Atmospheric Physics

Atmospheric Physics

Editor: Mary D'souza

RCallisto
Reference

www.callistoreference.com

Callisto Reference,
118-35 Queens Blvd., Suite 400,
Forest Hills, NY 11375, USA

Visit us on the World Wide Web at:
www.callistoreference.com

© Callisto Reference, 2017

ISBN: 978-1-63239-854-3 (Hardback)

The publisher's policy is to use permanent paper from mills that operate a sustainable forestry policy. Furthermore, the publisher ensures that the text paper and cover boards used have met acceptable environmental accreditation standards.

Trademark Notice: Registered trademark of products or corporate names are used only for explanation and identification without intent to infringe.

Printed in the United States of America.

Cataloging-in-publication Data

Atmospheric physics / edited by Mary D'souza.
    p. cm.
Includes bibliographical references and index.
ISBN 978-1-63239-854-3
1. Atmospheric physics. 2. Geophysics. 3. Atmospheric deposition. 4. Climatology. I. D'souza, Mary.
QC852 .A86 2017
551.5--dc23

Table of Contents

Preface

Atmospheric physics is a vital part of atmospheric sciences. It deals with the study of the atmosphere by applying the laws of physics. It aims at modeling the atmosphere of different planets, especially earth, by using radiation budget, fluid flow equations, energy transfer processes, etc. It uses the elements of cloud physics, climatology, spatial statistics, scattering theory, remote sensing, etc. to study the atmosphere and model weather systems. This book unravels the recent studies in the field of atmospheric physics. The objective of this text is to give a general view of the different areas of atmospheric physics and its applications. The various studies that are constantly contributing towards advancing technologies and evaluations of this field are also examined in detail. It is a compilation of chapters that discuss the most vital concepts and emerging trends in this field. This text is a resource guide for experts as well as students. It will help the readers in keeping pace with the rapid changes in this area.

The world is advancing at a fast pace like never before. Therefore, the need is to keep up with the latest developments. This book was an idea that came to fruition when the specialists in the area realized the need to coordinate together and document essential themes in the subject. That's when I was requested to be the editor. Editing this book has been an honour as it brings together diverse authors researching on different streams of the field. The book collates essential materials contributed by veterans in the area which can be utilized by students and researchers alike.

Each chapter is a sole-standing publication that reflects each author's interpretation. Thus, the book displays a multi-facetted picture of our current understanding of applications and diverse aspects of the field. I would like to thank the contributors of this book and my family for their endless support.

Editor

Atmospheric Reaction Systems as Null-Models to Identify Structural Traces of Evolution in Metabolism

Petter Holme[1,2]*, Mikael Huss[3,4], Sang Hoon Lee[1]

1 IceLab, Department of Physics, Umeå University, Umeå, Sweden, **2** Department of Energy Science, Sungkyunkwan University, Suwon, Korea, **3** Science for Life Laboratory Stockholm, Solna, Sweden, **4** Department of Biochemistry and Biophysics, Stockholm University, Stockholm, Sweden

Abstract

The metabolism is the motor behind the biological complexity of an organism. One problem of characterizing its large-scale structure is that it is hard to know what to compare it to. All chemical reaction systems are shaped by the same physics that gives molecules their stability and affinity to react. These fundamental factors cannot be captured by standard null-models based on randomization. The unique property of organismal metabolism is that it is controlled, to some extent, by an enzymatic machinery that is subject to evolution. In this paper, we explore the possibility that reaction systems of planetary atmospheres can serve as a null-model against which we can define metabolic structure and trace the influence of evolution. We find that the two types of data can be distinguished by their respective degree distributions. This is especially clear when looking at the degree distribution of the reaction network (of reaction connected to each other if they involve the same molecular species). For the Earth's atmospheric network and the human metabolic network, we look into more detail for an underlying explanation of this deviation. However, we cannot pinpoint a single cause of the difference, rather there are several concurrent factors. By examining quantities relating to the modular-functional organization of the metabolism, we confirm that metabolic networks have a more complex modular organization than the atmospheric networks, but not much more. We interpret the more variegated modular arrangement of metabolism as a trace of evolved functionality. On the other hand, it is quite remarkable how similar the structures of these two types of networks are, which emphasizes that the constraints from the chemical properties of the molecules has a larger influence in shaping the reaction system than does natural selection.

Editor: Matjaz Perc, University of Maribor, Slovenia

Funding: PH was supported by the Swedish Research Council and the World Class University program through National Research Foundation Korea funded by Ministry of Education, Science and Technology R31-2008-000-10029-0. The funders had no role in study design, data collection and analysis, decision to publish, or preparation of the manuscript.

Competing Interests: The authors have declared that no competing interests exist.

* E-mail: petter.holme@physics.umu.se

Introduction

Reaction systems are, at many levels of the universe, motors driving the creation of higher structure. From the metabolism in our bodies, via reactions in planetary interiors and atmospheres, to the nuclear reaction systems in stars; these are all systems shaped by the physical properties of constituents—the atoms and molecules. Among these systems, metabolism is special in the sense that its control has evolved by natural selection. But the physical properties of molecules and the relative abundance of elements constrain the evolution of this genetic control. Perhaps these constraints explain that very different reaction systems—reactions in planetary atmospheres and the organismal metabolism—share large-scale features (like the right-skewed probability distributions of degree, which roughly speaking reflects the number of molecules a molecule can react with) [1,2]. Still, as we will see, there are differences between these two types of systems and in this paper we will focus on what these differences are and what they can tell us of the evolution of metabolism. To put it short, we explore the idea that the reaction systems of planetary atmospheres can be null-models for studying metabolic networks in an evolutionary perspective.

The study of reaction-system topology (the set of all participating reactions) has long been restricted, by lack of data, to small subsystems. These systems, like e.g. the citric acid cycle of metabolism [3] or the carbon-nitrogen-oxygen cycle of stellar nuclear reactions [4] (two systems that were, coincidentally, both discovered in the mid-1930's), have been modeled in great detail with e.g. differential equations. It has, however, not until recently been possible to investigate the system-wide organization of any type of reaction system. Since about a decade, we do have methods to infer the entire set of reactions (again coincidentally) both in metabolism and planetary atmospheres. Still these datasets are so crude that our conclusions in this paper will be rather hypothetical in nature. On the encouraging side, however, the early conclusions mentioned above—that reaction network are right-skewed and fat-tailed [1,2]—still hold for contemporary datasets. If we go beyond the topology, even less is known. A full picture of reaction rates and concentrations for a traditional kinetic modeling is far into the future. One complication comes from the fact that metabolites (and also molecular species in atmospheres) are distributed heterogeneously in space [5] and sometimes so few in number that concentration based models do not apply. This means that when investigating the global organization of reaction systems, we will have to rely on graph-based analysis techniques for still some time. Even though graph-based methods need to discard much of the knowledge we have about reaction kinetics, one can still encode much information into the graph. The molecular species present determine the vertices of the network; the catalysts present define the reactions. But what should the edges represent? Should one also include separate vertex-types for reactions and catalysts? The fundamental trade-off is between a graph representation including more information and a simpler representation that suits a larger variety of analysis methods. Much of the recent development in the graph structure of reaction systems has focused on either adapting analysis techniques to complex and informative graph representations [6–9], or to find simple graph representations encoding as much relevant information as possible [10–12]. In this paper, we will focus more on the latter developments and study the topology of two simple graph representations: one *substance graph* where the vertices are molecular species and an edge represents that two vertices participate in the same reaction, and a *reaction graph* where vertices symbolize reactions and two vertices are linked if they share some molecular species. In addition to these representations we also study the reaction systems as a bipartite graph with two classes of vertices, one for reactions and one for molecular species with edges connecting substances to the reactions they participate in. (Note that this representation, although more informative, still means a reduction of the information from the entire reaction system since one no longer can see which reactants that need to be present for a reaction to occur, or which products that are produced.) We investigate several topological properties of such graphs from reaction systems of planetary atmospheres and organismal data sets. Apart from degree distributions, we study network modularity (reflecting how well a graph can be decomposed into dense sub-graphs that are relatively weakly interconnected), currency substances (abundant molecular species that can react with a broad spectrum of other substances) and degree correlations (if edges primarily go between vertices of similar degree, or if the degrees are unbalanced with many edges between high- and low-degree vertices).

Results

The different degree distributions of the human metabolic and Earth atmospheric networks

Since the degree of a vertex count the number of other vertices it interacts with, it is a fundamental network quantity. The high-degree vertices can, and in most situations will, interact with many other vertices. The early findings that reaction systems have fat-tailed degree distributions—i.e. most vertices interacts only with a few others while some interact with a number far larger than the average—points at a diversity of functions among the vertices. For the metabolism, the common interpretation is that the high-degree metabolites are supplying building blocks to metabolites with more specialized functions, and lower degree. For atmospheric reaction networks, the low-degree vertices typically correspond to more complex molecules. We start our comparison of planetary and metabolic reaction system by looking at the substance and reaction graphs of Earth's atmosphere and the human metabolism. In Fig. 1A, we show the degree distributions of the substance graphs of the human metabolism and Earth's atmospheric reaction system. These distributions are rather similar—peaked and right skewed with tails of about the same slope. The degree distributions of the reaction graph, seen in Fig. 1B, are strikingly different. The human reaction graph is skewed and fat-tailed like its substance graph (but with a smaller exponent), whereas the Earth reaction graph has a degree distribution of an entirely different functional form, suggesting a different organization. The graphs are too big, however, for layout programs to give a hint of a deeper explanation of this difference (Fig. 2). Indeed, it is difficult to single out a more fundamental quantity causing the differences in degree distributions, as we will see in the rest of this section.

In our quest for a more detailed explanation of the difference of degree distributions in Fig. 1, we look closer at the bipartite representations mentioned above. In Fig. 3 (panels A, B, E and F), we plot the probability distribution of bipartite degree K_i for the human metabolic (Figs. 3A and B) and Earth atmospheric (Figs. 3E and F) networks in the substance (Figs. 3A and E) and reaction (Figs. 3B and F) projections. (For the other data sets this information can be found in Figure S1 and S2.) For substances, the degree distributions are right skewed in a fashion similar to the substance graph of Fig. 1A. For reactions, the two types of reaction systems both show unimodal degree distributions. A slight difference is that the Earth data gives a left-skewed distribution while the human network is right-skewed. This also means that the bipartite reaction-degree distribution, for the human data, is radically different than the projected distribution of Fig. 1B. To understand this better, we can decompose the degrees of the projected networks into three quantities as follows (where the left-

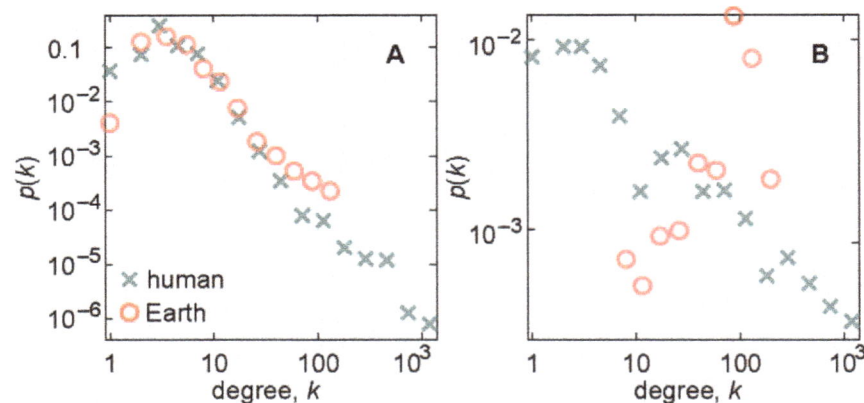

Figure 1. Degree distributions of substance and reaction graphs of the human metabolism and Earth's atmospheric reaction system. Panel A shows the probability mass-function of the degree of the substance graph of the reaction system of the Earth's atmosphere and the human metabolic networks. B shows the same as A, but for the reaction network. The similar behavior in A is drastically different in B. The plots are log-binned and plotted on double logarithmic scales.

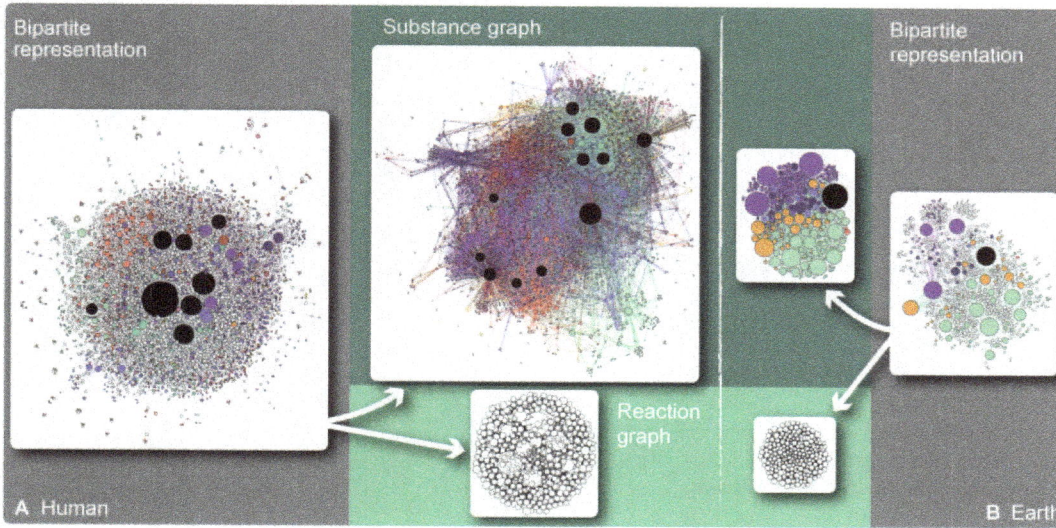

Figure 2. Ridiculograms of the human metabolism and Earth's atmospheric reaction system in bipartite, substance and reaction graph representations. The areas of the vertices are proportional to their degree. White vertices are reaction vertices; black vertices are currency vertices. For the other vertices the color represent different network modules. The colors of the edges are the same as their vertex of largest degree.

hand side is the degree of the projected network and the right-hand side quantities refer to the bipartite representations):

$$k_i = S_i - K_i - X_i = K_i(\kappa_i - 1) - X_i, \qquad (1)$$

where S_i is the sum of degrees of i's neighbors, K_i is i's degree, X_i is

the number of four-cycles that i is a part of, and κ_i is the average degree of i's neighbors. If there are few four-cycles in the bipartite network and there are no strong degree correlations (so κ_i can be assumed constant with respect to k_i), then i's degree in the bipartite network is a linear function of k_i (according to Eq. (1)). This is thus not the case for, at least, the metabolic reaction network where the k- and K-degree distribution, as mentioned, differs much. Indeed,

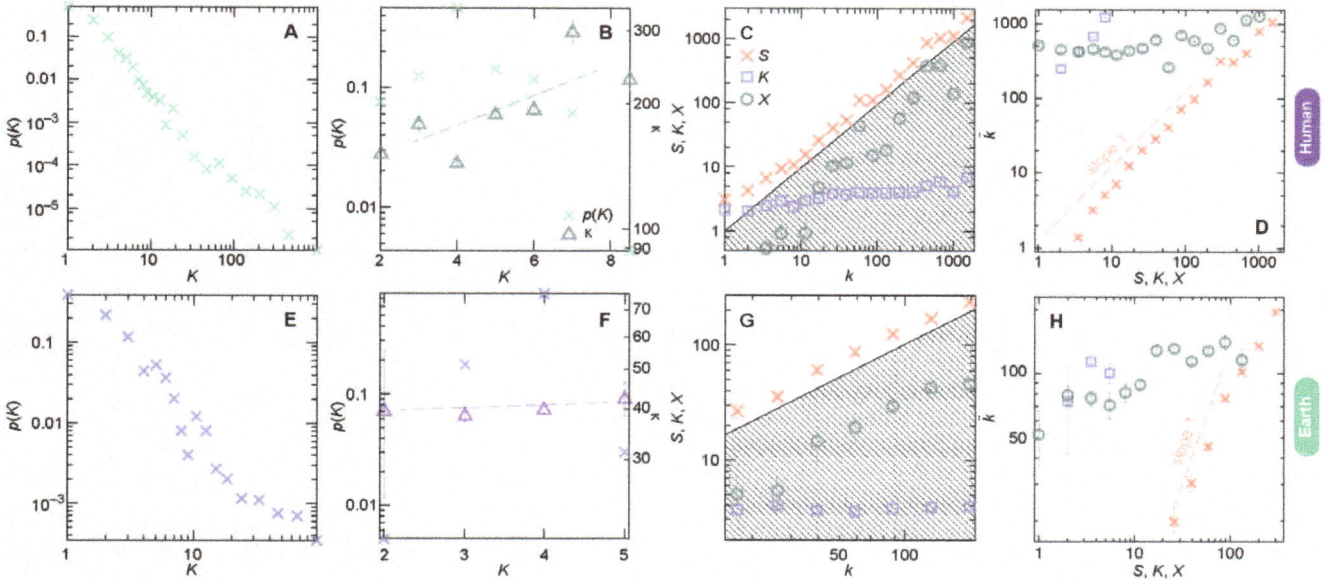

Figure 3. Deeper investigations of the degree distributions. Panel A displays the degree distribution of substances in a bipartite representation of the reaction system, i.e. the probability distribution of the number of reactions a substance participates in. Panel B shows the corresponding plot for reactions and also the average degree of neighbors. The dashed line is a linear-regression line to highlight the trend in κ. C and G displays the values of the three bipartite-network terms of k—S (the sum of the degrees of neighbors), K (the degree) and X (the number of four-cycles the vertex participates in). The diagonal line shows the k-value (so if you subtract the values of circles and squares from the values of crosses you would get this line). Panel D and H shows the average degrees $\bar{k}$ of nodes with certain values of the three terms that contribute to the degree in the projected reaction networks. $\bar{k}$ is averaged over logarithmic bins of S, K, and X values. The dashed line is a reference corresponding to a linear $\bar{k}$-dependence. Panels A–D are for the human metabolic reaction networks, E–H show the corresponding plots for the Earth atmospheric reaction networks.

in Fig. 3B we see a positive correlation between K and κ, stronger than the corresponding correlation for the Earth network in Fig. 3F (which is almost absent). This means that $S = K\kappa$ grows super-linearly with K so the tail of $p(K)$ gets stretched into the distribution of Fig. 1B. Here, we still assume that the number of four-cycles does not contribute to k significantly, which we justify below. This is justified to some extent in Fig. 3C (and 3G for the Earth network)—the k-scaling of S and X is similar, so $S - X$ scales like S (and thus the arguments above still hold). That S (and thus $S - X$) scales like X is also true for the atmospheric network (Fig. 3G), which explains that the shape of Fig. 1B is to a large degree determined by K (so the hump shape of $p(K)$ gives a hump-shaped $p(k)$). Another view of S, K and X is given in panels D and H where, we plot the average degrees of nodes given their S-, K- and X-values. We can see that, as expected, S is the best predictor of $\bar{k}$ (showing close to a linear relationships for the metabolic data, and a clear correlation for the atmospheric network). Another observation is that X shows more structure (apart from the scaling itself) in the metabolic network compared to the atmospheric network. This can perhaps be explained by the more pronounced modular structure of the metabolic network (that we will discuss further below). From Fig. 3D and H we also learn that $\bar{k}$ shows a strong positive K-dependence for the metabolic network, but not for Earth's atmospheric network. This is reflected in Figs. 3B and F too—since κ grows with K for the metabolic network, S and K will be positively correlated, and since $\bar{k}$ grows with S then it will also grow with K.

In summary, the difference between the degree distributions of the reaction graphs of the metabolic and atmospheric networks cannot be explained by one single feature of the original reaction system's topology. Instead it can be traced to a combination of the slightly different skewness of the distribution of a reaction's number of participating substances and the different correlation properties between the degree of a vertex and the average degree of its neighbors. In Figures S3 and S4, we plot the bipartite degree distributions of all the planets and organisms. Essentially, the conclusions for the Earth's atmospheric network extends to other planets, except that the data sets are smaller and the degree distributions does not have the same negative trend similar to power-laws.

Comparing degree distributions of planetary atmospheric and organismal metabolic networks

So far, we focused on finding lower-level causes for the degree distributions of projected networks of the human metabolic and Earth atmospheric networks. We now turn to the question how much these observations can be generalized to the other networks. To this end, we will use more rigorous methods for analyzing probability distributions than we used so far. We will analyze the data using methods from Ref. [11]. First, we test the hypothesis that degrees are power-law distributed by (roughly speaking, details in the Methods section) finding parameter values for the power-law distribution that fits the data best, then draw as many series of numbers from this distribution with the same size as the raw data and check the likelihood that the synthetic and real data come from the same distribution. We also check which is the most likely distribution generating that degree distribution—power-law or log-normal (a right-skewed distribution with a more narrow tail than a power-law that is visually similar to the Earth reaction graph of Figure 1B). The results of these measurements are shown in Table 1. As hypothesized above, the reaction graphs are unanimously inconsistent with power-laws. Of the substance graphs, only planetary atmospheric networks are consistent with

Table 1. Statistical tests of various types of degree distributions.

| | Atmospheres of planets and moons | | | | | | | | Metabolism of organisms | | | | | | | | |
	Earth	Venus	Titan	Titan 2	Mars	Jupiter	Io	Solar system	Human (KEGG)	Human (BiGG)	M. genitalium	S. cerevisiae (KEGG)	S. cerevisiae (BiGG)	E. coli	M. musculus	D. melanogaster	C. elegans
Substance graph Power-law?	Y	N	Y	Y	Y	Y	Y	Y	N	N	N	N	N	N	N	N	N
PL or LN	PL	LN	LN	LN	LN	LN	LN	LN	PL	PL	PL	PL	PL	PL	PL	PL	PL
Reaction Graph Power-law?	N	N	N	N	N	N	N	N	N	N	N	N	N	N	N	N	N
PL or LN	LN	LN	LN	LN	LN	LN	LN	LN	PL	LN	PL	PL	PL	PL	PL	PL	PL
Bipartite substances Power-law?	Y	Y	Y	Y	N	Y	Y	Y	N	Y	Y	N	Y	N	N	N	Y
PL or LN	PL	LN	LN	LN	LN	LN	LN	LN	PL	PL	PL	PL	PL	PL	PL	PL	PL

Statistics of the reactions in the bipartite representation are omitted since they are not fat-tailed. "Y" ("N") indicates that the data set is consistent (inconsistent) with the tested hypothesis. "PL" stands for "power-law" (i.e., testing for a power-law hypothesis); "LN" means "log-normal".

power-laws. This does not mean that it is fair to describe them as power-laws; especially since most of them fit better to a log-normal form. Since the planetary data sets are relatively small, the relative errors are larger and it is harder to refute the possibility of another functional form. The substance graphs are, on the other hand, closer to log-normals than power-laws. The reason is seen for the human metabolic network in Fig. 1 (and for the other datasets sets in Fig S1), that they are even more fat-tailed than a power-law—they have more vertices of highest degrees than the best-fitting power-law does. Thus they are even further from log-normals than power-laws. The reaction graphs are more similar to power-laws than log-normals for the metabolic networks, but the other way around for the planetary atmospheres, which is also in line with our observations. This study cannot, however, strengthen the observation that the substance graphs are similar to the metabolic networks except for Earth's network that falls into the same category as the metabolic networks. There are two possibilities— either the difference can be explained by a difference in sizes and that the other planetary atmospheres have to be measured by indirect methods, or the Earth network is radically different (more than just the sizes). Ref. [2] makes the latter hypothesis, and argues a difference from the influence on the biosphere on the Earth's atmosphere creates a visible difference. On the other hand, many reactions typical for Earth (e.g. involving molecular oxygen) are also present in the other datasets.

The substances' degrees in the bipartite representation do not separate the planetary and metabolic data so well (both types of datasets contain degree distributions consistent with power laws, and not). Similar to the observations in the detailed studies above, the projections to substance or reaction graphs create the difference. However, the planet-network distributions are more similar to log-normal than power-laws, whereas it is the other way around for the metabolic networks.

Modularity and currency metabolites

Biological systems are commonly described as modular—being composed of different subunits, or modules, which perform some specific task relatively independent of the rest of the system. Some modules are quite conspicuous—a cell is a prime example—but also more nebulous systems, like metabolism, are thought to consist of modules. If we treat all reactions equal (the essence of the

graph theoretic approach), then independence means that the connections within the network module should be denser than the connections out of the module. A module on a graph-level resolution of metabolism is thus equal to what is commonly known as a network cluster or community [13]. This is not quite the whole story however. The most abundant metabolites (like water, carbon dioxide and so on) do not put any restriction on the reactions, and would not contribute to the specialized function of a module. It is thus common to preprocess the graph by identifying such *currency metabolites* and removing them from the network, considering only a network of other less frequent molecular species that are more of bottlenecks in the metabolic machinery. There are methods to identify both network clusters and currency metabolites (described in the Methods section) from the topology of substance graphs. Although these definitions have been developed for metabolic networks, there is nothing that stops us from applying them to networks of planetary atmospheres. *A priori*, since atmospheric reaction system has not evolved through natural selection, we expect them to have less distinct modules and currency metabolites. This is indeed the case as can be seen in Fig. 4—there is a size-difference between the metabolic and atmospheric networks, but it is less pronounced than both the relative modularity and the number of currency vertices. Thus there seems to be a stronger tendency for the metabolic networks to be organized into modules supplied by currency vertices than the networks of planetary atmospheres.

Discussion

In this article, we have directly compared functionally informative network characteristics of metabolic reaction systems of a wide variety of organisms and the reaction systems of planets and moons of the solar system. One such quantity is degree—the number of other nodes a node interacts with. (Where "interact" is defined via the network in question.) In most types of networks, degree indicates the importance of a node, but in biochemical networks, where both low- and high-degree vertices can be essential for the cell's functionality, then degree rather separates chemical substances of different functionality—at least in meta-bolic substance networks, the high-degree vertices are typically light molecules that supply atoms and molecular groups to the functionally more specialized low-degree vertices [14]. For

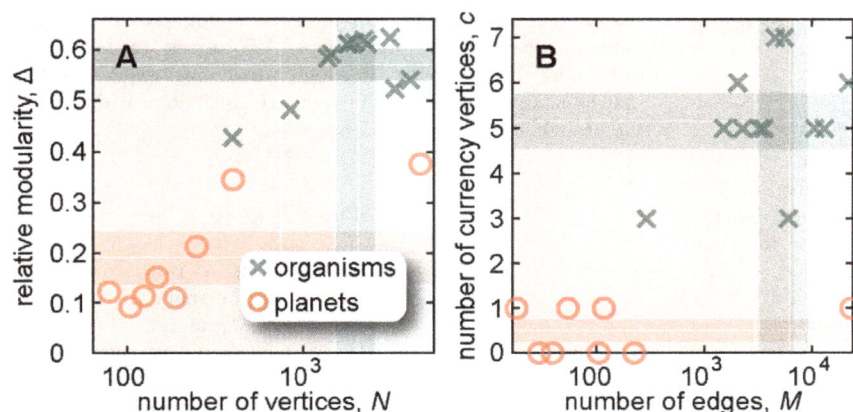

Figure 4. Relative modularity and the number of currency vertices separate networks of metabolism from networks of planetary atmospheres more than their sizes do. To show that the maximal relative modularity separates metabolism from reaction systems of planetary atmospheres, we display (panel A) the relative modularity Δ as a function of the number of vertices N. The shaded areas indicate the standard deviation and means of the respective quantities. Similarly, in B, we show another quantity related to the functional organization, the number of currency vertices c, as a function of the number of edges M in the network. Note that axes are linear and logarithmic respectively.

reaction networks one can assume a similar interpretation—high-degree vertices are reactions supporting many subsystems of the reaction system. All substance projections, for both atmospheric and metabolic networks, do indeed have relatively broad degree distributions. This supports the above-mentioned picture of functional differentiation by degree. Using statistical tests, we can separate organisms from planet fairly well. The networks of planetary atmospheres are typically consistent with power-laws, but the metabolic networks are not. The planetary networks are, however, statistically more similar to log-normal distributions, which suggests that the fact they are deemed consistent with power-laws is an effect that they are, on average, smaller than the metabolic systems (and thus does not provide enough data to give statistical significance).

We note that in the substance-network projection, the Earth atmospheric and human metabolic networks have rather similar degree distributions, but for the reaction-network projection the distributions are strikingly different. We investigate lower-level explanations for this observation in terms of degree distributions of a bipartite representation of the reaction system and degree correlations. It is however not easy to single out a low-level cause for this difference, rather it seems to be a combined effect of a slightly difference in the distribution of reaction-degrees and degree correlations in the bipartite representation.

When we look closer at quantities designed to characterize the modular functionality, we see higher network modularity and more currency metabolites in metabolic networks than atmospheric networks. On the other hand, the differences are not larger than that they can almost be explained by the sizes of the networks alone. Furthermore, fundamental structures such as the shape of some of the degree distributions are skewed in a qualitatively similar way. Our conclusion is thus that the main structure of metabolic networks is probably shaped by the same fundamental stoichiometric constraints as all chemical reaction systems, but there are also traces of evolution in the network structure of metabolism. At the same time the network-modular structure, the traces of evolution, is not so clear as the picture the analogy to engineering paints—there are more than a couple of in- and output terminals. Maybe the largest open question is not why metabolic networks are modular but why they are not more modular? How can we reconcile the logical picture of evolution operating by adding and deleting of modules with the modular-but-not-very-much-so picture of metabolic networks? We believe the approach we take in this paper, to use a natural system as a null-model for the metabolism can be fruitful.

Methods

Datasets for metabolic and chemical networks

Reaction sets for planetary atmospheres are described in Ref. [5], except the "solar system" data that was obtained from the UMIST database [15]. The metabolic networks come from the KEGG [16] and BiGG [17] database and are described in Ref. [7]. We select nine datasets from the KEGG and BiGG databases to match the number of planetary atmosphere datasets. To get a rough error estimate of sampling effects, we also analyze the human data both from BiGG and KEGG, and two independent datasets from Jupiter's atmosphere. Our selection criterion is that the datasets should be a diverse selection among the most well-studied model organisms.

Network representations

To choose the graph representation of a reaction system involves a trade-off between information content and usefulness.

One can use a complex representation with substances, catalysts and reactions as separate classes of vertices and directed edges representing the general direction of the matter flow. The advantage with such a representation is that all topological aspects of the reaction system are encoded into the graph. But the price for this is that there few general analysis methods can be applied to it; they would need to be modified, something that is not always possible. Alternatively, one chooses a simple-graph representation with one type of vertices and one type of (undirected) edges, without multiple edges or self-edges. Such a representation can be analyzed by a multitude of off-the-shelf methods. A disadvantage with simple graphs, except that they encode less information, is that there is no obvious way of reducing the reaction system to a simple graph. We choose a substance graphs as our main graph simple-graph representation. In such a graph one put an edge between all substances that can participate in the same reaction, so if the reaction $2H_2O \rightarrow 2H_2 + O_2$, would contribute with three edges—(H_2O, H_2), (H_2, O_2) and (O_2, H_2O)—to a substance graph. There is some evidence that substance graphs are good simple-graph representations of metabolic networks [11,18], but to the best of our knowledge, no corresponding studies for other categories of reaction systems. In addition, we use a reaction graph representation that is in some sense dual to the substance graphs—every reaction is a vertex in this network and two reactions that have a substance in common is connected.

Testing degree distributions

We use the approach in Clauset et al. [11] to test the degree distributions for the hypothesis that they follow power-laws. This method starts from the real data and obtains the exponent of a best-fitting power-law, α, by maximum likelihood estimation. Then one draws sets of random numbers, of the same cardinality as the original data, from the probability distribution

$$p_k = \begin{cases} \Lambda k^{-\alpha} & \text{if } 0 < k \leq k_{\max} \\ 0 & \text{otherwise} \end{cases} \qquad (2)$$

where Λ is a normalization constant. Finally, one use the Kolmogorov–Smirnov test statistics (the maximal difference, for all k-values, between the cumulative density functions of the real and synthetic data) to estimate the p-value of the hypothesis that the real data was drawn from p_k.

Ref. [19] also adapts a method by Vuong [20] to compare different heavy-tailed distributions. We use it to test which distribution of power-law and log-normal distribution functions that best fits our data. The log-normal distribution is defined by the probability density function

$$p_k = \frac{A'}{k} \exp\left[-a'(\ln k - \mu)^2\right] \qquad (3)$$

where A', a' and μ are positive constants (A' is a normalization factor, a' and μ are parameters giving the shape of the curve). Vuong's method takes the likelihoods, L_1 and L_2, of the two functional forms generating the observed data as its starting point. The method uses the result that $V = \ln(L_1/L_2)$ is normally distributed for large data sets to compute a p-value for the hypothesis that the data was generated by distribution 1 rather than distribution 2.

Network Modularity

The concept of network modularity, cluster, or community structure strives to capture the large-scale organization of networks into dense subnetworks that are relatively weakly interconnected [21]. There is no unique way of deriving a measure for network modularity or dividing a graph into such dense subgraphs; rather, there is a number of different methods each capturing some certain aspect of network modularity. The method in this work is based on the popular method of maximizing Newman and Girvan's Q-modularity. For this measure, one assume the graph is divided into a number of subgraphs and let e_{ij} be the fraction of all edges going between subgraph i and j, and defines

$$Q = \sum_i \left[e_{ii} - \left(\sum_j e_{ij} \right)^2 \right] \qquad (4)$$

A class of module-detection methods starts by assuming that the division maximizing Q is a sensible decomposition into subgraphs. Already from the Equation (4) one can see that edges within a subgraph give a positive contribution to Q, and edges between communities decrease Q. The advantages with this clustering algorithm are that Q is easy to interpret and closely matching the verbal definition of a network module above; and furthermore the maximal Q, $\hat{Q}$, is a crude measure of the network modularity of an entire graph. The two disadvantages with Q-maximization methods are the following. First, it fails to divide some subgraphs into what looks like obvious clusters. This is roughly speaking because the second sum compares a division i with all other divisions j, even if it does not matter (for a visually good clustering) if i and j are far apart [22]. Second, it is technically hard to find the maximizing division—Q is a very flat function (in sub-division space) near its maximum [23]. For our purpose these latter two objections are not so serious—there is no general biological argument that the modules that look like they can be further subdivided are not sensible clusters, and there is no need to find the actual subdivision into modules, we just want a good estimate of $\hat{Q}$, which we do have if we only get close to the mentioned plateau in subdivision space.

As a measure of the modularity of a graph, $\hat{Q}$, is not ideal. On one hand $\hat{Q}$ close to zero would mean a low modularity and $\hat{Q}$ close to one would imply modularity. On the other hand, the intermediate values depend on many factors regarded as more fundamental (like the number of vertices and edges and the degree distribution) than modularity. To compensate for such effects as much as possible we rather measure $\hat{Q}$ relative to the average value of $\hat{Q}$ in an ensemble, or null-model, of graphs (obtained by standard edge rewiring [24]) with the same sizes N and M and the same degrees as the substance graph G, but everything else random. So we define

$$\Delta = \hat{Q} - \bar{Q} \qquad (5)$$

where $\bar{Q}$ is the average of the maximal modularity over 1000 rewired graphs.

Currency vertices

The hubs in metabolic networks—e.g. H_2O, NADH, ATP and CO_2—are typically also the most abundant metabolites through-

out the cell. These are the workhorses of metabolism, supplying functional groups to proteins and other molecules with more specialized functions. Since these currency metabolites are present throughout the cell and do not put much of constraints on the reactions they participate in, one can learn more about the functionality of the network if one exclude them from the graph representation. The circumstance that they are common throughout the cell and participate in many reactions also means that they connect network modules and effectively lower the modularity. This observation, along with the fact they have a high degree, has been used as a definition of currency metabolites [10]. If one deletes vertices in order of their degree (starting from large degrees) and monitor Δ, then for metabolic networks, Δ typically first increase to a maximum and later decrease. Ref. [10] defines currency metabolites as those that give the largest Δ before Δ reached a value larger than in the original graph. This definition is general enough to apply to other reaction-system networks, and one can speak of *currency vertices* also for atmospheric or nuclear reaction systems [14].

Supporting Information

Figure S1 Degree distributions for the substance networks. The data is log-binned and plotted in log–log scale.

Figure S2 Degree distributions for the reaction networks. The data is log-binned and plotted in log–log scale.

Figure S3 Degree distributions for the substances in the bipartite representations. The data is log-binned and plotted in log–log scale.

Figure S4 Degree distributions for the reactions in the bipartite representations. The data is log-binned and plotted in log–log scale.

Figure S5 A plot corresponding to Fig. 3C, D, G and H for substance networks. Panels A and C display the values of the three terms of k—S, K and X. The diagonal line shows the k-value. Panels B and D show the average degrees $\bar{k}$ of nodes with certain values of the three terms that contribute to the degree in the projected networks. $\bar{k}$ is averaged over logarithmic bins of S, K, and X values. Panels A and B is data for the human network; C and D are the corresponding plots for the Earth atmospheric network.

Acknowledgments

The authors thank Andreea Munteanu for help with the data acquisition.

Author Contributions

Conceived and designed the experiments: PH MH. Analyzed the data: PH SHL. Wrote the paper: PH MH SHL.

References

1. Jeong H, Tombor B, Oltvai ZN, Barabási AL (2000) The large-scale organization of metabolic networks. Nature 407: 651–654.
2. Solé RV, Munteanu A (2004) The large-scale organization of chemical reaction networks in astrophysics. Europhys Lett 68: 170–176.
3. Krebs HA, Johnson (1937) The role of citric acid in intermediate metabolism in animal tissues. Enzymologia 4: 148–156.
4. Bethe HA (1939) Energy production in stars. Phys Rev 55: 434–456.
5. Yung YL, Demore WB (1999) Photochemistry of planetary atmospheres. New York: Oxford University Press.
6. Veeramani B, Bader JS (2010) Predicting functional associations from metabolism using bi-partite network algorithms. BMC Systems Biology 4: 95.
7. Miyake S, Takenaka Y, Matsuda H (2004) A graph analysis method to detect metabolic sub-networks based on phylogenetic profile. IEEE Computational Systems Bioinformatics Conference. pp 634–635.

8. Klamt S, Haus U-U, Theis F (2009) Hypergraphs and cellular networks. PLoS Comput Biol 5: e1000385.

9. Holme P, Huss M, Jeong H (2003) Subnetwork hierarchies of biochemical pathways. Bioinformatics 19: 532–538.

10. Huss M, Holme P (2007) Currency and commodity metabolites: Their identification and relation to the modularity of metabolic networks. IET Systems Biology 1: 280–288.

11. Holme P, Huss M (2010) Substance networks are optimal simple-graph representations of metabolism. Chinese Sci Bull 55: 3161–3168.

12. Serrano MA, Sagues F (2010) Network-based confidence scoring system for genome-scale metabolic reconstruction. e-print arXiv: 1008.3166.

13. Newman MEJ (2010) Networks: An Introduction. Princeton, NJ: Princeton University Press.

14. Holme P (2009) Signatures of currency vertices. J Phys Soc Jpn 78: 034801.

15. Woodall J, Agúndez M, Markwick-Kemper AJ, Millar TJ (2007) The UMIST database for astrochemistry. Astronomy and Astrophysics 466: 1197–1204.

16. Kanehisa M, Goto S (2000) KEGG: Kyoto Encyclopedia of Genes and Genomes. Nucleic Acids Res 28: 27–30.

17. Duarte NC, Becker SA, Jamshidi N, Thiele I, Mo ML, et al. (2007) Global reconstruction of the human metabolic network based on genomic and bibliomic data. Proc Natl Acad Sci USA 104: 1777–1782.

18. Holme P (2011) Metabolic robustness and network modularity: A model study. PLoS ONE 6, e16605.

19. Clauset A, Shalizi CR, Newman MEJ (2009) Power-law distributions in empirical data. SIAM Rev 51: 661–703.

20. Vuong QH (1989) Likelihood ratio tests for model selection and non-nested hypotheses. Econometrica 57: 307–333.

21. Fortunato S (2009) Community detection in graphs. Phys Rep 486: 75–174.

22. Fortunato S, Barthélemy M (2006) Resolution limit in community detection. Proc Natl Acad Sci USA 104: 36–41.

23. Good BH, de Montjoye Y-A, Clauset A (2010) The performance of modularity maximization in practical contexts. Phys Rev E 81: 046106.

24. Sneppen K, Maslov S (2002) Specificity and stability in topology of protein networks. Science 296: 910–913.

2

Can Hydrographic Data Provide Evidence That the Rate of Oceanic Uptake of Anthropogenic CO$_2$ Is Increasing?

William Carlisle Thacker*

University of Miami and Atlantic Oceanographic and Meteorological Laboratory, Miami, Florida, United States of America

Abstract

Predictions of the rate of accumulation of anthropogenic carbon dioxide in the Pacific Ocean near 32°S and 150°W based on the P16 surveys of 1991 and 2005 and on the P06 surveys of 1992 and 2003 underestimate the amount found in the P06 survey of 2009–2010, suggesting an increasing uptake rate. Assuming the accumulation rate to be constant over the two decades, analyses using all five surveys lead to upward revision of the rates based only on the first four. On the other hand, accumulation rates estimated for 2003–2010 are significantly greater than those for 1991–2003, again suggesting an increasing uptake rate. In addressing this question it is important to acknowledge the limitations of the repeat hydrography and consequent uncertainties of estimated accumulation rates.

Editor: Inés Álvarez, University of Vigo, Spain

Funding: This work has been supported by the Physical Oceanography Division of National Oceanic and Atmospheric Administration's Atlantic Oceanographic and Meteorological Laboratory and by the Office of Naval Research Award N00014-101-0498. The funders had no role in study design, data collection and analysis, decision to publish, or preparation of the manuscript.

Competing Interests: The author has declared that no competing interests exist.

* E-mail: cthacker@rsmas.miami.edu

Introduction

A recent paper [1] introduced a two-regression method for estimating the rate of accumulation of anthropogenic carbon dioxide within a limited region of the ocean. The first regression serves as a filter, removing the variability of the observed total inorganic carbon that should be present in the absence of anthropogenic sources while leaving the anthropogenic signal in the residuals together with unexplained variability that might be regarded as noise. The second regression fits a linear function of time to the residuals to account for a systematic temporal trend that can be identified as the anthropogenic signal, with the slope of the trend line providing an estimate of the accumulation rate. This method was illustrated using data obtained from repeated surveys along 32.5°S (P06) and along 150°W (P16) in the South Pacific within the period from 1991 to 2005. The rates of accumulation were estimated for the region surrounding the intersection of the survey tracks for two layers, one centering on 200 dbar and the other on 400 dbar. For each layer, six different sets of regressors were used to account for the natural variability in the region and accumulation rates were estimated from each model's residuals.

With no accumulation before the industrial revolution, the rate had to have increased from zero, and it is reasonable to expect that it is still increasing. The simplest argument is that the increasing amounts of atmospheric CO$_2$ provide more for the ocean to take up. The observation-based fact that carbon appears to be increasing faster in the 200 dbar layer than in the deeper 400 dbar layer [1] also supports the notion of accelerating accumulation, as the anthropogenic signal takes longer to reach the deeper layer. The objective of this study is to see whether the addition of the more recent data from 2009–2010 allows for observational evidence of accelerating rates within the separate layers.

The new data are found to have a greater amount of anthropogenic carbon than would be expected given the accumulation rates inferred from the four previous surveys. While the excess is relatively small and within the range of unexplained variability, the fact that it is seen consistently for all models used to filter out the natural environmental variability might be taken as an indication of an increasing rate of accumulation of anthropogenic carbon. Similarly, revised analyses of the data from all five surveys, regardless of the filtering, lead to upwardly revised accumulation rates. And rates estimated using data from the first through third surveys and separately using data from the third through fifth surveys indicate substantial increases regardless of how natural environmental variability is treated.

Data

The data considered here, which are freely available from CLIVAR, are subsets of P06 and P16 surveys that were obtained in the vicinity of the intersection of the two survey lines as illustrated in figure 1. Because evidence of accumulating anthropogenic CO$_2$ is expected to be strongest in the upper ocean, the study focuses on data within two upper-ocean layers, 125–275 dbar and 275–525 dbar, as illustrated in figures 2 and 3. The shallower layer has a stronger anthropogenic signal but greater variability due to near-surface processes, while the deeper layer has a smaller signal and less environmental noise. If the signal-to-noise ratio is large enough, there is a possibility of detecting a changing rate of accumulation of anthropogenic carbon.

The number of samples available to this study, which are listed in Table 1, varies from survey to survey with the 1991 P16 survey providing fewest data and the 2009–2010 P06 survey providing the most. Consequently, variability over latitude is more poorly

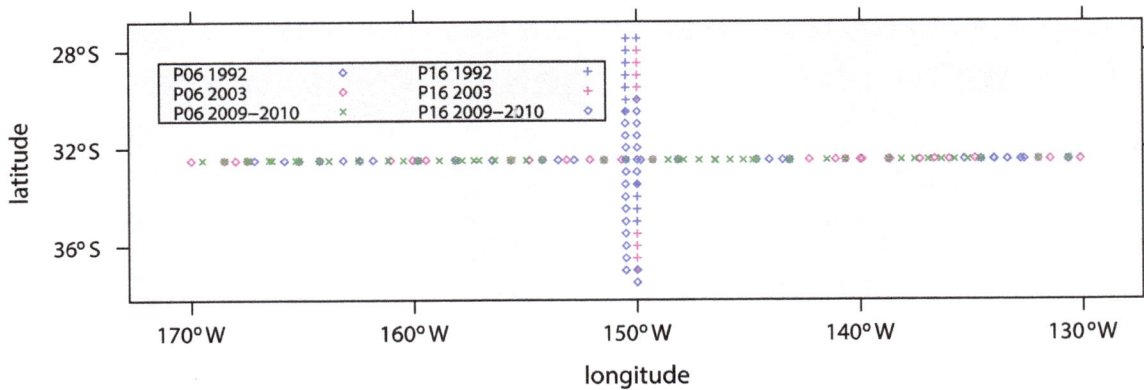

Figure 1. Locations of stations contributing data to this study.

sampled than variability over longitude. Descriptions of the data can be found in [2–7].

Local averages of the observed amounts of total inorganic carbon for each survey provide a time series that might serve to indicate an increase in the region. While such volume averages are clearly more stable than the individual point observations, as they serve to remove some of the natural variability that might be confounded with an anthropogenic signal, the limited nature of the sampling brings into question how representative of the natural variability of the region over the two decades of this study is the variability seen at the time of the surveys. This argues for combining data from multiple surveys so that the analysis might be based on as much of the variability as possible.

By working with the surveys collectively it is possible to exploit the co-variability of inorganic carbon with other measured variables in an attempt to identify and to remove the consequences of horizontal and vertical shifts in the water mass properties and of mixing process in the upper ocean, so that remaining systematic

temporal changes might be attributed to anthropogenic causes alone. Although co-variability of carbon with other environmental variables can mitigate this concern, there is still the issue of whether the sample is sufficient to capture this co-variability. As these are the only data available, it is necessary to proceed while recognizing their limitations.

Models

The models listed in Table 2 provide alternative characterizations of the natural variability of dissolved inorganic carbon through its co-variability with the regressors. They are exactly the same models as those discussed when the two-regression method was introduced [1]. The default model m0, which has only the intercept term, makes no attempt to account for non-anthropogenic variability other than through its averaging of the carbon observations. Because near-surface process stratify water according to its density, all of the other models have potential density as a regressor. In addition model m2, m3, m4, and m5 have one or

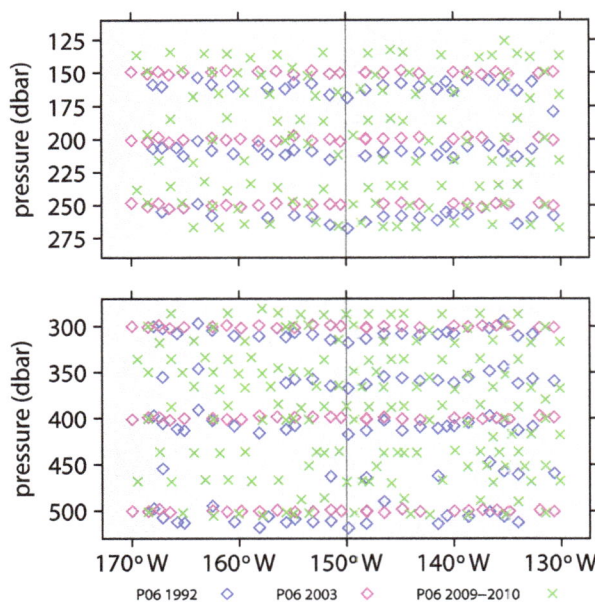

Figure 2. Locations of data from P06 surveys. The vertical lines indicates where P06 and P16 cross. Blue indicates data from 1992; magenta, 2003; green, 2009 and 2010

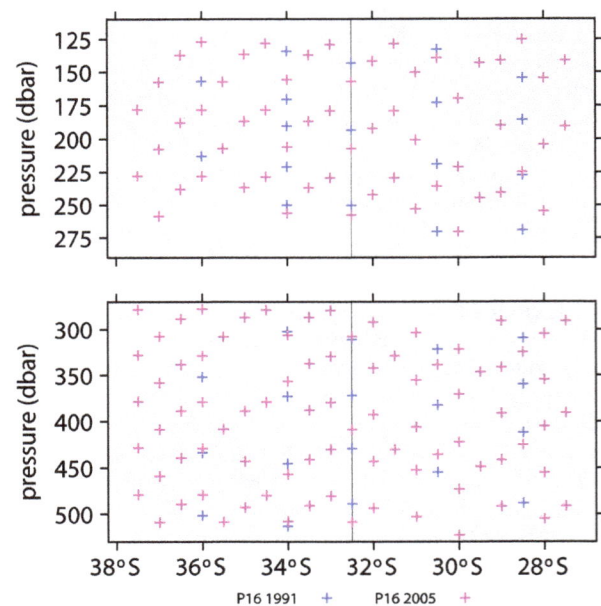

Figure 3. Locations of data from P16 surveys. The vertical lines indicates where P06 and P16 cross. Blue indicates data from 1991; magenta, 2005.

Table 1. Number of samples in study region.

	200 dbar	400 dbar
1991 P16	18	18
1992 P06	**68**	**94**
2003 P06	**74**	**77**
2005 P16	57	83
2009–10 P06	**115**	**154**

more additional regressors to account for variability due to biogeochemical processes. These models are fitted separately to the data in the 200 dbar layer and to those in the 400 dbar layer, as the two layers experience different variability.

After being fitted to the data from the first four surveys, each of these models characterizes the non-anthropogenic part of the variability of total dissolved inorganic carbon as being linearly proportional to the variability of the regressors. Each model's residuals contain its version of the anthropogenic signal plus noise. Because the intercept terms contain the regional average of inorganic carbon from these four surveys (2077.3 for 200 dbar and 2122.4 for 400 dbar), that average is missing from the residuals. If the regression is to be regarded as filtering out variability without removing the mean, the means should be added back to the residuals.

Figure 4 shows the filtered data from these surveys as blue circles. Because the means have been restored, the panels for the default models, m0 at 200 and 400 dbar, show the actual measured values for dissolved inorganic carbon. The other panels show the data after they have been adjusted by the indicated model to remove natural variability. Note that, for all of the non-default models in both layers, the spread of the adjusted values is considerably less than that of the measured values, indicating that each of these models does indeed explain a sizable fraction of the variance. The filtered data in each panel are somewhat different, as each model provides a slightly different view of non-anthropogenic variability. With much better sampling, more regressors might safely be used and a better view of the natural variability might be possible. But given the available data, differences between models provide an indication of the uncertainties associated with the filtering.

Because of the limited nature of the sampling provided by the repeated surveys, overfitting is a concern. Using too many regressors to account for carbon's environmental co-variability

can limit a models ability to filter new data. Additional regressors can confuse the idiosyncrasies of the earlier data with meaningful information, causing the model to expect those same idiosyncrasies to manifest in data from the most recent survey. Thus, models m3, m4, and especially m5 are more likely to be overfitting the data than the other models.

To indicate the rate at which anthropogenic carbon has been accumulating during the temporal span of the first four surveys, a linear function of time has been fitted to the filtered values, as is illustrated in figure 4. The accumulation rates given by the slopes of the regression lines are listed in Table 3 for each model. Note that all methods for modeling natural variability yield higher accumulation rates than the those provided by m0. Also, the estimated rates indicate faster accumulation in the 200 dbar layer than in the 400 dbar layer. As these rate estimates were determined from the data from the first four surveys, they should be associated with the mean date of those observations – 1999.524 in the 200 dbar layer and 1999.424 in the 400 dbar layer. (To compute the fraction of year corresponding to month and day, observational dates were first converted to Julian days.).

Since each model characterizes the natural environmental variability differently, model-to-model differences in accumulation rate should be regarded as an indication of the uncertainty of the rate estimates stemming from the uncertainty of the environmental variability. Fitting the models to one pair of surveys and predicting for the other pair provided a quantitative measure of the accuracies of the rate estimates in an earlier study [1]. Here, a similar measure is provided by the ability to predict the anthropogenic component of the carbon measured in the most recent survey using models fitted to the four earlier surveys.

Predictions

After being fitted to the data from four earlier surveys, the models of Table 2 can be used with the values for their regressors measured during the most recent survey to remove the environmental variability from the new inorganic carbon data. These adjusted values are shown along with those from the earlier surveys in figure 4. Note that the regression line, which was fitted to the filtered data from first four surveys, does pass close to the center of the filtered data from the most recent survey. This indicates that, whichever model of environmental co-variability is used to account for natural variability, the following temporal regression does a fair job at projecting the rate at which carbon has been accumulating.

But each regression line passes below the blue cross marking the mean of the adjusted data. In other words, there is a consensus among models in both layers that more carbon has accumulated than should be expected based on the rates estimated from the earlier surveys. The amount that it exceeds expectations depends on the model used to account for natural environmental variability. Table 4 list the mean differences between the data filtered by the models and corresponding values predicted by the regression lines. This table also lists the errors associated with those differences, which include the standard error of the mean differences and the standard error of the temporal prediction at the mean date of the survey combined as the square root of the sum of the squares.

The largest excess in both layers is for the default models where nothing is done to account for natural variability. On the other hand, data for the 200 dbar layer that have been filtered by model m1, which uses co-variability of carbon with potential density to characterize natural variability, are quite well centered on the regression line, and the tabulated error indicates that the

Table 2. Regressors for the models used to account for environmental variability are indicated by crosses.

	1	density	nitrate	silicate	aou
m0	x				
m1	x	x			
m2	x	x	x		
m3	x	x	x	x	
m4	x	x	x		x
m5	x	x	x	x	x

The intercept term is indicated by 1, apparent oxygen utilization by aou.

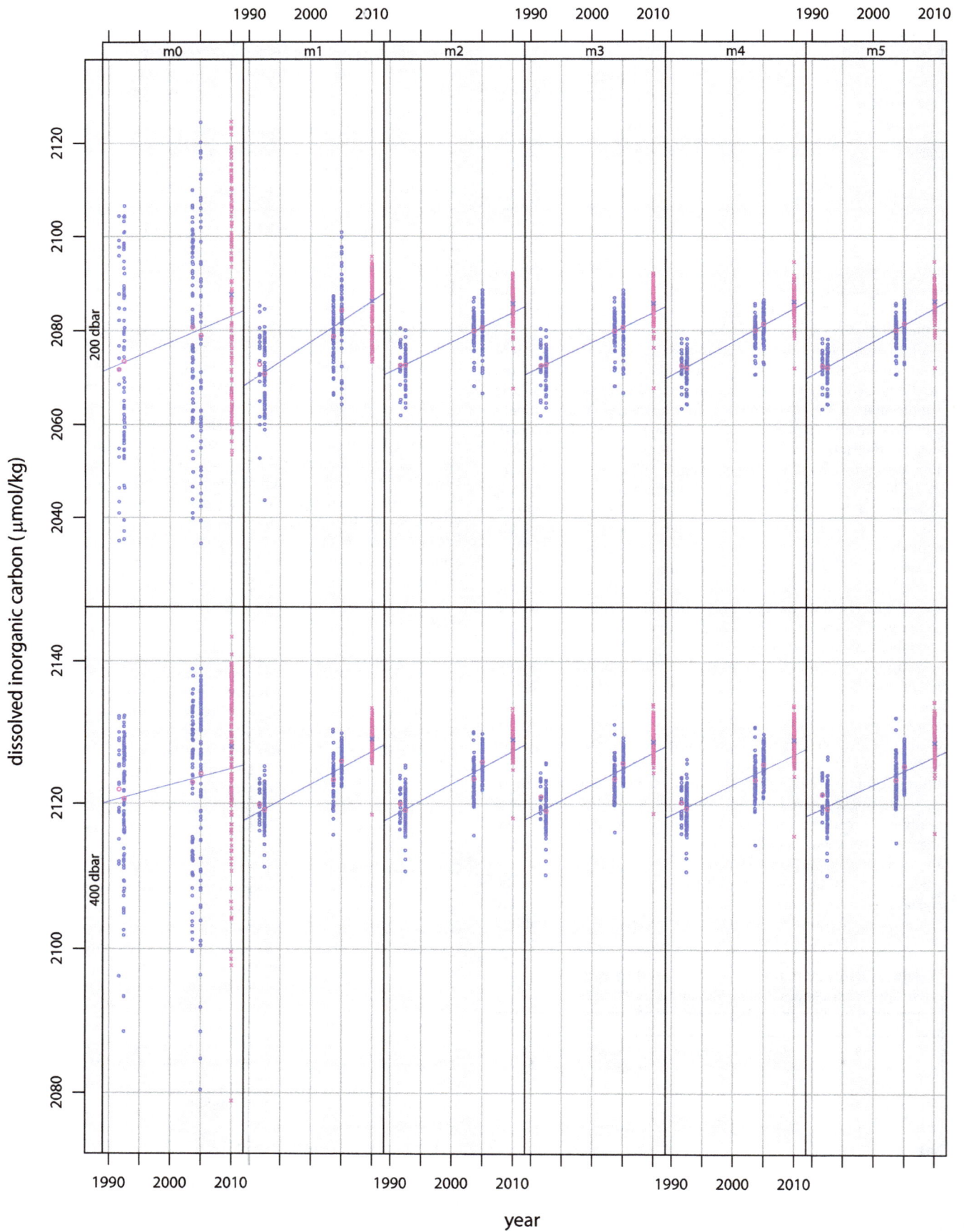

Figure 4. Projections based on the first four surveys. Blue circles indicate data from the first four surveys with environmental variability removed using the model indicated in the panel label. The regression lines have been fitted to these data. Magenta crosses indicate data from the most recent survey. Overlays in contrasting colors indicate the means of the filtered data for each of the five surveys.

extrapolated accumulation rate can be considered a perfect prediction. In contrast, model m2, which uses co-variability with both nitrate and potential density, leads to an underestimation of 1.90 μmol/kg of accumulated anthropogenic carbon; as the accumulation rate in the upper layer resulting from m2 filtering was estimated to be 0.628 μmol/kg/yr, the excess accumulation is equivalent to what would have been expected three years later. This excess accumulation is three times larger than the error of 0.64 μmol/kg. Similarly, except for m2 in the 200 dbar layer, the excess accumulation for all filtering models is greater than the tabulated error. The fact that nine of the ten models clearly underestimate the anthropogenic signal in the new data suggests that the accumulation rates in both layers might be increasing.

To gauge the size of the excess accumulation of anthropogenic carbon, the distances of the blue crosses from the regression lines can be compared with the similar offsets of the magenta circles marking the means of the data from the previous surveys. While the excess accumulation is generally larger, it can nevertheless be regarded as being of similar magnitude and as providing support for the accumulation rate being constant. The difficulty in deciding whether the rate of accumulation actually has been increasing or whether it simply should be revised upward in light of the new observations is clearly due to the paucity of the available observations.

Revised Analyses Using All Five Surveys

The most recent survey provides additional information about the nature of the environmental variability as well as about accumulated carbon from anthropogenic CO_2, so it is useful to redo the analysis using data from all five surveys. Figure 5 illustrates the results. Qualitatively each panel looks quite similar to its counterpart in figure 4. Even the groupings of points are similar. A close look reveals that the blue regression lines, which have been fitted to data from all five surveys are much closer to the fifth-survey means than are their counterparts in figure 4. Clearly, the points have moved and the slopes of the regression lines are different.

Table 5 lists the revised rates. Note that for each of the models in each of the layers the revised rates are larger than those based on only the first four surveys, which are listed in Table 3, but the amounts of increase are not the same for all models. In particular, using silicate to help to characterize environmental variability in

the upper layer now allows the m3 rate to be distinguished from m2 rate and the m5 rate from the m4 rate. Also note that the standard errors are smaller than for the previous estimates of the accumulation rates due to two factors: the increase in the number of samples and the increase in the interval over which the slope has been determined. The fact that the various models yield different results for the rates illustrates the difficulty in accounting for natural environmental variability. Consequently, model to model differences in the rates should be regarded as a measure of uncertainty. In the 200 dbar layer the accumulation rate is likely to be in the range of 0.7–0.8 μmol/kg/yr, and in the 400 dbar layer, 0.5–0.55 μmol/km/yr.

Split Interval Analyses

After accounting for natural variability seen in the data from all five surveys, it is possible to check whether or not anthropogenic carbon has been accumulating at a constant rate over the two decades by separately fitting trend lines to filtered data from the first three surveys and from the last three surveys. In doing so, the data from the 2003 P06 survey contribute to both fits. The results are indicated in figure 5 by the green lines, where dashes correspond to fits to the 1st, 2nd, and 3rd clusters of points and dots to fits to the 3rd, 4th, and 5th. In all panels the slopes determined from the more recent surveys can be seen to be larger than their counterparts from the earlier surveys.

The slopes of these regression lines, which provide estimates of the accumulation rates for the two periods, are listed in Table 6. All estimates of accumulation rates clearly increase from the first period to the second. The standard errors are all larger than those of Tables 3 and 5 due to the fits being determined by fewer data and over shorter intervals. Nevertheless, the differences between the earlier and later estimates appear appreciable when compared to the sizes of the errors. T-tests show all increases in accumulation rates to be significant at the 95% level or higher, except for the case with no adjustment for background variability in 200 dbar layer. Similarly, while the exceptionally low 1991–2003 rate for the 400 dbar m0 case leads to an extremely high level of significance for the increase, it might be better viewed as an

Table 3. Accumulation rates inferred from data from first 4 surveys (1992–2005).

	200 dbar	400 dbar
m0	0.557 (0.235)	0.231 (0.111)
m1	0.855 (0.081)	0.459 (0.025)
m2	0.628 (0.045)	0.461 (0.024)
m3	0.628 (0.045)	0.443 (0.024)
m4	0.705 (0.036)	0.418 (0.026)
m5	0.705 (0.036)	0.393 (0.027)

Units are μmol/kg/yr. Standard errors are indicated in parentheses.

Table 4. Mean differences between the anthropogenic component of the 2009–2010 data and their counterparts predicted by extrapolating the accumulation rates inferred from the first 4 surveys.

	200 dbar	400 dbar
m0	4.52 (3.40)	3.09 (1.59)
m1	0.07 (1.13)	1.76 (0.34)
m2	1.90 (0.64)	1.67(0.33)
m3	1.92 (0.64)	1.55 (0.34)
m4	1.43 (0.51)	2.05 (0.37)
m5	1.44 (0.51)	1.96 (0.39)

Units are μmol/kg. Errors indicated in parentheses include both the standard error of the mean distance from the regression line and the standard error of the regression line at the mean date of the survey.

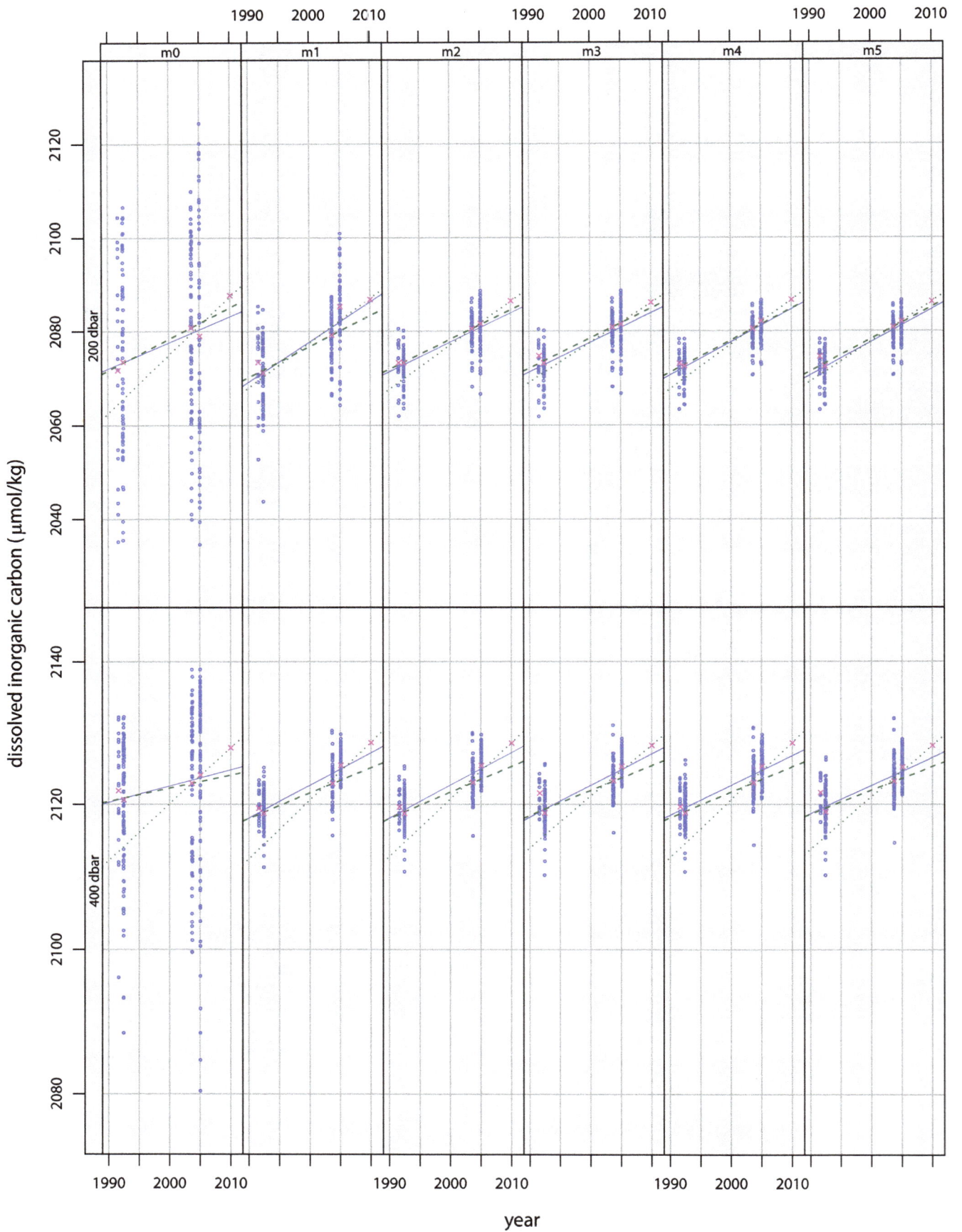

Figure 5. Analysis of data from all five surveys. Environmental variability has been removed using the model indicated in the panel label. Filtered data are indicated in blue. Magenta indicates the means of the filtered data for each survey. Solid black lines indicates fits to all 5 surveys, green dashed lines, fits to surveys 1–3, and green dotted lines, fits to surveys 3–5.

Table 5. Accumulation rates inferred from data from all 5 surveys (1992–2010).

	200 dbar	400 dbar
m0	0.780 (0.162)	0.386 (0.075)
m1	0.864 (0.052)	0.547 (0.016)
m2	0.728 (0.030)	0.543 (0.016)
m3	0.695 (0.032)	0.502 (0.017)
m4	0.778 (0.024)	0.537 (0.016)
m5	0.745 (0.026)	0.492 (0.018)

Units are μmol/kg/yr. Standard errors are indicated in parentheses.

Table 6. Accumulation rates before and after the 3rd survey.

	200 dbar		400 dbar	
	1991–2003	2003–2010	1991–2003	2003–2010
m0	0.67 (0.26)	1.24 (0.45)	0.19 (0.13)	0.79 (0.22)
m1	0.66 (0.08)	0.95 (0.14)	0.36 (0.03)	0.81 (0.04)
m2	0.63 (0.05)	0.94 (0.09)	0.37 (0.03)	0.79 (0.04)
m3	0.64 (0.05)	0.84 (0.09)	0.35 (0.04)	0.72 (0.05)
m4	0.68 (0.04)	0.95 (0.07)	0.36 (0.03)	0.80 (0.04)
m5	0.69 (0.05)	0.85 (0.07)	0.33 (0.04)	0.72 (0.05)

Adjustments for non-anthropogenic variability is based on data from all 5 surveys. Units are μmol/kg/yr. Standard errors are indicated in parentheses. Note that all treatments of natural variability indicate accelerating rates of accumulation in both layers.

indication of a problem associated with ignoring carbon's natural variability.

Discussion

In the upper ocean in the neighborhood of 32.5°S and 150°W an anthropogenic signal of increasing dissolved inorganic carbon is clearly seen. This is true whether or not any attempt is made to account for the effects of natural variability through carbon's co-variability with other environmental observables. Exploiting this co-variability leads to a quantitative indication of the uncertainty of the accumulation rate.

When added to the earlier data, those from the most recent survey provide upwardly revised accumulation rates. The nearly linear alignment of the means from the five surveys suggests that the accumulation rate has been nearly steady for two decades. However the fact that rates based on the first four surveys underestimate the anthropogenic component of carbon measured during the fifth survey suggests that the rate has been increasing. This conclusion is also supported by the split interval analyses, which show a clear increase in rates. However, the uncertainty in rates stemming from our limited ability to characterize the natural

variability is of a similar magnitude as the increase in rates, so this conclusion should be taken with caution.

These results apply only to this small region in the southeast Pacific. It would be interesting to see the results for other regions where multiple surveys provide sampling that is comparable to or better than what was available for this study. Further consensus of an accelerating accumulation in other regions would reinforce the notion that anthropogenic carbon's accumulation rate is increasing.

Acknowledgments

The many people who have spent much time and effort at sea to obtain the data used in this study deserve our thanks. These data as well as those from other repeated surveys can be downloaded at http://www.clivar.org/resources/data/clivar-carbon-and-hydrographic-sections.

Author Contributions

Conceived and designed the experiments: WCT. Analyzed the data: WCT. Wrote the paper: WCT.

References

1. Thacker WC (2012) Regression-based estimates of the rate of accumulation of anthropogenic CO$_2$ in the ocean: A fresh look. Marine Chemistry 132–133: 44–55, doi:10.1016/j.marchem2012.02.004.
2. Takahashi T, Goddard JG, Rubin S, Chipman DW, Sutherland SC, et al. (1995) Carbon Dioxide, Hydrographic, and Chemical Data Obtained in the Central South Pacific Ocean (WOCE Sections P17S and P16S) During the TUNES-2 Expedition of the R/V Thomas Washington, July-August, 1991. ORNL/CDIAC-86 NDP-054, Carbon Dioxide Information Analysis Center, Oak Ridge National Laboratory, U.S. Department of Energy, Oak Ridge, Tennessee.
3. Rubin S, Goddard JG, Chipman DW, Takahashi T, Sutherland SC, et al. (1998) Carbon Dioxide, Hydrographic, and Chemical Data Obtained in the South Pacific Ocean (WOCE Sections P16A/P17A,P17E/P19S, and P19C, R/V Knorr, October 1992-April 1993. ORNL/CDIAC-109 NDP-065, Carbon Dioxide Information Analysis Center, Oak Ridge National Laboratory, U.S. Department of Energy, Oak Ridge, Tennessee.
4. Johnson KM, Haines M, Key RM, Neill C, Tilbrook B, et al. (2001) Carbon Dioxide, Hydrographic, and Chemical Data Obtained During the R/V Knorr Cruises 138-3, -4, and -5 in the South Pacific Ocean (WOCE Sections P6E, P6C, and P6W, May 2 - July 30, 1992). ORNL/CDIAC-132 NDP-077, Carbon

Dioxide Information Analysis Center, Oak Ridge National Laboratory, U.S. Department of Energy, Oak Ridge, Tennessee.
5. Uchida H, Fukasawa M, Murata A (2005) WHP P6, A10, I3/I4 REVISIT DATA BOOK Blue Earth Global Expedition 2003 (BEAGLE2003) Volumes 1,2. Japan Agency for Marine-Earth Science and Technology.
6. Feely RA, Sabine CL, Millero FJ, Langdon C, Dickson AG, et al. (2008) Carbon Dioxide Hydrographic, and Chemical Data Obtained During the R/Vs Roger Revelle and Thomas G. Thompson Repeat Hydrography Cruises in the Pacific Ocean: CLIVAR CO2 Sections P16S 2005 (6 Janury -19 February 2005) and P16N 2006 (13 February - 30 March, 2006). ORNL/CDIAC-155 NDP-090, Carbon Dioxide Information Analysis Center, Oak Ridge National Laboratory, U.S. Department of Energy, Oak Ridge, Tennessee.
7. Wanninkhof R, Millero F, Swift J, Carlson C, McNichol A, et al. (2010) Carbon Dioxide, Hydrographic, and Chemical Data Obtained During the R/V Melville Cruise in the Pacific Ocean on CLIVAR Repeat Hydrography Section P06 2009 (November 21, 2009 - February 11, 2010). Technical report, Carbon Dioxide Information Analysis Center, Oak Ridge National Laboratory, U.S. Department of Energy, Oak Ridge, Tennessee. http://cdiac.ornl.gov/ftp/oceans/CLIVAR/P06_2009/.

3

Predicting Insect Migration Density and Speed in the Daytime Convective Boundary Layer

James R. Bell[1]*, **Prabhuraj Aralimarad**[2], **Ka-Sing Lim**[1], **Jason W. Chapman**[1,3]

1 Department of Agro-Ecology, Rothamsted Research, Harpenden, Hertfordshire, United Kingdom, **2** Department of Entomology, University for Agricultural Sciences, Raichur, Karnataka, India, **3** Environment and Sustainability Institute, University of Exeter, Penryn, Cornwall, United Kingdom

Abstract

Insect migration needs to be quantified if spatial and temporal patterns in populations are to be resolved. Yet so little ecology is understood above the flight boundary layer (i.e. >10 m) where in north-west Europe an estimated 3 billion insects km^{-1} month^{-1} comprising pests, beneficial insects and other species that contribute to biodiversity use the atmosphere to migrate. Consequently, we elucidate meteorological mechanisms principally related to wind speed and temperature that drive variation in daytime aerial density and insect displacements speeds with increasing altitude (150–1200 m above ground level). We derived average aerial densities and displacement speeds of 1.7 million insects in the daytime convective atmospheric boundary layer using vertical-looking entomological radars. We first studied patterns of insect aerial densities and displacements speeds over a decade and linked these with average temperatures and wind velocities from a numerical weather prediction model. Generalized linear mixed models showed that average insect densities decline with increasing wind speed and increase with increasing temperatures and that the relationship between displacement speed and density was negative. We then sought to derive how general these patterns were over space using a paired site approach in which the relationship between sites was examined using simple linear regression. Both average speeds and densities were predicted remotely from a site over 100 km away, although insect densities were much noisier due to local 'spiking'. By late morning and afternoon when insects are migrating in a well-developed convective atmosphere at high altitude, they become much more difficult to predict remotely than during the early morning and at lower altitudes. Overall, our findings suggest that predicting migrating insects at altitude at distances of ≈100 km is promising, but additional radars are needed to parameterise spatial covariance.

Editor: Zoe G. Davies, University of Kent, United Kingdom

Funding: Rothamsted Research is a national institute of bioscience strategically funded by the UK Biotechnology and Biological Sciences Research Council (BBSRC). PA's sabbatical at Rothamsted was funded by the University of Agricultural Sciences, Raichur. The funders had no role in study design, data collection and analysis, decision to publish, or preparation of the manuscript.

Competing Interests: The authors have declared that no competing interests exist.

* E-mail: james.bell@rothamsted.ac.uk

Introduction

Ecologists have argued that universal patterns should emerge from ecosystems despite their apparent complexity [1,2]. Particularly, patterns should be evident at given spatial and temporal resolutions because individuals within populations either interact individually in some way (parasitism, competition, predation etc) to produce a signal, or the community responds to an exogenous driver such as climate change that cascades through trophic levels [3–7]. In this paper we consider mixed populations of numerous insect species migrating at high altitudes (i.e. 150–1200 m above ground level (a.g.l.)) and ask the simple question, can average aerial densities and displacement speeds of 1.7 million insect migrants over the last decade be predicted using reductive linear models?

At and above 100 m a.g.l., the great majority of the daytime fauna in northern Europe comprises Homoptera (primarily aphids), small flies, beetles and parasitic Hymenoptera (parasitoids) [8]. Other groups, such as spiders and mites, can also be found at these heights [9]. Because convection currents circulate small arthropods around in the atmosphere, the fauna is not dissimilar to that captured by suction traps at a much lower height of 12.2 m

a.g.l. in which parasitoids, flies and aphids dominate a broad range of invertebrate taxa [10–11]. These species are either beneficial to agriculture (e.g. aphid biocontrol agents such as parasitoids, hoverflies, carabids and ladybirds), are pests or disease vectors (most aphids and some mites and flies), or contribute substantially to biodiversity [8,11].

Periodicity is a feature of flying insects which tend to have quite predictable diel activities, causing fluxes throughout the 24-hour period [12–14]. These diel fluxes are largely in response to temperature and light intensity changes, the effects of which are evident during dawn, the middle of the day, and at dusk, when pronounced density peaks are apparent [13–17]. In fine weather, insects take advantage of updrafts produced in unstable convective atmospheres and are then transported downwind at speeds often greatly in excess of their self-powered flight speeds [16,18–20]. Day-active species generally descend before dusk, but on some occasions when air-temperatures remain particularly warm at high-altitude daytime species have been known to remain aloft [12,21]. However, the night migratory flights are usually restricted to a different set of nocturnal insects, and these often become concentrated into layers as a result of the more stable atmosphere in the nocturnal boundary layer which allows insects to rapidly traverse hundreds of kilometres in fast-moving winds [13,18,20,22–25]. We do not consider nocturnal movements any further here however, but instead concentrate on the daytime phase which is less well understood.

With such a diverse fauna in the air and with each species or group seemingly conditioned to have discrete flight behaviours that include preferences for height, time and meteorological conditions, it might be expected that general patterns would be elusive. Indeed this is true within the flight boundary layer [FBL] below about 10 m [26]. Above this height, 'noisy' density profiles give way to two of the most compelling patterns. Firstly, log density of insects declines close to a linear fashion with log height, generating a negative slope that emerges as a result of changes in atmospheric stability [12,27]. The slope declines rapidly when the boundary layer is stable, and is markedly shallower if the atmosphere becomes convective [27]. This is an important finding because it indicates that empirical models should be able to relate measures of atmospheric stability, notably temperature and wind speed, to patterns of migration. The log density-log height relationship encapsulates insect populations in the vertical plane and may even be universal above a given height.

In the horizontal plane, Taylor's Power Law is manifest and simply describes log-transformed variance in abundance as a function of log transformed mean abundance [28]. The law was developed from measurements of insect density in the air in collaboration with C.G. Johnson in the 1950s [12]. Their earlier collaboration fuelled later entomological work on mean-variance relationships from aphids caught above the FBL [29–31]. In this horizontal plane at heights of 12.2 m, Bell et al. [32] recently described the abundance-occupancy relationship for 170 species of aphids migrating over the United Kingdom, showing that the occupancy and continuity (persistence) of aphids is a function of their log abundance which generates sigmoidal curves with varying lower and upper asymptotes.

In this paper we attempt to model insect densities and their speeds measured by radar at altitudes between 150–1200 m to test a biological hypothesis concerning the period when solar heating of the ground produces rising thermals that generate convective plumes. Our motivation is purely to reveal migration predictability that can be generalised at the community level. Our hypotheses are that exogenous drivers are linearly related to average aerial densities and displacement speeds. We use a 10-year dataset, the longest continuous time series of high-flying diurnal insects in the world, to model 1.7 million radar-detected insect targets that have masses between 5–700 mg. The novelty of this work is that whilst there are hundreds of studies of insects and spiders migrating within their FBL (i.e. usually ~0.5–10 m; [9,26,33–34]) there are few empirical non-invasive surveys of day-flying insects within the convective boundary layer [CBL] above the level of the FBL (Ecological: [13,19,23]; Meteorological: [35–36]). These latter studies either focus on a time-series of just a few days when either particular meteorological, or insect phenomena, were apparent or are purely meteorological. Thus, our study is unique in providing a long time series analysis with mixed meteorological and temporal effects on insect densities and displacement speeds.

Materials and Methods

Deriving Data From Vertical-looking Entomological Radar (VLR)

Data were derived from daily radar observations at two altitude ranges (150–300 m and 600–1200 m a.g.l.) using Rothamsted Research's vertical-looking radar (VLR) located in Harpenden (51°49′N; 0°22′W; Fig. 1) [15]. Target insects >5 mg migrating

through the radar beam are automatically and individually-resolved on a cycle of 5 minutes, once every 15 minutes, 24 h a day using a novel iterative procedure based on components of their complex Fourier transformations [37,38]. The process yields the horizontal speed, displacement direction, orientation and three radar scattering terms of the target and also calculates the distance of closest approach to the beam's axis of rotation and the time that this point was reached. All these parameters are then used to create a simulated signal and the correlation between the simulation and the actual radar return provides a quantitative estimate of how well our model has described the target [37]. It should be noted that targets are modelled as 'insects' as they do not have unique cross-sectional areas that would allow a species identification. It has been well established through aerial netting that these 'insects' collectively comprise diurnally-active flies, beetles, and true bugs, while the occasional large insects (>50 mg) are most likely butterflies, dragonflies and grasshoppers [13,15–16]. In the process of the procedure, the radar algorithms also filter out non-biological bodies such as raindrops, radar 'chaff' and aerial detritus, as well as bird and bat targets [39]. Ballooning spiders, flying aphids and other micro-insects fall below the minimum mass threshold (≈2 mg) for radar detection and are thus not included in our analyses [8,15]. Further, we excluded any target that had a mass greater than 700 mg as these are unlikely to be an insect [24]. Target insects were automatically logged and stored in a database ready for extraction.

VLR Data

We split daily data into 'early morning' (06:00–10:00 GMT) and 'late morning-afternoon' (10:00–18:00 GMT) to provide a contrast in potential atmospheric conditions between sunrise and a period around midday when the thermal input was likely to be

Figure 1. Location of the vertical-looking radars. The radars at Rothamsted, Hertfordshire (shaded) and Chilbolton, Hampshire (white-filled) are shown. The arrow indicates that the Euclidean distance between the sites is 104 km.

much higher (Fig. 2). Firstly, we sought to ascertain which meteorological variables predict migration intensity and speed at the lower altitude range (150–300 m) during the 'early morning'. Concomitantly, we asked if the same meteorological predictors were still relevant to insects flying at a much higher altitude (600–1200 m) during the late morning-afternoon period, hereafter denoted as the 'late period', when a large number of insects were known to be airborne and at altitude when air temperatures were high (Fig. 2). In both cases, we modelled the total density of insects and their average displacement speed (i.e. their speed in relation to the ground) as separate responses.

A single value for both density and speed was derived for each early and late period (during each day) over five months (May–September) for 10 years (2002 to 2011). All density values are expressed as counts per 10^7 m^3 and all displacement speeds as m s^{-1}. Aerial densities were corrected for the volume of air sampled, automatically taking account of wind speed. Only days with >10 radar-detected insects in the relevant altitude and time range were included in the analyses. Associated with early and late periods were a set of temporal factors: each period within each radar day was associated with a given week number, a month and year. It is important to note that there are normally few insects at high altitude in the early morning period but instead there is a progression of insects, known as a discharge, from the lower altitudes to the upper altitudes as convection develops toward midday. These high flyers then usually 'fall out' as nightfall approaches, most landing before sunset [12]. Hence, our experimental structure cannot be a fully crossed design, but instead it is fractional and reflects the biological nature of migration in the CBL.

The Rothamsted VLR was then compared with our other VLR, situated at the Chilbolton Facility for Atmospheric and Radio Research, Hampshire (2004 onwards) (51°09′N; 1°26′W; Fig. 1) in an attempt to derive parameters on the predictability of insect densities and displacement speeds in southern Britain. The radars are assumed to be independent because insect migrants are highly unlikely to travel between the sites due to large-scale circulation patterns. Data from Chilbolton was cross-tabulated to meet data requirements described above for Rothamsted (i.e. altitude, mass, period).

Meteorological Data

Air temperature and wind speeds at relevant altitudes were estimated using the operational mesoscale version of the UK Met Office's numerical weather prediction model, named the 'Unified Model' (hereafter referred to as 'MetUM'; see http://www.metoffice.gov.uk/research/modelling-systems/unified-model). For our analyses, we used MetUM mean outputs from 190 m and 770 m for the 'early morning' and 'late morning-afternoon' periods respectively.

Statistical Analysis

Data exploration. Given that high-altitude daytime insect densities and their associated displacement speeds are little understood, we use some basic models to examine their structure. Lorenz curves and associated statistics (i.e. Gini and asymmetry co-efficients) were used to examine the change in inequalities in each response [40]. Briefly, if all densities and velocities were the same the Gini coefficient describing the size of area of inequality from the Lorenz curve would be zero – that is, the relative mean difference would be the same. Additionally, bias can be examined using the axis of symmetry which has the coordinates (1, 0) to (0, 1). If S>1, the point where the Lorenz curve is parallel with the line of equality will also be above the axis of symmetry.

Correspondingly, if S<1, the point where the Lorenz curve is parallel to the line of equality is thus below the axis of symmetry. For both Gini and asymmetric co-efficients, 95% bootstrap confidence intervals were produced. Pearson moment correlations were used to express changes in density as a function of increasing displacement speeds.

Regression models with meteorological variables. The total density of insects in the air and displacement speeds of those insects once airborne were studied using generalized linear mixed effect models (GLMMs). GLMM used the method by Schall [41] in which models were fitted using penalised quasi-likelihood [42]. Early period densities; late period densities; early period displacement speeds and late period displacement speeds were the four responses that were considered. The fixed effects were temperature and wind, at the relevant heights to the response. The dispersion parameter (Ø) was estimated from the residual mean square which should approximate unity. We sought to minimize the difference from unity by removing redundant random effects. The maximal random effects structure was the product of all the effects (i.e. year*month*week) which allows for there to be exceptional weeks or months as separate interacting terms as well as individual single effects. It was quickly established, however, that over the decade of the time series that month was a redundant effect in the models which inflated Ø. Year.week was used as a single random effect and gave the closest approximation to a dispersion parameter (Ø) of unity for all models: this model infers that only the combined effect of year.week has a profound effect on the models, and, as additive terms year and week are not needed.

Within GLMMs $log(\mu_{ijk})$ is the expected response (on the link scale: log or identity), that insects are subject to prevailing weather, where i is the unit for j response (displacement speed or density) in k period (early, late and in which the height of the radar is synonymous with period and thus not crossed) and temperature ($l = -1,0,1,...27$) and wind ($m = 0,1,2,...21$) are the fixed effects of interest. The random effect b included only the interaction between year ($x = 1,2,3,...10$) and week ($z = 1,2,3...21$)) and estimated the variance component. Thus, the general form of the maximal model was:

$$\mu_{ijk} = \beta_0 + \beta_1 Temperature_l + \beta_2 Wind_m$$
$$+ \beta_3 (Temperature x Wind)_{lm} + b(Year_x.Week_z)$$

Distributional assumptions differed between total density of insects and their average velocity. Due to the very large variance heterogeneities and overdispersion in the errors, total density fluxes were modelled using the negative binomial with an aggregation parameter of unity and a logarithmic link [43]. Velocity measurements were instead modelled with a normal distribution and an identity link. We used reverse model selection procedures in our search for the parsimonious model. The Wald statistic assessed the contributions of individual terms in the fixed model: if a term produced non-significant values of P>0.05, the variable(s) was subsequently removed from the model, and the model updated until all terms had significance values P<0.05. Velocity measurements did not have very large variance heterogeneities but were instead modelled with a normal distribution and an identity link, producing very similar parameters to that of a linear mixed-effects model (not shown).

Throughout the text, subscripts e and l refer to the 'early period' and 'late period' respectively. These two subscripts are then combined with the notation for displacement speeds (V) and

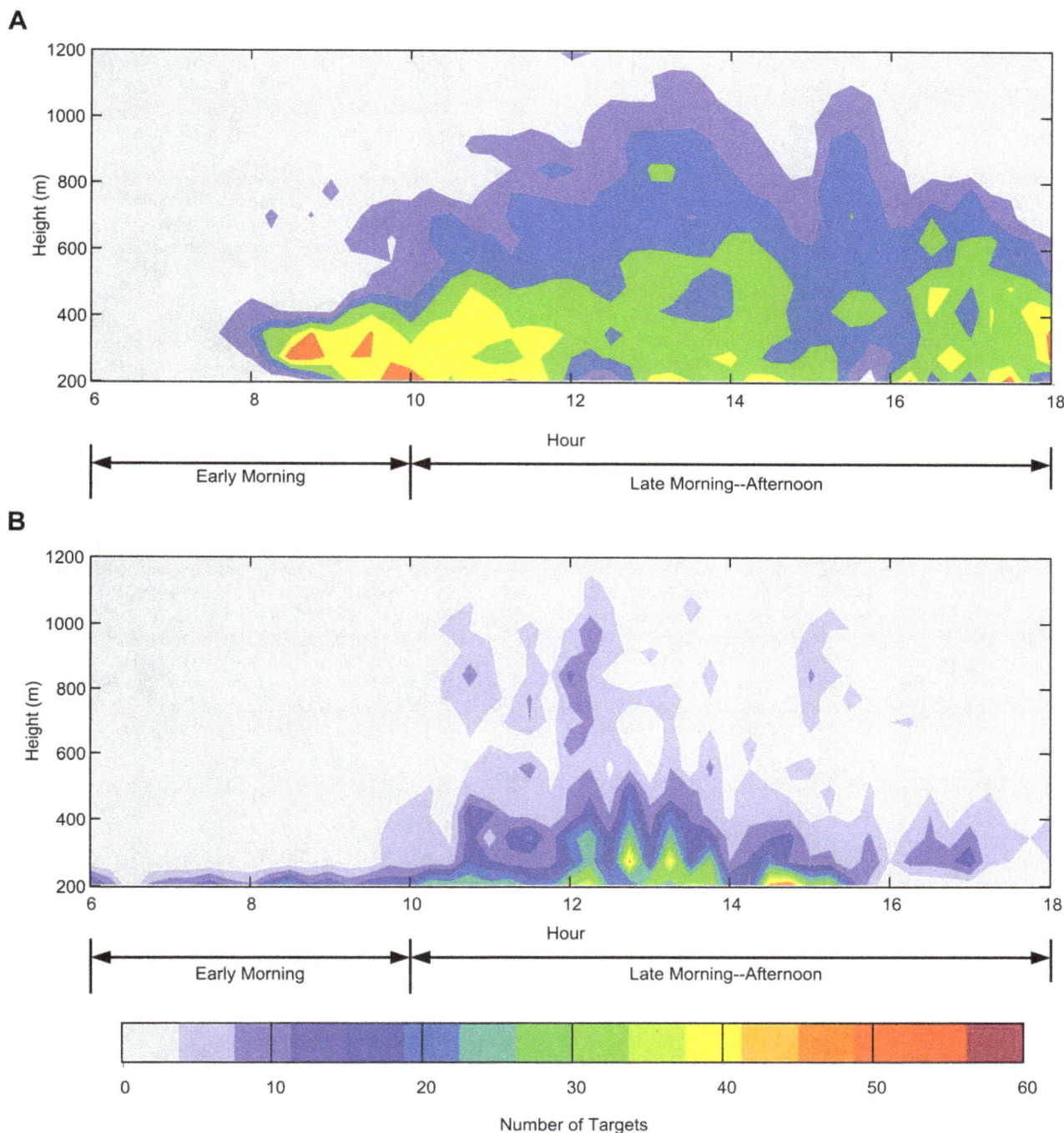

Figure 2. Time/height plots for numbers of insects recorded by a VLR at Cholbolton, Hampshire, UK. The colour scale bar refers to the number of individually-resolvable insects detected by the radar at each sampling height in each 5-minute period. The X-axis shows time of day (GMT), and the 'early' and 'late' analysis periods are indicated. A. Insect densities on a warm day (05 September 2004), when air temperatures at 10 m, 150 m and 600 m were 24.7°C, 22.8°C and 18.5°C respectively. B. Insect densities on a much cooler day (15 September 2004), when air temperatures at 10 m, 150 m and 600 m were 15.4°C, 13.2°C and 8.8°C respectively. Substantial density was constrained to time-periods and altitudes where air temperatures were relatively warm. Air temperatures were obtained from the UK Met Office's 'Unified Model' [14].

density (D) responses in the regression models or to parameters (i.e G, S, r and mean) in data exploration section.

Regression models between sites. Discrete regression models for both densities and speeds were developed for Chilbolton using ordinary least-squared regression models (OLS). The Rothamsted VLR was the response and Chilbolton

was the explanatory variable. In these models the sole aim was to establish how predictable VLR data were from a separation distance of 104 km away. We consider insects directly and do not consider analysing the residuals of the model to look at insects having taken account of the prevailing weather. The body of evidence suggests that the convective plumes that the insects are using to remain aloft are scaled at much finer levels (<5 km) than the distance between the radars ([19], and see discussion later). Speeds were log transformed and densities square-root transformed prior to analyses.

Results

Data Exploration

Insect displacement speeds tended to be close to symmetrical ($S_e = 0.9540$; 95% CI: 0.916, 0.997; $S_l = 0.9797$; 95% CI: 0.923, 1.050) and near to the line of equality ($G_e = 0.1679$; 95% CI: 0.159, 0.175; $G_l = 0.1446$; 95% CI: 0.134, 0.154). All the indications are that the distribution of displacement speeds, irrespective of time of day or flight altitude, was not skewed but uniform because both the relative mean difference between all values was broadly even and the distribution was closely aligned to the axis of symmetry (Fig. 3). As might be expected, at a higher altitude and during a later period of the day, insects travelled faster on average (mean = 14.29 m s^{-1}, se ±0.18) than those individuals earlier in the day and at lower altitude (mean = 7.83 m s^{-1}, se ±0.08).

Insect densities tended to depart radically from the line of equality ($G_e = 0.6037$; 95% CI: 0.588, 0.625; $G_l = 0.5456$; 95% CI: 0.510, 0.582), indicating that distributions were dominated by comparatively few but large values. Early morning densities were particularly asymmetrical about their distribution ($S_e = 0.8786$; 95% CI: 0.844, 0.919), although this asymmetry dissipated out by the later period ($S_l = 1.049$; 95% CI: 0.989, 1.094). Approximately, 50% of the insect density values above the median contained only ≈13% of the overall insect densities for both early and late periods (Fig. 3), suggesting that there are occasions when insect densities are super abundant but such events are uncommon. The mean

densities in the early morning are a particularly extreme case (mean = 2219, se ±99.69 per 10^7 m^3) in which the range is vast (min = 15.37; max = 16,409 per 10^7 m^3) although values remain modest even by the 75% percentile ($Q_2 = 3146$ per 10^7 m^3). Later on in the day, the range of insect densities is relatively small in comparison to the early period (min = 9.74; max = 1064 per 10^7 m^3).

The relationship between displacement speeds and densities was negative: as speeds increase towards the maximum for the period (early max = 17.78 m s^{-1}; late max = 27.48 m s^{-1}), insect densities decline. This decline is twice as fast in the morning than in the afternoon ($r_e = -0.2562$ P = <0.001; $r_l = -0.1444$; P = 0.002).

Regression models

The evidence for an interaction between temperature and wind that would then influence insect displacement speed or migration intensity was absent for both periods (temp.wind P>0.05: V_e [F = 0.52, ddf = 709.0, P = 0.470]; V_l [F = 1.78, ddf = 328.6, P = 0.183]; D_e [F = 2.15, ddf = 756.4, P = 0.143]; D_l [F = 0.31, ddf = 439, P = 0.575]). Instead, for insect displacement speeds it was possible to generate a very simple model that described how wind increased displacement speeds for early and late periods respectively (V_e[F = 1290.36, ddf = 736.2, P<0.001]; V_l[F = 593.48, ddf = 335.0, P<0.001]). In this model there is no requirement to include a temperature component (V_e [F = 0.68, ddf = 750.2; P = 0.411], V_l [F = 1.22, ddf = 384.4, P = 0.271]).

Both wind (D_e [F = 263.36; ddf = 742.5; P = <0.001]; D_l [F = 17.31; ddf = 428.0; P = <0.001]) and temperature (D_e [F = 240.19; ddf = 740.7; P = <0.001]; D_l [F = 41.96; ddf = 433.9; P = <0.001]) have a highly significant effect on insect densities which fall with increasing wind (early: −0.18 ±se 0.01; late: −0.05 ±se 0.01) and rise with increasing temperatures (early:

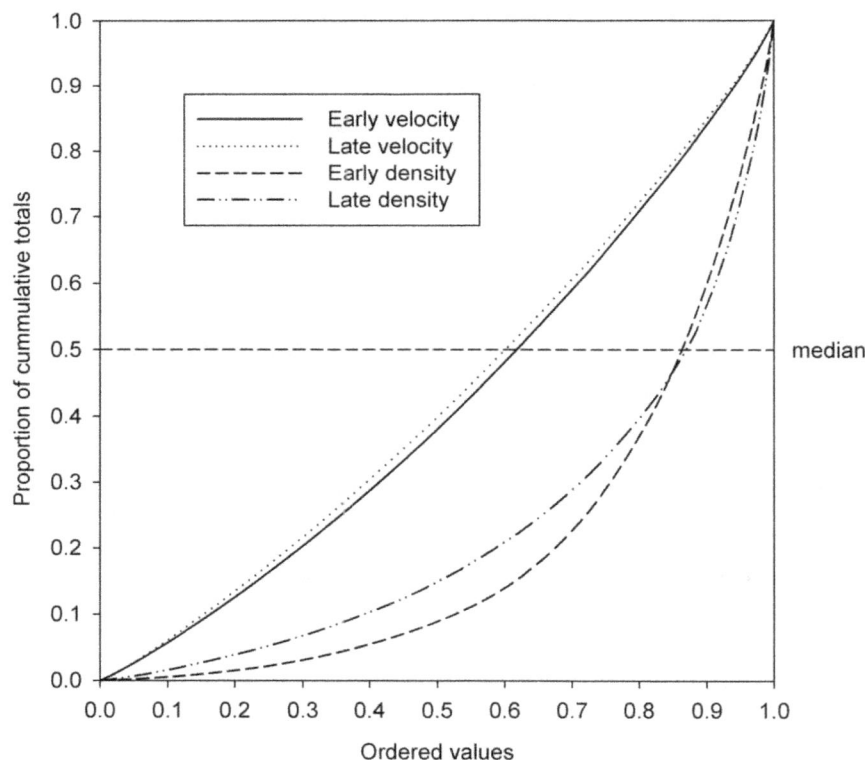

Figure 3. Lorenz curves for insect displacement speeds and densities by period.

0.212 $\pm$se<0.01; late: 0.09 $\pm$se 0.01). Notably, the rate of change is faster during the early period compared to the late for both fixed effects.

Predicting Rothamsted VLR dynamics remotely

For all regression models, Chilbolton was highly significant at predicting the dynamics of the prevailing insect displacement speeds (V_e [$F_{1,550} = 557.15$; P<0.01; $r^2 = 0.502$]; V_1 [$F_{1,340} = 564.19$; P<0.01; $r^2 = 0.623$]) and densities (D_e [$F_{1,550} = 513.87$; P<0.01; $r^2 = 0.48.2$]; D_1 [$F_{1,341} = 101.05$; P<0.01; $r^2 = 0.226$]) at Rothamsted, 104 km away. Insect displacement speeds have very similar slopes throughout the day (V_e $\beta = 0.623\pm0.026$; V_1 $\beta = 0.692\pm0.029$) (Fig. 4A, 4B), but densities differ markedly (D_e $\beta = 0.886\pm0.039$; D_1 $\beta = 0.231\pm0.023$). Further, as evidenced from the models, when insects are migrating in a convective atmosphere that is well developed during the late morning and afternoon, the migrating population becomes more difficult to predict even though the numbers are considerably smaller (Fig. 4C, 4D).

Discussion

To date the relationship between migrating aerial insects across space is little understood, mostly due to the technicalities of concurrently measuring the aerial biomass without significant bias. We found that $\approx$50% of the variance in radar-detected insect densities can be explained by a predictor radar $\approx$100 km away (Figs. 1 and 4A, 4B, 4C, 4D) which bodes well for models that aim to produce forecasts of flying insects. The size of the variance captured was hypothesized as being a function of the difference in landscape between these sites, particularly the surface characteristics (water, soil, vegetation and the built environment). For example, there is substantially more development around Rothamsted which sits on the boundary between urban and rural environments, compared to Chilbolton which is distinctly more rural. These surface characteristics are well known to determine the scale at which individual convection plumes operate [44–46]. In the horizontal plane, the size of the plume is known to be scaled to quite a fine resolution, much less than the Euclidean distance between the sites we studied (i.e. 105 km). For example, Kitchen and Caughey [47] suggest that the scaling lies somewhere between 100 m to 1 km although Bénard-type convective cells are 3–4 km across [48, Don Reynolds *pers. comm.*] which is not dissimilar to the size of more general plumes reported by Geerts and Miao [19]. From a bottom-up perspective, landscape heterogeneities that have an effect on the atmosphere tend to be scaled at distances greater than this, perhaps up to 5 km [46,49].

Thus, the scale at which landscape surface varies will directly impact the convection process and our models reflect that implicitly. Deforestation or wide-scale agricultural management over large areas for example, may expose the soil surface to greater thermal input and stronger convection as a result [44,50], which may promote migration. Topography also plays an important role: towards the west coast and on upland sites in the UK, surface winds are characterised by their high speeds [51] which may depress the onset of migration in these regions. Variation in both topography and surface characteristics determine what happens in the vertical plane too. Here, thermal plumes rise in the CBL in response to increasing surface temperatures and these convective updrafts carry warm air which cools, slowing in the rate of ascent as it rises to heights of up to $\approx$1–2 km until it finally loses its identity [48].

Given all these convective complexities, our finding that $\approx$50% of the variance in radar-detected insect densities can be explained

by a predictor radar $\approx$100 km away (Figs. 1 and 4A, 4B, 4C, 4D), is not only somewhat surprising, but it also highlights a need for more spatially replicated data to establish scaling rules and determine levels of spatial synchrony. We acknowledge that this is beyond the limits of our study but what we can conclude is that the best predictors for an entomological radar will be a radar that is sited in a landscape with the same configurational and compositional components [52]. Any prediction of the aerial biomass over space and time is, at the very least, a three-way interaction between the convective landscape, habitat and differences in individual behaviour (e.g. phenology/circadian rhythms). This is perhaps why we are better able to estimate speeds of the animals in transit than total densities because speed can be reduced to pure physics whereas densities cannot. Whilst diurnal insect displacement speeds we recorded agreed well with previous studies using radar [13], they correlate strongly with the actual speed of the wind at altitude [35–36]. For example, the mean displacement speed of the carabid beetle *Notiophilus biguttatus* between 150 to 408 m was 7.3 m s^{-1}, and the mean wind speed was 4.8 m s^{-1} [53], indicating that the mean air speed of the migrating *N. biguttatus* was small and around 2.5 m s^{-1}.

Speed and density were shown to be negatively related. Weak-flying, small-bodied (<10 mm) insects not only decline with increasing height [27] but we show that densities fall with increasing wind speed and rise with increasing temperatures (Fig. 2). The rate of change is fastest in the morning when the atmospheric stability is in a non-equilibrium state [48] yielding occasions, albeit rarely, when insect densities are super abundant at altitudes between 150–300 m (i.e. $\approx$16,000 per 10^7 m^3 period^{-1}). Considering all early mornings studied, the majority yield only modest densities, much less than 3,500 insects (75% percentile) which is a product of non-ideal take-off conditions and concomitantly weaker thermals in a windier atmosphere.

The interplay between speed and density is an interesting facet that may have arisen from turbulence which redistributes the fauna laterally. Alternatively, the fall in numbers as speeds increase may simply indicate that for an increasing proportion of the potential aerial fauna, conditions are simply not suitable for flight which is either not initiated or curtailed [12]. Flight conditions, particularly temperature-insect relationships at altitude, have been described formally since Johnson [12]. More recently Wood et al. [23] established that warmer days are associated with more aerial migration and that the minimum threshold was around 13–14°C. Hotter days are associated with more frequent, longer-lasting and deep convective plumes allowing more insects to be spread through the CBL (Fig. 2). Interestingly, our results also show that at a higher altitude and during a later period of the day, insects travelled horizontally twice as fast on average (14.29 m s^{-1}, se $\pm$0.18) than those individuals earlier in the day and at lower altitude (7.83 m s^{-1}, se $\pm$0.08), which is a function of wind speed and consistent with transport by updrafts (i.e. thermals that are driven by convection) within the CBL [23].

Revisiting our original hypotheses regarding the prediction of average aerial densities and displacement speeds and also in light of the complex meteorology discussed so far, we found that insect densities become much more difficult to predict remotely, especially by the afternoon. This is most likely because during fine weather in summer, morning may often begin with the erosion of any stable nocturnal temperature inversion and by 11am convection may increase to heights of 750–900 m in the UK depending on location, wind and solar input [54]. In response to these thermals, there may be large numbers of invertebrates that have taken advantage of the updrafts and have risen to $\approx$800 m by late morning [13] —the weather now becomes increasingly

A

B

C

D

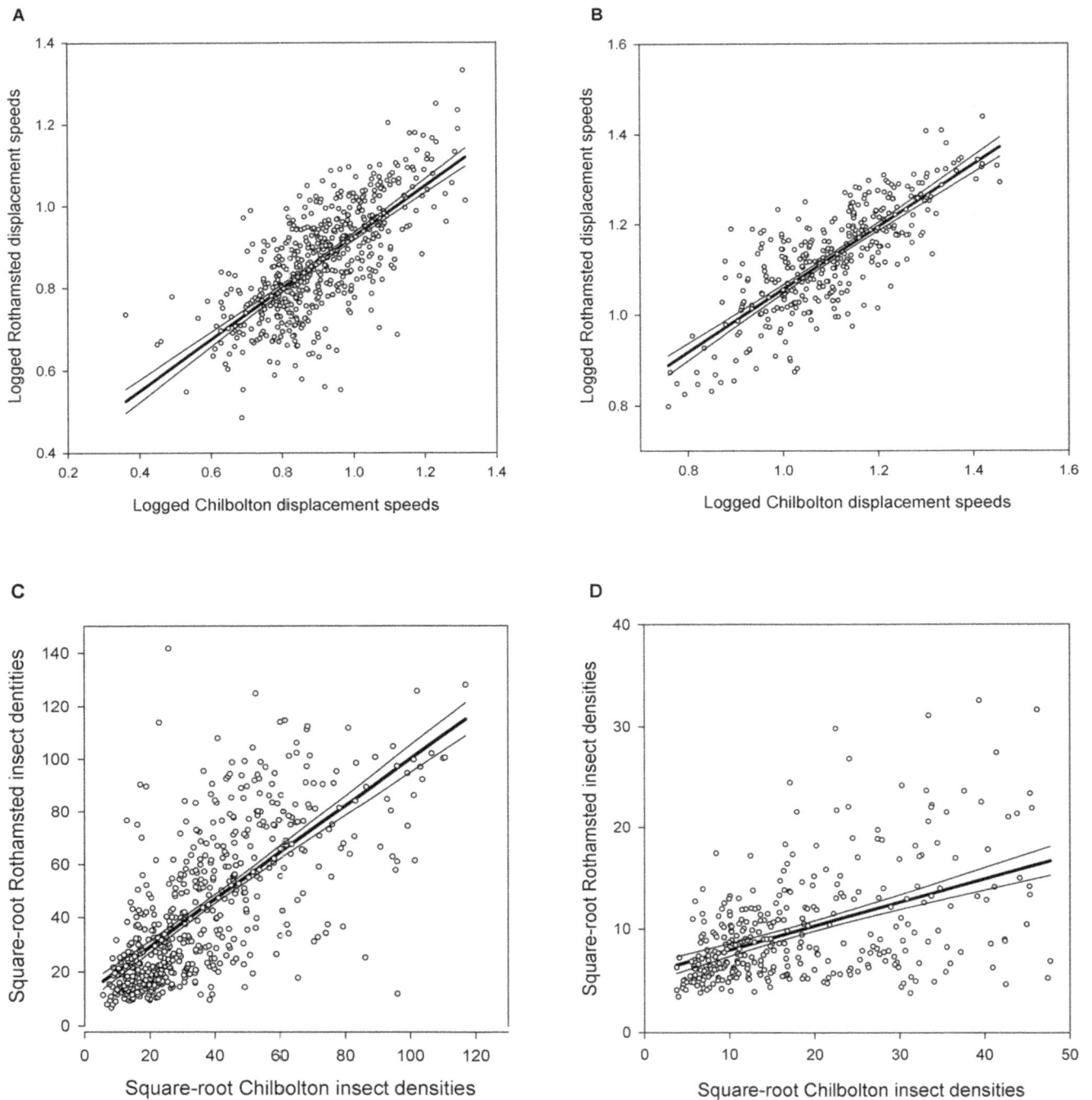

Figure 4. Linear regressions. Fitted and observed relationships with 95% confidence intervals between Chilbolton and Rothamsted logged displacement speeds by period: A. early period; B. late period, and, Chilbolton and Rothamsted square-root transformed insect densities by period: C. early period, D. late period.

heterogeneous and varying in spatial extent. Within the plumes, wind speeds are driven by convection wherein 80% of the turbulence is explained by thermals with updrafts reaching 1–2 m s^{-1} in the mixed layer below the cloud base, but averaging 0.5 m s^{-1} [36,47,55]. Above the base of the cloud when convection has reached its highest extent by mid-afternoon, the atmospheric boundary layer may reach heights of 1–2 km in which wind speeds approach 5 m s^{-1} at the top of the clouds

[23,36,47]. But by afternoon, the way in which the numbers of insects are then redistributed in these plumes becomes quite difficult to predict remotely at scales greater than the plumes themselves (i.e. ≈5 km). Thus it is clear that entomologically the dynamics are not consistent across space and that significant spatial covariance may be ultimately elusive at regional scales.

Finally, it seems a widely held misconception that the act of migrating is a risky redistribution strategy, given the heights individuals achieve. As Johnson [12] states, the cost associated with moving large distances is low at altitude because most insects are vagile even well above 1000 m where the atmosphere is quite cold. The real cost of migration is deferred because whilst the actual cost of transit is low, the great majority of insects may experience the cost of moving during and after deposition [56] thus advocating a mixed Evolutionary Stable Strategy for long distance movements. For example, Ward et al. [57] showed that after migrating, less than 1% of the bird cherry-oat aphid populations were able to find hosts. However, larger insect migrants are known to utilise wind currents in a highly efficient manner, thus reducing the probability of being dispersed to unsuitable habitats [20,24,58].

Conclusions

Insect migration studies have made major contributions to ecology [3,12,28–32] and have been complemented by more general studies on effects of surface landscape heterogeneity at various scales [50] and the meteorology itself [35–36]. Migration studies are important because of the sheer abundance of insects that use the CBL to move. For example, Chapman et al. [15] established that for the southern U.K. the size of the migrating populations is equivalent to 3 billion insects km^{-1} $month^{-1}$ comprising pests, beneficial insects and other species that contribute to biodiversity. Indeed it is because of the very nature of migrating pests and beneficial insects that it is prescient that we understand their dynamics to better deliver food security and future pest control. We have shown that it is possible to generalize about the aerial fauna over a long time series which is a significant step forward in the ecology of aerial migrants. Meteorological mechanisms, principally related to wind speed and temperature, drive variation in density and displacements speeds with increasing altitude. A strong linear relationship was apparent between sites separated by ≈ 100 km, 10-fold larger than previously expected (i.e. 1–5 km), which will go some way in meeting the challenge of forecasting migration.

Acknowledgments

We thank Sue Welham, Don Reynolds, Mark Mallott and Andy Reynolds who have all been generous with their knowledge and expertise. We are grateful to those who have spent time developing the VLR in the early years, particularly Alan Smith and Joe Riley.

Author Contributions

Conceived and designed the experiments: JRB PA JWC. Performed the experiments: KSL JWC. Analyzed the data: JRB. Contributed reagents/materials/analysis tools: KSL JWC. Wrote the paper: JRB.

References

1. Bascompte J, Sole RV (1995) Rethinking complexity: Modelling spatiotemporal dynamics in ecology. TREE 10: 361–366.
2. Storch D, Gaston KJ (2004) Untangling ecological complexity on different scales of space and time. Basic Appl Ecol 5: 389–400.
3. Hanski I (1998) Metapopulation dynamics. Nature 396: 41–49.
4. Gaston KJ (2000) Global patterns in biodiversity. Nature 405: 220–227.
5. Gaston KJ, Blackburn TM, Greenwood JJD, Gregory RD, Quinn RM, et al. (2000) Abundance-occupancy relationships. J Appl Ecol 37: 39–59.
6. McGill BJ, Etienne RS, Gray JS, Alonso D, Anderson MJ, et al. (2007) Species abundance distributions: moving beyond single prediction theories to integration within an ecological framework. Ecol Lett 10: 995–1015.
7. Thackeray SJ, Sparks TH, Frederiksen M, Burthe S, Bacon PJ, et al. (2010) Trophic level asynchrony in rates of phenological change for marine, freshwater and terrestrial environments. Global Change Biol 16: 3304–3313.
8. Chapman JW, Reynolds DR, Smith AD, Smith ET, Woiwod IP (2004) An aerial netting study of insects migrating at high-altitude over England. Bull Entomol Res 94: 123–136.
9. Bell JR, Bohan DA, EM Shaw, Weyman GS (2005) Ballooning dispersal using silk: world fauna, phylogenies, genetics and models. Bull Entomol Res 46: 69–114.
10. Benton TG, Bryant DM, Cole L, Crick HQP (2002) Linking agricultural practice to insect and bird populations: a historical study over three decades. J Appl Ecol 39: 673–87.
11. Shortall CR, Moore A, Smith E, Hall MJ, Woiwod IP, et al. (2009) Long-term changes in the abundance of flying insects. Insect Conserv Divers 2: 251–260.
12. Johnson CG (1969) Migration and Dispersal of Insects by Flight. London: Methuen.
13. Reynolds DR, Smith AD, Chapman JW (2008) A radar study of emigratory flight and layer formation at dawn over southern Britain. Bull Entomol Res 98: 35–52.
14. Wood CR, Reynolds DR, Wells PM, Barlow JF, Woiwod IP, et al. (2009) Flight periodicity and the vertical distribution of high-altitude moth migration over southern Britain. Bull Entomol Res 99, 525–535.
15. Chapman JW, Reynolds DR, Smith AD (2003) High-altitude insect migration monitored with vertical-looking radar. Bioscience 53: 503–511.
16. Chapman JW, Drake VA, Reynolds DR (2011) Recent insights from radar studies of insect flight. Ann Rev Entomol 56: 337–356.
17. Reynolds AM, Reynolds DR, Smith AD, Chapman JW (2010) Orientation cues for high-flying nocturnal insect migrants: do turbulence-induced temperature and velocity fluctuations indicate the mean wind flow? PloS One 5: e15758.
18. Drake VA, Farrow RA (1988) The influence of atmospheric structure and motions on insect migration. Ann Rev Entomol 33: 183–210.
19. Geerts B, Miao Q (2005) Airborne radar observations of the flight behavior of small insects in the atmospheric convective boundary layer. Environ Entomol 34: 361–377.
20. Chapman JW, Klaassen RHG, Drake VA, Fossette S, Hays GC, et al. (2011) Animal orientation strategies for movement in flows. Curr Biol 21: R861–R870.
21. Reynolds DR, Chapman JW, Edwards AS, Smith AD, Wood CR, et al. (2005) Radar studies of the vertical distribution of insects migrating over southern Britain: the influence of temperature inversions on nocturnal layer concentrations. Bull Entomol Res 95: 259–274.
22. Drake VA (1984) The vertical distribution of macroinsects migrating in the nocturnal boundary layer: a radar study. Bound-Lay Meteorol 28: 353–74.
23. Wood CR, O'Connor EJ, Hurley RA, Reynolds DR, Illingworth AJ (2009) Cloud-radar observations of insects in the UK convective boundary layer. Meteorol Appl 16: 491–500.
24. Chapman JW, Nesbit RL, Burgin LE, Reynolds DR, Smith AD, et al. (2010) Flight orientation behaviors promote optimal migration trajectories in high-flying insects. Science 327: 682–685.
25. Reynolds AM, Reynolds DR, Smith AD, Chapman JW (2010) A single wind mediated mechanism explains high altitude 'non-goal oriented' headings and layering of nocturnally migrating insects. Proc Roy Soc B 277: 765–772.
26. Taylor LR (1960) The distribution of insects at low levels in the air. J Anim Ecol 29: 45–63.
27. Reynolds AM, Reynolds DR (2009) Aphid aerial density profiles are consistent with turbulent advection amplifying flight behaviours: abandoning the epithet 'passive'. Proc. Roy. Soc B 276: 137–143.
28. Taylor LR (1961) Aggregation, variance and mean. Nature 189: 732–735.
29. Taylor LR, Woiwod IP (1980) Temporal stability as a density-dependent species characteristic. J Anim Ecol 49: 209–224.
30. Taylor LR, Woiwod IP, Perry JN (1980) Variance and the large scale spatial stability of aphids, moths and birds. J Anim Ecol 49: 831–854.
31. Taylor LR, Perry JN, Woiwod IP, Taylor AR (1988) Specificity of the spatial power-law exponent in ecology and agriculture. Nature 332: 721–722.
32. Bell JR, Taylor MS, Shortall CR, Welham SJ, Harrington R (2012) The trait ecology and host plants of aphids and their distribution and abundance over the United Kingdom. Global Ecol Biogeog 20: 405–415.
33. Reynolds AM, Bohan DA, Bell JR (2007) Ballooning dispersal in arthropod taxa: conditions at take-off. Biol Lett 3: 237–240.
34. Byers JA (2011) Analysis of vertical distributions and effective flight layers of insects: three-dimensional simulation of flying insects and catch at trap heights. Environ Entomol 40: 1210–1222.
35. Achtemeier GL (1991) The use of insects as tracers for "clean air" boundary-layer studies by Doppler radar. J Atmos Oceanic Technol 8: 746–764.
36. Chandra AS, Kollias P, Giangrande SE, Klein SA (2010) Long-term observations of the convective boundary layer using insect radar returns at the SGP ARM climate research facility. J Climate 23: 5699–5714.
37. Smith AD, Riley JR, Gregory RD (1993) A method for routine monitoring of the aerial migration of insects by using a vertical-looking radar. Phil Trans Roy Soc B340, 393–404.
38. Smith AD, Reynolds DR, Riley JR (2000) The use of vertical-looking radar to continuously monitor the insect fauna flying at altitude over southern England. Bull Entomol Res 90: 265–277.

39. Chapman JW, Smith AD, Woiwod IP, Reynolds DR, Riley JR (2002) Development of vertical-looking radar technology for monitoring insect migration. Comput Electron Agric 35: 95–110.

40. Damgaard C, Weiner J (2000) Describing inequality in plant size or fecundity. Ecology 81: 1139–1142.

41. Schall R (1991) Estimation in generalized linear models with random effects. Biometrika 78: 719–727.

42. Bolker BM, Brooks ME, Clark CJ, Geange SW, Poulsen JR, et al. (2009) Generalized linear mixed models: a practical guide for ecology and evolution. TREE 24: 127–135.

43. Zuur AF, Ieno EN, Walker NJ, Saveliev AA, Smith GM (2009) Mixed Effects Models and Extensions in Ecology with R. London, Springer.

44. Betts AK, Ball JH, Beljaars ACM, Miller MJ, Viterbo PA (1996) The land surface-atmosphere interaction: A review based on observational and global modeling perspectives. J Geophys Res 101: 7209–7225.

45. Cleugh HA, Grimmond CSB (2001) Modelling regional scale surface energy exchanges and CBL growth in a heterogeneous, urban-rural landscape. Bound-Lay Meteorol 98: 1–31.

46. Weaver CP, Avissar R (2001) Atmospheric disturbances caused by human modification of the landscape. B Am Meteorol Soc 82: 269–281.

47. Kitchen M, Caughey SJ (1981) Tethered-balloon observations of the structure of small cumulus clouds. Quart J Roy Meteorol Soc 107: 853–874.

48. Ahrens CD (2000) Meteorology today. London: Sixth edn. Brooks/Cole.

49. Letzel MO, Raasch S (2003) Large eddy simulation of thermally induced oscillations in the convective boundary layer. J Atmos Sci 60: 2328–2341.

50. Nathan R, Sapir N, Trakhtenbrot A, Katul GG, Bohrer G, et al. (2005) Long-distance biological transport processes through the air: can nature's complexity be unfolded in silico? Divers Distrib 11 131–137.

51. Palutikof J, Holt T, Skellern A (1997) Wind, resource and hazard. In: Hulme M, Barrow E, editors. Climates of the British Isles. Present, past and future. London, Routledge. pp. 220–242.

52. Fahrig L, Baudry J, Brotons L, Burel FG, Crist TO, et al. (2011) Functional landscape heterogeneity and animal biodiversity in agricultural landscapes. Ecol Lett 14: 101–112.

53. Chapman JW, Reynolds DR, Smith AD, Riley JR, Telfer MG, et al. (2005) Mass aerial migration in the carabid beetle Notiophilus biguttatus. Ecol Entomol 30: 264–272.

54. Moores WH, Caughey SJ, Readings CJ, Milford JR, Mansfield DA, et al. (1979) Measurements of boundary layer structure and development over SE England using aircraft and tethered balloon instrumentation. Quart J Roy Meteorol Soc 105, 397–421.

55. Rotach MW, Gryning S-E, Tassone C (1996) A two-dimensional Lagrangian stochastic dispersion model for daytime conditions. Quart J Roy Meteorol Soc 122: 367–389.

56. DeAngelis DL, Wolkowicz GSK, Lou Y, Jian YX, Novak M, et al. (2011) The effect of travel loss on evolutionarily stable distributions of populations in space. Am Nat 178:15–29.

57. Ward SA, Leather SR, Pickup J, Harrington R (1998) Mortality during dispersal and the cost of host-specificity in parasites: how many aphids find hosts? J Anim Ecol 67: 763–773.

58. Chapman JW, Bell JR, Burgin LE, Reynolds DR, Pettersson LB, et al. (2012) Seasonal migration to high latitudes results in major reproductive benefits in an insect. PNAS 109: 14924–14929.

Hydroclimate Variations in Central and Monsoonal Asia over the Past 700 Years

Keyan Fang[1,2]*, Fahu Chen[3]*, Asok K. Sen[4], Nicole Davi[5], Wei Huang[3], Jinbao Li[6], Heikki Seppä[1]

1 Key Laboratory of Humid Subtropical Eco-geographical Process (Ministry of Education), College of Geographical Sciences, Fujian Normal University, Fuzhou, Fujian province, China, **2** Department of Geosciences and Geography, University of Helsinki, Helsinki city, Helsinki, Finland, **3** Key Laboratory of Western China's Environmental Systems (MOE), Lanzhou University, Lanzhou, Gansu province, China, **4** Richard G. Lugar Center for Renewable Energy and Department of Mathematical Sciences, Indiana University, Indianapolis, Indiana, United States of America, **5** Tree-Ring Lab, Lamont-Doherty Earth Observatory, Columbia University, Palisades, New York, United States of America, **6** Department of Geography, University of Hong Kong, Hong Kong city, Hong Kong, China

Abstract

Hydroclimate variations since 1300 in central and monsoonal Asia and their interplay on interannual and interdecadal timescales are investigated using the tree-ring based Palmer Drought Severity Index (PDSI) reconstructions. Both the interannual and interdecadal variations in both regions are closely to the Pacific Decadal Oscillation (PDO). On interannual timescale, the most robust correlations are observed between PDO and hydroclimate in central Asia. Interannual hydroclimate variations in central Asia are more significant during the warm periods with high solar irradiance, which is likely due to the enhanced variability of the eastern tropical Pacific Ocean, the high-frequency component of PDO, during the warm periods. We observe that the periods with significant interdecadal hydroclimate changes in central Asia often correspond to periods without significant interdecadal variability in monsoonal Asia, particularly before the 19th century. The PDO-hydroclimate relationships appear to be bridged by the atmospheric circulation between central North Pacific Ocean and Tibetan Plateau, a key area of PDO. While, in some periods the atmospheric circulation between central North Pacific Ocean and monsoonal Asia may lead to significant interdecadal hydroclimate variations in monsoonal Asia.

Editor: Eryuan Liang, Chinese Academy of Sciences, China

Funding: This research is supported by National Basic Research Program of China (2012CB955301), National Science Foundation of China (41210002), the Nordic top-level research initiative CRAICC (Cryosphere-atmosphere interactions in a changing Arctic climate) and the National Science Foundation fellowship (NSF AGS-PRF1137729). The funders had no role in study design, data collection and analysis, decision to publish, or preparation of the manuscript.

Competing Interests: The authors have declared that no competing interests exist.

* Email: kujanfang@gmail.com (KF); fhchen@lzu.edu.cn (FC)

Introduction

Hydroclimate changes in central and monsoonal Asia have generated much concern due to their close linkages with water resources, critical for increasing population and economical development [1,2]. Central and monsoonal Asia are strongly influenced by the westerlies and Asian summer monsoon. Hydroclimate changes in the two regions were out-of-phase or anti-phase on glacial-interglacial, millennial and centennial timescales [3,4,5]. Possible mechanisms related to this inverse relationship include the different responses of hydroclimate changes to external forcings (e.g. orbital changes and solar irradiance), the boundary conditions (e.g. ice volume), the internal ocean-atmospheric feedbacks (e.g. North Atlantic Oscillation (NAO) and Atlantic Multidecadal Oscillation (AMO)) and the regional topographic features (e.g. Tibetan Plateau), which can modulate the strength of the Asian summer monsoon and westerlies [3,4,5]. However, hydroclimate variations in central and monsoonal Asia at the interannual and interdecadal timescales and their relationships remain unclear [6,7].

Tree-ring records are very useful for investigating the spatio-temporal climate changes extending before the industrial era on interannual and interdecadal timescales due to their annual resolution and wide spatial distribution, although the long-term variations on timescales longer than a century may not be well preserved due to the "segment length curse" problem in the developments of tree-ring chronology [8,9]. The Palmer Drought Severity Index (PDSI) is a hydroclimate metric of the moisture deficiency relative to the normal condition and is scaled to have equal mean across space [10,11]. Tree rings have been successfully used to reconstruct the PDSI to establish the PDSI atlas for the past 700 years over the entire monsoonal Asia and most of the central Asia, i.e. the Monsoon Asia Drought Atlas (MADA) [9]. In this study, we used the MADA to provide a comprehensive investigation of the hydroclimate variability of central and monsoonal Asia and their relationships.

Data and Methods

Climate data

The study region ($60°E–160°E$, $10°S – 60°N$) covers regions of topographic complexity from the Tibetan Plateau to the low-lying regions such as the basins and river valleys (Figure 1). Precipitation generally decreases from the coast to the inland since the monsoon strength decays according to its distance to ocean [3]. The driest regions are the low-lying inland regions, e.g. the Taklimakan desert in Tarim Basin and the wettest region is the southern windward boundary of the Himalayan Mountains. Instrumental PDSI data were derived from a global dataset with $2.5° \times 2.5°$ gridding from 1870 to 2005 [11]. The MADA consists of 534 grid cells of summer (June-July-August) PDSI records in central and monsoonal Asia, reconstructed from 327 tree-ring chronologies [9], most of which start since 1300 (504 grids). The point-by-point regression method was employed to locate the nearby tree-ring chronologies according to a search radius [9]. The PDSI data in regions with sparse tree-ring chronologies were reconstructed from remote regions and thus may contain some biases. For example, there are limitations in recovering the drought patterns in southern China, central Inner Mongolia and south central Mongolia where limited tree-ring chronologies are available [12]. The MADA is normally distributed with extremely dry (<-4) and wet (>4) PDSI values accounting for only a small percentages [12]. This indicates that the reconstructions are reasonable for spatiotemporal comparisons [13]. All the PDSI reconstructions were detrended by fitting a 150-year spline to remove the centennial or multi-centennial variations and highlight the interdecadal variations.

Analytical methods

We investigate the percentage of areas affected by drought over the past 700 years, i.e. the Drought Area Index (DAI). The DAI is calculated as a ratio between the number of PDSI grids exceeding a threshold value and the sum of the PDSI grids of a given year. That is, high DAI indicates enlarged dry areas, and vice versa. This index has been applied in tree-ring based reconstructions in western North America [14]. The DAI pays special attention on the spatial features of the hydroclimate changes. As is known, the long-term climate changes in the tree-ring based reconstructions may be removed in the curve-fitting standardization process. While the spatial features of the DAI can show some long-term changes even if the low-frequency variations of the reconstructions for individual grids are removed. That is, the long-frequency information of the spatial coverage of the droughts can be better recovered relative to the amplitudes of the reconstructed droughts using the curve-fitting standardization method. In this study, we tested the use of the threshold values of lower than -1 and -2 that result in similar results and we thus only shown the DAI variations according to the threshold value lower than -1. We herein calculated the DAI for the arid central Asia and the monsoonal Asia [3], separately, due to the different climate regimes between them (Figure 1). The spectral properties of these time series were detected using the multi-taper method (MTM) that is particularly efficient for short time series [15]. We applied the wavelet analysis based on a Morlet function to transform the DAI series into a time-frequency profile [16,17]. We additionally calculate the Cross Wavelet Transform (XWT) and Wavelet Coherence (WTC) of the Pacific decadal oscillation (PDO) and DAI in central and monsoonal Asia. The XWT identifies the common power, whereas WTC gives a measure of local correlation between two time series in the time-frequency plane. The relative phase between the two time series can also be discerned from the XWT or WTC. In order to further investigate the frequency-dependent relationships between climate patterns, a reciprocal pair of low-pass and high-pass Gaussian filters were also

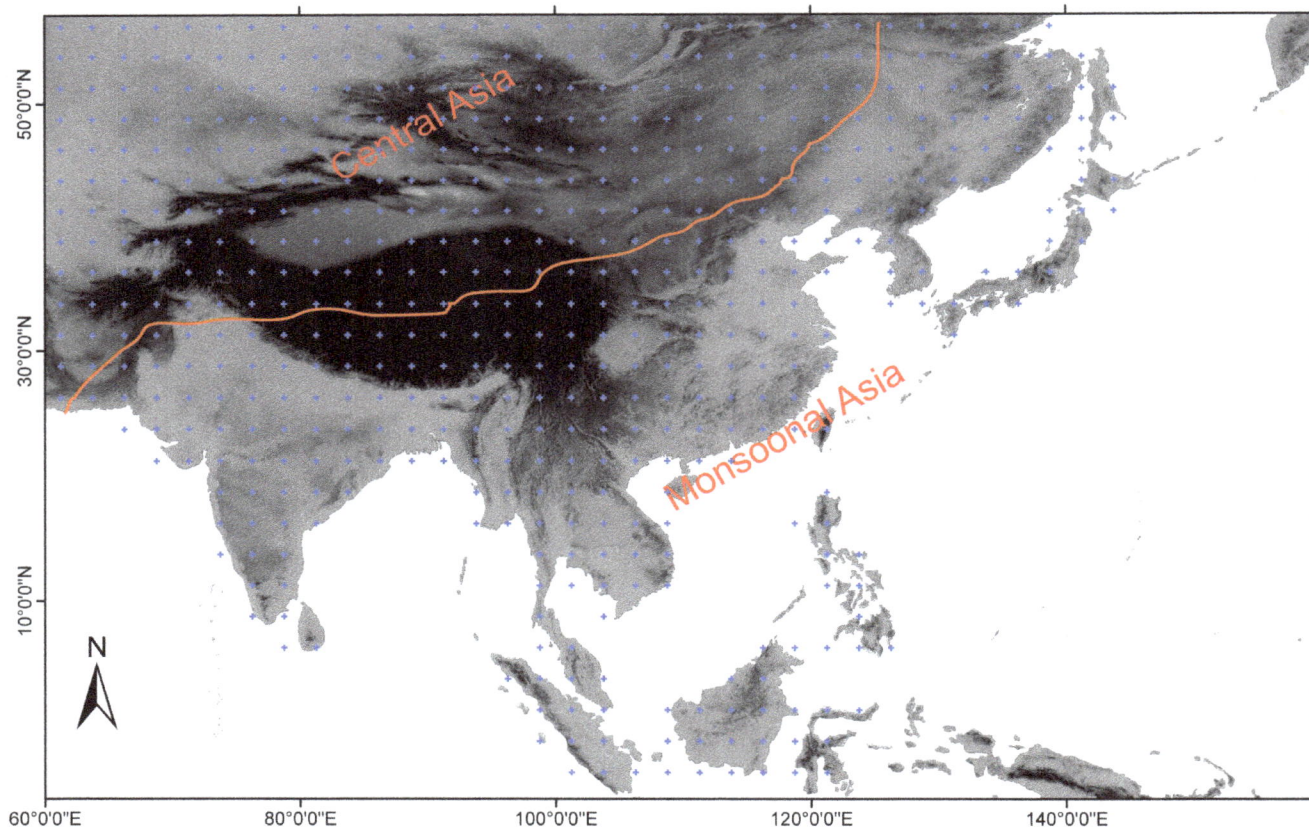

Figure 1. Location of the PDSI grids and the central and eastern Asia is delimited according to Chen et al. (2010).

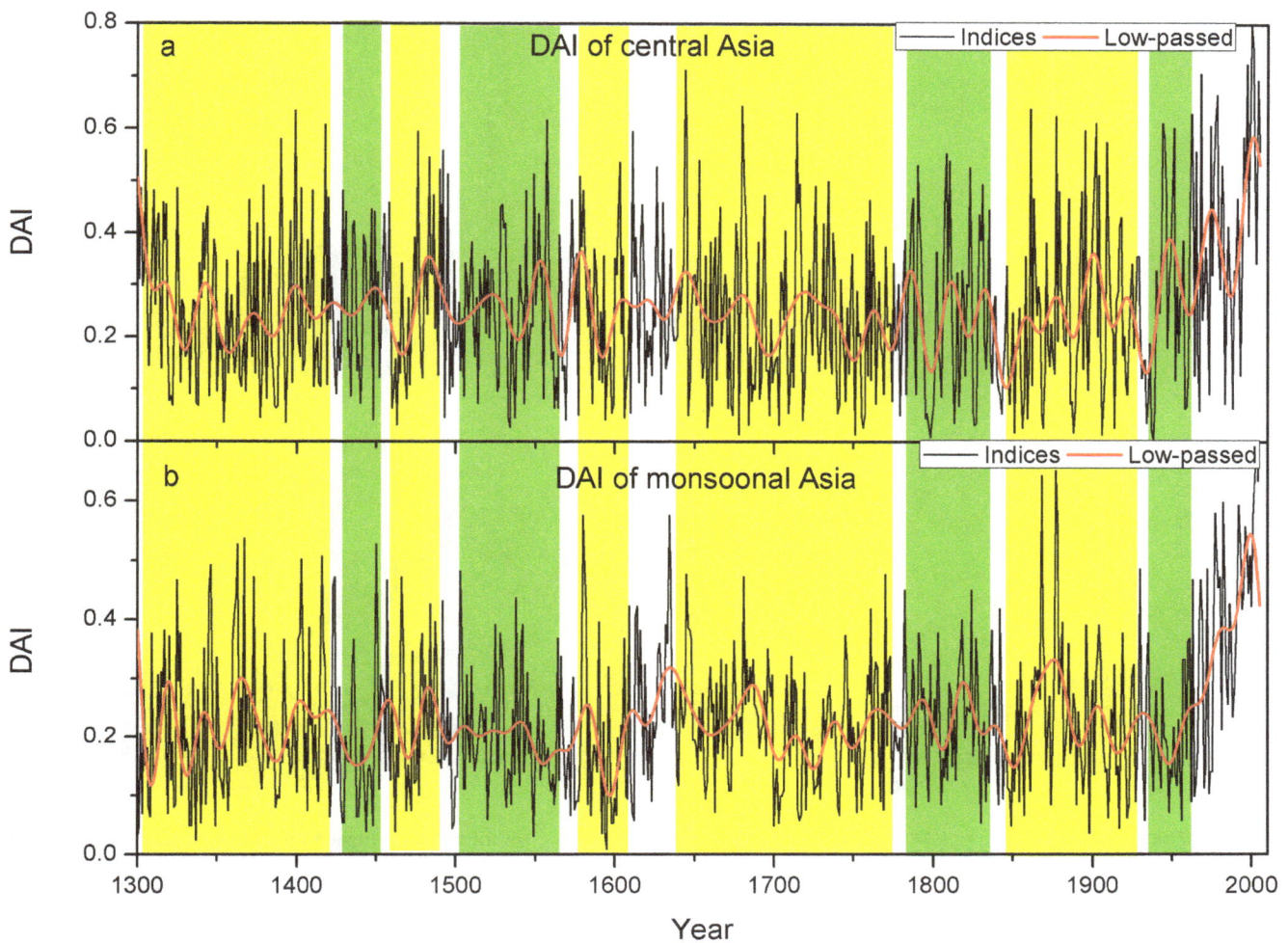

Figure 2. The drought area index (DAI) and the (red bold line) low-passed (lower than the cutoff frequency of 0.02) DAI for (a) central and (b) monsoonal Asia. The yellow shaded area indicates the in-phase decadal variations and the green shaded area indicates the anti-phase decadal variations between them.

employed to separate low-frequency climate signals greater than 8 years from the high-frequency variations shorter than 8 years.

To explore the linkages between hydroclimate and sea surface temperature (SST), we applied the singular value decomposition (SVD) method [18] to detect the leading pattern between the reconstructed PDSI and the Extended Reconstruction Sea Surface Temperature (ERSST.v3b) [19] in the Indian and Pacific oceans (herein 20S-60N; 40E-90W) during the common period from 1854 to 2005. The SVD is an efficient method to identify coupled modes from the cross-covariance matrix between the two fields (herein reconstructed PDSI and ERSST) according to their importance (the portion of explained variances) [18]. In addition, we calculated the spatial correlations between MADA and the instrumental PDO indices (http://jisao.washington.edu/pdo/) [20] and the reconstructed PDO indices from tree rings [21].

Hydroclimate variations in central and monsoonal Asia

DAI and its spectral properties. The prominent feature of the DAI series is the increasing trend of the dry area since the 1950s (Figure 2). This trend coincides with the drying trend since the 1950s [6,22] and may be exaggerated due to the potential overestimation of the evapotranspiration under the global

warming trend [11,23]. As revealed by the MTM analyses (Figure 3a, 3b), significant spectral peaks at around 2–3 years in both central and monsoonal Asia may be dominated by tropospheric biennial oscillation (TBO) [24], which has been widely documented in many tree-ring based hydroclimatic reconstructions [25,26,27,28,29,30]. Other significant cycles of 3–7 years in both study regions might be related to El Niño-Southern Oscillation (ENSO). The interdecadal cyclic pattern of 22–24, 26–27 and 46–56 years can be linked with the PDO [31]. Potential relationships with ENSO and PDO will be further discussed below. The significant 22-year cycle might also be associated with Hale's solar cycle, which is also evidenced in a previous reconstruction in northeastern Tibetan Plateau [27].

Relationships between hydroclimate in central and monsoonal Asia. According to the wavelet analyses, shifts in frequency bands between the presence and absence of the interdecadal changes for central Asia were identified in the 1460s, 1590s, 1770s, 1820s, and 1850s (Figure 3c). Similarly, in monsoonal Asia, we detected shifts at interdecadal timescale in the 1440s, 1550s, 1700s, 1810s and 1920s (Figure 3d). Significant interdecadal variations during the 1460s–1590s and 1770s–1820s were found in central Asia, while the interannual variations

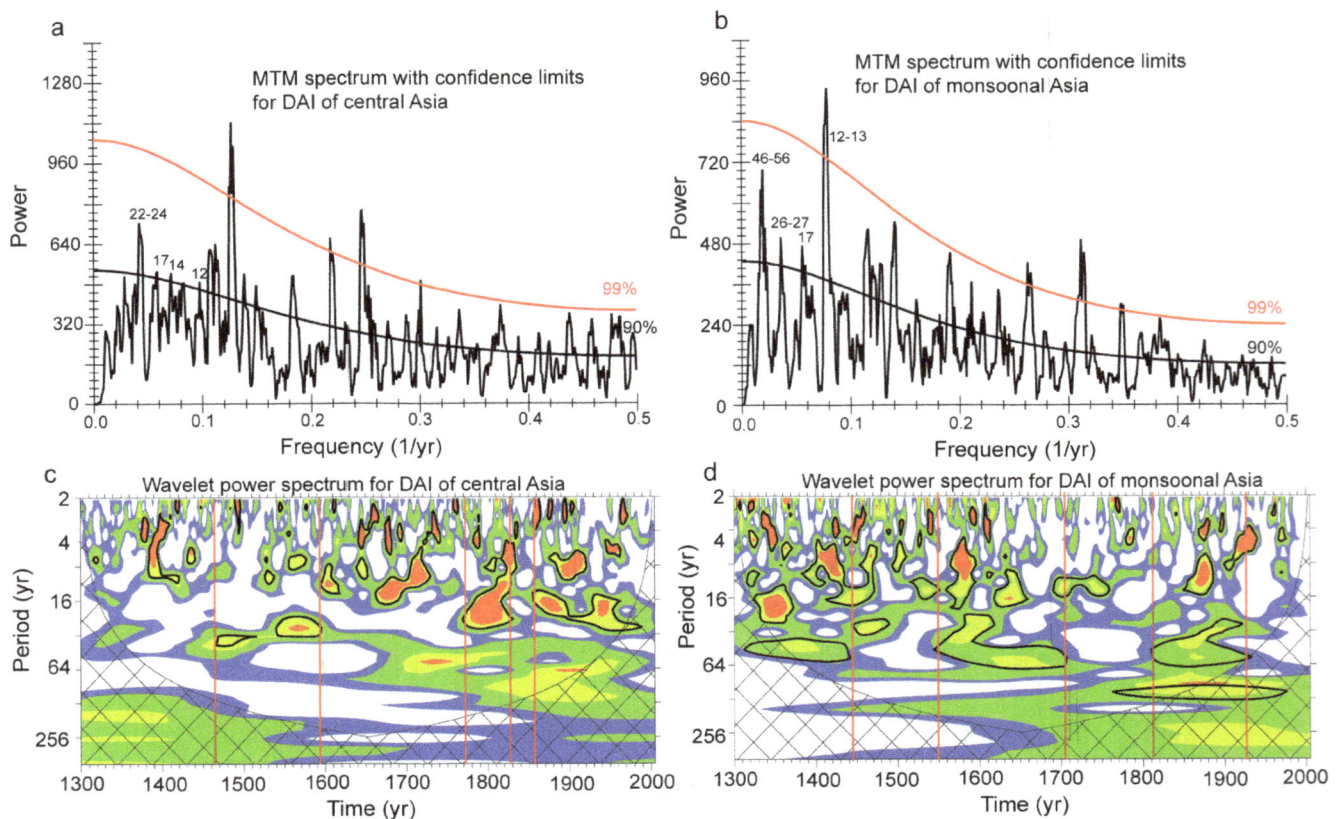

Figure 3. The multi-taper method (MTM) spectra of the drought area index (DAI) of the (a) central Asia and the (b) monsoonal Asia, as well as the wavelet spectra of the DAI of the (c) central Asia and the (d) monsoonal Asia.

dominated monsoonal Asia over the corresponding periods of the 1440s–1550s and the 1700s–1810s. On the other hand, significant interannual variations were found before the 1460s, 1590s–1770s, and 1820s–1850s for central Asia, which corresponds to significantly interdecadal variations over corresponding periods before the 1440s, 1550s–1700s and 1810s–1920s in monsoonal Asia (Figure 3). Some frequency shifts of the hydroclimate changes appear to be triggered by extreme climate conditions. For example, the frequency shift at 1450s–1460s corresponds with a dry period in much of the study region (Figure 2), particularly in northeastern Tibetan Plateau [27,29,30,32]. The different dominant frequency properties were more conspicuous in the early periods, particularly before the 19th century. Most of the long-term tree-ring chronologies are located in the central Asia and the marginal areas of monsoonal Asia [9], suggesting that different frequency properties might focus on central Asia and the marginal areas of monsoonal Asia. Increased number of tree-ring chronologies after 19th century (Figure S1) can be associated with the reconstruction of increased spatial features, such as for the monsoon-dominated Indian subcontinent with only two chronologies available in recent centuries. Further reconstructions of hydroclimate changes with more old chronologies of relatively even distribution may help clarify this. Although our results do not indicate anti-phase variations in the mean hydroclimate conditions between central and monsoonal Asia at the interannual or interdecadal timescale, there is evidence of the inverse shifts of the dominant frequency properties of hydroclimate changes occurring at intervals from some decades to a century. Different

frequency properties between central and monsoonal Asia suggest different hydroclimate regimes between two regions.

Since the DAI displayed frequency-dependent variations, we further calculated the 51-year running correlations between the DAI in central and monsoonal Asia at interannual and interdecadal timescales (Figure S2). The relationships between DAI in central and monsoonal Asia is unstable through time. We further calculated the WTC between DAI in both regions (Figure S3). The DAI variations at interdecadal timescale correlate with each other better than at interannual timescale. This agrees with the visual comparisons between DAI in central and monsoonal Asia (Figure 2). Significant correlations between DAI in both regions at interdecadal timescale correspond to peaks in the running correlations of the low-passed data (Figures S1 and S2). Unstable relationships between DAI in central and monsoonal Asia may be because the coupled ocean-atmospheric patterns dominating these regions vary through time. We therefore explore the linkages between DAI variations in both regions with the coupled ocean-atmosphere pattern in the following section.

Linkages to the coupled ocean-atmosphere patterns

PDO and hydroclimate changes. As for the first leading SVD mode between MADA and SST, significant (0.01 level) positive heterogeneous correlation with MADA are found in central Asia (around 30–40°N, 70–90°E), and are surrounded by horseshoe-like negative correlations in monsoonal Asia of eastern Mongolia, eastern China and Indochina (Figure 4a). Significant positive correlations with SST are seen in the eastern equatorial Pacific and oceans near western North America and negative

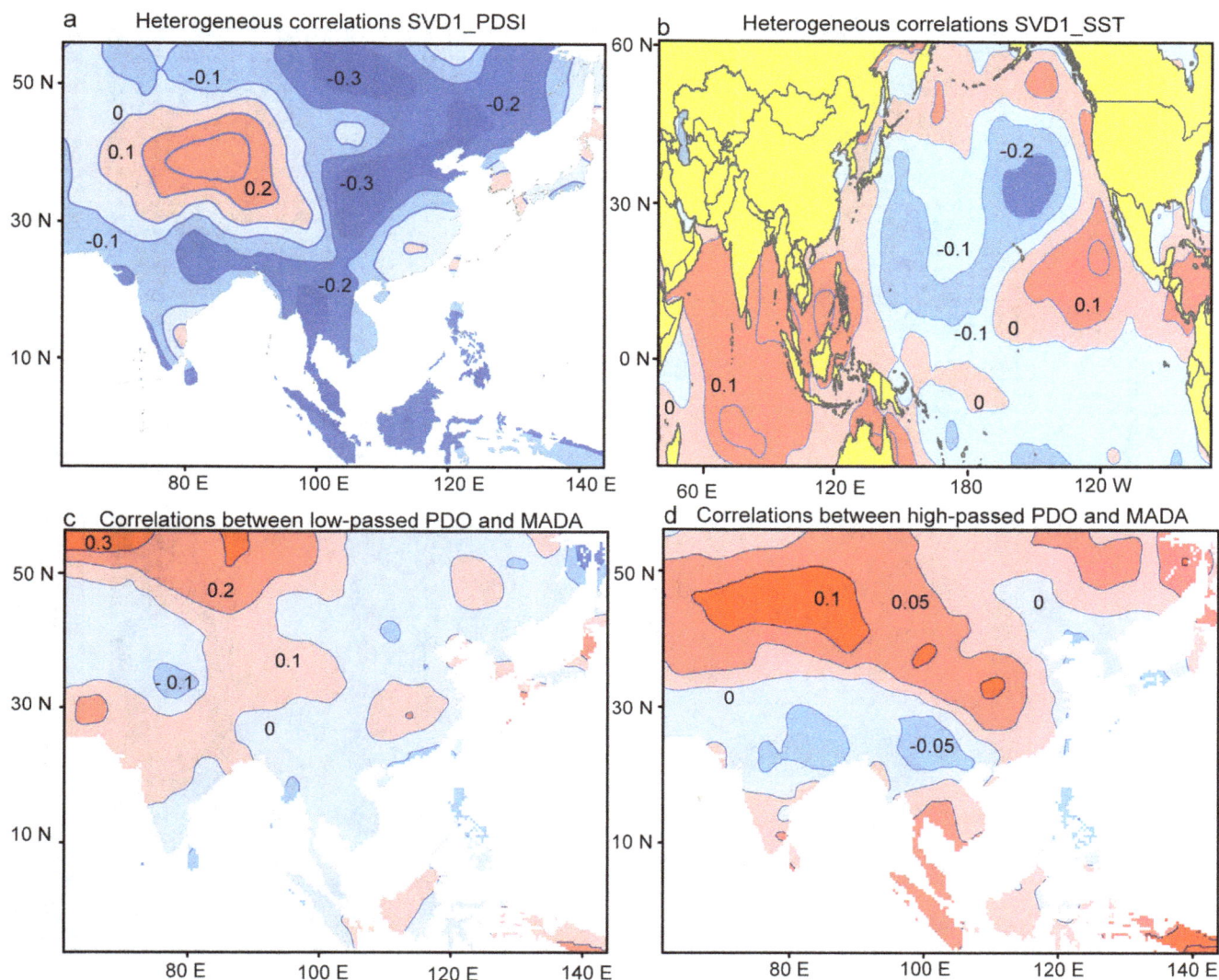

Figure 4. Heterogeneous correlations of the first leading SVD mode for the (a) reconstructed PDSI and (b) extended SST during the common period 1854–2005 and the (c) correlations between the low-passed PDO indices and the Monsoon Asia Drought Atlas (MADA) and the (d) correlations between the high-passed PDO and MADA over their common period period 1300–1996.

correlations are located in the north Pacific (Figure 4b), corresponding to a positive phase of the PDO [20]. Positive correlations are also found in much of the Indian Ocean. This indicates that a positive (negative) PDO and warm (cold) Indian Ocean SST lead to pluvial (dry) conditions in central Asia and dryness (wetness) in monsoonal Asia. The close linkage to PDO with strong interdecadal variations may account for the above findings that the DAI in central and monsoonal Asia are higher correlated at the interdecadal timescale. Therefore our later discussions focus on the relationships with PDO. We further calculated the correlations between PDO and MADA at the interannual (Figure 4c) and interdecadal (Figure 4d) timescales. At both timescales, the spatial correlation pattern is similar to the heterogeneous map in Figure 4a, except that the negative correlations in monsoonal Asia are less significant. In addition, higher correlations are found over the tropical Asia at interannual timescale. The most significant, positive correlations since 1300 are found for central Asia, indicating that the relationships between PDO and

hydroclimate variations in central Asia are most robust through time.

To further detect the frequency-dependent relationships between PDO and hydroclimate through time in these two regions, we calculated the XWT and WTC between PDO and DAI in central and monsoonal Asia. There is significant common power between DAI in central Asia and PDO on interdecadal timescales, particularly from 1770 to 1820, and after 1850 (Figure 5a). These significant common powers correspond to significant interdecadal variations of DAI in the 1770s–1820s and since the 1850s. The WTC plot (Figure 5c) also indicates significant correlations between PDO and DAI in central Asia in the late 15th century and since the 1850s on interdecadal timescale, which account for the significant interdecadal variations from the 1460s to 1590s and since 1850s in central Asia. There are significant correlations on interannual timescale before the 14th century and from middle 17th century to 18th century, and since the 1850s, which correspond to the significantly interannual variations of the hydroclimate changes in central Asia. The good

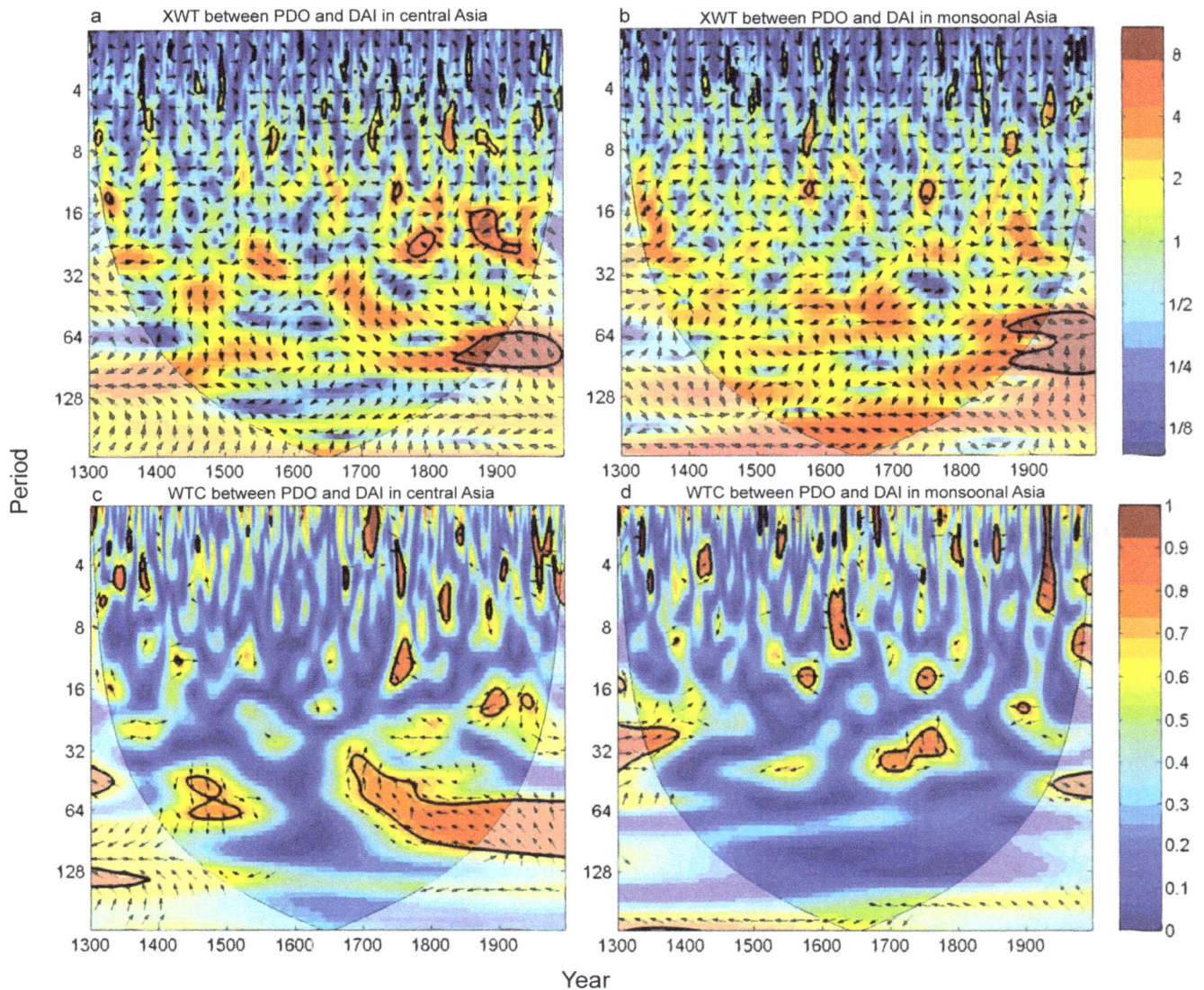

Figure 5. The cross wavelet transform (XWT) between PDO and DAI of (a) central and (b) monsoonal Asia, and the squared wavelet coherence (WTC) between PDO and DAI of (c) central and (d) monsoonal Asia. The significance level ($p<0.05$) is indicated by thick contours. The left arrows indicate anti-phase relationship, and right arrows indicate in-phase relationship between PDO and DAI.

matches of the significant common powers and correlations between PDO and hydroclimate changes add in proofs to the close linkages between PDO and hydroclimate changes in central Asia. We also found correspondence between PDO and DAI in monsoonal Asia (Figure 5b and 5d). For example, significant correlation at the interdecadal timescale in the 14th century may be related to the significant interdecadal variations prior to the 1450s. However, in comparison, the common power and local correlation between DAI in monsoonal Asia and PDO appears to be less significant.

We calculated the correlations between PDO and MADA over different subperiods generally corresponding to the different levels of solar irradiance [33]. Consistent with the above findings, we found that the correlations between PDO and MADA are more stable in central Asia. Positive correlations between PDO and PDSI in central Asia are significant in all periods except for the two periods with the minimum solar irradiance (Figure 6), i.e. the Spörer Minimum from 1460–1550 and the Maunder Minimum

from 1645–1715 [33], and thus low temperature [34]. There is seemingly paradox as the interdecadal signal in central Asia is not conspicuous in warm periods. This is because the correlations during the subperiods are more likely to represent the relationships on interannual timescale as the interannual variations of these tree-ring based reconstructions in short subperiods generally account for larger variance. At interannual timescale, the DAI in central Asia is more conspicuous in warm periods. Taken together, significant interdecadal (interannual) hydroclimate changes in central Asia are modulated by the interdecadal (interannual) variations of PDO. The linkages between PDO and hydroclimate in central Asia is more conspicuous in warm periods high solar irradiance at the interannual timescale. While the relationships between PDO and DAI in monsoonal Asia is not stable.

Mechanisms for the interannual and interdecadal changes. The PDO is closely associated with ENSO and is considered the reddening signal (low-frequency) of ENSO with its

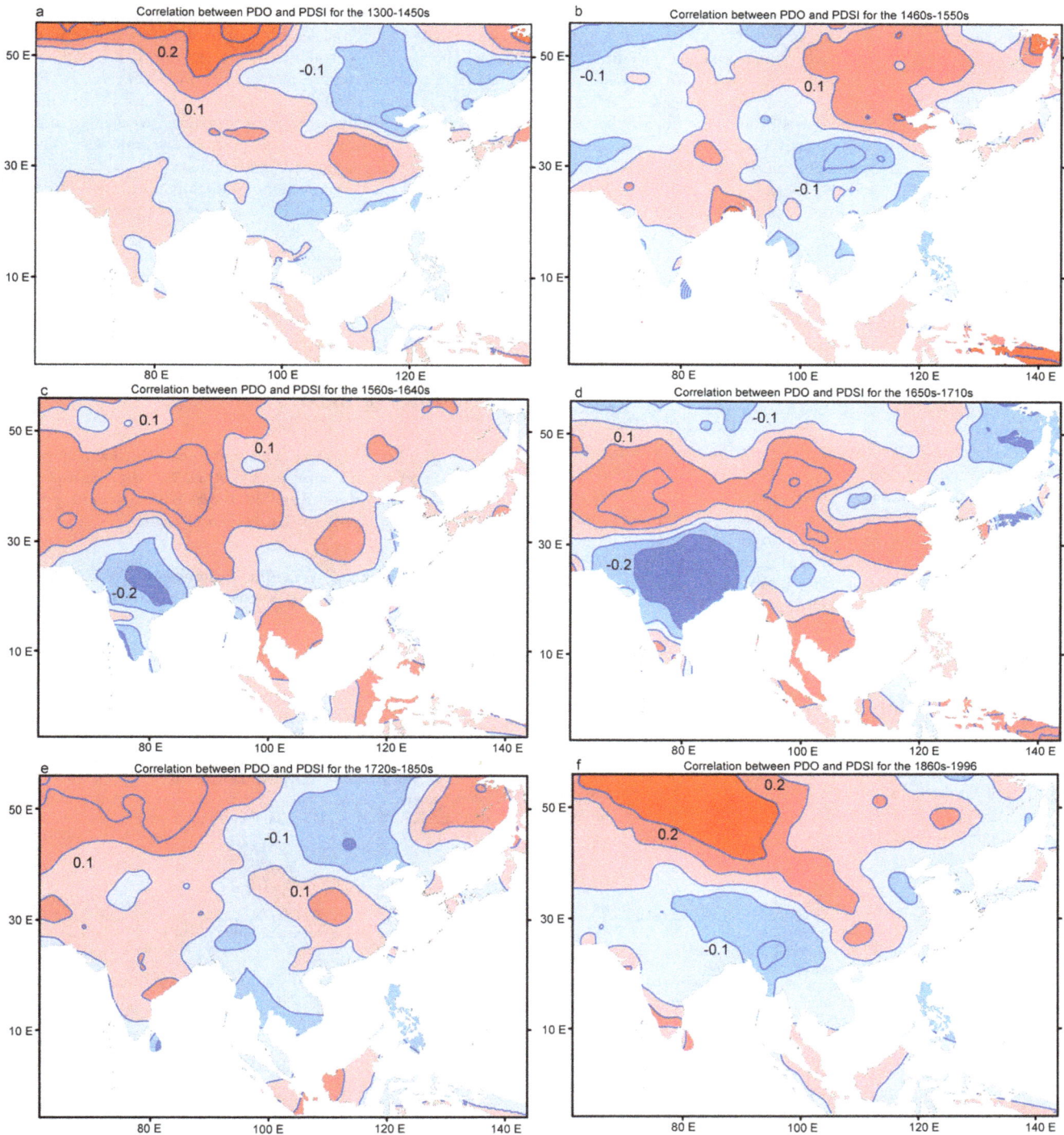

Figure 6. Correlations between the reconstructed PDO series and the reconstructed PDSI for the periods (a) 1300–1450s, (b) 1460s–1550s, (c) 1560s–1640s, (d) 1650s–1710s, (e) 1720s–1850s and (f) 1860s–1996.

interannual variability focusing on the tropical ocean and the interdecadal variability focusing on the north central Pacific Ocean [20,35]. Therefore we treat the PDO-hydroclimate linkages at interannual timescale mainly via the tropical ocean similar as the ENSO-hydroclimate teleconnections, and the interdecadal PDO-hydroclimate linkages mainly via the North Pacific Ocean. On the interannual timescale, hydroclimate changes in central Asia can be influenced by propagating moisture from the Indian Ocean via the Iran Plateau [7,36]. This northward moisture transportation can be enhanced by the anticyclone pattern in the Arabian Sea and the cyclone-like anomalies in the Iran Plateau and a trough in central Asia [36]. The anticyclone pattern over Arabian Sea can be strengthened when the SST of the tropical eastern Pacific Ocean is above average [37]. Since the variability of the tropical eastern Pacific Ocean tends to be strong (weak) in warm (cold) periods [38], the interannual variability of the hydroclimate changes in central Asia tends to be strengthened (weakened) in warm (cold) periods.

On the interdecadal timescale, the PDO-hydroclimate relationships may be linked via the atmospheric circulation between Tibetan Plateau and central North Pacific Ocean. An upward atmospheric circulation over Tibetan Plateau and a downward limb in the central North Pacific Ocean can lead to an increase in SST over the central North Pacific Ocean [39,40], which corresponds to a negative phase of the PDO [20]. On the other hand, a rising motion over Tibetan Plateau can contribute to the drying tendency of central Asia by a subsiding flow of dry air [41,42]. That is, during the positive (negative) phase of PDO, cold (warm) central North Pacific Ocean can be associated with downward (rising) atmospheric anomalies and thus more (less) air from Tibetan Plateau can be descended over central North Pacific Ocean, which leads to weak (enhanced) subsidences and thus wet (dry) climate over central Asia. This accounts for the significant interdecadal variations of hydroclimate changes in central Asia. On the other hand, PDO may be linked to monsoonal Asia instead of Tibetan Plateau, leading to significant interdecadal variations in monsoonal Asia. That is, a rising motion over monsoonal Asia when the monsoon is strong corresponds to a downward motion on central North Pacific Ocean and a high SST and thus a negative PDO. The negative correlations between PDO and hydroclimate in monsoonal Asia generally agree with the previous findings that suggested the negative correlation between PDO and the strength of Asian summer monsoon [7,25,35].

Conclusions

We calculated DAI for central and monsoonal Asia and investigated their relationships and linkages with coupled ocean-atmosphere patterns. The presence (absence) of interdecadal variability in central Asia often correspond to periods without significant interdecadal variability in monsoonal Asia, particularly before the 19th century, and vice versa, indicating different hydroclimate regimes between the two regions. This may be due

to their linkages with PDO. The relationship between PDO and hydroclimate in central Asia is more robust than in monsoonal Asia. The positive correlations between the PDO and hydroclimate in central Asia are more robust in warm periods. This is because the interannual PDO variability, similar to the interannual ENSO variability, tends to be enhanced (weakened) in warm (cold) periods, resulting in high (low) interannual variability in central Asia during the warm (cold) periods prior to the industrial era. The significant interdecadal hydroclimate changes in central Asia are likely linked to the PDO via the atmospheric circulation between Tibetan Plateau and the central North Pacific Ocean. The significant interdecadal hydroclimate changes in monsoonal Asia may be linked to the atmospheric circulation between monsoonal Asia and the central North Pacific Ocean.

Supporting Information

Figure S1   The time-varying number of tree-ring chronologies used in MADA.

Figure S2   The 51-year running correlations between DAI in central and monsoonal Asia for the high-passed and low-passed data derived from a reciprocal pair of low-pass and high-pass Gaussian filters.

Figure S3   Squared wavelet coherence (WTC) of the drought area index (DAI) between central and monsoonal Asia. The arrows pointing left (right) indicate anti-phase (in-phase) relationship. The areas with significance ($p<0.05$) relationships are indicated by thick contours.

Author Contributions

Conceived and designed the experiments: KF FC. Performed the experiments: KF FC HS. Analyzed the data: KF ND AS. Contributed reagents/materials/analysis tools: KF AS. Contributed to the writing of the manuscript: KF FC AS JL HS WH. Contributed the software used herein: AS.

References

1. Huang J, Guan X, Ji F (2012) Enhanced cold-season warming in semi-arid regions. Atmospheric Chemistry and Physics 12: 4627–4653.
2. Feng S, Fu Q (2013) Expansion of global drylands under a warming climate. Atmospheric Chemistry and Physics 13: 10081–10094.
3. Chen F, Yu Z, Yang M, Ito E, Wang S, et al. (2008) Holocene moisture evolution in arid central Asia and its out-of-phase relationship with Asian monsoon history. Quaternary Sci Rev 27: 351–364.
4. Chen FH, Chen JH, Holmes J, Boomer I, Austin P, et al. (2010) Moisture changes over the last millennium in arid central Asia: a review, synthesis and comparison with monsoon region. Quaternary Science Review 29: 1055–1068.
5. An Z, Colman SM, Zhou W, Li X, Brown ET, et al. (2012) Interplay between the Westerlies and Asian monsoon recorded in Lake Qinghai sediments since 32 ka. Scientific reports 2: doi:10.1038/srep00619
6. Fang K, Davi N, Gou X, Chen F, Cook E, et al. (2010) Spatial drought reconstructions for central High Asia based on tree rings. Climate Dynamics 35: 941–951.
7. Fang K, Gou X, Chen F, Li J, Zhou F, et al. (2011) Covariability between tree-ring based precipitation reconstructions in Northwest China and sea surface temperature of Indian and Pacific oceans. Climate Research 49: 17–27.
8. Cook E, Briffa KR, Meko DM, Graybill DA, Funkhouser G (1995) The'segment length curse'in long tree-ring chronology development for palaeoclimatic studies. Holocene 5: 229–237.
9. Cook E, Anchukaitis KJ, Buckley BM, D'Arrigo RD, Jacoby GC, et al. (2010) Asian Monsoon Failure and Megadrought During the Last Millennium. Science 328: 486–489.
10. Palmer WC (1965) Meteorological drought. Washington, D.C.
11. Dai A, Trenberth KE, Qian T (2004) A global dataset of Palmer Drought Severity Index for 1870–2002: Relationship with soil moisture and effects of surface warming. J Hydrometeorol 5: 1117–1130.

12. Fang K, Gou X, Chen F, Davi N, Liu C (2012) Spatiotemporal drought variability for central and eastern Asia over the past seven centuries derived from tree-ring based reconstructions. Quaternary International: doi.org/10.1016/j.quaint.2012.1003.1038
13. Wells N, Goddard S, Hayes MJ (2004) A self-calibrating palmer drought severity index. Journal of Climate 17: 2335–2351.
14. Cook E, Woodhouse CA, Eakin CM, Meko DM, Stahle DW (2004) Long-term aridity changes in the western United States. Science 306: 1015.
15. Mann ME, Lees JM (1996) Robust estimation of background noise and signal detection in climatic time series. Climatic Change 33: 409–445.
16. Narisma GT, Foley JA, Licker R, Ramankutty N (2007) Abrupt changes in rainfall during the twentieth century. Geophysical Research Letters 34: L06710, doi:06710.01029/02006GL028628
17. Torrence C, Compo GP (1998) A practical guide to wavelet analysis. Bulletin of the American Meteorological Society 79: 61–78.
18. Bretherton CS, Smith C, Wallace JM (1992) An intercomparison of methods for finding coupled patterns in climate data. Journal of Climate 5: 541–560.
19. Smith TM, Reynolds RW, Peterson TC, Lawrimore J (2008) Improvements to NOAA's historical merged land-ocean surface temperature analysis (1880–2006). Journal of Climate 21: 2283–2296.
20. Mantua NJ, Hare SR (2002) The Pacific decadal oscillation. Journal of Oceanography 58: 35–44.
21. MacDonald GM, Case RA (2005) Variations in the Pacific Decadal Oscillation over the past millennium. Geophys Res Lett 32: L08703, doi:08710.01029/02005GL022478
22. Li J, Cook ER, Chen F, Davi N, D'Arrigo R, et al. (2009) Summer monsoon moisture variability over China and Mongolia during the past four centuries. Geophysical Research Letters 36: L22705 doi:22710.21029/22009GL041162
23. Sheffield J, Wood EF, Roderick ML (2012) Little change in global drought over the past 60 years. Nature 491: 435–438.

24. Meehl GA, Arblaster JM (2010) The tropospheric biennial oscillation and Asian-Australian monsoon rainfall. J Climate 15: 722–744.

25. Fang K, Gou X, Chen F, Li J, D'Arrigo R, et al. (2010) Reconstructed droughts for the southeastern Tibetan Plateau over the past 568 years and its linkages to the Pacific and Atlantic Ocean climate variability. Clim Dyn 35: 577–585.

26. Gou X, Chen F, Yang M, Gordon J, Fang K, et al. (2008) Asymmetric variability between maximum and minimum temperatures in Northeastern Tibetan Plateau: Evidence from tree rings. Science in China Series D: Earth Sciences 51: 41–55.

27. Gou XH, Deng Y, Chen FH, Yang MX, Fang KY, et al. (2010) Tree ring based streamflow reconstruction for the Upper Yellow River over the past 1234 years. Chinese Science Bulletin 55: 4179–4186.

28. Liu Y, Bao G, Song H, Cai Q, Sun J (2009) Precipitation reconstruction from Hailar pine (Pinus sylvestris var. mongolica) tree rings in the Hailar region, Inner Mongolia, China back to 1865 AD. Palaeogeography, Palaeoclimatology, Palaeoecology 282: 81–87.

29. Shao X, Huang L, Liu H, Liang E, Fang X, et al. (2005) Reconstruction of precipitation variation from tree rings in recent 1000 years in Delingha, Qinghai. Sci China 48: 939–949.

30. Zhang QB, Cheng GD, Yao TD, Kang XC, Huang JG (2003) A 2,326-year tree-ring record of climate variability on the northeastern Qinghai-Tibetan Plateau. Geophys Res Lett 30: 1739–1742.

31. Mantua NJ, Hare SR, Zhang Y, Wallace JM, Francis RC (1997) A Pacific interdecadal climate oscillation with impacts on salmon production. B Am Meteorol Soc 78: 1069–1079.

32. Liu Y, An Z, Ma H, Cai Q, Liu Z, et al. (2006) Precipitation variation in the northeastern Tibetan Plateau recorded by the tree rings since 850 AD and its relevance to the Northern Hemisphere temperature. Science in China Series D: Earth Sciences 49: 408–420.

33. Eddy JA (1976) The maunder minimum. Science 192: 1189–1202.

34. Mann ME, Zhang Z, Rutherford S, Bradley RS, Hughes MK, et al. (2009) Global signatures and dynamical origins of the Little Ice Age and Medieval Climate Anomaly. Science 326: 1256–1260.

35. Ma ZG (2007) The interdecadal trend and shift of dry/wet over the central part of North China and their relationship to the Pacific Decadal Oscillation (PDO). Chinese Science Bulletin 52: 2130–2139.

36. Yang LM, Zhang QY (2007) Circulation characteristics of interannual and interdecadal anomalies of summer rainfall in north Xinjiang. Chinese Journal of Geohysics 50: 412–419 (in Chinese with English abstract).

37. Mariotti A (2007) How ENSO impacts precipitation in southwest central Asia. Geophysical Research Letters 34: L16706.

38. Li J, Xie SP, Cook ER, Huang G, D'Arrigo R, et al. (2011) Interdecadal modulation of El Nino amplitude during the past millennium. Nature Climate Change 1: 114–118.

39. Nan S, Zhao P, Yang S, Chen J (2009) Springtime tropospheric temperature over the Tibetan Plateau and evolutions of the tropical Pacific SST. J Geophys Res 114: D10104, doi:10110.11029/12008JD011559

40. Zhao P, Zhu Y, Zhang R (2007) An Asian–Pacific teleconnection in summer tropospheric temperature and associated Asian climate variability. Clim Dyn 29: 293–303.

41. Broccoli AJ, Manabe S (1992) The effects of orography on midlatitude Northern Hemisphere dry climates. Journal of Climate 5: 1181–1201.

42. Sato T, Kimura F (2005) Impact of diabatic heating over the Tibetan Plateau on subsidence over northeast Asian arid region. Geophysical Research Letters 32: L05809.

5

Saharan Dust Deposition May Affect Phytoplankton Growth in the Mediterranean Sea at Ecological Time Scales

Rachele Gallisai[1]*, **Francesc Peters**[1], **Gianluca Volpe**[2], **Sara Basart**[3], **José Maria Baldasano**[3,4]

1 Departament de Biologia Marina i Oceanografia, Institut de Ciències del Mar, CSIC, Barcelona, Spain, **2** Istituto di Scienze dell'Atmosfera e del Clima, Roma, Italy, **3** Earth Sciences Department, Barcelona Supercomputing Center-Centro Nacional de Supercomputación, BSC-CNS, Barcelona, Spain, **4** Environmental Modelling Laboratory, Technical University of Catalonia, Barcelona, Spain

Abstract

The surface waters of the Mediterranean Sea are extremely poor in the nutrients necessary for plankton growth. At the same time, the Mediterranean Sea borders with the largest and most active desert areas in the world and the atmosphere over the basin is subject to frequent injections of mineral dust particles. We describe statistical correlations between dust deposition over the Mediterranean Sea and surface chlorophyll concentrations at ecological time scales. Aerosol deposition of Saharan origin may explain 1 to 10% (average 5%) of seasonally detrended chlorophyll variability in the low nutrient-low chlorophyll Mediterranean. Most of the statistically significant correlations are positive with main effects in spring over the Eastern and Central Mediterranean, conforming to a view of dust events fueling needed nutrients to the planktonic community. Some areas show negative effects of dust deposition on chlorophyll, coinciding with regions under a large influence of aerosols from European origin. The influence of dust deposition on chlorophyll dynamics may become larger in future scenarios of increased aridity and shallowing of the mixed layer.

Editor: Tomoya Iwata, University of Yamanashi, Japan

Funding: The work was supported by the following: JAE-Predoc fellowship from the Spanish Scientific Research Council (CSIC), RG; Aerosol deposition and ocean plankton dynamics (CTM2011-23458), Spanish Ministry of Science and Innovation project, http://www.idi.mineco.gob.es/, RG and FP; Estructura de la materia orgánica en el océano costero: implicaciones biogeoquímicas y ecológicas (CTM2009-09352), Spanish Ministry of Science and Innovation project, http://www.idi.mineco.gob.es/, FP; Grup d'Oceanografia Mediterrània (2009/SGR/588), Generalitat de Catalunya project, http://www10.gencat.cat/agaur_web/AppJava/catala/index.jsp, RG and FP; La ricerca italiana per il mare, Italian Ministry of Education, University and Research, http://www.ritmare.it/, GV; Acoplamiento on-line de un modulo completo de aerosoles multicomponente al modelo atmosferico global regional NMMB (CGL2010-19652), Spanish Ministry of Science and Innovation project, http://www.idi.mineco.gob.es/, SB and JMB; Supercomputación and e-ciencia project (CSD2007-0050) from the Consolider-Ingenio 2010 program of the Spanish Ministry of Economy and Competitivity, http://www.idi.mineco.gob.es/, SB and JMB; and Severo Ochoa program (SEV-2011-00067) from Spanish Ministry of Economy and Competitivity, http://www.idi.mineco.gob.es/, SB and JMB. The funders had no role in study design, data collection and analysis, decision to publish, or preparation of the manuscript.

Competing Interests: The authors have declared that no competing interests exist.

* Email: gallisai@icm.csic.es

Introduction

Aerosols have major impacts on weather and climate regulations [1,2] and even on crop production [3]. Atmospheric desert dust may travel large distances from its source and has been proposed to have ocean production regulation effects over geological times scales [4]. The Mediterranean Sea (hereafter Med) atmosphere is subject to the continuous injection of Saharan and Middle East mineral dust particles [5]. The deposition of these mineral particles supply numerous macro and micro- nutrients to the ocean surface [6–14] and some authors consider it as the major source of "new" nutrients [15] for system production.

Calculations show that the atmospheric input of nutrients in the Med is of the same magnitude as riverine inputs [16–18], thus playing a significant role in the regulation of the nutrient balance of the basin at decadal or longer time scales [19,20]. The contribution of atmospheric deposition can be especially important and efficient in oligotrophic environments such as the Med, which has a marked stratification period and a pronounced nutrient limitation [21]. The deposition of some of these soluble compounds on surface waters may influence biological production, at least during certain events [9,22,23]. Dust deposition spreads over vast areas and dilutes into the water column often preventing the potential effects on system production to be unequivocally detected at ecological time scales. Experiments and observations in low nutrient – low chlorophyll areas have so far shown mixed results [24–26]. Reasons may include a tremendous variability in dust nutrient bioavailability content [27–29] and a relatively small increase of the background nutrient concentration when vertical mixing is active and represents the major source of nutrients [21] as well as a rapid transfer of increased primary production to other trophic levels and a variety of plankton community structures and physiological states.

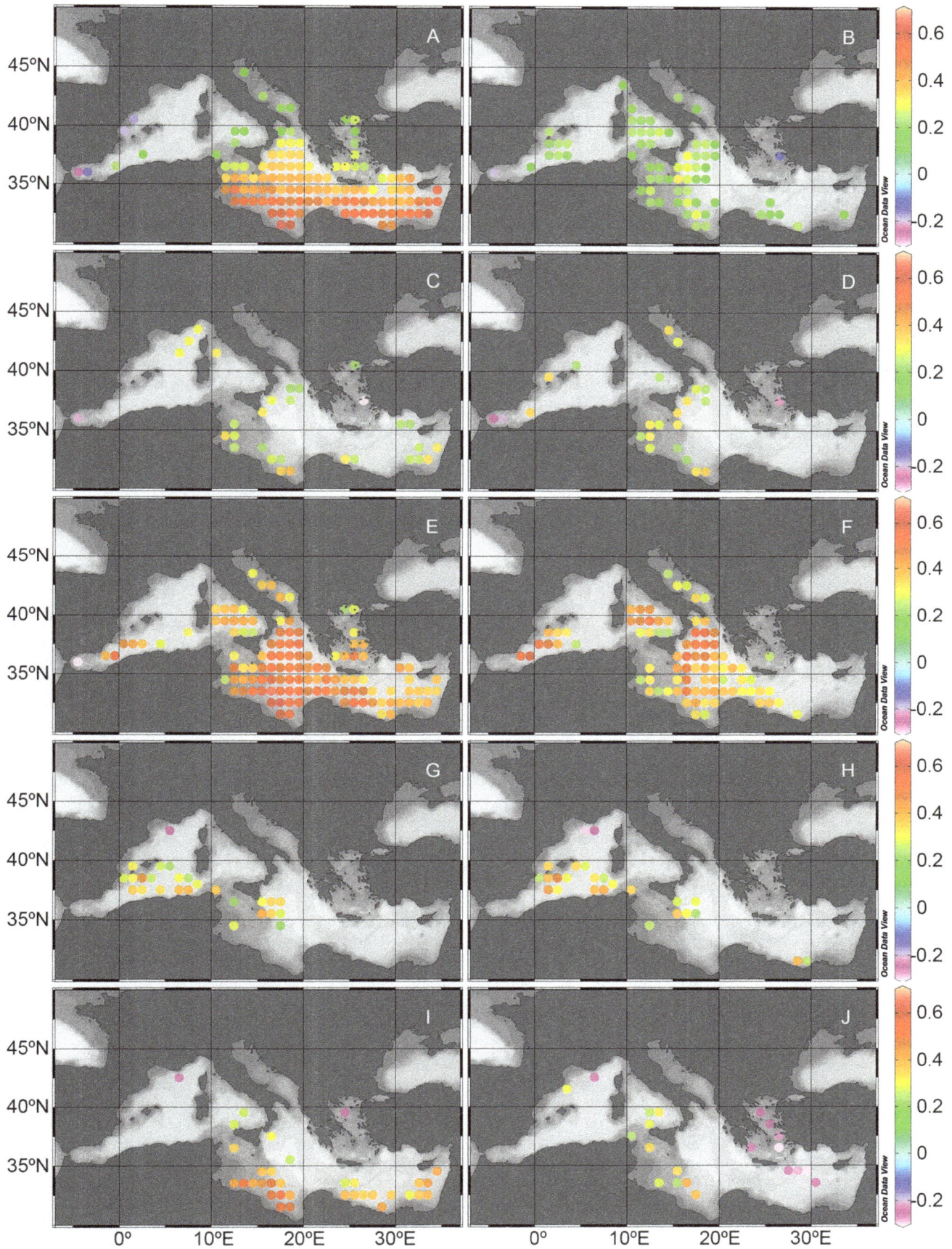

Figure 1. Correlation between chlorophyll concentration and dust deposition. Statistically significant (p<0.05) correlation coefficient (r) between chlorophyll concentration and dust deposition (left panels) and between the seasonally detrended chlorophyll concentration and the seasonally detrended dust deposition (right panels) for the whole time series and for different seasons. Panels: a, b) annual; c, d) winter (January to March); e, f) spring (April to June); g, h) summer (July to September) and i, j) autumn (October to December).

Given the episodic nature of dust events, an additional complication may reside in the human capacity of detecting the dust event deposition with sufficient space-time resolution in order to build a statistically significant dust event database. Previous attempts used satellite-derived aerosol optical thickness (AOT) as a proxy of dust in the atmosphere to infer the deposition events [26,30,31]. Dust generally travels from several hundreds to thousands of meters high in the atmosphere, making this approach not quantitatively adequate for discerning between transport and deposition. Deposition is measured *in situ* at a few terrestrial (mainland and islands) sites, which are extremely valuable for ground truth validation but are dependent on local conditions, making generalizations hard to draw especially towards the open ocean. Here, we employ a state-of-art atmospheric transport and deposition model, the BSC-DREAM8b model [32], which has been validated [33–35] and gives the power of having aerosol deposition data over the whole Med basin with daily temporal resolution. A previous study showed the potential positive effects of dust deposition on SeaWiFS-derived chlorophyll (Chl) in the Med [32]. However, in the Med, the used NASA OC4v4 algorithm falls far short to retrieve Chl with accuracy smaller than 100%, casting doubts on the relationships found. Here we extend this approach by relating deposition to SeaWiFS Chl using the Med-specific algorithm MedOC4 [36]. When we think of dust deposition, we tend to think about very large events, those that are obvious in true color images or that we recognize because we find our cars covered with red dust, but the truth is that, to some extent, there is Saharan dust in the atmosphere over the Mediterranean almost continuously and deposition does not occur only during large events but also when atmospheric aerosol concentrations are not so high. Thus, rather than focusing on single events or experiments, we take a correlational approach using an 8-year data time series in order to find relationships between Chl dynamics and dust deposition over the Mediterranean Sea.

Methods

Chlorophyll data

SeaWiFS HRPT Level-1A data (2000–2007) were collected at the Istituto di Scienze dell'Atmosfera e del Clima of Rome, Italy, and processed up to Level-3 using the MedOC4 regional algorithm (http://www.myocean.eu/web/69-myocean-interactive-catalogue.php) [36]. This algorithm takes into account the peculiar blue-green ratio of Med waters. Level-3 Chl data, with a native 1 km resolution, were $\log_{10}$-transformed averaged, over a period of eight days, and regridded over the $1°$ resolution grid of the basin (179 cells, see Table S1). A previous study showed that it is recommended not to use 8-d averages when computing correlation analysis between Chl and dust events [26]. To account for the possible contamination by atmospheric dust mimicking chlorophyll, here, before averaging over the period of eight days, the quality of the entire Chl dataset was carefully checked by i) applying all the SeaDAS Level-2 processing masks and flags (http://oceancolor.gsfc.nasa.gov/VALIDATION/flags.html), ii) removing all isolated pixels, iii) removing all pixels exceeding 3 standard deviations within a moving box of 3×3 pixels, and iv) by applying a median filter over all remaining good pixels. This procedure increases the confidence level on data quality, with the only shortcoming of reducing the

number of observations with respect to the NASA standard processing. The time series of daily observations was temporally binned into periods of 8 days. This results into 45 bins up to the 360th day of the year. The last bin was computed with the remaining 5 days, and in the case of leap years, with the remaining 6 days. The climatic mean is then calculated across years for each of the natural 8-d time periods.

Dust deposition

For the present study, a dust deposition simulation from the BSC-DREAM8b model (http://www.bsc.es/earth-sciences/mineral-dust/catalogo-datos-dust) model [33,37] was used for the period between 1 January 2000 and 31 December 2007, over the Med basin. BSC-DREAM8b tracks mineral dust particles from their sources in the Sahara and Middle East regions. Output, after being $\log_{10}$-transformed, was provided for the same space and time resolution as for chlorophyll. A low cut-off threshold $(10^{-8} \text{ Kg m}^{-2} \text{ d}^{-1})$ is applied to the numerical deposition output from BSC-DREAM8b since the dataset showed numerically correct but physically unrealistic low value spikes [32]. The model main features were described in detail in Pérez et al. [37] and Basart et al. [35]. It has been used for dust forecasting and as a dust research tool in North Africa and the Med [32,38–40]. Several studies have checked its performance [33,41], concerning both the horizontal and vertical extent of the dust plumes in the Med Basin. The model daily evaluation with near-real time observations is conducted at the Barcelona Supercomputing Center, and includes satellite data (MODIS and MSG) and AERONET sun photometers. BSC-DREAM8b has also been validated and tested over longer time periods in the European region [34,42,43] and against measurements at source regions [44].

Aerosol Optical Thickness

AOT at 865 nm data were derived from SeaWiFS radiometer measurements and they were downloaded from the Giovanni database (http://gdata1.sci.gsfc.nasa.gov/daac-bin/G3/gui.cgi?instance_id=ocean_8day). We acquired 8-d averaged, 9 km resolution product from 2000 to 2007. Similarly to Chl and deposition data, AOT data were $\log_{10}$-transformed and regridded over the $1°$ resolution grid of the basin, with the same temporal binning. This was done for the same 179 $1° \times 1°$ cells as for chlorophyll. It should be noted that AOT contains information of total aerosol particles in the atmosphere, not only of particles from Saharan origin. However, over much of the Mediterranean Sea most particles are indeed of Saharan origin [45].

Statistical analyses

Pearson's correlation coefficient (r) was calculated between chlorophyll concentration, modeled dust deposition and AOT time series for each grid cell. Significance was considered at $p < 0.05$ using Student's t-test. In addition, the degrees of freedom used for significance testing were adjusted to take into account the possible presence of autocorrelation in the time series. The number of effective independent observations, N^*, were calculated as described in Pyper el al. [46]. Correlations were computed both for the entire series and for each season. The same analyses were performed after seasonally detrending the data by subtracting the

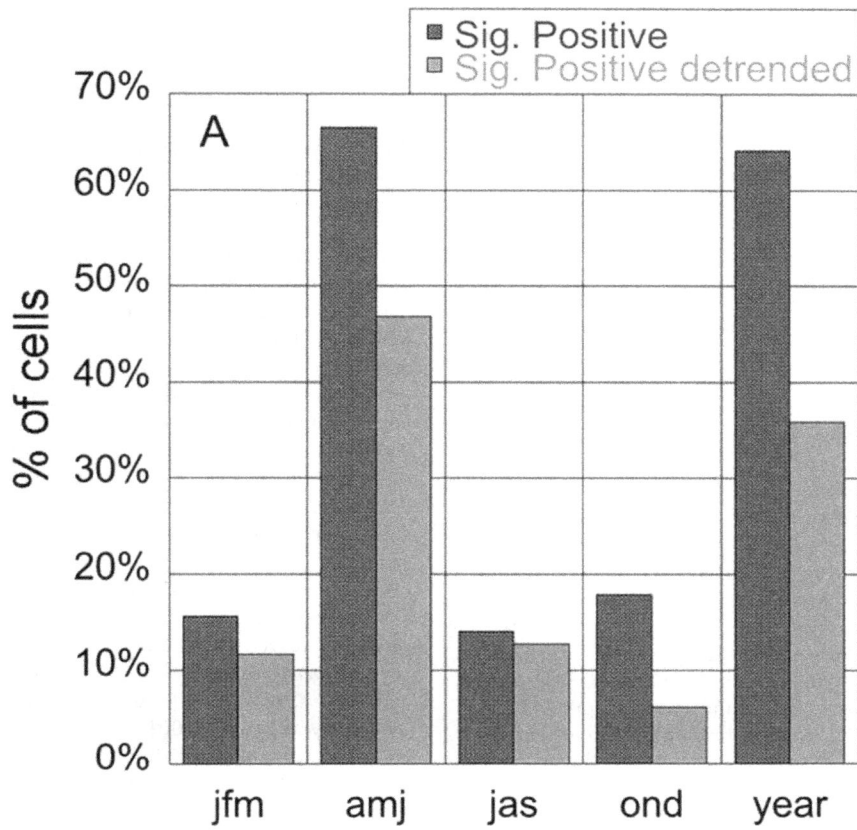

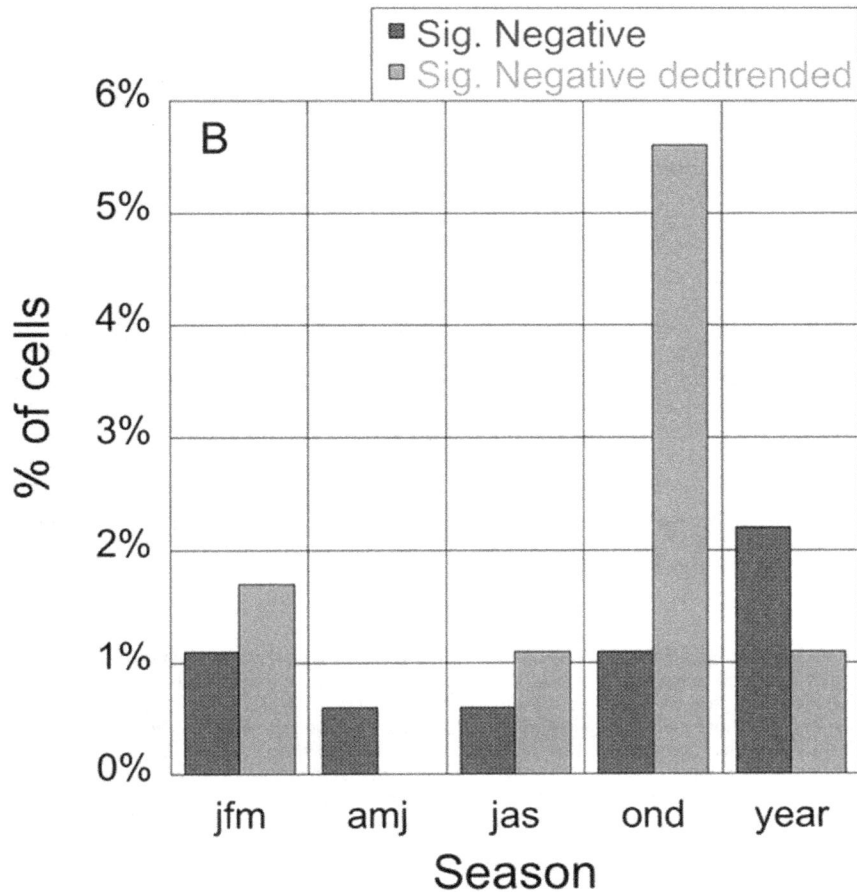

Figure 2. Percentage of cells showing significant correlations between chlorophyll and deposition. Left panel: positive correlations. Right panel: negative correlations. Red bars represent non seasonally detrended data and blue bars seasonally detrended data.

climatic mean at each time series data point. The r^2 of the correlation in a cell is the variance explained by the correlation in that cell. The minimum, maximum and average variance-explained values (expressed in %variability) were calculated for the population of cells with a $p < 0.05$.

Results and Discussion

We have found statistically significant positive correlations between surface chlorophyll and mineral dust deposition in large areas of the Med, covering 64% of the analyzed surface and located mainly in the Central and Eastern basins (Fig. 1) and with a clear south to north gradient in correlation intensity from 0.63 to 0.12. Significant negative correlations (r from -0.15 to -0.25) are observed in only 4 cells located in the Alboran Sea and in the eastern coast of Spain. Positive correlations can be found during all seasons, although it is in spring when we see the largest effects with correlations ranging from 0.22 to 0.65 mainly in the Central, Eastern and Southwestern Med. The Western and Central Med also show regions with positive correlations in summer, while in autumn there are some areas affected in the Central and Eastern Med. Most of the Med phytoplankton variability (>80%) is well explained by the variability of the mixed layer depth [47], and especially the winter-spring mixing bringing nutrient-rich deep waters to the surface. Thus, at least part of the explained variability between our deposition and chlorophyll time series must be due to the partial matching of the annual cycles of both variables.

The relationship between the seasonally detrended data of chlorophyll and dust deposition, that represents more of a response of short-term chlorophyll peaks to dust outbreaks, is somewhat weaker in intensity and in area covered. Largest positive correlations are found in the Central Med (from 0.13 to 0.32) (Fig. 1). Again, it is in spring where the largest impacted area is found, mainly in the Central Med and extending into the Eastern Med and Southwestern Med with r ranging from 0.24 to 0.58. The Western Med shows the largest area affected in summer. This is not surprising given that the seasonal dust event frequency peaks during spring in the Central-Eastern Med, and during summer over the Central-Western basin [26]. Seasonally detrended data tend to slightly increase the number of cells showing significant negative correlations and decrease the number of significantly positive correlated cells (Fig. 2). It is in autumn when we see the largest number of negatively correlated cells (6% of analyzed surface) and located mainly in the Aegean Sea and extending southeasterly of Crete.

For the seasonally detrended data, we checked that the correlation values were not caused by chance. We generated synthetic seasonally detrended chlorophyll time series with the observed mean and standard deviation for each cell. Correlations were computed with dust deposition model outputs, and the process repeated 100 times (Fig. 3). The observed significant correlations were compared to the distribution of the synthetic correlations for each cell, and in all cases they were statistically different with an $\alpha < 0.001$ and a power $(1-\beta)$ undistinguishable from 1. This confirmed the non-spurious nature of the relationships between dust deposition and non-seasonal chlorophyll time series.

Bulk Saharan dust deposition over the Med is not straightforwardly related to dust travelling in the atmosphere (Fig. S1). Meteorological conditions and wind patterns at different times of the year often have large amounts of dust (AOT) travelling at altitude with little deposition [48–50]. AOT and dust deposition

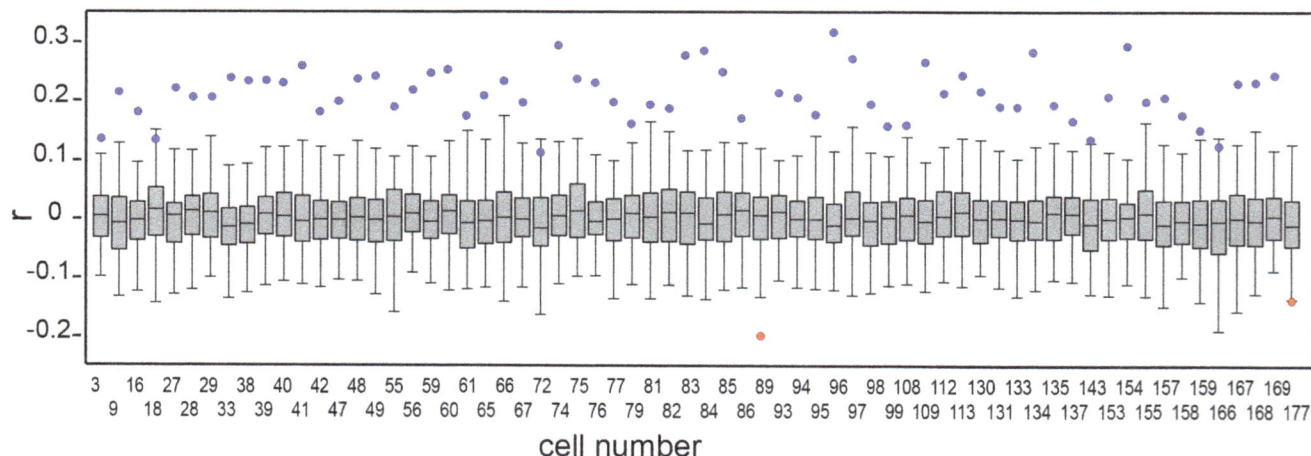

Figure 3. Analysis of the chance of significant correlations being spurious. Comparison between the box plots of the distribution of correlation coefficients between synthetic seasonally detrended chlorophyll time series and seasonally detrended dust deposition model outputs ($N = 100$) and the actual observed correlation between the seasonally detrended chlorophyll and the seasonally detrended dust deposition model outputs (dots). Data is shown only for those cells showing significant ($p < 0.05$) observed correlations. Dots in blue represent significantly positive correlations and red significantly negative correlations. Box plots show the median, the grey box englobing all data between the 25 and 75 percentiles, and the range between the smallest and largest values that are not outliers. Starting from the detrended data of the cells that show statistically significant correlations between detrended chlorophyll and deposition (Fig. 1b), synthetic Chl time series with the same mean and standard deviation (normal distribution) as the original detrended chlorophyll time series, were computed for each cell. The correlation between these synthetic Chl time series and the modelled dust deposition were computed. For each cell, this process was repeated 100 times, and the probability distribution functions (PDFs) of the correlations were then obtained and presented as box plots.

Table 1. Percentage of observed chlorophyll variability explained by modeled dust deposition.

Season	Non-detrended data				Seasonally detrended data			
	N	Min	Max	Average	n	Min	Max	Average
Annual	115	1.4	40.2	16.4	64	1.3	10.1	4.7
Winter	28	4.3	15.1	7.8	21	4.4	12.6	8.5
Spring	119	5.0	41.6	19.1	84	5.7	33.3	15.0
Summer	25	4.2	19.6	10.2	23	4.7	21.3	11.1
Autumn	32	6.5	26.9	15.6	11	5.4	16.5	9.6

Number of cells (n) with significantly ($p<0.05$) positive correlations. Minimum (min), maximum (max) and average percentage of chlorophyll variability explained in the significantly positive cells.

show positive correlation in the Western Med (Fig. S2.) especially in spring and summer with correlated areas shifting depending on the season. The Eastern sub-basin presents the highest correlations in spring (from 0.23 to 0.51) and the Central Med (Tyrrhenian Sea, Sicily channel and Dardanelle strait) in autumn. Once the data are seasonally detrended, deposition events are more related to AOT events, both when the whole series is considered and when the data are analyzed for the different times of the year (Fig. S2). With respect to non-detrended data, seasonally detrended data show main increases in correlation and correlated area in the Central Med for most of the year, as well as in the Eastern Med in autumn. Some overall hotspots appeared in the Alboran and in the Tyrrhenian Sea and around Crete, where the correlations ranged from 0.36 to 0.48.

The annual cycles of chlorophyll and AOT do not match (Fig. S1). The maximum chlorophyll concentrations occur in winter and minima coincide with the summer months. On the contrary, the highest AOT is found in summer and the minimum in autumn. Overall, AOT and chlorophyll (Fig. S3) show no significant correlations in the Med, except for some areas near the African coasts, where the correlation is negative (from −0.28 to −0.36). While no correlations are evident between AOT and chlorophyll there are significant correlations between seasonally detrended AOT and chlorophyll data (Fig. S3). A plume of higher correlation, with r-values between 0.33 and 0.39, appear in the northern part of Cyrenaica region with an extension up to the south of Italy. The best match between both series was found in summer (Fig. S3). Volpe et al. [36] ground truthed the chlorophyll satellite estimates with in situ measurements and concluded that the atmospheric correction was appropriate. In addition, we compared the data from the chlorophyll measurements at the DYFAMED station (1998–2007) with SeaWiFS estimates corrected with a regional algorithm giving a slope of ~1 ($\log DYF = 0.0129 + 1.0497 \cdot \log SW$; Adjusted $R^2 = 0.68$; N = 91; $p<0.001$). Moreover DYFAMED chlorophyll was unrelated to AOT, providing further evidence of the independence between satellite measurements of chlorophyll and AOT. Aerosols travelling over a certain area are not necessarily depositing. When a deposition event is occurring, it should coincide with high aerosol content in the air (AOT), thus if we find relationships between dust deposition events and non-seasonal chlorophyll peaks it is also logical to expect that chlorophyll is related to AOT, while the non-detrended AOT data show little or no relationship.

As mentioned before, the largest positive correlations between dust deposition and chlorophyll occur around the Central and Eastern Med. Calculations [8,20,51] and experiments [11,20,24] tell us that aerosol deposition effects on primary production should be small in most situations and thus we do not expect African dust deposition in general to explain a large portion of chlorophyll variability. Accordingly, positive significant correlations between mineral dust deposition of Saharan origin and chlorophyll do explain only a 1 to 10% (average 5%) of chlorophyll variability for seasonally detrended and a 1 to 40% (average 16%) for non-detrended data although it may be higher for certain seasons (Table 1). It should be noted that the explained variability does not provide direct information of the magnitude of chlorophyll impacted.

Winter shows overall the lowest significantly positive correlations, while spring presents the highest. This is to be expected since the entrance of new nutrients should be mostly due to seasonal winter overturning and mixing of nutrient-rich deep waters with upper ocean surface waters, through a number of physical processes that increase vertical diffusion at certain moments. But even at times when nutrient concentrations are expected to be relatively high in the water, low concentrations and strong imbalances between N and P are often observed [52,53], opening windows of opportunity for the nutrients from atmospheric deposition to have an impact in the sustainment of phytoplankton production. We can only speculate on the positive cause-effect relationship between aerosol deposition and chlorophyll in the Med at certain times. Terrestrial inputs through major rivers occur mainly in the Western Med [54], and atmospheric inputs may dominate nutrient supply at certain times [10,55]. Phosphorus (P) limitation alleviation has often been invoked [8,56] as the surface waters of the Med are among the most P-limited in the world [57]. Although aerosols show a disproportionally large ratio of nitrogen to phosphorus [18], potentially only exacerbating P-limitation, they do carry an amount of P that could be used by phytoplankton and bacteria, especially in spring and summer when the concentrations of this element in surface waters of the open Med are at their lowest. Guieu et al. [10] calculated that, if P is considered the limiting element for phytoplankton growth, atmospheric deposition could account for chlorophyll increases of ca. $0.2 \ \mu g \ L^{-1}$ in the upper mixed layer for a single large deposition event or for the average total deposition during the summer-stratified period. The Central and Eastern Med do not show the typical spring phytoplankton bloom and have been defined as no blooming areas [58]. The ultra-oligotrophic conditions [59] found in these areas should make them most responsive to external nutrient supplies. As is the case for high nutrient – low chlorophyll areas, micronutrients such as iron from aerosols have also been proposed to stimulate Med phytoplankton production under certain situations [60], albeit addition experiments have not shown a direct increase in dissolved iron (Fe) [61]. Fe in the mixed layer of the Mediterranean is found at concentrations from 0.13 to 2.7 nM [62,63]. It seems though that Fe is, relative to the needs of plankton, in excess with respect to P in the Mediterranean [64]. Nevertheless, in a system where all elements are relatively scarce, responses to the combination of elements arriving through aerosol deposition, may be very complex, with elements becoming successively limiting in a chained reaction. Ridame et al. [65] found stimulation of nitrogen (N) fixation in dust pulse experiments, in general related to a primary alleviation of P-limitation. In their Central Med experiment though, they found high N-fixation stimulation unrelated to P- or Fe-limitation, further showing the complexity of the processes involved and the potential spatial and temporal variability. An initial stimulation of heterotrophic bacteria [25,66] should not be discarded since these organisms have a potential advantage at low nutrient concentrations owing to their high surface to volume ratio. Secondarily, released nutrients from recycling could then stimulate phytoplankton processes. Contrary to the Eastern Med showing the lowest nutrient concentrations in the Mediterranean [59], the Central Med was found somewhat more responsive to dust deposition in the present study. Pey et al. [5] mention the Central Med as a transitional area, receiving a higher frequency of dust outbreaks than similar latitudes in the Western and Eastern Med. Additionally, the dust source areas are not homogeneous. The Libyan Desert is the main source of dust for the Central Med while the Eastern Med receives dust from Libya and from the Middle East [67]. Thus, positive correlations between dust deposition and surface chlorophyll seem to arise from the combination of areas of low nutrient concentrations with the right nature, timing and frequency of dust outbreaks.

Negative relationships between dust deposition and chlorophyll have been related to metal (mainly Cu but also Al) inhibition of phytoplankton growth [13,68]. The toxicity of Cu in reducing phytoplankton growth rate has been shown in laboratory experiments (see [68] and references therein). A recent correlation study between chlorophyll and metals from onshore-measured aerosols in the Northwest Med shows negative relationships in the area under northerly wind (Tramontane) conditions [13]. These winds favor the transport of anthropogenic aerosols from Europe to the Med. Although most Cu pulses are anthropogenically derived, pulses originating in Africa showed effects on chlorophyll undistinguishable from those originating locally [13]. A reduction in chlorophyll growth of up to 20% can be seen along the French and Spanish coasts. In addition, Jordi et al. [13] argue that since Cu toxicity seems to be taxon specific, the summer phytoplankton community with a predominance of nanoflagellates over the less sensitive diatoms, is more vulnerable to atmospheric deposition. This is an area where we also see some negative correlations between the modeled deposition and chlorophyll. We only track Saharan mineral dust, while some of the high load of metals may be more related to local anthropogenic sources. Most of the large deposition events in the Northwest Med come in the form of wet deposition [69]. In our model, the deposition field only originates from Saharan and Middle East dust transport and does not account for local anthropogenic aerosol sources, but rain washes out the entire atmospheric column aerosol loading, no matter the origin. Results from our correlation analysis agree with previous more detailed local studies [13]. We also see a negative relationship between deposition and chlorophyll, both seasonally detrended, mainly in autumn in the Aegean region (Fig. 1). This area is affected by long-range transport of air pollutants from Eastern Europe [70] but it is also heavily impacted by anthropogenic emissions generated in Athens and Istanbul [71,72]]. A high-density population together with a massive number of vehicles, many of them still using non-catalytic or old technology diesel engines, contributed to exceed the EU annual aerosol limit. The amount of Cu in these aerosols is high with an annual mean concentration between 0.013 and $0.22 \ \mu g \ m^{-3}$ ([73] and references therein). An estimated dry deposition flux of Cu over the sea ranges then between 22 and $380 \ \mu g \ Cu \ m^{-2} \ d^{-1}$ surpassing the threshold limit for Cu to inhibit phytoplankton growth according to [13] and [68].

Conclusion

Desert dust storm events seem to be increasing in frequency and intensity [14,69,72–76] in the last decades, due to human activities and climate forcing. This means that the presence of aerosols over the Med is likely to increase with future aridity. Thus, it is important to understand basin level patterns in the response of Med biogeochemistry to aerosol deposition. Only a few studies [26,30] have tried to analyze the potential links between aerosols in the air column and chlorophyll for the entire Med basin, with non definitive results. In this study, we use a modeled actual aerosol deposition product and show both positive and negative significant correlations with chlorophyll dynamics in certain areas and times of the year. Mineral dust from North Africa and the Middle East correlates to chlorophyll in large areas of the Med Sea. This is especially true for the Central and Eastern Med sub-basins, where Saharan dust deposition dynamics matches that of chlorophyll, particularly during spring. Here the atmospheric input may be an intrinsic part of the annual ecosystem dynamics. In terms of large dust outbreaks, chlorophyll best relates to aerosol deposition in the Central Med, extending both into the Eastern and Southwestern Med (Fig. S3). Some areas of the Western Med and Aegean Sea show negative correlations between chlorophyll and deposition in accordance with some recent findings of toxicity brought by metals in aerosols. As expected, dust deposition does not explain an overall large amount of chlorophyll variability since the main ecosystem production driver in the Med is the vertical mixing of nutrients from deep waters. Variability related to carbon to chlorophyll ratios, the consumption of biomass with a varying degree of coupling and the variable settling of primary production, are all additional sources of surface chlorophyll variability that we could not account for in our correlations and thus add to the noise. No matter how small significant correlations are, they are not distributed randomly in space and coincide with independent estimates that follow the same trend. Thus, albeit the mechanisms that affect chlorophyll through aerosol deposition cannot be pinpointed and may be indirect, non-unique, and dependent on local spatio-temporal conditions, our study shows a clear potential for effects at ecological scales. These effects should become more important in a future scenario with increased aerosols over the Med and a shallower mixed layer depth, owing to increased temperatures, over which aerosols may leach out nutrients.

Supporting Information

Figure S1 Seasonal average values of chlorophyll concentration, dust deposition and aerosol optical thickness. Average chlorophyll concentration (left panels). Average dust deposition (central panels) and average aerosol optical thickness (right panels) for different seasons. Winter (a, b, c), spring (d, e, f), summer (g, h, i) and autumn (j, k, l).

Figure S2 Correlation between dust deposition and aerosol optical thickness. Statistically significant (p<0.05) correlation coefficient (r) between dust deposition and aerosol optical thickness (left panels) and between seasonally detrended dust deposition and seasonally detrended aerosol optical thickness (right panels) for the whole time series and for different seasons. Panels: a, b) annual; c, d) winter (January to March); e, f) spring (April to June); g, h) summer (July to September) and i, j) autumn (October to December).

Figure S3 Correlation between chlorophyll concentration and aerosol optical thickness. Same as Fig. S2 but for chlorophyll concentration versus aerosol optical thickness.

Table S1 Geographical coordinates for the 1°×1° grid cells analyzed in this study. Coordinates refer to the central point of the cell. Latitudes are all North. Positive longitudes are East and negative longitudes West.

Acknowledgments

The Goddard Earth Sciences Data and Information Services Center (GES DISC), the SeaWiFS mission scientists and associated NASA personnel for the production of the data used in this study and for the development and maintenance of SeaDAS software. The Barcelona Supercomputing Center (BSC) for hosting the BSC-DREAM8b simulations on the Mare Nostrum Supercomputer. The DYFAMED (CNRS-INSU) Observatoire Océanologique de Villefranche-sur-mer (http://www.obs-vlfr.fr/dyfBase/) for providing *in situ* chlorophyll measurements. The developers of Ocean Data View (http://odv.awi.de) for their useful tool.

Author Contributions

Conceived and designed the experiments: RG FP. Performed the experiments: RG FP. Analyzed the data: RG FP GV SB JMB. Contributed reagents/materials/analysis tools: RG FP GV SB JMB. Contributed to the writing of the manuscript: RG FP GV SB JMB.

References

1. Booth BBB, Dunstone NJ, Halloran PR, Andrews T, Bellouin N (2012) Aerosols implicated as a prime driver of twentieth-century North Atlantic climate variability. Nature 484: 228–232.
2. Creamean JM, Suski KJ, Rosenfeld D, Cazorla A, DeMott PJ, et al. (2013) Dust and Biological Aerosols from the Sahara and Asia Influence Precipitation in the Western U.S. Science 339: 1572–1578.
3. Liu X, Zhang Y, Han W, Tang A, Shen J, et al. (2013) Enhanced nitrogen deposition over China. Nature 494: 459–462.
4. Jaccard SL, Hayes CT, Martinez-Garcia A, Hodell DA, Anderson RF, et al. (2013) Two Modes of Change in Southern Ocean Productivity Over the Past Million Years. Science 339: 1419–1423.
5. Pey J, Querol X, Alastuey A, Forastiere F, Stafoggia M (2013) African dust outbreaks over the Mediterranean Basin during 2001–2011: PM10 concentrations, phenomenology and trends, and its relation with synoptic and mesoscale meteorology. Atmospheric Chemistry and Physics 13: 1395–1410.
6. Bonnet S, Guieu C, Chiaverini J, Ras J, Stock A (2005) Effect of atmospheric nutrients on the autotrophic communities in a low nutrient, low chlorophyll system. Limnology and Oceanography 50: 1810–1819.
7. Bergametti G, Remoudaki E, Losno R, Steiner E, Chatenet B, et al. (1992) Source, transport and deposition of atmospheric phosphorus over the northwestern Mediterranean. Journal of Atmospheric Chemistry 14: 501–513.
8. Ridame C, Guieu C (2002) Saharan input of phosphate to the oligotrophic water of the open western Mediterranean sea. Limnology and Oceanography 47: 856–869.
9. Markaki Z, Oikonomou K, Kocak M, Kouvarakis G, Chaniotaki A, et al. (2003) Atmospheric deposition of inorganic phosphorus in the Levantine Basin, eastern Mediterranean: Spatial and temporal variability and its role in seawater productivity. Limnology and Oceanography 48: 1557–1568.
10. Guieu C, Loÿe-Pilot MD, Benyahya L, Dufour A (2010) Spatial variability of atmospheric fluxes of metals (Al, Fe, Cd, Zn and Pb) and phosphorus over the whole Mediterranean from a one-year monitoring experiment: Biogeochemical implications. Marine Chemistry 120: 164–178.
11. Pulido-Villena E, Rerolle V, Guieu C (2010) Transient fertilizing effect of dust in P-deficient LNLC surface ocean. Geophysical Research Letters 37: L01603.
12. Herut B, Krom MD, Pan G, Mortimer R (1999) Atmospheric input of nitrogen and phosphorus to the Southeast Mediterranean: Sources, fluxes, and possible impact. Limnology and Oceanography 44: 1683–1692.
13. Jordi A, Basterretxea G, Tovar-Sanchez A, Alastuey A, Querol X (2012) Copper aerosols inhibit phytoplankton growth in the Mediterranean Sea. Proceedings of the National Academy of Sciences of the United States of America 109: 21246–21249.
14. Goudie AS, Middleton NJ (2001) Saharan dust storms: nature and consequences. Earth-Science Reviews 56: 179–204.
15. Ternon E, Guieu C, Ridame C, L'Helguen S, Catala P (2011) Longitudinal variability of the biogeochemical role of Mediterranean aerosols in the Mediterranean Sea. Biogeosciences 8: 1067–1080.
16. Guieu C, Martin JM, Thomas AJ, Elbazpoulichet F (1991) Atmospheric versus river inputs of metals to the Gulf of Lions: Total concentrations, partitioning and fluxes. Marine Pollution Bulletin 22: 176–183.
17. Ludwig W, Bouwman AF, Dumont E, Lespinas F (2010) Water and nutrient fluxes from major Mediterranean and Black Sea rivers: Past and future trends and their implications for the basin-scale budgets. Global Biogeochemical Cycles 24: GB0A13.
18. Markaki Z, Loÿe-Pilot MD, Violaki K, Benyahya L, Mihalopoulos N (2010) Variability of atmospheric deposition of dissolved nitrogen and phosphorus in the Mediterranean and possible link to the anomalous seawater N/P ratio. Marine Chemistry 120: 187–194.
19. Bethoux JP, Morin P, Ruiz-Pino DP (2002) Temporal trends in nutrient ratios: chemical evidence of Mediterranean ecosystem changes driven by human activity. Deep-Sea Research Part Ii-Topical Studies in Oceanography 49: 2007–2016.
20. Herut B, Zohary T, Krom MD, Mantoura RFC, Pitta P, et al. (2005) Response of East Mediterranean surface water to Saharan dust: On-board microcosm experiment and field observations. Deep-Sea Research Part Ii-Topical Studies in Oceanography 52: 3024–3040.
21. Estrada M (1996) Primary production in the northwestern Mediterranean. Scientia Marina 60: 55–64.
22. Guerzoni S, Chester R, Dulac F, Herut B, Loÿe-Pilot MD, et al. (1999) The role of atmospheric deposition in the biogeochemistry of the Mediterranean Sea. Progress in Oceanography 44: 147–190.
23. Morales-Baquero R, Pulido-Villena E, Reche I (2006) Atmospheric Inputs of Phosphorus and Nitrogen to the Southwest Mediterranean Region: Biogeochemical Responses of High Mountain Lakes. Limnology and Oceanography 51: 830–837.

24. Romero E, Peters F, Marrasé C, Guadayol T, Gasol JM, et al. (2011) Coastal Mediterranean plankton stimulation dynamics through a dust storm event: An experimental simulation. Estuarine, Coastal and Shelf Science 93: 27–39.

25. Lekunberri I, Lefort T, Romero E, Vázquez-Domínguez E, Romera-Castillo C, et al. (2010) Effects of a dust deposition event on coastal marine microbial abundance and activity, bacterial community structure and ecosystem function. Journal of Plankton Research 32: 381–396.

26. Volpe G, Banzon VF, Evans RH, Santoleri R, Mariano AJ, et al. (2009) Satellite observations of the impact of dust in a low-nutrient, low-chlorophyll region: Fertilization or artifact? Global Biogeochem Cycles 23: GB3007.

27. Carbo P, Krom MD, Homoky WB, Benning LG, Herut B (2005) Impact of atmospheric deposition on N and P geochemistry in the southeastern Levantine basin. Deep-Sea Research Part II: Topical Studies in Oceanography 52: 3041–3053.

28. Herut B, Collier R, Krom MD (2002) The role of dust in supplying nitrogen and phosphorus to the Southeast Mediterranean. Limnology and Oceanography 47: 870–878.

29. Baker AR, French M, Linge KL (2006) Trends in aerosol nutrient solubility along a west-east transect of the Saharan dust plume. Geophysical Research Letters 33: L07805.

30. Cropp RA, Gabric AJ, McTainsh GH, Braddock RD, Tindale N (2005) Coupling between ocean biota and atmospheric aerosols: Dust, dimethylsulphide, or artifact? Global Biogeochemical Cycles 19: GB4002.

31. Gabric AJ, Cropp R, Ayers GP, McTainsh G, Braddock R (2002) Coupling between cycles of phytoplankton biomass and aerosol optical depth as derived from SeaWiFS time series in the Subantarctic Southern Ocean. Geophysical Research Letters 29: 1112.

32. Gallisai R, Peters F, Basart S, Baldasano JM (2012) Mediterranean basin-wide correlations between Saharan dust deposition and ocean chlorophyll concentration. Biogeosciences Discussion 9: 28.

33. Pérez C, Nickovic S, Baldasano JM, Sicard M, Rocadenbosch F, et al. (2006) A long Saharan dust event over the western Mediterranean: Lidar, Sun photometer observations, and regional dust modeling. Journal of Geophysical Research-Atmospheres 111: D15214.

34. Basart S, Pay MT, Jorba O, Pérez C, Jinez-Guerrero P, et al. (2012) Aerosols in the CALIOPE air quality modelling system: Evaluation and analysis of PM levels, optical depths and chemical composition over Europe. Atmospheric Chemistry and Physics 12: 3363–3392.

35. Basart S, Perez C, Nickovic S, Cuevas E, Baldasano JM (2012) Development and evaluation of the BSC-DREAM8b dust regional model over Northern Africa, the Mediterranean and the Middle East. Tellus Series B-Chemical and Physical Meteorology 64: 18539.

36. Volpe G, Santoleri R, Vellucci V, d'Alcala MR, Marullo S, et al. (2007) The colour of the Mediterranean Sea: Global versus regional bio-optical algorithms evaluation and implication for satellite chlorophyll estimates. Remote Sensing of Environment 107: 625–638.

37. Pérez C, Nickovic S, Pejanovic G, Balcasano JM, Özsoy E (2006) Interactive dust-radiation modeling: A step to improve weather forecasts. Journal of Geophysical Research-Atmospheres 111: D16206.

38. Amiridis V, Kafatos M, Perez C, Kazadzis S, Gerasopoulos E, et al. (2009) The potential of the synergistic use of passive and active remote sensing measurements for the validation of a regional dust model. Annales Geophysicae 27: 3155–3164.

39. Alonso-Perez S, Cuevas E, Perez C, Querol X, Baldasano JM, et al. (2011) Trend changes of African airmass intrusions in the marine boundary layer over the subtropical Eastern North Atlantic region in winter. Tellus, Series B: Chemical and Physical Meteorology 63: 255–265.

40. Pay MT, Jiménez-Guerrero P, Jorba O, Basart S, Querol X, et al. (2012) Spatio-temporal variability of concentrations and speciation of particulate matter across Spain in the CALIOPE modeling system. Atmospheric Environment 46: 376–396.

41. Papanastasiou DK, Poupkou A, Katragkou E, Amiridis V, Melas D, et al. (2010) An Assessment of the Efficiency of Dust Regional Modelling to Predict Saharan Dust Transport Episodes. Advances in Meteorology 2010: 154368.

42. Jiménez-Guerrero P, Pérez C, Jorba O, Baldasano JM (2008) Contribution of Saharan dust in an integrated air quality system and its on-line assessment. Geophysical Research Letters 35: L03814.

43. Pay MT, Piot M, Jorba O, Gassó S, Gonçalves M, et al. (2010) A full year evaluation of the CALIOPE-EU air quality modeling system over Europe for 2004. Atmospheric Environment 44: 3322–3342.

44. Haustein K, Pérez C, Baldasano JM, Müller D, Tesche M, et al. (2009) Regional dust model performance during SAMUM 2006. Geophysical Research Letters 36: L03812.

45. Barnaba F, Gobbi GP (2004) Aerosol seasonal variability over the Mediterranean region and relative impact of maritime, continental and Saharan dust particles over the basin from MODIS data in the year 2001. Atmos. Chem. Phys. 4: 2367–2391.

46. Pyper BJ, Peterman RM (1998) Comparison of methods to account for autocorrelation in correlation analyses of fish data. Can. J. Fish. Aquat. Sci. 55: 2127–2140.

47. Volpe G, Nardelli BB, Cipollini P, Santoleri R, Robinson IS (2012) Seasonal to interannual phytoplankton response to physical processes in the Mediterranean Sea from satellite observations. Remote Sens. Environ. 117: 223–235.

48. Papayannis A, Amiridis V, Mona L, Tsaknakis G, Balis D, et al. (2008) Systematic lidar observations of Saharan dust over Europe in the frame of EARLINET (2000–2002). J. Geophys. Res. (Atmos.) 113: D10204.

49. Mona L, Amodeo A, Pandolfi M, Pappalardo G (2006) Saharan dust intrusions in the Mediterranean area: Three years of Raman lidar measurements. J. Geophys. Res. (Atmos.) 111: D16203.

50. Gobbi GP, Angelini F, Barnaba F, Costabile F, Baldasano JM, et al. (2013) Changes in particulate matter physical properties during Saharan advections over Rome (Italy): a four-year study, 2001–2004. Atmos. Chem. Phys. 13: 7395–7404.

51. Eker-Develi E, Kideys AE, Tugrul S (2006) Role of Saharan dust on phytoplankton dynamics in the northeastern Mediterranean. Mar. Ecol. Prog. Ser. 314: 61–75.

52. Diaz F, Raimbault P, Boudjellal B, Garcia N, Moutin T (2001) Early spring phosphorus limitation of primary productivity in a NW Mediterranean coastal zone (Gulf of Lions). Mar. Ecol. Prog. Ser. 211: 51–62.

53. Rahav E, Herut B, Levi A, Mulholland MR, Berman-Frank I (2013) Springtime contribution of dinitrogen fixation to primary production across the Mediterranean Sea. Ocean Sci. 9: 489–498.

54. Struglia MV, Mariotti A, Filograsso A (2004) River discharge into the Mediterranean Sea: Climatology and aspects of the observed variability. J. Clim. 17: 4740–4751.

55. Durrieu de Madron X, Guieu C, Sempéré R, Conan P, Cossa D, et al. (2011) Marine ecosystems' responses to climatic and anthropogenic forcings in the Mediterranean. Prog. Oceanogr. 91: 97–166.

56. Izquierdo R, Benítez-Nelson CR, Masqué P, Castillo S, Alastuey A, et al. (2012) Atmospheric phosphorus deposition in a near-coastal rural site in the NE Iberian Peninsula and its role in marine productivity. Atmos. Environ. 49: 361–370.

57. Marty JC, Chiaverini J, Pizay MD, Avril B (2002) Seasonal and interannual dynamics of nutrients and phytoplankton pigments in the western Mediterranean Sea at the DYFAMED time-series station (1991–1999). Deep-Sea Res. Part II. 49: 1965–1985.

58. D'Ortenzio F, Ribera D'Alcalà M (2009) On the trophic regimes of the Mediterranean Sea: A satellite analysis. Biogeosciences 6: 139–148.

59. Pujo-Pay M, Conan P, Oriol L, Cornet-Barthaux V, Falco C, et al. (2011) Integrated survey of elemental stoichiometry (C, N, P) from the western to eastern Mediterranean Sea. Biogeosciences 8: 883–899.

60. Bonnet S, Guieu C (2006) Atmospheric forcing on the annual iron cycle in the western Mediterranean Sea: A 1-year survey. J. Geophys. Res. (Oceans) 111: C09010.

61. Wagener T, Guieu C, Leblond N (2010) Effects of dust deposition on iron cycle in the surface Mediterranean Sea: results from a mesocosm seeding experiment. Biogeosciences 7: 3769–3781.

62. Sarthou G, Jeandel C (2001) Seasonal variations of iron concentrations in the Ligurian Sea and iron budget in the Western Mediterranean Sea. Mar. Chem. 74: 115–129.

63. Guieu C, Bozec Y, Blain S, Ridame C, Sarthou G, et al. (2002) Impact of high Saharan dust inputs on dissolved iron concentrations in the Mediterranean Sea. Geophys. Res. Lett. 29: 1911.

64. Guieu C, Dulac F, Desboeufs K, Wagener T, Pulido-Villena E, et al. (2010) Large clean mesocosms and simulated dust deposition: a new methodology to investigate responses of marine oligotrophic ecosystems to atmospheric inputs. Biogeosciences 7: 2765–2784.

65. Ridame C, Le Moal M, Guieu C, Ternon E, Biegala IC, et al. (2011) Nutrient control of N-2 fixation in the oligotrophic Mediterranean Sea and the impact of Saharan dust events. Biogeosciences 8: 2773–2783.

66. Pulido-Villena E, Wagener T, Guieu C (2008) Bacterial response to dust pulses in the western Mediterranean: Implications for carbon cycling in the oligotropic ocean. Global. Biogeochem. Cy. 22: GB1020.

67. Gaetani M, Pasqui M (2012) Synoptic patterns associated with extreme dust events in the Mediterranean Basin. Reg. Environ. Change.: 1–14.

68. Paytan A, Mackey KRM, Chen Y, Lima ID, Doney SC, et al. (2009) Toxicity of atmospheric aerosols on marine phytoplankton. Proc. Natl. Acad. Sci. USA. 106: 4601–4605.

69. Avila A, Peñuelas J (1999) Increasing frequency of Saharan rains over northeastern Spain and its ecological consequences. Sci. Total Environ. 228: 153–156.

70. Lelieveld J, Berresheim H, Borrmann S, Crutzen PJ, Dentener FJ, et al. (2002) Global air pollution crossroads over the Mediterranean. Science 298: 794–799.

71. Kanakidou M, Mihalopoulos N, Kindap T, Im U, Vrekoussis M, et al. (2011) Megacities as hot spots of air pollution in the East Mediterranean. Atmos. Environ. 45: 1223–1235.

72. Querol X, Alastuey A, Pey J, Cusack M, Perez N, et al. (2009) Variability in regional background aerosols within the Mediterranean. Atmos. Chem. Phys. 9: 4575–4591.

73. Theodosi C, Grivas G, Zarmpas P, Chaloulakou A, Mihalopoulos N (2011) Mass and chemical composition of size-segregated aerosols (PM1, PM2.5,

PM10) over Athens, Greece: local versus regional sources. Atmos. Chem. Phys. 11: 11895–11911.

74. Ganor E, Osetinsky I, Stupp A, Alpert P (2010) Increasing trend of African dust, over 49 years, in the eastern Mediterranean. J. Geophys. Res. (Atmos.) 115: D07201.

75. Goudie AS (2009) Dust storms: Recent developments. J. Environ. Manage. 90: 89–94.

76. Mahowald NM, Kloster S, Engelstaedter S, Moore JK, Mukhopadhyay S, et al. (2010) Observed 20th century desert dust variability: impact on climate and biogeochemistry. Atmos. Chem. Phys. 10: 10875–10893.

6

Land Surface Reflectance Retrieval from Hyperspectral Data Collected by an Unmanned Aerial Vehicle over the Baotou Test Site

Si-Bo Duan[1,2,3], Zhao-Liang Li[3,4]*, Bo-Hui Tang[1], Hua Wu[1], Lingling Ma[5], Enyu Zhao[2,5], Chuanrong Li[5]

1 State Key Laboratory of Resources and Environment Information System, Institute of Geographic Sciences and Natural Resources Research, Chinese Academy of Sciences, Beijing, China, **2** University of Chinese Academy of Sciences, Beijing, China, **3** Laboratoire des sciences de l'ingenieur, de l'informatique et de l'imagerie, Université de Strasbourg, Centre National de la Recherche Scientifique, Illkirch, France, **4** Key Laboratory of Agri-informatics,Ministry of Agriculture/Institute of Agricultural Resources and Regional Planning, Chinese Academy of Agricultural Sciences, Beijing, China, **5** Earth Observation Technology Application Department, Academy of Opto-Electronics, Chinese Academy of Sciences, Beijing, China

Abstract

To evaluate the in-flight performance of a new hyperspectral sensor onboard an unmanned aerial vehicle (UAV-HYPER), a comprehensive field campaign was conducted over the Baotou test site in China on 3 September 2011. Several portable reference reflectance targets were deployed across the test site. The radiometric performance of the UAV-HYPER sensor was assessed in terms of signal-to-noise ratio (SNR) and the calibration accuracy. The SNR of the different bands of the UAV-HYPER sensor was estimated to be between approximately 5 and 120 over the homogeneous targets, and the linear response of the apparent reflectance ranged from approximately 0.05 to 0.45. The uniform and non-uniform Lambertian land surface reflectance was retrieved and validated using *in situ* measurements, with root mean square error (RMSE) of approximately 0.01–0.07 and relative RMSE of approximately 5%–12%. There were small discrepancies between the retrieved uniform and non-uniform Lambertian land surface reflectance over the homogeneous targets and under low aerosol optical depth (AOD) conditions (AOD = 0.18). However, these discrepancies must be taken into account when adjacent pixels had large land surface reflectance contrast and under high AOD conditions (e.g. AOD = 1.0).

Editor: Kimberly Patraw Van Niel, University of Western Australia, Australia

Funding: This work was supported by the Hi-Tech Research and Development Program of China (863 Plan Program) under Grant 2012AA12A302 and by the State Key Laboratory of Resources and Environment Information System under Grant 088RA801KA. Mr. Si-Bo Duan is financially supported by the China Scholarship Council for his stay in LSIIT, France. The funders had no role in study design, data collection and analysis, decision to publish, or preparation of the manuscript.

Competing Interests: The authors have declared that no competing interests exist.

* E-mail: lizhaoliang@caas.cn

Introduction

Hyperspectral data in the solar-reflective region (0.4–2.5 μm) has been collected since the mid-1980 s [1]. Hyperspectral remote sensing is increasingly being used in a wide range of applications, including geology, agriculture, forestry, and ecology [2–4].

An adequate pre-processing of hyperspectral data is a mandatory prerequisite to extract quantitative information about the land surface from hyperspectral data. Radiometric calibration is an important process in the pre-processing of hyperspectral data. The radiometric calibration of airborne hyperspectral sensors is usually performed in the laboratory. However, the radiometric performance of these sensors can be reduced by the significant stresses generated during their transport, installation, and/or data acquisition [5]. Therefore, the radiometric calibration coefficients determined in the laboratory may not be appropriate for data acquired during the flight. Vicarious calibration methods are often used to produce a new set of radiometric calibration coefficients to replace those derived in the laboratory [6], [7]. For airborne hyperspectral sensors, a feasible vicarious calibration method is reflectance-based test site calibration [8], [9]. To perform a test site calibration for airborne hyperspectral sensors, portable or permanent reference reflectance targets must be deployed over the test sites. In addition, *in situ* measurements of target reflectance and atmospheric properties during the flight are required to predict the at-sensor radiances [10].

Besides radiometric calibration, quality assessment is also a key step in the pre-processing of hyperspectral data. The signal-to-noise ratio (SNR) is an important criterion for characterizing the quality of hyperspectral data. Accurate evaluation of the SNR is crucial to quantitatively analyze the data, and a high SNR is required to optimize the use of the data [11]. Therefore, bands with particularly low SNR must be discarded. Image-based SNR estimation is a feasible method to assess the quality of hyperspectral data [12]. Several methods have been developed to perform image-based SNR estimation [13], [14].

After the pre-processing of hyperspectral data, accurate removal of atmospheric absorption and scattering effects is required to extract land surface reflectance from remotely sensed data. The atmospheric absorption and scattering effects in remotely sensed data can be corrected by a number of physical-based methods [1]. In addition to atmospheric absorption and scattering effects, the adjacency effect must be considered during the retrieval of land surface reflectance from hyperspectral data. The magnitude of this

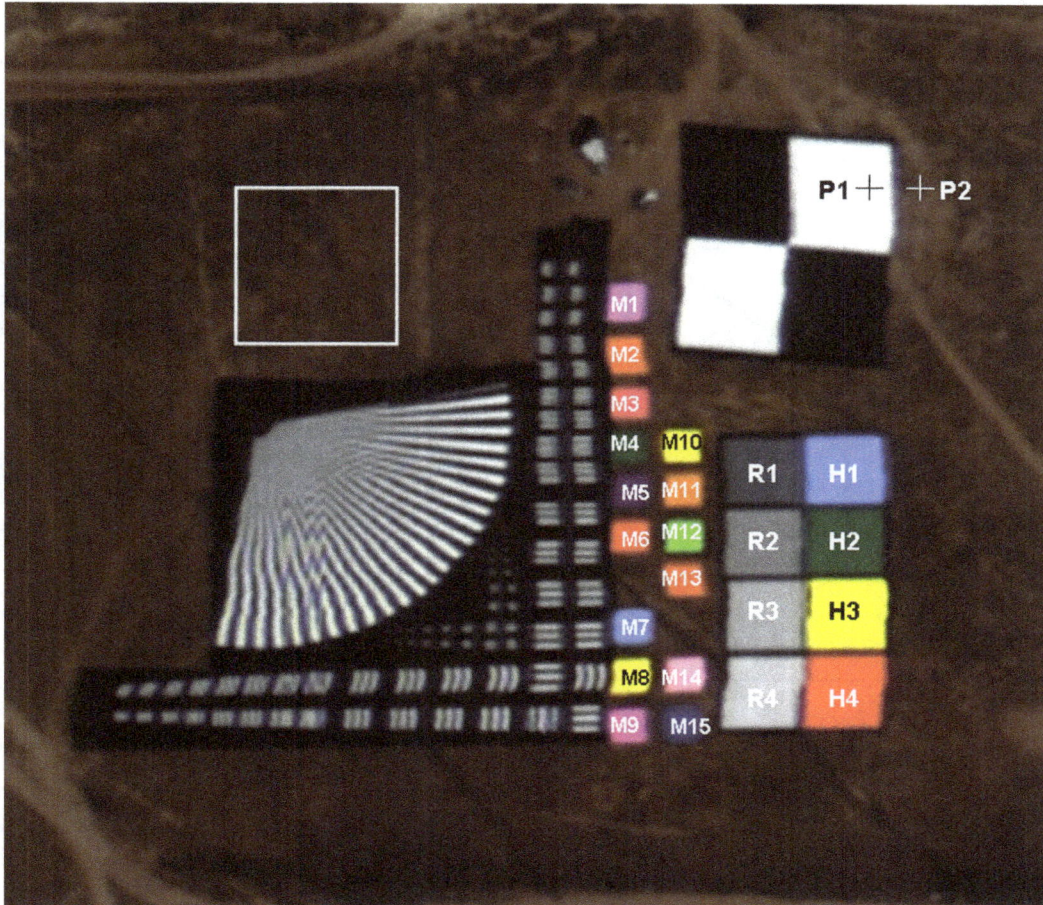

Figure 1. A subset image extracted from hyperspectral data acquired over the test site on 3 September 2011 at 06:42 UTC. The locations of the 23 targets (R1–R4, H1–H4, and M1–M15) are displayed in the image. A bare area highlighted in a white rectangle is used to perform the signal-to-noise ratio estimation. The two pixels labeled as P_1 and P_2 are used to demonstrate the discrepancy between the uniform and non-uniform Lambertian land surface reflectance.

effect directly depends on atmospheric turbidity and surface heterogeneity [15]. Therefore, the adjacency effect is the most intricate problem that must be solved when removing atmospheric effects from hyperspectral data [16].

To validate land surface reflectance derived from airborne hyperspectral data, *in situ* measurements must be collected. *In situ* measurements are used to evaluate the performance of the retrieval algorithms of land surface reflectance. The accuracies of the retrieval parameters are characterized by comparing the values of the retrieval parameters with the *in situ* measurements. The accuracies to which the retrieved values match the *in situ* measurements are used to further improve the performance of the retrieval algorithms of land surface reflectance. The objectives of this study are 1) to assess the radiometric performance of a new hyperspectral sensor onboard an unmanned aerial vehicle (UAV) and 2) to validate land surface reflectance retrieval from airborne hyperspectral data using *in situ* measurements.

Test Site and Data

1. Test site. To evaluate the in-flight performance of a new hyperspectral sensor onboard an UAV, a comprehensive field campaign was conducted over the Baotou test site (Inner Mongolia, China: 40.88°N, 109.53°E) on 3 September 2011. The Baotou test site is located in a rural area, is surrounded by

agricultural parcels, and has an average ground elevation of approximately 1.3 km above sea level (ASL). The test site receives little precipitation and has a high percentage of cloud-free days. The area has a continental climate that is characterized by four seasons and a large diurnal temperature variation. The yearly average temperature is 6–7°C, and the average annual rainfall is 200–250 mm.

A number of portable reference reflectance targets were deployed over the test site. Figure 1 shows a subset image extracted from data acquired by the hyperspectral sensor onboard the UAV on 3 September 2011 at 06:42 UTC. The targets denoted as R1–R4, H1–H4, and M1–M15 in Figure 1 are used in this study. Targets R1–R4 and H1–H4 are 15 m×15 m in size, while targets M1–M15 are 7 m×7 m in size. Targets R1–R4, which have nominal surface reflectance of 0.2, 0.3, 0.4, and 0.5, respectively, are used to perform the radiometric calibration of the hyperspectral sensor. Targets H1–H4 and M1–M15 are employed to evaluate the accuracies of the land surface reflectance retrieved from the hyperspectral data.

2. UAV-HYPER sensor. The hyperspectral sensor, which was developed by the Changchun Institute of Optics, Fine Mechanics and Physics, Chinese Academy of Sciences, was installed on an UAV operated by the Research Institute of Unmanned Flight Vehicle Design, Beihang University, China. Hereafter, the hyperspectral sensor is referred to as UAV-HYPER.

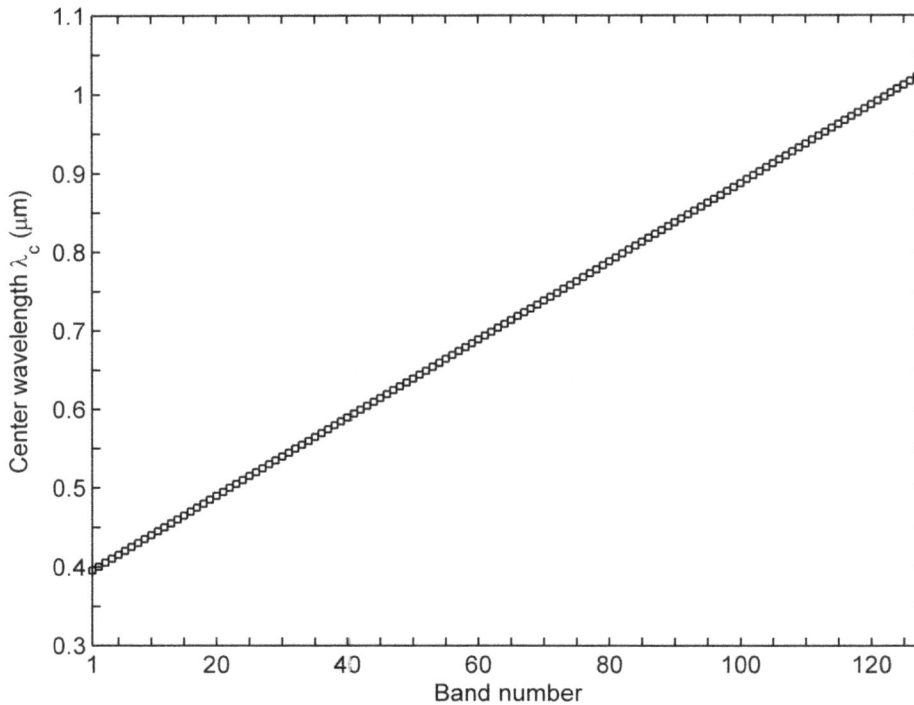

Figure 2. Center wavelengths λ_c of 128 bands of the UAV-HYPER sensor.

The UAV-HYPER sensor is a pushbroom scanner that utilizes linear CCD arrays. The main characteristics of the UAV-HYPER sensor are presented in Table 1. During the campaign, the operational altitude of the UAV-HYPER was approximately 3.5 km above ground level (AGL), which gives a spatial resolution of approximately 0.7 m at nadir. The UAV-HYPER image has an across-track sampling of 1024 pixels, which gives a swath width of approximately 0.7 km. The spectral response functions of the UAV-HYPER sensor are simulated using Gaussian functions with the center wavelengths and band widths that were measured during the laboratory calibration. The center wavelengths of 128 bands of the UAV-HYPER sensor are shown in Figure 2.

3. In situ measurements. *In situ* measurements of the 23 targets (R1–R4, H1–H4, and M1–M15) were carried out to collect the surface reflectance spectra with a SVC HR-1024 field portable spectroradiometer at the time of the UAV-HYPER data acquisition. The spectroradiometer has 1024 channels that cover the spectral range from 350 to 2500 nm. A reference measurement was collected with a white Spectralon reference panel before and after each target measurement. The spectra were measured in absolute radiance mode at nadir. The raw spectra of each target were scaled with the reference measurements to produce reflectance spectra. Five measurements of each target were averaged to yield a representative reflectance spectrum. The averaged reflectance spectra of each of the 23 targets are shown in Figure 3. Because the wavelength range of the UAV-HYPER sensor is in the 0.4–1.03 μm region and the sensitivity of the Si detector of the SVC spectroradiometer is reduced around 1 μm, only the reflectance spectra in the wavelength range of 0.4–0.95 μm are plotted in Figure 3.

In addition to the surface reflectance measurements of the targets, aerosol optical depth (AOD) and columnar water vapor (CWV) were also collected with an automatic CIMEL CE318 sunphotometer. The sunphotometer has nine channels at nominal wavelengths of 340, 380, 440, 500, 670, 870, 936, 1020, and 1640 nm. Measurements at 936 nm were used to derive the CWV [17] with the coefficients simulated by MODTRAN [18]. The AOD at 550 nm was derived from the other channels using the Ångström law. Detailed information on the method used to retrieve the AOD can be found in [19]. The measured values of the AOD at 550 nm (AOD@550) and the CWV at the time of the UAV-HYPER data acquisition are 0.18 and 1.7 g cm^{-2}, respectively.

Radiometric Performance of the UAV-HYPER Sensor

1. SNR estimation. Some bands of the UAV-HYPER data have low SNR values. A method based on local means and local

Table 1. Main characteristics of the UAV-HYPER sensor.

Parameter	Requirement
Instantaneous field of view	0.2 mrad
Field of view	11.5°
Pixel per line	1024
Spectral range	350–1030 nm
Spectral resolution	5 nm
Spatial resolution	1 m @ 5 km flight altitude AGL
Number of bands	128
Swath width	1 km @ 5 km flight altitude AGL
Digitization	12 bits
Signal-to-noise ratio	>100:1

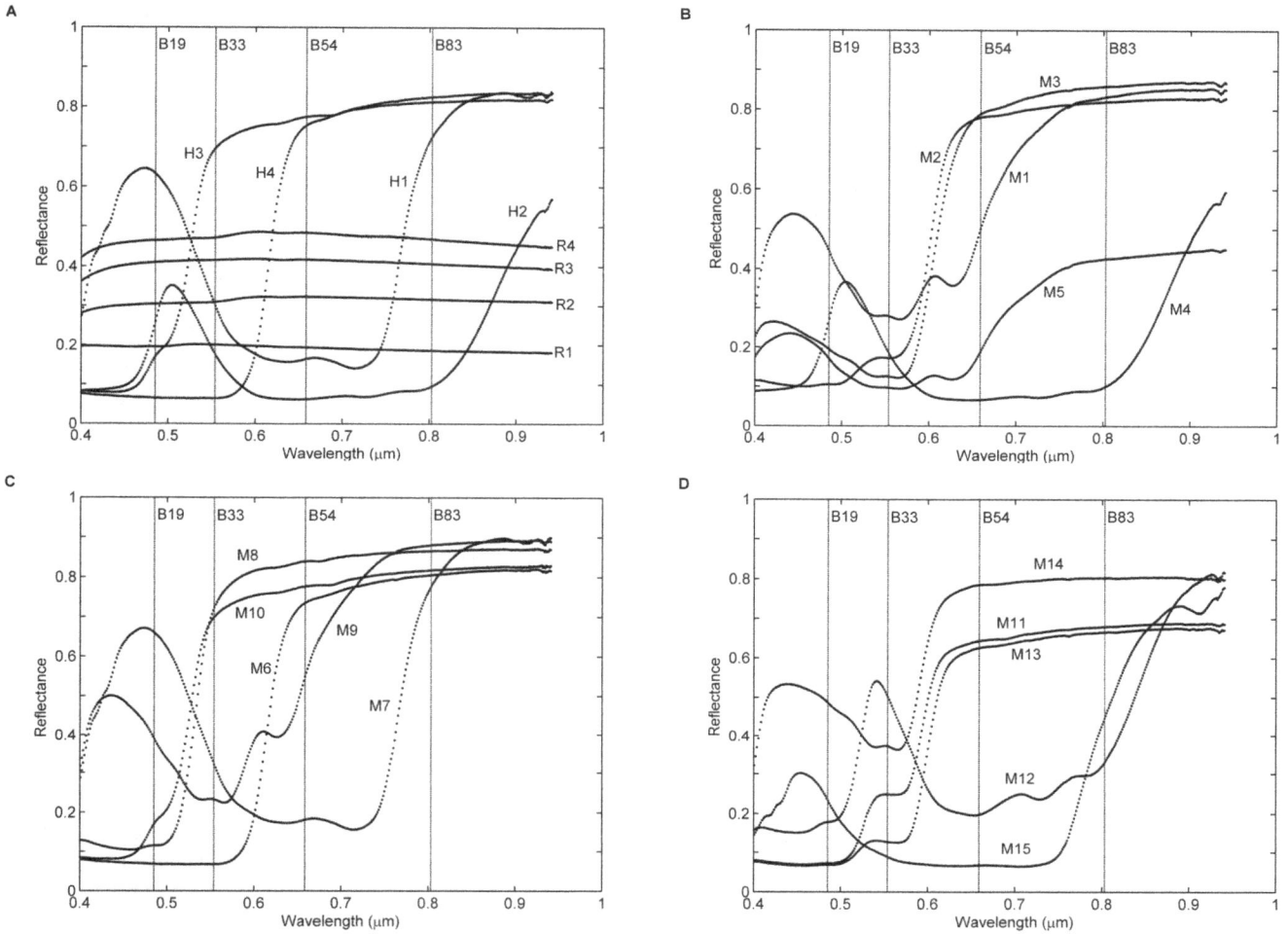

Figure 3. *In situ* **surface reflectance spectra of the 23 targets (R1–R4, H1–H4, and M1–M15) in the wavelength range of 0.4–0.95 μm.** The four dashed vertical lines denote the positions of the center wavelengths of four selected bands (19, 33, 54, and 83) that represent the blue, green, red, and near-infrared bands, respectively.

standard deviations of small imaging blocks is used to estimate the SNR from the UAV-HYPER data. A 3×3 pixel window is chosen as the block size, and the SNR is calculated as the ratio of the average signal to the average noise of the UAV-HYPER data. The detailed procedure to estimate the SNR from data acquired with imaging spectrometers can be found in [14].

Figure 4 shows the SNR estimated using a bare area (50×50 pixels) shown in Figure 1. The SNR values are in the range of 4–110. Low SNR can be found in the first and last bands of the UAV-HYPER sensor. For comparison, the SNR is also estimated using targets R1–R4 and H1–H4. An area of 10×10 pixels is extracted from the center of each of the eight targets to estimate the SNR. The SNR values of the eight targets are then averaged to yield a single averaged SNR. The results are also shown in Figure 4. The SNR estimated using the target area range from approximately 5 to 120. Low SNR values can also be found in the first and last bands of the UAV-HYPER data. Except for the bands with low SNR, the SNR values estimated using the target area are slightly greater than those estimated using the bare area. This is because the target area is more homogeneous than the bare area. Therefore, the SNR of the target area is more suitable to characterize the quality of the UAV-HYPER data. To minimize

the effect of low SNR, the bands with SNR values lower than 40 are discarded. Therefore, only bands 13–108, with SNR values between 40 and 120, are used in the following analysis.

2. Radiometric calibration of the UAV-HYPER sensor. Radiometric calibration coefficients generally differ from the laboratory pre-flight values due to in-flight changes in instrument behavior, such as optical defocusing or distortion of the dispersed radiation on the detector arrays. The procedure that converts the digital number (DN) to the at-sensor radiance L_{sensor} according to the radiometric calibration coefficients is called radiometric calibration and can be given by:

$$L_{sensor} = gain \times DN + offset \qquad (1)$$

where *gain* and *offset* are the radiometric calibration coefficients.

To determine the radiometric calibration coefficients, the at-sensor radiance L_{sensor} is calculated using Equation (2) assuming a uniform Lambertian surface:

$$L_{sensor} = L_p + \frac{\rho_{UL} F_d \left[e^{-\tau/\mu_v} + t_d(\mu_v) \right]}{\pi (1 - \rho_{UL} S)} \qquad (2)$$

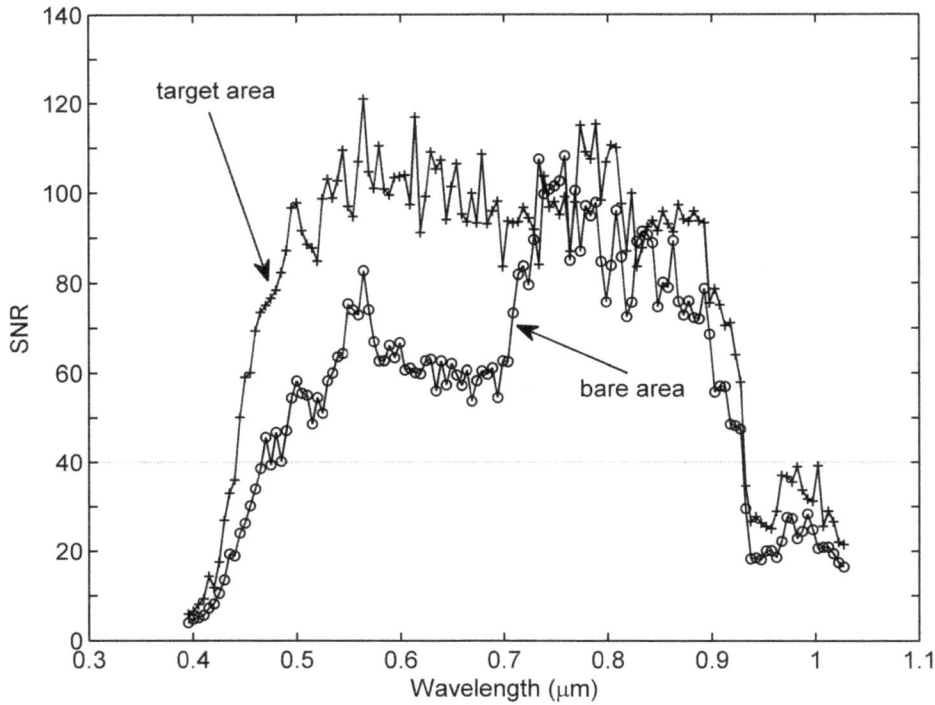

Figure 4. Signal-to-noise ratio estimated using the bare area shown in Figure 1 and the target area including targets R1–R4 and H1–H4. The bands with signal-to-noise ratio values lower than 40 are discarded in this study.

where L_p is the atmospheric path radiance, ρ_{UL} is the uniform Lambertian land surface reflectance, F_d is the total solar flux at ground level, $\mu_v = \cos(\theta_v)$ is the cosine of the view zenith angle θ_v, $e^{-\tau/\mu_v}$ and $t_d(\mu_v)$ are the direct and diffuse transmittances in the viewing direction, and S is the spherical albedo of the atmosphere.

In this study, the apparent reflectance corresponding to the at-sensor radiance L_{sensor} calculated using Equation (1) is denoted as ρ_{app}, while the at-sensor reflectance corresponding to the at-sensor radiance L_{sensor} simulated using Equation (2) is referred to as ρ_{sim}.

Targets R1–R4 are used to determine the radiometric calibration coefficients of the UAV-HYPER sensor. The DNs are averaged over a 3×3 pixel window, which is extracted from the center of each of the four targets. The at-sensor radiances L_{sensor} are calculated using Equation (2) in conjunction with five atmospheric parameters (L_p, S, F_d, $e^{-\tau/u_v}$, and $t_d(\mu_v)$) simulated by MODTRAN and the measured surface reflectance shown in Figure 3. The input parameters to MODTRAN for the radiative transfer calculations are a mid-latitude summer atmosphere, rural aerosol with AOD@550 of 0.18, CWV of 1.7 g cm^{-2}, flight altitude (FA) of 4.8 km ASL, ground elevation (GE) of 1.3 km ASL, solar zenith angle (SZA) of 44.1°, viewing zenith angle (VZA) of 2.5°, and relative azimuth angle (RAA) of 42.5°/137.5°. The radiometric calibration coefficients (*gain* and *offset*) are obtained by a least squares regression from Equation (1) using the DNs and the corresponding at-sensor radiances L_{sensor} of the four targets (R1–R4). Figure 5 shows a flowchart of the radiometric calibration procedure of the UAV-HYPER sensor.

Four bands (band 19 centered at 0.485 μm, band 33 at 0.554 μm, band 54 at 0.659 μm, and band 83 at 0.803 μm) are arbitrarily selected to represent the blue, green, red, and near-infrared bands, respectively. The positions of the center wavelengths of the four bands are shown in Figure 3. The at-sensor radiances L_{sensor} as a function of the DNs for the four bands are shown in Figure 6. The linear response of the UAV-HYPER sensor is good for the four bands, with R^2 of approximately 1 and root mean square error (RMSE) of approximately 1 W/(m^2 sr μm). The other bands have similar performance, which is not shown.

To further demonstrate the linear response range of the UAV-HYPER sensor, Figure 7 shows the apparent reflectance ρ_{app} versus the simulated at-sensor reflectance ρ_{sim} for the 19 targets (H1–H4 and M1–M15) in bands 13–108. Different symbols with different colors represent different targets. As shown in Figure 7, ρ_{app} matches ρ_{sim} well in the apparent reflectance range of approximately 0.05–0.45. The result illustrates that the linear response of the UAV-HYPER sensor is good in this apparent reflectance range. Nevertheless, ρ_{app} does not correspond well to ρ_{sim} when ρ_{app} is greater than approximately 0.45. This may be due to the non-linear response of the UAV-HYPER sensor beyond $\rho_{app} = 0.45$ and/or errors in the *in situ* measurements.

To evaluate the accuracies of the radiometric calibration of the UAV-HYPER sensor, the RMSE and relative RMSE (RRMSE) between ρ_{app} and ρ_{sim} for targets H1–H4 and M1–M15 in bands 13–108 are calculated according to Equations (3) and (4):

$$RMSE_k = \sqrt{\frac{1}{N} \sum_{i=1}^{N} \left(\rho_{app}^{k,i} - \rho_{sim}^{k,i} \right)^2} \qquad (3)$$

$$RRMSE_k = \frac{RMSE_k}{\frac{1}{N} \sum_{i=1}^{N} \rho_{sim}^{k,i}} \qquad (4)$$

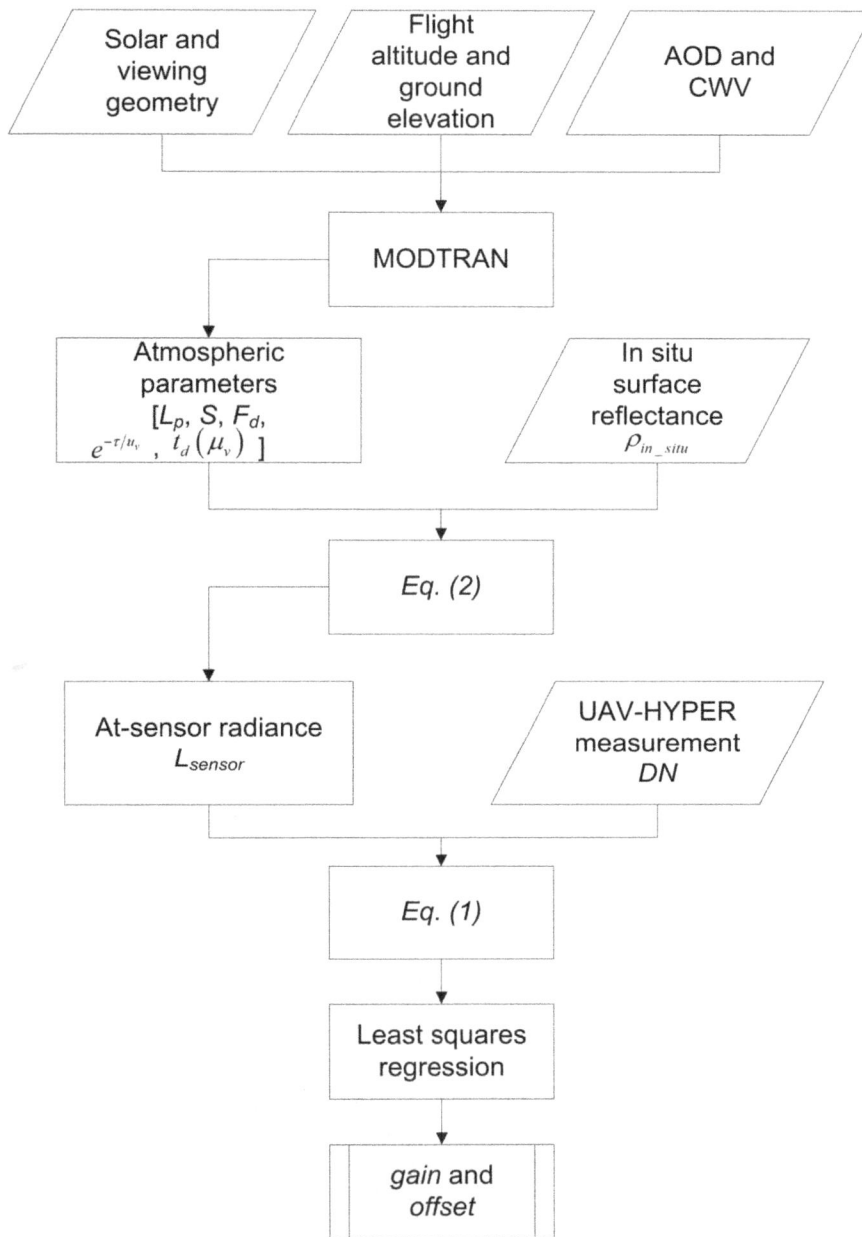

Figure 5. Flowchart of the radiometric calibration procedure of the UAV-HYPER sensor.

where k is the band number, i is the target number, and $N = 19$ (targets H1–H4 and M1–M15).

Figure 8A displays the RMSE and RRMSE values between ρ_{app} and ρ_{sim} calculated from targets H1–H4 and M1–M15 in bands 13–108. The RMSE values are between approximately 0.01 and 0.06. The larger RMSE values occur in the near-infrared region, where most of the apparent reflectance is beyond the linear response range of the UAV-HYPER sensor. Conversely, the smaller RMSE values occur in the visible range, where most of the apparent reflectance is in the linear response range of the UAV-HYPER sensor. The RRMSE values are between approximately 4% and 10%. The larger RRMSE values are approximately 10% and occur in the first and last bands of the UAV-HYPER sensor.

The smaller RRMSE values are approximately 4% and are approximately 0.54 μm.

To further examine the RMSE and RRMSE values between ρ_{app} and ρ_{sim} in the linear response range of the UAV-HYPER sensor, apparent reflectance greater than 0.45 in bands 13–108 were discarded. The results are shown in Figure 8B. The range of RMSE values is approximately 0.005 to 0.03, which is less than those in Figure 8A. The RRMSE values are between approximately 3% and 9%, which are slightly less than those in Figure 8A.

Methodology

1. Retrieval of uniform lambertian land surface reflectance. The compilation of a large atmospheric look-up

A

L_{sensor}=0.3396xDN-26.49
RMSE=1.33 W/(m^2 sr μm)
R^2=1

B

L_{sensor}=0.1623xDN-17.99
RMSE=1.08 W/(m^2 sr μm)
R^2=1

C

L_{sensor}=0.1140xDN-11.59
RMSE=1.12 W/(m^2 sr μm)
R^2=1

D

L_{sensor}=0.1314xDN-10.75
RMSE=0.85 W/(m^2 sr μm)
R^2=1

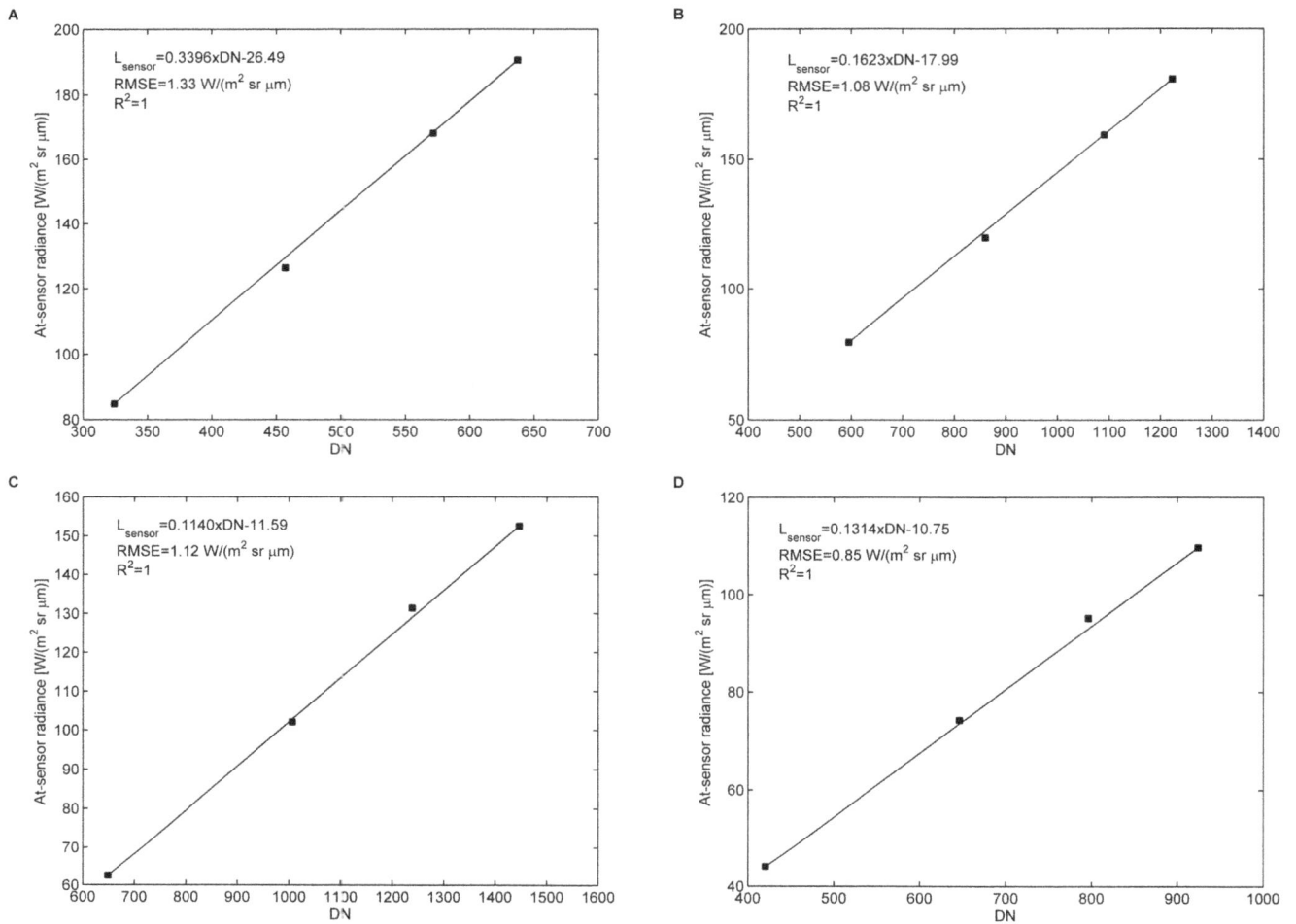

Figure 6. At-sensor radiances L_{sensor} as a function of the DN for bands 19, 33, 54, and 83. λ_c is the center wavelength.

table (LUT) is useful in deriving land surface reflectance from airborne hyperspectral data, especially for the operational atmospheric processing of large volumes of data [20]. MOD-TRAN is used to establish the atmospheric LUT because of its high accuracy and fine spectral resolution. A mid-latitude summer atmospheric model is selected. Multiple scattering is calculated using the scaled DIScrete Ordinate Radiative Transfer (DISORT) option of MODTRAN with eight streams. The carbon dioxide (CO$_2$) mixing ratio of the atmosphere is set to 380 parts per million by volume (ppmv). Due to its low spatial and temporal variations, the total ozone column content is fixed at 0.33 atm-cm for ground at sea level. The rural aerosol model is selected to represent aerosol in areas that are not strongly affected by urban or industrial sources. The radiative transfer calculations are performed using the default MODTRAN 5 cm^{-1} atmospheric database.

Six free parameters are selected as inputs in the atmospheric LUT: AOD@550, CWV, FA, GE, SZA, and RAA. Due to the small FOV of the UAV-HYPER sensor (FOV = 11.5°), VZA is fixed at 2.5° for the establishment of the atmospheric LUT. An AOD@550 range of 0.05–1.5 is used to characterize clean to very turbid atmospheric conditions. A CWV range of 0.1–5 g cm^{-2} represents a normal range for a mid-latitude summer atmosphere. The maximum flight altitude of the UAV is 7 km. The ground elevation at the Baotou test site is between 0 and 2.5 km. SZA

ranges from 0° to 70° with an increment of 10°, and RAA ranges from 0° to 180° with an increment of 30°. The breakpoint positions in the atmospheric LUT for the six input parameters are presented in Table 2. The number of breakpoints describing each dimension in the atmospheric LUT is selected as a trade-off between sufficient sampling and LUT size. Given a certain set of inputs, the values of the atmospheric parameters are calculated through linear interpolation in the six directions of the parameter space [21]. The atmospheric LUT gives five atmospheric parameters as outputs: L_p, S, F_d, $e^{-\tau/u_v}$, and $t_d(\mu_v)$. Detailed information on calculating the atmospheric parameters can be found in [22].

Once the atmospheric parameters are determined, ρ_{UL} can be calculated from L_{sensor} by inverting Equation (2) on a pixel-by-pixel basis:

$$\rho_{UL} = \frac{\pi\left(L_{sensor} - L_p\right)}{\pi\left(L_{sensor} - L_p\right)S + F_d[e^{-\tau/\mu_v} + t_d(\mu_v)]} \quad (5)$$

The procedure for retrieving ρ_{UL} is shown in Part 1 of Figure 9.

2. Retrieval of non-uniform lambertian land surface reflectance. For the case of a non-uniform Lambertian surface,

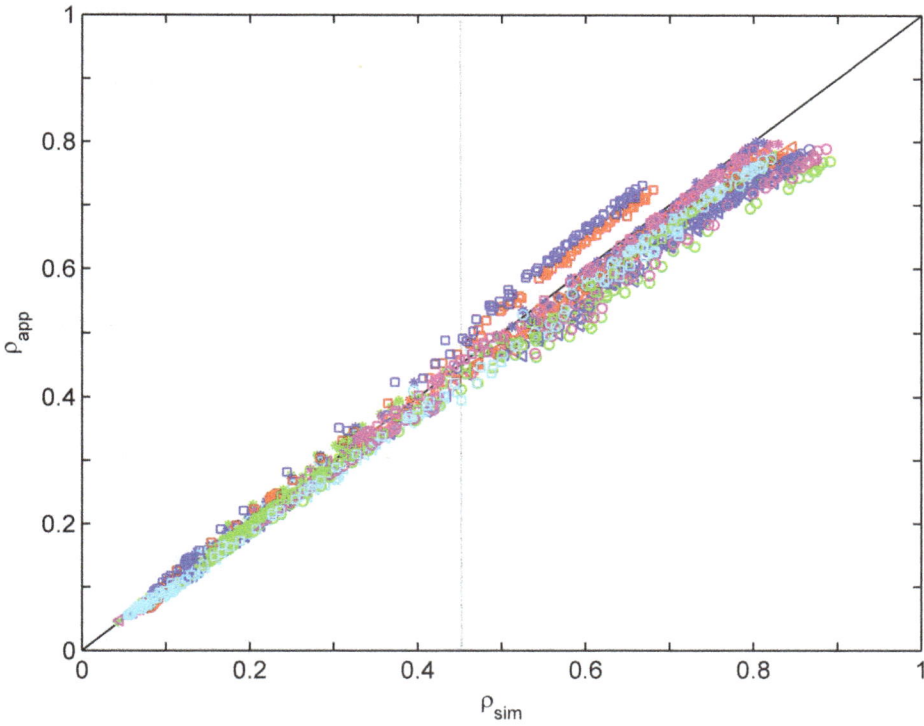

Figure 7. Comparison of the apparent reflectance ρ_{app} with the simulated at-sensor reflectance ρ_{sim} for targets H1–H4 and M1–M15 in bands 13–108. Different symbols with different colors represent different targets.

Equation (2) can be rewritten as:

$$L_{sensor} = L_p + \frac{F_d\left[e^{-\tau/\mu_v}\rho_{NUL} + t_d(\mu_v)\langle\rho\rangle\right]}{\pi(1 - \langle\rho\rangle S)} \qquad (6)$$

The non-uniform Lambertian land surface reflectance ρ_{NUL} can then be calculated from L_{sensor} by inverting Equation (6) on a pixel-by-pixel basis:

$$\rho_{NUL} = \frac{1 - \langle\rho\rangle S}{1 - \rho_{UL}S}\left(1 + \frac{t_d(\mu_v)}{e^{-\tau/\mu_v}}\right)\rho_{UL} - \frac{t_d(\mu_v)}{e^{-\tau/\mu_v}}\langle\rho\rangle \qquad (7)$$

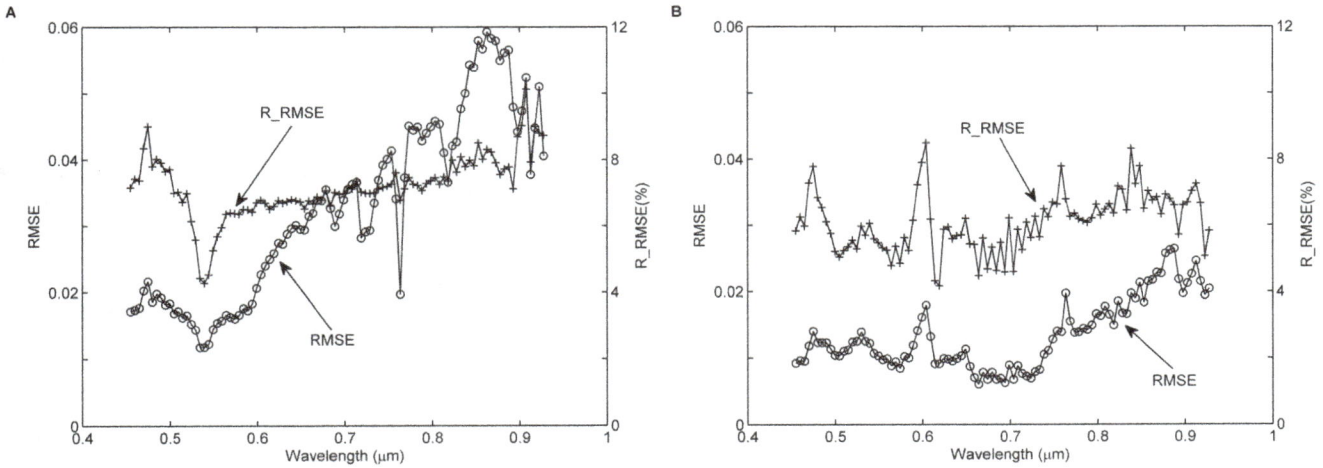

Figure 8. Root mean square error (RMSE) and relative RMSE (RRMSE) values of apparent reflectance as a function of the wavelength for targets H1–H4 and M1–M15. A: RMSE and RRMSE values between the apparent reflectance ρ_{app} and the simulated at-sensor reflectance ρ_{sim} for targets H1–H4 and M1–M15 in bands 13–108. B: Same as Figure 8A, but the apparent reflectance greater than 0.45 in bands 13–108 has been discarded due to the non-linear response of the UAV-HYPER sensor.

Table 2. Breakpoint positions in the atmospheric LUT for the six input parameters.

Parameter*	#1	#2	#3	#4	#5	#6	#7	#8
AOD@550	0.05	0.1	0.3	0.6	1.0	1.5	–	–
CWV (g cm^{-2})	0.1	0.5	1.5	2.5	3.5	5.0	–	–
FA (km)	1	2	3	4	5	6	7	–
GE (km)	0	0.5	1.0	1.5	2.0	2.5	–	–
SZA (°)	0	10	20	30	40	50	60	70
RAA (°)	0	30	60	90	120	150	180	–

*AOD@550: aerosol optical depth at 550 nm, CWV: columnar water vapor, FA: flight altitude, GE: ground elevation, SZA: solar zenith angle, and RAA: relative azimuth angle.

where the average reflectance of the surrounding $\langle\rho\rangle$ can be weighted by an atmospheric point spread function (PSF) that takes into account the contribution of the surrounding region according to the distance from the target [23]:

$$\langle\rho\rangle = \int\limits_{-\infty}^{+\infty}\int\limits_{-\infty}^{+\infty} PSF(x,y)\rho(x,y)dxdy \qquad (8)$$

According to [24], the unnormalized atmospheric PSF is defined as:

$$PSF_{unnorm}(i,j) = \sum_{h=BOA}^{at-sensor} P(\theta,\lambda)\Omega(i,j)e^{-(\tau_a\sec\theta+\tau_b)}\Delta\tau_h \quad (9)$$

where h is the atmospheric layer height, BOA is the bottom of atmosphere, θ is the scattering angle, λ is the wavelength, P is the scattering phase function, Ω is the solid angle subtended by the unit cross section as seen by the (i, j)th surrounding pixel, τ_a is the atmospheric optical depth from the surrounding pixel to the atmospheric layer height h, τ_b is the atmospheric optical depth from the atmospheric layer height h to the sensor, and $\Delta\tau_h$ is the atmospheric optical depth at the atmospheric layer height h.

To calculate the weight of each surrounding pixel, it is necessary to normalize the atmospheric PSF (i.e. the atmospheric PSF must integrate to unity). Assuming that the atmosphere is homogeneous within each atmospheric layer, $\Delta\tau_h$ can be canceled, leaving [24]:

$$PSF(i,j) = \frac{\sum_{h=BOA}^{at-sensor} P(\theta,\lambda)\Omega(i,j)e^{-(\tau_a\sec\theta+\tau_b)}}{\sum_{i=1}^{WS}\sum_{j=1}^{WS}\sum_{h=BOA}^{at-sensor} P(\theta,\lambda)\Omega(i,j)e^{-(\tau_a\sec\theta+\tau_b)}} \quad (10)$$

where WS is the moving window size, which depends on the pixel size, the atmospheric parameters, the spectral band, and the spatial frequencies of the image itself [25].

The average reflectance of the surrounding $\langle\rho\rangle$ can then be calculated in the discrete form of Equation (8), namely:

$$\langle\rho\rangle = \sum_{i=1}^{WS}\sum_{j=1}^{WS} PSF(i,j)\rho(i,j) \qquad (11)$$

To determine the window size WS, an iterative method is used to calculate the $\langle\rho\rangle$ difference between two successive iterations 'm' and '$m+1$'. If the difference is less than δ (e.g. the noise equivalent reflectance of the sensor at that wavelength), the iterative process is stopped. Otherwise, the iteration procedure goes back to recalculate $\langle\rho\rangle$, and the order 'm' is increased by 1. The iteration procedure is shown in the Part 2 of Figure 9. The outputs of the iteration procedure are the final WS image and the initial ρ_{NUL} image.

In theory, the reflectance ρ on the right-hand side of Equation (11) should be the actual reflectance; however, the actual reflectance is not available at this stage. Therefore, an iteration procedure is used to reduce the error introduced by replacing the actual reflectance with ρ_{NUL} [26]. The iteration procedure is shown in Part 3 of Figure 9. The output of the iteration procedure is the final ρ_{NUL} image.

Results and Discussion

1. Results of Uniform Lambertian Land Surface Reflectance Retrieval

The *in situ* land surface reflectance measurements of the 19 targets (H1–H4 and M1–M15) are used to evaluate the accuracies of the atmospheric correction of the UAV-HYPER data. A 3×3 pixel window is selected from the center of each of the 19 targets to yield the average surface reflectance. The uniform Lambertian land surface reflectance ρ_{UL} derived from Equation (5) is compared with the apparent reflectance ρ_{app} and the *in situ* surface reflectance ρ_{in_situ} for targets H1–H4 in bands 13–108 in Figure 10. The absorption effects of oxygen (0.76 μm) and water vapor (0.82 μm) are clearly observed in ρ_{app} but have been nearly removed in ρ_{UL}. These results demonstrate that the spectral shift of the UAV-HYPER sensor is small around the oxygen absorption feature centered at 0.76 μm. However, small dips can still be found at approximately 0.76 μm in targets H3 and H4; this is most likely because the oxygen concentration given in the radiative transfer calculations is lower than the actual conditions. ρ_{UL} generally agrees closely with ρ_{in_situ} for targets H1–H4. However, large discrepancies are present in some bands; these may be caused by radiometric calibration errors of the UAV-HYPER sensor and/or the radiative transfer calculations.

Figure 11 shows ρ_{UL} versus ρ_{in_situ} for targets H1–H4 and M1–M15 in bands 19, 33, 54, and 83. The results show that ρ_{UL} generally agrees well with ρ_{in_situ} in these four bands, with R^2 values of 0.992, 0.997, 0.991, and 0.977 and RMSE values of 0.022, 0.018, 0.034, and 0.051, respectively. However, they do not match well in the high reflectance conditions, as is shown in Figures 11C and D.

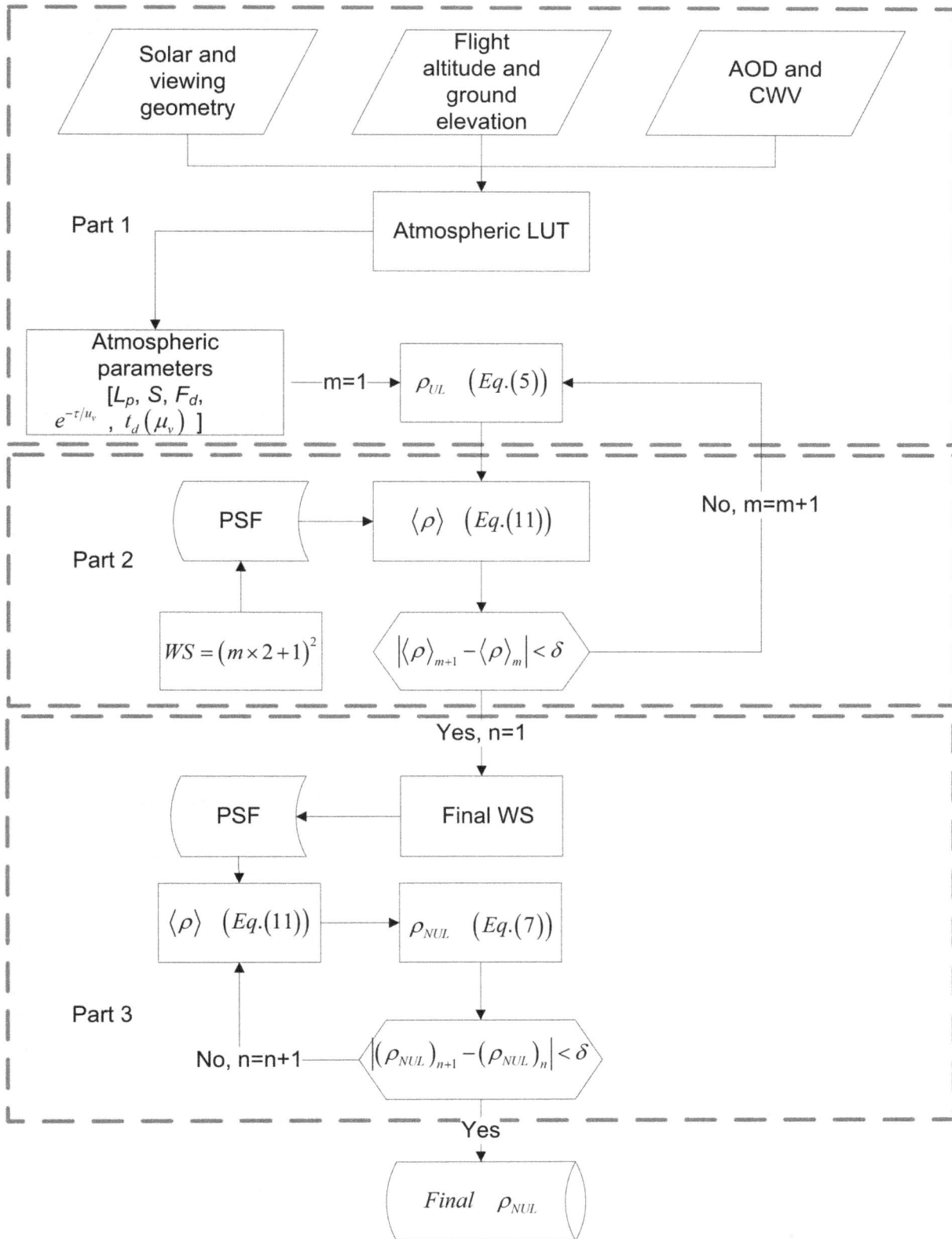

Figure 9. Flowchart of the atmospheric correction procedure for the UAV-HYPER data. Part 1 is used to derive the uniform Lambertian surface reflectance ρ_{UL} using the atmospheric look-up table (LUT). Part 2 is used to determine the window size (WS) of the atmospheric point spread function (PSF) by calculating the $\langle \rho \rangle$ difference between two successive iterations. Part 3 is used to determine the final non-uniform Lambertian surface reflectance ρ_{NUL} by calculating the ρ_{NUL} difference between two successive iterations.

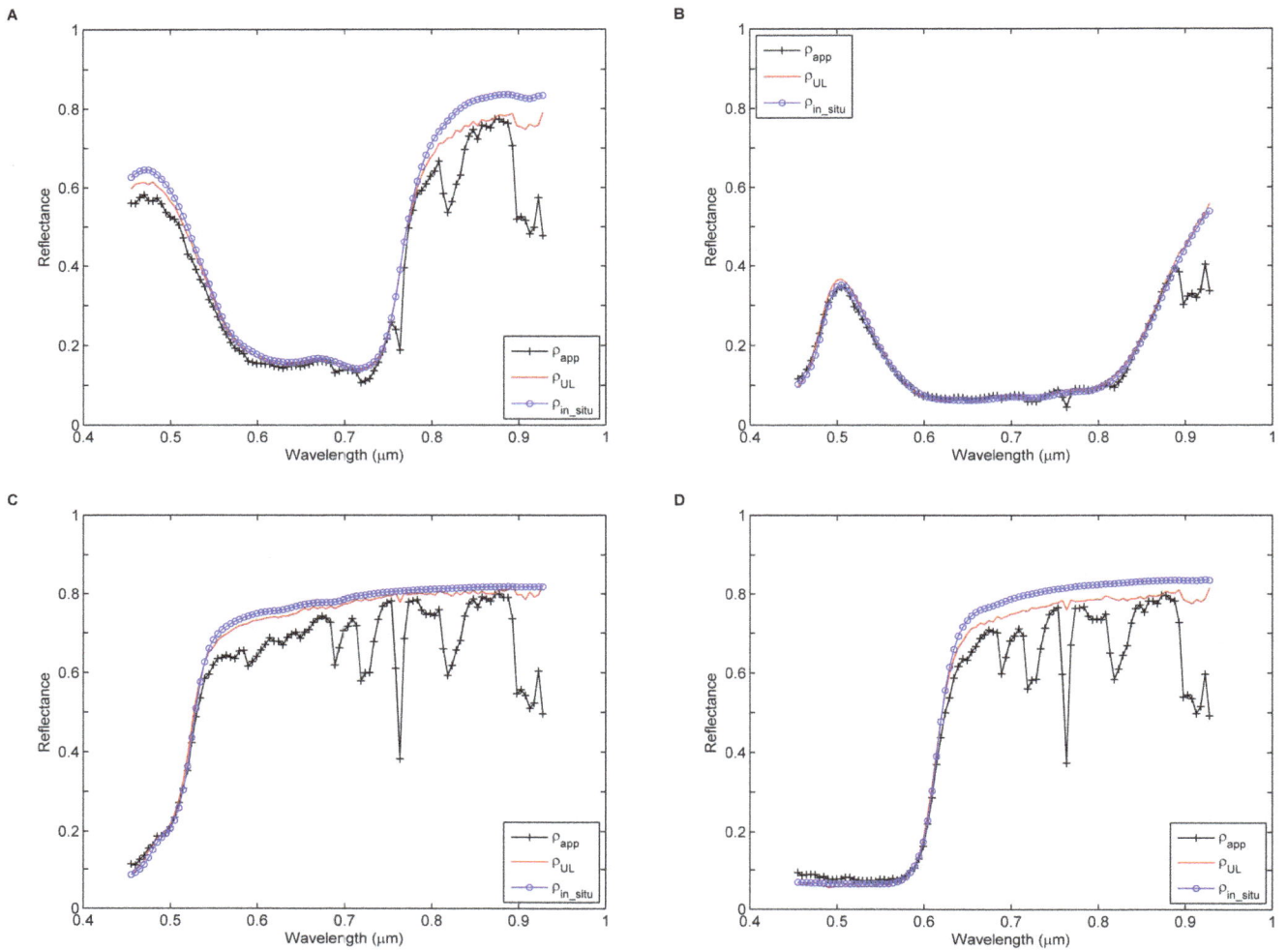

Figure 10. Comparison of the uniform Lambertian surface reflectance ρ_{UL} **derived using Equation (5) with the apparent reflectance** ρ_{app} **and the** *in situ* **surface reflectance** ρ_{in_situ} **for targets H1–H4 in bands 13–108.**

To further analyze these results, Figure 12 shows ρ_{UL} versus ρ_{in_situ} for targets H1–H4 and M1–M15 in bands 13–108. Different symbols with different colors represent different targets. The results show that ρ_{UL} does not match ρ_{in_situ} well when ρ_{in_situ} is greater than approximately 0.5. This discrepancy is believed to be mainly caused by the large errors of the radiometric calibration of the UAV-HYPER sensor due to its non-linear response when the surface reflectance is greater than approximately 0.5 and/or by measurement errors of ρ_{in_situ}. Nevertheless, ρ_{UL} matches ρ_{in_situ} well in the surface reflectance range between approximately 0.05 and 0.5, which demonstrates the good accuracy of land surface reflectance retrieval from the UAV-HYPER data in the linear response range of the UAV-HYPER sensor.

2. Results of Non-uniform Lambertian Land Surface Reflectance Retrieval

Figure 13 shows the relative errors between ρ_{UL} as well as the non-uniform Lambertian land surface reflectance ρ_{NUL} derived from Equation (7) and ρ_{in_situ} for targets H1–H4 in bands 13–108. The relative errors are less than 10% in most bands. Compared with Figure 10, large relative errors occur in the bands with

wavelengths less than approximately 0.5 μm and surface reflectance less than approximately 0.4. This occurs because a small absolute difference for a low surface reflectance may lead to a large relative error. There is no evident improvement and difference in terms of the relative errors between ρ_{NUL} and ρ_{UL} for targets H1–H4. Two reasons can explain these findings. One reason is that the AOD was relatively low at the time of the UAV-HYPER data acquisition (AOD = 0.18). The other reason is that the targets are large and have relatively homogeneous surface reflectance. Both these effects lead to a small difference between ρ_{NUL} and ρ_{UL}.

Two pixels, labeled as P_1 and P_2 in Figure 1, are selected to calculate the discrepancy between ρ_{UL} and ρ_{NUL}. As shown in Figure 14, the surface reflectance of pixel P_1 ($\rho_{UL} \approx 0.6$) is greater than that of pixel P_2 ($\rho_{UL} < 0.4$ for all bands). Therefore, ρ_{UL} of pixel P_1 is lower than its actual surface reflectance because photons escaping from the FOV of the UAV-HYPER sensor are not counterbalanced by those coming from the surrounding pixels (e.g. pixel P_2). In contrast, ρ_{UL} of pixel P_2 is greater than its actual surface reflectance because more photons come from the surrounding pixels (e.g. pixel P_1) than escape from the FOV of the UAV-HYPER sensor. As shown in Figure 14, because the

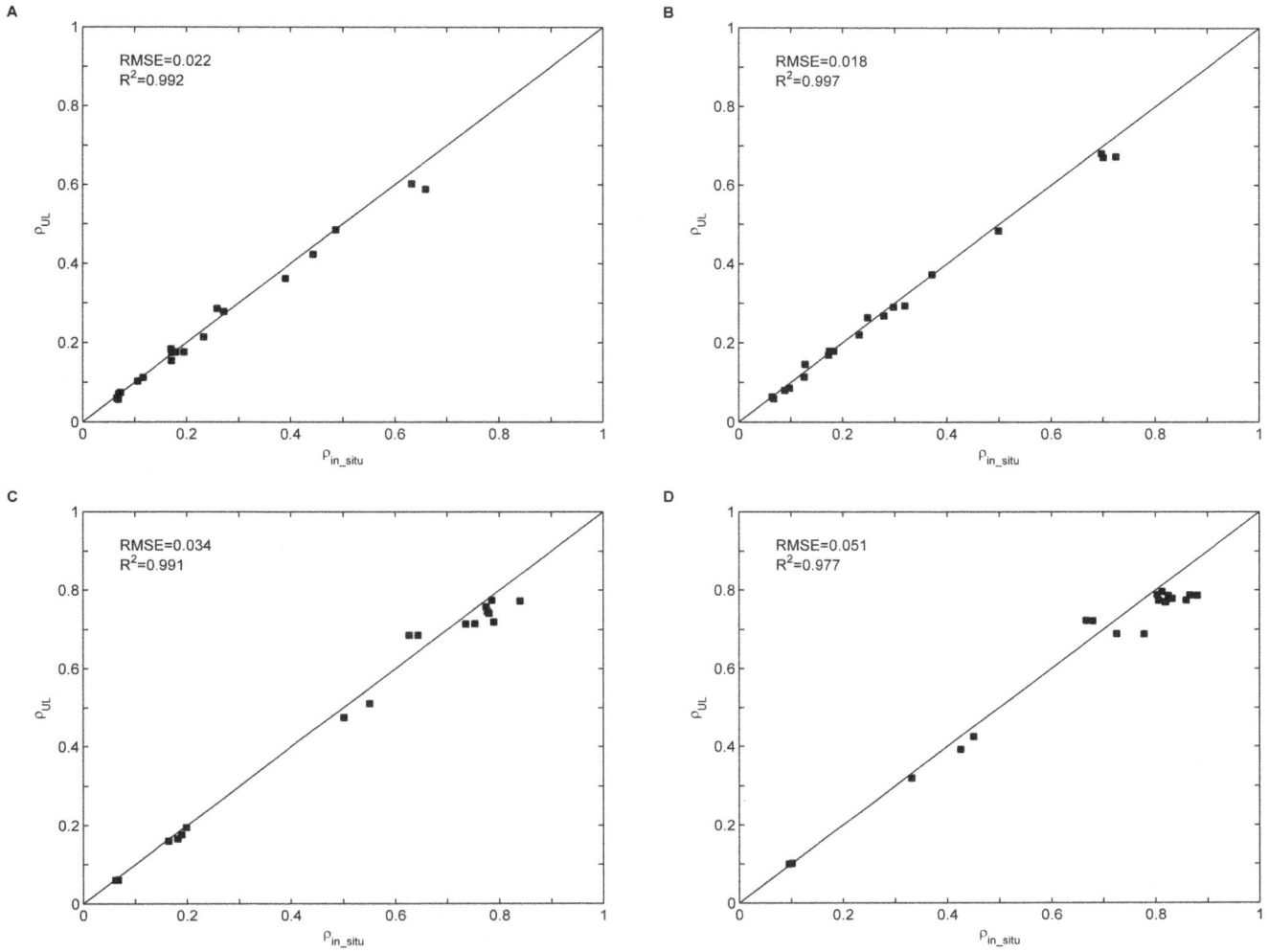

Figure 11. Comparison of the uniform Lambertian surface reflectance ρ_{UL} derived using Equation (5) with the *in situ* surface reflectance ρ_{in_situ} for targets H1–H4 and M1–M15 in bands 19, 33, 54, and 83. λ_c is the center wavelength of each of the four bands.

surface reflectance of pixel P_1 is greater than that of its surrounding pixels, $\rho_{NUL} > \rho_{UL}$ for the pixel P_1. Furthermore, the surface reflectance difference $\Delta\rho$ ($\Delta\rho = \rho_{NUL} - \rho_{UL}$) for pixel P_1 decreases as wavelength increases because the effect of atmospheric scattering in the near-infrared region is less than in the visible region, where scattering from atmospheric aerosols dominates. In contrast, $\rho_{NUL} < \rho_{UL}$, and absolute $\Delta\rho$ decreases as wavelength increases for pixel P_2.

To further demonstrate the impact of AOD on the discrepancy between ρ_{UL} and ρ_{NUL}, four AOD@550 values (0.05, 0.3, 0.5, and 1.0) are used to calculate the surface reflectance differences $\Delta\rho$ for pixels P_1 and P_2. The AOD@550 values of 0.05, 0.3, 0.5, and 1.0 represent clear, slightly turbid, turbid, and very turbid atmospheric conditions, respectively. To simulate the at-sensor radiances of the UAV-HYPER sensor, the other input parameters for the radiative transfer calculations are the same as those used to perform the atmospheric correction of the UAV-HYPER data. Furthermore, the image of ρ_{NUL} is used as a reference image of surface reflectance. The average reflectance $\langle\rho\rangle$ of each pixel is simulated by Equation (11) using the atmospheric PSF and the reference image of surface reflectance, and the at-sensor radiances

for the four AOD@550 values are simulated using Equation (6). The images of ρ_{UL} and ρ_{NUL} for the four AOD@550 values are then derived from the corresponding at-sensor radiances using Equations (5) and (7), respectively.

The surface reflectance differences $\Delta\rho$ for the four AOD@550 values are shown in Figure 15. For pixel P_1, $\Delta\rho$ decreases as wavelength increases for each AOD@550 value. In addition, $\Delta\rho$ increases as the AOD@550 value increases for each band. The maximum $\Delta\rho$ for the case of AOD@550 = 1.0 can reach 0.035 in the blue region and is approximately seven times larger than that for the case of AOD@550 = 0.05. The relative surface reflectance difference $\Delta\rho\%$ ($\Delta\rho\% = \Delta\rho/\rho_{NUL} \times 100\%$) varies from approximately 1% in the blue region to approximately 0% in the near-infrared region for the case of AOD@550 = 0.05, while $\Delta\rho\%$ varies from approximately 6% in the blue region to approximately 2% in the near-infrared region for the case of AOD@550 = 1.0. For pixel P_2, $\Delta\rho$ is negative, and its absolute value $|\Delta\rho|$ decreases as wavelength increases for each AOD@550 value. In addition, $|\Delta\rho|$ increases as the AOD@550 value increases for each band. The maximum $|\Delta\rho|$ for the case of AOD@550 = 1.0 reaches 0.08 in the blue region and is approximately eight times larger than that

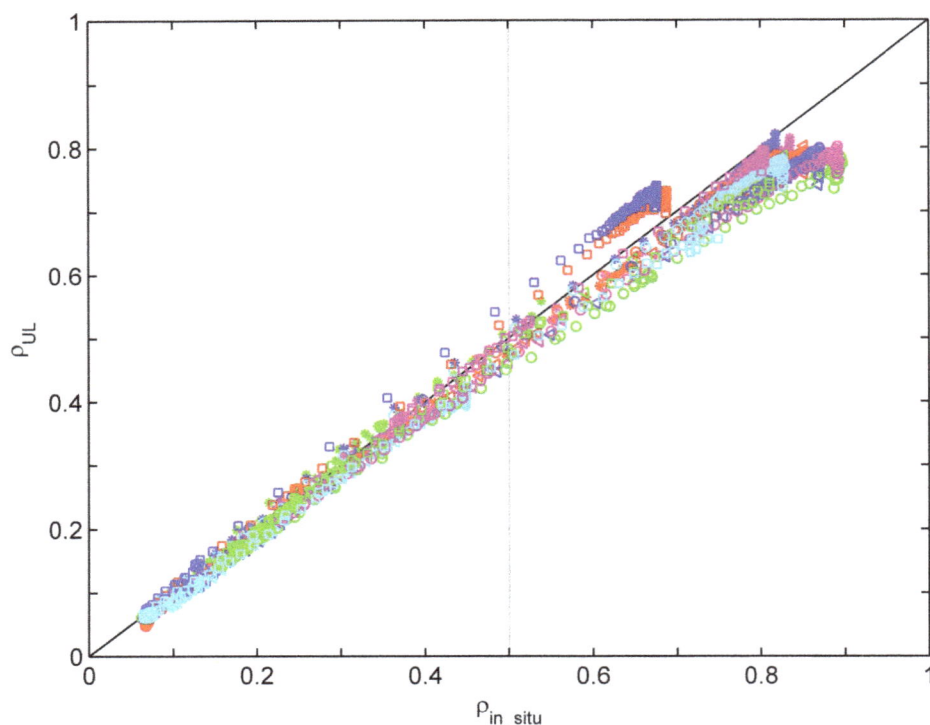

Figure 12. Comparison of the uniform Lambertian surface reflectance ρ_{UL} derived from Equation (5) and the *in situ* surface reflectance ρ_{in_situ} for targets H1–H4 and M1–M15 in bands 13–108. Different symbols with different colors represent different targets.

for the case of AOD@550 = 0.05. The relative surface reflectance difference $\Delta\rho\%$ varies from approximately −10% in the blue region to approximately 0% in the near-infrared region for the case of AOD@550 = 0.05, while $\Delta\rho\%$ varies from approximately −80% in the blue region to approximately −7% in the near-infrared region for the case of AOD@550 = 1.0.

The accuracies of land surface reflectance retrieval are evaluated in terms of the RMSE and RRMSE values between ρ_{NUL} and ρ_{in_situ} for targets H1–H4 and M1–M15 in bands 13–108. The results are shown in Figure 16A. The RMSE values range between approximately 0.01 and 0.07, while the R-RMSE values are between approximately 5% and 12%. The largest RMSE value occurs in the near-infrared region, while the smallest RMSE value occurs in the visible range. In contrast, the largest RRMSE value occurs in the visible region, while the smallest RRMSE value occurs in the near-infrared range. The accuracies of ρ_{UL} are similar to those of ρ_{NUL}, which are not shown in this study.

To further demonstrate the RMSE and RRMSE values between ρ_{NUL} and ρ_{in_situ} for the 19 targets in the linear response range of the UAV-HYPER sensor, surface reflectance greater than 0.5 in the bands 13–108 are discarded. The results are shown in Figure 16B. The RMSE values are between approximately 0.005 and 0.05, which is obviously smaller than those shown in Figure 16A. However, large RMSE values also occur in the near-infrared region. The RRMSE values are between approximately 4% and 10%, which is slightly less than those shown in Figure 16A.

Conclusions

The radiometric performance of the UAV-HYPER sensor was assessed in terms of SNR and the accuracy of the radiometric calibration. The SNR values estimated using the homogeneous targets were between approximately 5 and 120. The linear response of the UAV-HYPER sensor was found in the apparent reflectance range of approximately 0.05 and 0.45, while a non-linear response was observed for apparent reflectance greater than approximately 0.45. The accuracies of the radiometric calibration of the UAV-HYPER sensor were evaluated with RMSE of approximately 0.01–0.06 and RRMSE of approximately 4%–10%.

The retrieved uniform Lambertian land surface reflectance match the *in situ* surface reflectance well in the land surface reflectance range of approximately 0.05 to 0.5. There is a small difference between the retrieved uniform and non-uniform Lambertian land surface reflectance over the homogeneous targets and under low AOD conditions. The results demonstrate that the discrepancy between the uniform and non-uniform Lambertian land surface reflectance can be neglected under homogeneous surface and low AOD conditions. However, the discrepancy is up to 0.08 in the blue region when adjacent pixels have large land surface reflectance contrast and under high AOD conditions (e.g. AOD = 1.0). Therefore, this discrepancy should be taken into account under these conditions. The accuracies of land surface reflectance retrieval were evaluated using the *in situ* measurements with RMSE of approximately 0.01–0.07 and RRMSE of approximately 5%–12%.

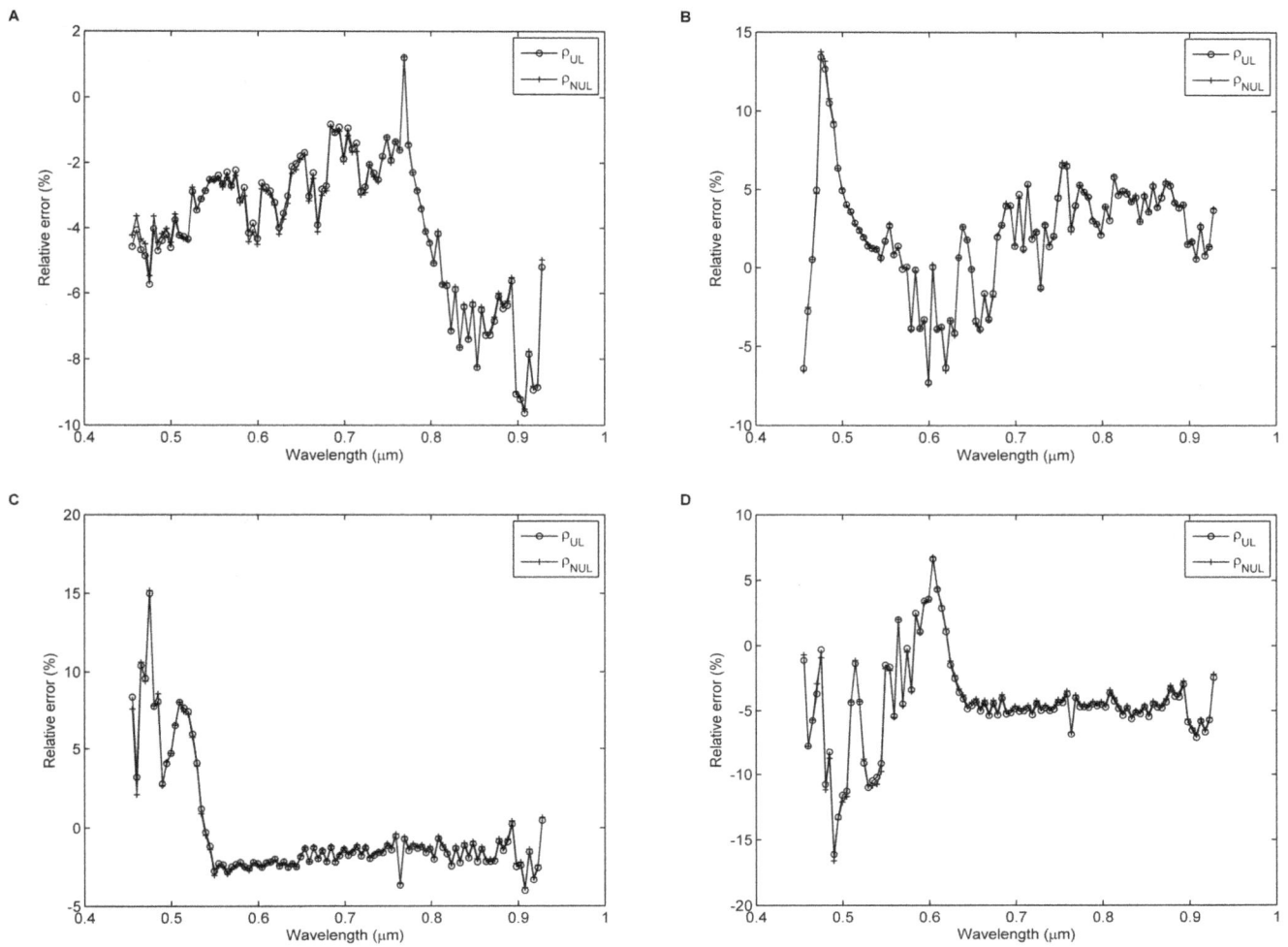

Figure 13. Relative errors of surface reflectance as a function of the wavelength for targets H1–H4. Relative errors between the uniform Lambertian surface reflectance ρ_{UL} derived from Equation (5) and the non-uniform Lambertian surface reflectance ρ_{NUL} derived from Equation (7) and the *in situ* surface reflectance ρ_{in_situ} for targets H1–H4 in bands 13–108.

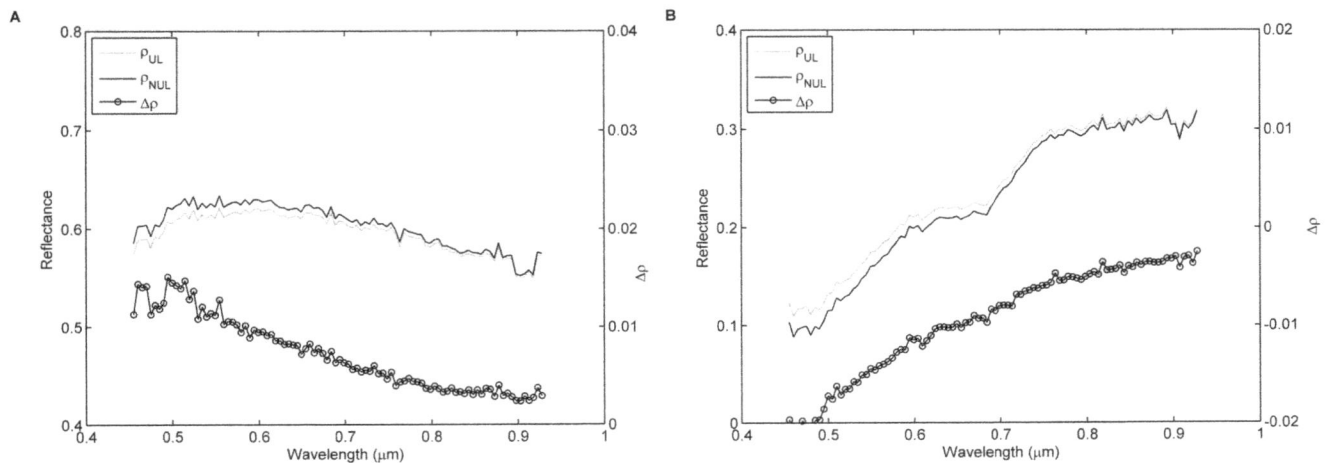

Figure 14. Comparison of the uniform Lambertian surface reflectance ρ_{UL} derived from Equation (5) and the non-uniform Lambertian surface reflectance ρ_{NUL} derived from Equation (7) for pixels (A) P_1 and (B) P_2. $\Delta\rho$ is the surface reflectance difference between ρ_{UL} and ρ_{NUL} ($\Delta\rho = \rho_{NUL} - \rho_{UL}$).

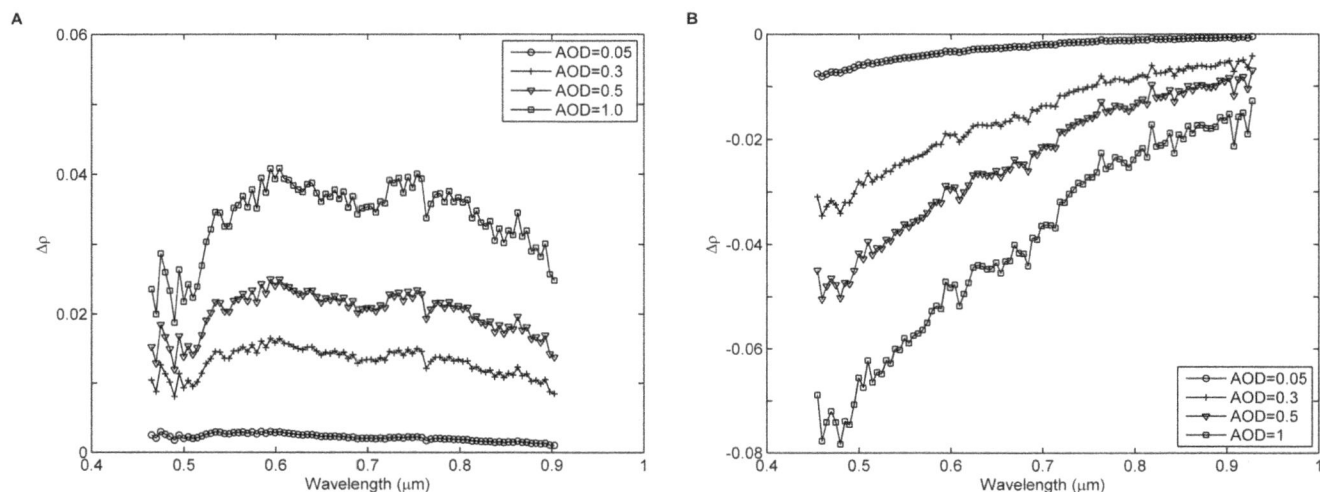

Figure 15. Surface reflectance differences $\Delta\rho$ ($\Delta\rho = \rho_{NUL} - \rho_{UL}$) **as a function of the wavelength for the four AOD@550 values for pixels (A)** P_1 **and (B)** P_2.

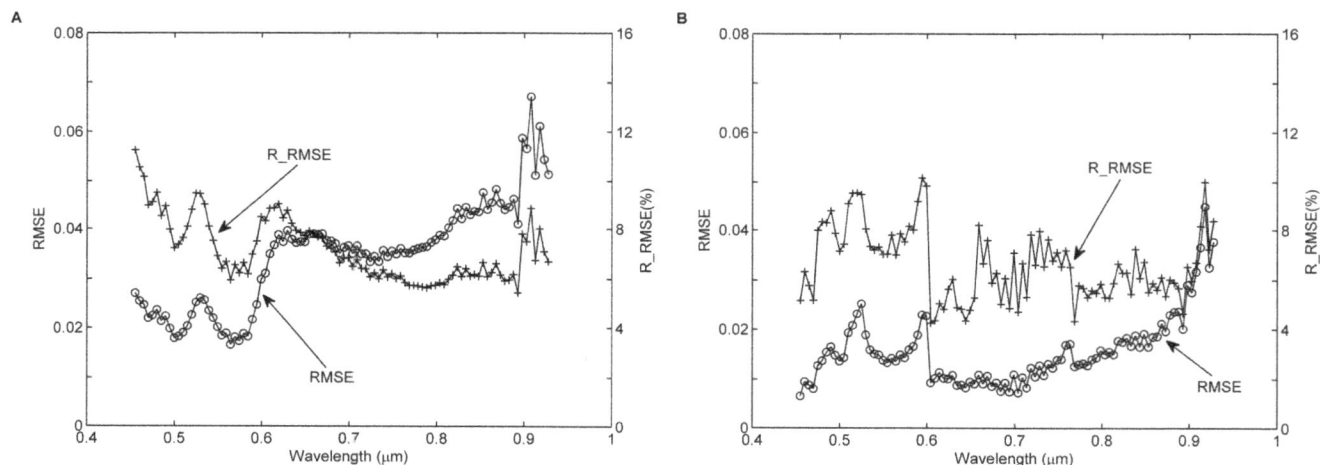

Figure 16. Root mean square error (RMSE) and relative RMSE (RRMSE) values of surface reflectance as a function of the wavelength for targets H1–H4 and M1–M15. A: RMSE and RRMSE values between the non-uniform Lambertian surface reflectance ρ_{NUL} derived using Equation (7) and the *in situ* surface reflectance ρ_{in_situ} for targets H1–H4 and M1–M15 in bands 13–108. B: Same as Figure 16A, but surface reflectance greater than 0.5 in bands 13–108 is excluded.

Acknowledgments

The authors would like to thank Yaokai Liu, Yonggang Qian, and Ning Wang at the Academy of Opto-Electronics, Chinese Academy of Sciences, and Kun Wang at the Institute of Geographic Sciences and Natural Resources Research, Chinese Academy of Sciences for the field work.

Author Contributions

Conceived and designed the experiments: SBD ZLL CRL. Performed the experiments: EYZ LLM. Analyzed the data: BHT HW. Contributed reagents/materials/analysis tools: SBD ZLL. Wrote the paper: SBD ZLL.

References

1. Gao BC, Montes MJ, Davis CO, Goetz AFH (2009) Atmospheric correction algorithms for hyperspectral remote sensing data of land and ocean. Remote Sens Environ 113: S17–S24.
2. Goetz AFH (2009) Three decades of hyperspectral remote sensing of the Earth: A personal view. Remote Sens Environ 113: S5–S16.
3. Li ZL, Wu H, Wang N, Qiu S, Sobrino JA, et al. (2013) Land surface emissivity retrieval from satellite data. Int J Remote Sens 34: 3084–3127.

4. Li ZL, Tang BH, Wu H, Ren HZ, Yan GJ, et al. (2013) Satellite-derived land surface temperature: Current status and perspectives. Remote Sens Environ 131: 14–37.
5. Secker J, Staenz K, Gauthier RP, Budkewitsch P (2001) Vicarious calibration of airborne hyperspectral sensors in operational environments. Remote Sens Environ 76: 81–92.
6. Brook A, Ben-Dor E (2011) Supervised vicarious calibration of hyperspectral remote sensing data. Remote Sens Environ 115: 1543–1555.

7. Green RO, Pavri BE, Chrien TG (2003) On-orbit radiometric and spectral calibration characteristics of EO-1 Hyperion derived with an underflight of AVIRIS and in situ measurements at Salar de Arizaro, Argentina. IEEE Trans Geosci and Remote Sens 41: 1194–1203.

8. Markelin L, Honkavaara E, Peltoniemi J, Ahokas E, Kuittinen R, et al. (2008) Radiometric calibration and characterization of large format digital photogrammetric sensors in a test field. Photogramm Eng Remote Sens 74: 1487–1500.

9. Markelin L, Honkavaara E, Hakala T, Suomalainen J, Peltoniemi J (2010) Radiometric stability assessment of an airborne photogrammetric sensor in a test field. ISPRS J Photogramm Remote Sens 65: 409–421.

10. Honkavaara E, Peltoniemi J, Ahokas E, Kuittinen R, Hyyppä J, et al. (2008) A permanent test field for digital photogrammetric systems. Photogramm Eng Remote Sens 74: 95–106.

11. Richter R, Schläpfer D, Müller A (2011) Operational atmospheric correction for imaging spectrometers accounting for the smile effect. IEEE Trans Geosci and Remote Sens 49: 1772–1780.

12. Atkinson PM, Sargent IM, Foody GM, Williams J (2005) Interpreting image-based methods for estimating the signal-to-noise ratio. Int J Remote Sens 26: 5099–5115.

13. Curran PJ, Dungan JL (1989) Estimation of signal-to-noise: A new procedure applied to AVIRIS data. IEEE Trans Geosci and Remote Sens 27: 620–628.

14. Gao BC (1993) An operational method for estimating signal to noise ratios from data acquired with imaging spectrometers. Remote Sens Environ 43: 23–33.

15. Liang SL, Fang HL, Chen MZ (2001) Atmospheric correction of Landsat ETM+ land surface imagery. I. Methods. IEEE Trans Geosci and Remote Sens 39: 2490–2498.

16. Semenov AA, Moshkov AV, Pozhidayev VN, Barducci A, Marcoionni P, et al. (2011) Estimation of normalized atmospheric point spread function and restoration of remotely sensed images. IEEE Trans Geosci and Remote Sens 49: 2623–2634.

17. Bruegge CJ, Conel JE, Green RO, Margolis JS, Holm RG, et al. (1992) Water vapor column abundance retrievals during FIFE. J Geophys Res 97: 18759–18768.

18. Halthore RN, Eck TF, Holben BN, Markham BL (1997) Sun photometric measurements of atmospheric water vapor column abundance in the 940-nm band. J Geophys Res 102: 4343–4352.

19. Estellés V, Utrillas MP, Martínez-Lozano JA, Alcántara A, Alados-Arboledas L, et al. (2006) Intercomparison of spectroradiometers and Sun photometers for the determination of the aerosol optical depth during the VELETA-2002 field campaign. J Geophys Res 111: D17207. doi: 10.1029/2005JD006047.

20. Guanter L, Richter R, Kaufmann H (2009) On the application of the MODTRAN4 atmospheric radiative transfer code to optical remote sensing. Int J Remote Sens 30: 1407–1424.

21. Guanter L, Estellés V, Moreno J (2007) Spectral calibration and atmospheric correction of ultra-fine spectral and spatial resolution remote sensing data. Application to CASI-1500 data. Remote Sens Environ 109: 54–65.

22. Verhoef W, Bach H (2003) Simulation of hyperspectral and directional radiance images using coupled biophysical and atmospheric radiative transfer models. Remote Sens Environ 87: 23–41.

23. Vermote EF, Tanre D, Deuze JL, Herman M, Morcette JJ (1997) Second simulation of the satellite signal in the solar spectrum, 6S: An overview. IEEE Trans Geosci and Remote Sens 35: 675–686.

24. Sanders LC, Schott JR, Raqueño R (2001) A VNIR/SWIR atmospheric correction algorithm for hyperspectral imagery with adjacency effect. Remote Sens Environ 78: 252–263.

25. Richter R (1990) A fast atmospheric correction algorithm applied to Landsat TM images. Int J Remote Sens 11: 159–166.

26. Putsay M (1992) A simple atmospheric correction method for the short wave satellite images. Int J Remote Sens 13: 1549–1558.

7

Remote Sensing of Atmospheric Optical Depth Using a Smartphone Sun Photometer

Tingting Cao, Jonathan E. Thompson*

Department of Chemistry & Biochemistry, Texas Tech University, Lubbock, Texas, United States of America

Abstract

In recent years, smart phones have been explored for making a variety of mobile measurements. Smart phones feature many advanced sensors such as cameras, GPS capability, and accelerometers within a handheld device that is portable, inexpensive, and consistently located with an end user. In this work, a smartphone was used as a sun photometer for the remote sensing of atmospheric optical depth. The top-of-the-atmosphere (TOA) irradiance was estimated through the construction of Langley plots on days when the sky was cloudless and clear. Changes in optical depth were monitored on a different day when clouds intermittently blocked the sun. The device demonstrated a measurement precision of 1.2% relative standard deviation for replicate photograph measurements (38 trials, 134 datum). However, when the accuracy of the method was assessed through using optical filters of known transmittance, a more substantial uncertainty was apparent in the data. Roughly 95% of replicate smart phone measured transmittances are expected to lie within ±11.6% of the true transmittance value. This uncertainty in transmission corresponds to an optical depth of approx. ±0.12–0.13 suggesting the smartphone sun photometer would be useful only in polluted areas that experience significant optical depths. The device can be used as a tool in the classroom to present how aerosols and gases effect atmospheric transmission. If improvements in measurement precision can be achieved, future work may allow monitoring networks to be developed in which citizen scientists submit acquired data from a variety of locations.

Editor: João Miguel Dias, University of Aveiro, Portugal

Funding: The authors would like to thank the State of Texas/Texas Tech University for financial support of this project through new investigator start-up funding. The funders had no role in study design, data collection and analysis, decision to publish, or preparation of the manuscript.

Competing Interests: The authors have declared that no competing interests exist.

* E-mail: jon.thompson@ttu.edu

Introduction

In recent years smart phones have become ubiquitous in society. These devices offer unique platforms for remote monitoring applications as they contain advanced processors and communication ability, are equipped with sophisticated sensors (cameras, accelerometers, and GPS), and are consistently located with an end user. In principle, the platform offers the potential of massively parallel sensing of environmental pollutants, inexpensive medical diagnosis, and unmatched utility for a variety of associated social and epidemiological studies. While it is unclear if this potential will ever be fully realized, several groups have already begun exploring the idea. Insurers have begun scoring consumers driving patterns based on acceleration and speed data collected with their smart phones [1]. A University group has used behavioral patterns assessed from data collected with smart phones to monitor physiological and mental health of a test group of students [2]. Smith *et al.* [3] have reported a simple cell-phone microscope and spectroscope fabricated from common laboratory supplies. A spectroscope and operating software has also recently been developed for pedagogical purposes [4]. In the research laboratory, Zhu *et al.* have used quantum dot luminescence as a probe to detect *E. coli* with cell phones [5]. Zhu *et al.* also have used a cell-phone to count fluorescently tagged cells flowing within a microfluidic channel [6]. In addition, Delaney *et al.* suggest cell phones coupled with electrogenerated chemiluminescence detec-

tion may offer options for rapid medical diagnosis in developing nations [7].

In this work we consider whether smart phone sun photometry is a viable option for remote sensing of atmospheric aerosols and gases. Our laboratory has an interest in developing single-point measurements to monitor the concentration and optical properties of airborne particulates (aerosols) [8–15] since particulate matter (PM) in earth's atmosphere degrades visibility, can affect climate, and human health. Particles in the micrometer size range effectively scatter electromagnetic radiation, so optical attenuation methods are suitable for estimating particle mass loading. However, single-point measurements can be costly to operate and maintain, require trained personnel to operate, and cannot provide air-column integrated information. In addition, information about the spatial pattern of aerosol loading cannot be obtained via one fixed sensor. In contrast, a network of smartphone-based sun photometers operated by citizen-scientists could offer air-column-integrated information at many locations simultaneously. Indeed, recent efforts have been directed towards this goal (see http://ispex.nl/en/; accessed 10/31/2013).

Sun photometers measure the irradiance of sunlight within narrow wavelength bands that reaches earth's surface. The top-of-the-atmosphere (TOA) solar irradiance can be estimated through construction of a Langley plot, which is a calibration procedure that relates known changes in atmospheric path length (air mass) with observed changes in signals. Once the TOA intensity is known, atmospheric column transmittance (T) and integrated optical depth can be calculated. Changes of optical depth mostly depend on the quantity of light removed through scattering and absorption by aerosol particles, which is the reason why optical depth of cloudy days will be quite different from clear days. Previous authors have used sun photometry measurements to describe aerosol scattering or to measure the concentration of atmospheric gases such as O_3 or water vapor [16–19]. In addition, measurements at multiple wavelengths has allowed retrieval of several aerosol characteristics including, size distributions, complex refractive index, and single-scatter albedo [20,21]. In principle, if sun photometry data collected with a smart phone could quickly be sent to a central location and automatically processed, environmental monitoring could be conducted in nearly real-time on a continental-to-global scale. However, the quality of such data sets would be directly linked to the performance characteristics of the smart phone sun photometers used.

It is the purpose of this work to explore the use of several band pass filters and the camera of a smart phone as a sun photometer. We present data on the accuracy and precision of measurements achieved with a typical smart phone (iPhone 4) and use the phone for estimating direct surface solar irradiance in several wavelength bands. While the work described within focuses on an initial proof-of-concept and characterization study, a network of smart phone based sun photometers could be a valuable resource for monitoring atmospheric PM mass loadings, air quality, or UV index. Initial results suggest the smart-phone sensor could be a valuable tool for reporting air quality in areas heavily affected by pollution. The device is also attractive for use in an educational setting.

Materials and Methods

2.1 Location, Solar Zenith Angle, and Airmass Estimation

All data was collected at ground level on the campus of Texas Tech University at Lubbock, TX (33.5778° N, 101.8547° W) during June - September 2012. The cosine of the solar zenith angle at any time/date for our location was computed by using the NOAA solar position calculator (http://www.esrl.noaa.gov/gmd/grad/solcalc/azel.html). Air mass was estimated as the inverse of this value. This estimate is known to be less accurate when the sun is near the horizon (air mass >10). Also, measurements made near sunrise/sunset are also subject to a higher uncertainty since air mass changes very rapidly at these times, and collecting data for a trial requires a finite amount of time. This introduces uncertainty into the Langley plot analysis described below.

2.2 Smartphone and Application Software

An iPhone 4 (Apple) was used for taking all measurements. The phones 8-bit, 5-megapixel iSight camera was used. Through experimentation, we found the smart phone camera produced some measureable response to light from a wavelength of approx. 200 nm to at least 1064 nm. In many Apps that run the camera, the ISO and shutter speed automatically adjusts based on ambient light level. This is clearly not acceptable for quantitative measurements so an application named 645 PRO (available in App store and Jag.gr) was used to take photographs since it allows the user control over variables such as exposure, ISO and focus. In order to avoid saturation of the 8-bit sensor, the maximum signal counts for each RGB channel must be <255. To achieve this we used neutral density filters to reduce the sunlight intensity and prevent potential damage to the camera. Then, the ISO of the camera was locked at 80 and the exposure time was at 1/

10,000 sec. The white balance and auto focus was also locked. Also, Q-mode was chosen in the App to bypass the film mode. Through experimentation we found it was necessary to make sure the App was completely closed (not running in background) prior to starting it again for use. Failure to do this produces unreliable quantitative results, but the exact nature of this error is not fully understood. (Figure S1 in File S1 depicts the apparatus used for solar irradiance measurements.).

2.3 Optical Filtering & Laboratory Transmission Experiments

Three band-pass filters with the color of yellow (590±2 nm), green (520±10 nm) and blue (450±10 nm) were used to spectrally select sunlight by placing them directly in front of the iPhone camera lens. We refer to these filters as "sun filters." Again, these sun filters were used to spectrally select the sun light. Neutral density filters were also used to attenuate the sunlight reaching the sensor. The transmission spectra of the sun filters are shown in Figure 1. The transmittances of the assembled filters (interference filters + neutral density) ranged between 0.1–0.4%. These low transmittances were needed to keep the cell phone camera sensor on-scale. Additional filters were also used to verify the accuracy of the sun photometer measurements made with the cell phone. For these experiments, four colored glass filters and four colored plastic filters were used. These filters are referred to as "test filters" and are essentially used to simulate aerosol loads (changes in transmittance). The test filters were used since their transmittances were fixed and known (or could be measured with a UV-VIS spectrophotometer) at the wavelengths of measurement. The glass test filters were a 400 nm long pass filter (FGL400), a 495 nm long pass filter (FGL495), a 550 nm short pass filter (FGL550), and a band pass filter (FGB37). The glass filters were obtained from Thor Labs, and a transmission spectrum for each filter is available at www.thorlabs.com. The colored plastic filters (LEE FILTERS) were named "straw tint", "chocolate", "dark steel blue", and "surprise pink". The transmission spectrum for each colored filter can be found at http://www.leefilters.com. The accepted percent transmittances of the test filters varied from near 0 to 92% at the test wavelengths. By comparing the smart phone measured transmittances for the test filters with their accepted transmittances, we can assess accuracy of the smart phone measurement. Smartphone measured transmittance was determined by taking photographs of the sun with and without a given test filter. These images corresponded to the test sample and spectroscopic blank (100% T), respectively. These experiments were conducted using the sun as the spectroscopic source, and a FGS 900 nm short pass filter was used to help eliminate near IR light which if present can lead to erroneous transmittance values. Since the atmospheric conditions can change during measurement it was decided that additional laboratory experiments should be performed in which a fiber-coupled UV-VIS light source (Hammamatsu, L10290) was used as the spectroscopic source. Details of the light source can be found in product information available at http://jp.hamamatsu.com/products/light-source/pd032/L10290/index_en.html. Light from the source was passed through the sun filters and into the smart phone camera. Again, the test filters provided samples of known transmission to test accuracy. It was anticipated the well-controlled laboratory setup would produce more accurate and precise measurements of test filter transmission, however, this was not achieved. Standard deviations of absolute errors (smart phone T – accepted T) were on the order of 0.06 or 6% transmittance. This is very similar to what was observed when the sun was the spectroscopic source. During experiments, we also found the order in which the filters were stacked to be significant. This is likely due

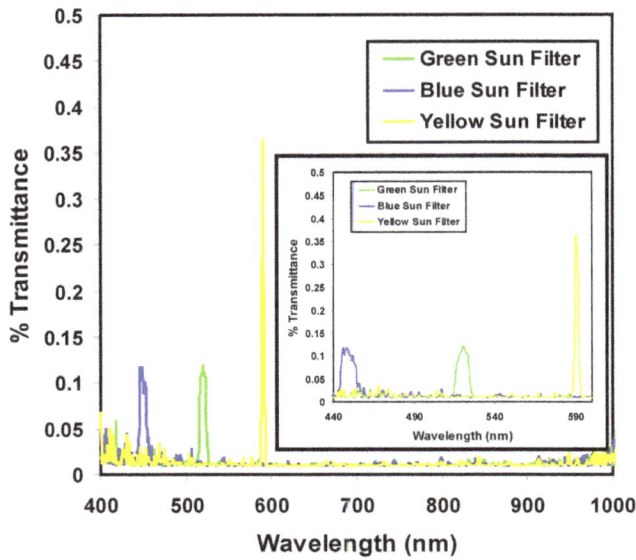

Figure 1. Plot of % Transmittance vs. wavelength for the 3 sun filters used in this study. Neutral density filters were used to attenuate the light to <1% of original value to provide "on-scale" measurements.

to etaloning off multiple surfaces. The optimal order was when the sun filter was placed first in the optical path, followed by the IR blocking filter, and then neutral density and test filters.

2.4 Image Analysis & Data Processing

All photos were imported into a personal computer and analyzed with Image J software (http://rsbweb.nih.gov/ij/). This allowed RGB count levels for each image pixel to be determined. For this procedure, the image file was opened and cropped to an area that depicted the center of the sun. An area was selected and the mean RGB intensities determined. This data was used without any further corrections for the subsequent calculations. Other areas in the photographs that did not depict the sun appeared very dark (e.g. black) and had very low signal counts (generally <<2 cts) so there was not a significant dark signal. If present and unaccounted for, dark counts could cause error in quantitative analysis. For accuracy experiments conducted with the test filters, the mean RGB values for blanks were assumed as 100% transmittances, and the percent transmittances of all test filters were calculated according to the ratio of the RGB channel values when the test filter was in the beam compared to the blank values. For ambient measurements of optical depth, the observed smart phone sun photometer signal counts ($S_{observed}$ - 0 - 256 counts) were determined in Image J and then ratioed to the expected top-of-the-atmosphere (S_{TOA}) signal values for each channel obtained via the Langley analysis described in section 3.2. This is the expected signal that would be obtained if the sensor was placed at the top of the atmosphere. Taking the ratio of $S_{observed}/S_{TOA}$ resulted in calculation of atmospheric transmission. Measured atmospheric optical depth (τ) was then computed via:

$$\tau = - \ln \left(\frac{S_{observed}}{S_{TOA}} \right)$$

For the stratus cloud data the Angstrom exponent was determined by plotting the observed optical depths vs. wavelength of measurement (using all 3 wavelengths) and fitting the data to a power law. The exponent value was rejected from the data pool if the fit to power law yielded an $R^2 < 0.75$. Most R^2 values encountered were >0.9.

Results and Discussion

3.1. Accuracy and Precision of Smartphone Measurements

The precision of replicate measurements was assessed by studying the variability in replicate blank measurements with only the sun filters present in the optical path. The image-to-image reproducibility was good, with an observed relative standard deviation of 1.2% in observed counts (28 trials with 145 total data points). From this we can estimate a minimum detectable transmission change as 3 times this value or a $\Delta T = 3.5\%$. This corresponds to an optical depth of roughly 0.04. To put this in perspective, the National Oceanic and Atmospheric Administration (NOAA) states typical aerosol optical depths for the United States is approximately 0.1–0.15 [22]. These values suggest smart phone sun photometry should be a viable option.

To investigate further we have performed experiments to assess the linearity and accuracy of the smart phone response. To accomplish this we have used the smart phone to make transmittance measurements of several optical filters of known transmission. We can then compare measured and accepted values directly. Results of this experiment are shown in Figure 2. As observed, the best-fit line determined by orthogonal distance regression has a slope close to 1 and an intercept near zero. This suggests the smart phone measured transmittance is in agreement with the accepted values and no large error is present in the measurement on average. However, measurement data is clearly scattered above and below the 1:1 line and this suggests significant variability/imprecision. We originally believed this scatter may be due to variability of aerosols in the atmosphere that we could not account for. However, the laboratory experiment using a lamp as a light source was not able to improve the measurement precision compared to solar measurements. To analyze this further we have prepared a histogram of the differences between iPhone measured filter transmittances and the accepted values (also presented in figure 2). As observed, most measured values are within 5% of the accepted transmittance, although a very slight bias (smart phone slightly high) may be apparent in the data. The standard deviation of the differences (iPhone%T – Accepted%T) was 0.058. If we assume the data is Gaussian distributed and 0.058 is representative of the standard deviation, we can define a 2s criteria for defining uncertainty. This criteria and the standard distribution leads to the conclusion that roughly 95% of replicate smart phone measured transmittances should lie within ±11.6% of the true value. This converts into an uncertainty in optical depth of roughly 0.12–0.13. We advocate these values for the uncertainty associated with the measurements made with the smart phone. This uncertainty is comparable to the mean aerosol optical depths for the continental United States cited previously, and is significantly poorer performance (by at least 1 order of magnitude) than commercially available sun photometers. For instance, the handheld Microtops II sunphotometer (solarlight.com) offers a precision of 1–2%. Also, the AERONET program estimates an overall uncertainty in optical depth of 0.01–0.02 for the research grade sunphotometers they employ (CE-318 CIMEL) [23]. Thus, it appears the smart phone sensor we describe here is best suited for environments in which optical depth is relatively high. This could correspond to

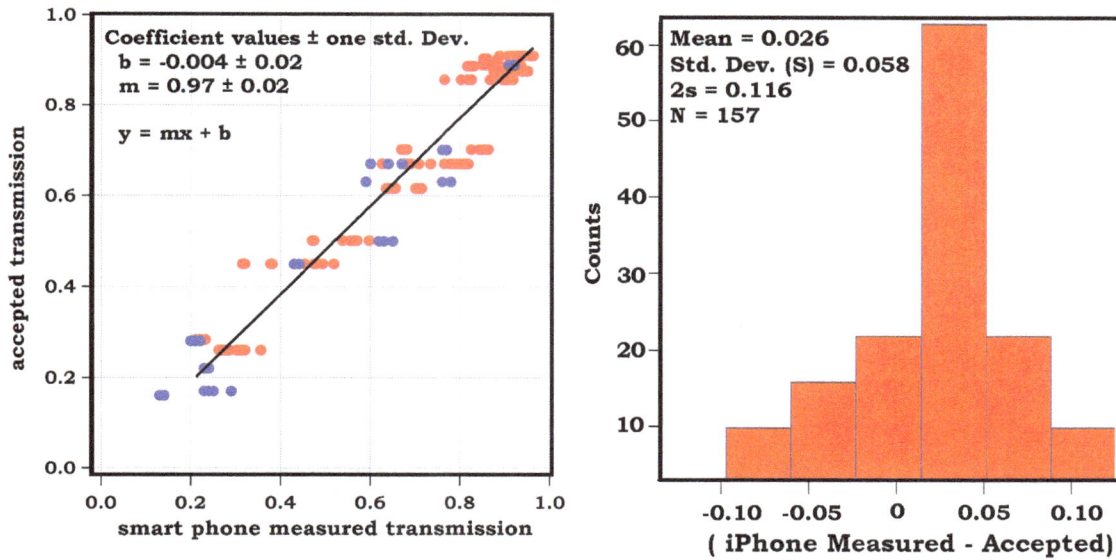

Figure 2. Evaluation of Accuracy. *Left* – Plot of accepted transmittance of a series of optical filters *vs.* smart phone measured transmittance using the sun (red) and a lamp (blue) as light source. The accepted filter transmittances were determined by either product literature or by using a spectrophotometer. The solid best-fit line was determined by orthogonal distance regression. The slope of the line is close to unity and the intercept indistinguishable from zero indicating good agreement between the measurements. *Right* – Histogram of differences between smart phone measured and accepted transmittances for sun data. A slight positive bias is reflected in the data. The calculated standard deviation (s) of differences between measurements was 0.058, yielding a 2s uncertainty of ±0.116 or 11.6% T.

applications in which users rapidly report significant dust events or use the smart phone sun photometer to monitor air quality in highly polluted urban centers.

3.2. Computation of Langley Plots

A significant problem in resolving integrated atmospheric optical depth is determining the signal a sensor would measure at the top-of-the-atmosphere (spectroscopic blank or 100% T). This is usually accomplished through an ingenious method known as Langley extrapolation [24–25]. In this approach, the detector signal for solar irradiance in a given wavelength band is logged at earth's surface throughout a day during which the sky is clear, humidity is low, and conditions are constant during the experiment. The solar zenith angle changes throughout the day, meaning the path length through the atmosphere differs from sunrise to sunset. This leads to differences in air mass (m, *e.g.* path length). When the sun is directly overhead (zenith angle = 0), the air mass is equal to m = 1. At dusk and dawn, air mass can exceed 15. If attenuation is caused by a homogeneous atmosphere, following the Beer-Lambert law we can write:

$$I(\lambda) = I_0(\lambda)e^{-\tau(\lambda)_{TOT}m}$$

Where $I(\lambda)$ and $I_0(\lambda)$ are irradiances at a given wavelength at the surface and the top of the atmosphere, m is air mass, and $\tau(\lambda)_{TOT}$ is the optical depth *per-unit-airmass* at this wavelength. Rearranging Eqn. 2 to create a new calibration yields:

$$lnI(\lambda) = \ln I_0(\lambda) - \tau(\lambda)_{TOT}m$$

Equation 3 suggests a linear relationship exists between $ln\ I(\lambda)$ and air mass (m) if all other variables are held constant. If the

detector response is linearly related to surface irradiance, then a plot of *ln* (observed signal counts) *vs.* air mass (m) should yield a line. The y-intercept of this line represents $ln\ I_0 (\lambda)$ which can yield the expected signal at the top of the atmosphere in a specific wavelength band. Figure 3 illustrates Langley plots for the cell phone sun photometer on the blue, green, and yellow channels. In order to generate these plots photos were taken every hour from sunrise to sunset on several days during July 30 – August 4 2012. These days featured clear and cloudless skies with temperature maxima of 35–40°C. Light winds from the southwest at 5–10 mph were typical. Air mass was computed from our locations geographic coordinates as described in the methods section. As observed in the figure, plots illustrating linear trends were obtained (R^2 approx. 0.9). The slopes of the best-fit lines were −0.25, −0.23, and −0.19 for the blue, green and yellow channels, respectively. This ordering is consistent with the enhancement in Rayleigh scattering at blue wavelengths; however, the observed slopes are not quantitatively consistent with the well-known λ^{-4} trend since these slopes are indicators of the total optical depth per unit airmass. Clearly, attenuation processes involve both gas and aerosol scattering (aerosol scattering does not follow the λ^{-4} trend). Also, the camera response may vary with wavelength.

3.3. Monitoring Changes in Optical Depth

After constructing the Langley plots for each wavelength, we turned our attention to monitoring the optical depth of the atmosphere using the smart phone device and sun filters. The afternoon of September 17 2012 provided an excellent opportunity to monitor solar attenuation since a high-altitude and thin stratus cloud layer was present at our location. Photographs of the sun were sequentially acquired through each of the three interference filters. Indicated signal counts were ratioed to the top-of-the-atmosphere signal obtained via the Langley analysis and optical depth computed from the resulting atmospheric transmittance. Measured optical depths were then plotted in time as

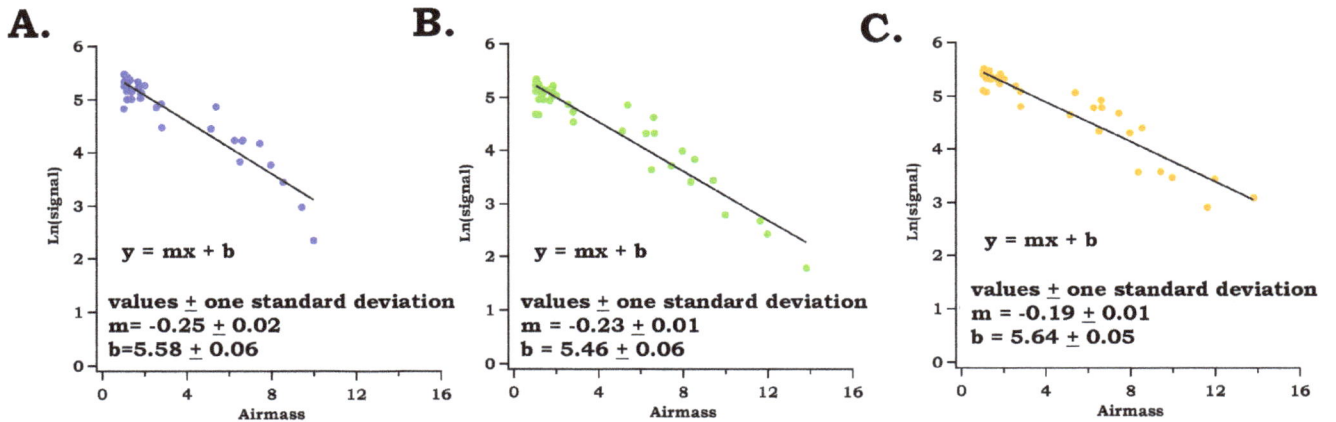

Figure 3. Langley Plots (ln signal counts *vs.* air mass) for the blue (A), green (B), and yellow (C) sun filters. The slope of the best-fit lines represents optical loss per unit air mass while the intercept describes the signal expected at the top-of-the-atmosphere (TOA). Best-fit lines were determined by orthogonal distance regression.

reported in figure 4. Figure 4 also plots observed Angstrom extinction exponents (AEE) values in time. The Angstrom exponent (α) is an empirically derived value used to describe optical attenuation with wavelength [26]. The basic premise is that the trend in atmospheric optical attenuation with wavelength follows a power law of the form:

$$\tau_\lambda = \tau_{\lambda 0}\left(\frac{\lambda}{\lambda_0}\right)^{-\alpha}$$

Where the τ terms represent optical depths at two different wavelengths, and λ and λ_0 represent the two wavelengths

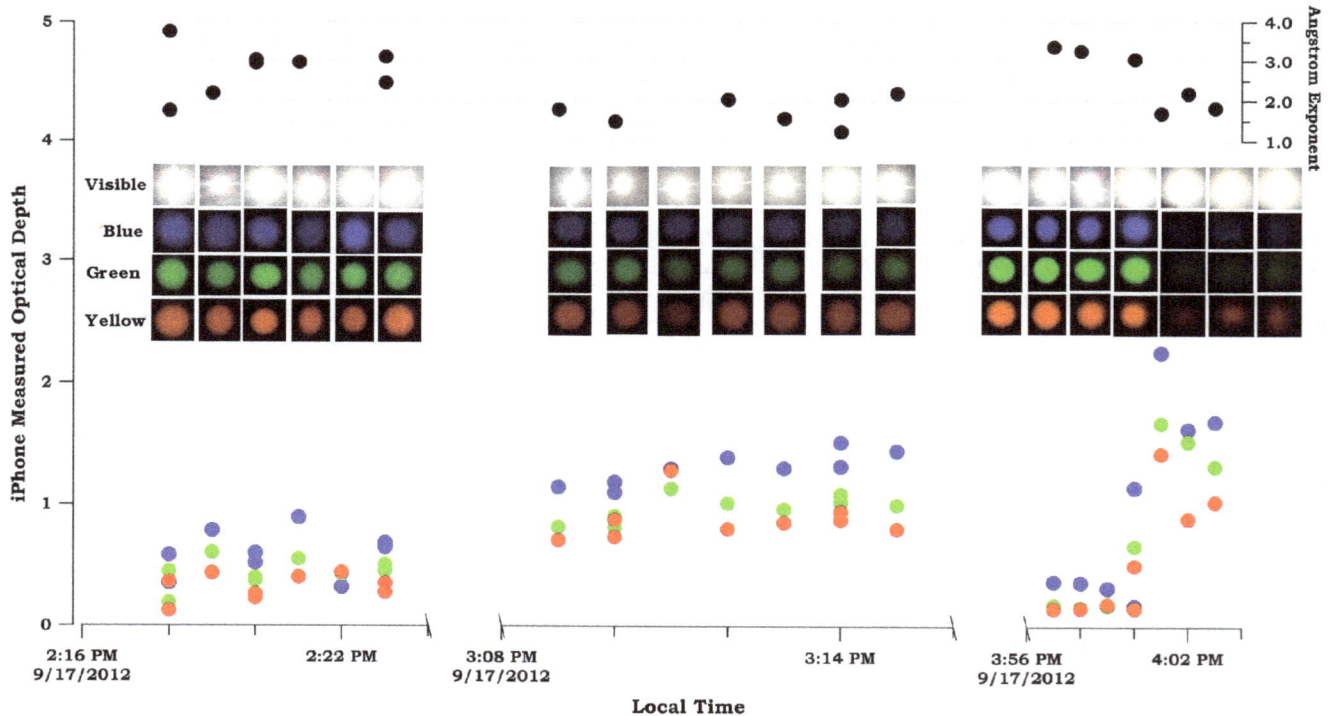

Figure 4. Monitoring atmospheric optical depth and Angstrom exponent at Lubbock, TX during the afternoon of September 17 2012 using the smart phone sun photometer on the blue, green, and yellow channels. This afternoon featured a very thin, high-altitude stratus cloud layer that periodically blocked the sun and increased optical depth. This effect is particularly noticeable near 4:00 PM local time when a rapid and large increase in optical depth was observed. This change was accompanied by a reduction in Angstrom exponent by approx. 1 unit which suggests larger particles contribute to the increase in optical attenuation. Angstrom exponents reported consider the effect of both gases and particles. No measurements are made on the "visible" channel – this is only included to provide the reader with a visual reference.

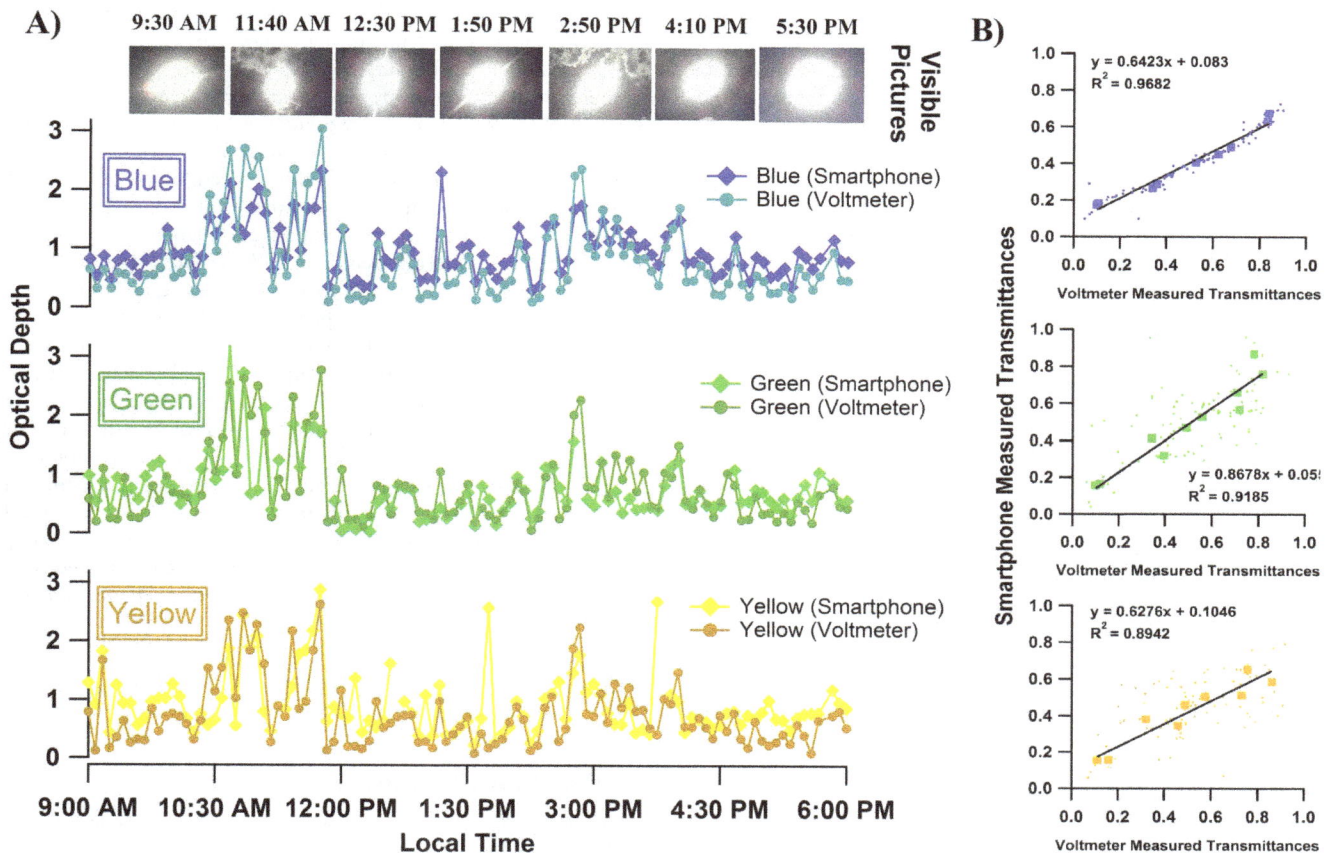

Figure 5. Optical depths measured with smartphone sun photometer and reference device. Panel (A) – Optical depths at Lubbock, TX from 9:00 AM to 6:00 PM on September 1 2013. Measurements were made with both smart phone sun photometer and reference method on blue, green and yellow channels. On this day, optical depths shifted up and down because of the existence of clouds in the sky as seen in the photographs. (B) – Comparison of transmittance measurements using smartphone and reference method. The y-axis shows atmospheric transmittance measured by the smartphone while x-axis shows transmittance measured by reference method. Small dots illustrate all measurements made with the two devices. Larger square markers are median values for different bins. Best-fit lines were determined by orthogonal distance regression from medians.

considered. The Angstrom exponent (α) essentially describes how rapidly attenuation changes with wavelength. It can reach a maxium value of $\alpha = 4$ for pure Rayleigh scatterers (gas molecules) and can be near zero (or even negative) for very large particles [27,28]. This is in fact why the clear sky appears blue but a overcast sky appears white or grey. Considering Figure 4 we can easily see the influence of the stratus clouds on observed optical depth around 4:00 PM local time as the optical depth increases from <0.5 to 1.5 over a few minutes. Interestingly, the Angstrom exponent observed suddenly decreases from approx. 3 to about 2 simultaneously. This decrease in average α is consistent with micron-sized particles such as water drops or ice influencing optical attenuation. In addition, for the middle and left panel of Fig. 4, we see highest Angstrom exponents were generally observed when optical depth was lowest. This is consistent with airborne particles being responsible for increasing the change in optical depth sensed.

3.4 Comparison of Sun Photometer with a Reference Device

Unfortunately, we did not have access to a commercial sun photometer during our study to compare measurements. Instead, a comparison measurement was made using a Si biased detector (Thorlabs, DET36A) integrated with a voltmeter (GB Instruments,

GDT-11) and the three interference filters (sun filters) described earlier. The field of view for the reference device was approx. 5 degrees. Linear response for this sensor was verified and Langley plots were constructed separately for this detector (see Figure S3 in File S1 and Fig. S4 in File S1) so that optical depths could be measured and compared for both the reference method, and smartphone. Measurements with both smartphone and reference method were taken on September 1 2013 from 9:00 AM to 6:00 PM local time. On that day, clouds were present in Lubbock, TX that periodically blocked the sun. Optical depth changes for both the smartphone and reference method are plotted in Figure 5. To ensure the accuracy of measurements, we performed an additional Langley calibration with the smartphone during August 22–24 2013 (see results in Figure S2 in File S1). Trends of optical depths measured with smartphone were generally consistent with the result from the reference device, for blue, green and yellow channels. During the whole day, optical depths shifted up and down showing the effects of the cloud layers in the sky. Clouds reflect and may absorb light from the sun, resulting in reduction of solar irradiance onto the surface of the earth. During 10:30 AM – 12:00 PM and 2:40 to 3:00 PM, optical depths went up to 2–3, indicating the existence of optically thick clouds in the atmosphere. These results were in agreement with the visible pictures shown at the top of Figure 5A indicating clouds were present. In Figure 5B, we plot the transmittance results obtained from both devices to directly compare them. Small dots illustrate all measured data points for the two devices. The larger square markers represent the median values when data was sorted by value and binned. From the best-fit lines determined by linear regression, we got the slopes of 0.64, 0.86 and 0.63 (R^2 approx. 0.9) and intercepts near zero. The data of Fig. 5B suggests a systematic underestimation of transmittance for the smartphone. The exact cause of this is unclear. The underestimation could be the lack of accuracy for the device itself; however, asynchronous measurements can also lead to variation between the two devices. Nonetheless, the measurements produced by the two separate devices are proportional to one – another.

Conclusions

A common smart phone has been adapted for use as a sun photometer. The uncertainty in optical depth achieved is at least on the order of 0.12. This is approximately an order of magnitude poorer performance compared to research grade devices, but this disadvantage is offset to a degree by the ubiquity of smartphones. The measurement uncertainty is also similar to the average aerosol optical depths for the continental United States. These results suggest improvement in quantitative performance will be required before the full potential of the smart phone sun photometer can be reached for environmental research. Nonetheless, this work has demonstrated the concept may be viable – particularly during periods of high aerosol loading such as in dust events or in polluted urban centers where aerosol optical depths often exceed 0.12. We have clearly demonstrated a change in observed optical depth during the transit of a thin cloud through the optical path between the sun and an observer. This increase in optical depth corresponded with a decrease in Angstrom extinction exponent. This would be expected for optical attenuation by large water or ice particles of a cloud.

Further investigations should focus on defining the limit of spectroscopic precision practically achievable with the on-board camera. The 8-bit resolution of the cameras plays a role in defining this limit, however, our current measurement uncertainty is >20 fold worse than what we might expect based on the 8-bit resolution alone. Significant performance improvement could be achieved by building a stand – alone spectroscopic sensor that would plug into the smart phone, however, this would significantly decrease user friendliness and could add significant additional expense to a monitoring network. In addition to the hardware limitations, new smart phone application software (an app) should be written to serve the specific purposes of the experiment. Data analysis was time consuming and complicated. In principle, it should be possible to author an app that could collect data and analyze results rapidly in the field. This could include use of geospatial data and the clock capability to rapidly determine air mass. Geospatial data could also be used for rapid and automated spatial mapping of results obtained. In summary, creating a network of smart phone sun photometers for real time chemical and aerosol monitoring is possible provided measurement uncertainty can be reduced and a suitable app created that maximizes measurement precision while automating data analysis.

Author Contributions

Conceived and designed the experiments: JET TC. Performed the experiments: TC. Analyzed the data: TC. Contributed reagents/materials/analysis tools: TC JET. Wrote the paper: JET TC.

References

1. Aviva Inc. Available: http://www.aviva.co.uk/ratemydrive/?intcmp = rmd-cppod. Accessed 22 Aug 2012.
2. Madan A, Cebrian M, Lazer D, Pentland A (2010) Social Sensing to Model Epidemiological Behavior Change. Proceedings of ACM Ubicomp 2010, Copenhagen, Denmark.
3. Smith ZJ, Chu K, Espenson AR, Rahimzadeh M, Gryshuk A, et al. (2011) Cell-Phone-Based Platform for Biomedical Device Development and Education Applications. PLoS ONE, 6(3): e17150. doi:10.1371/journal.pone.0017150.
4. Scheeline A (2010) Teaching, Learning, and Using Spectroscopy with Commercial, Off-the-Shelf Technology. Applied Spectros. 64(9): 256A–268A.
5. Zhu H, Sikora U, Ozcan A (2012) Quantum dot enabled detection of E. coli using a cell-phone. Analyst, 137: 2541–2544.
6. Zhu H, Mavandadi S, Coskun AF, Yaglidere O, Ozcan A (2011) Optofluidic Fluorescent Imaging Cytometry on a Cell Phone. Anal. Chem. 83(17): 6641–6647.
7. Delaney JL, Hogan CF, Tian J, Shen W (2011) Electrogenerated Chemiluminescence Detection in Paper-Based Microfluidic Sensors. Anal. Chem. 83(4): 1300–1306.
8. Thompson JE, Spangler H (2006) Tungsten Source Integrated Cavity Output Spectroscopy (W-ICOS) for the Determination of Ambient Atmospheric Extinction Coefficient. Applied Optics 45(11): 2465–2473.
9. Thompson JE, Myers K (2007) Cavity Ring-Down Lossmeter using a Pulsed Light Emitting Diode Source and Photon Counting. Measurement Science & Technology 18: 147–154.
10. Thompson JE, Barta N, Policarpio D, DuVall R (2008) Development of a Fixed Frequency Aerosol Albedometer. Optics Express 16: 2191–2205.
11. Thompson JE, Smith BW, Winefordner JD (2002) Monitoring Atmospheric Extinction Through Cavity Ringdown Turbidity. Anal. Chem. 74: 1962–1967.
12. Thompson JE, Smith BW, Winefordner JD (2003) Atmospheric Aerosol Measurements by Cavity Ringdown Turbidimetry. Aerosol Sci. & Technol. 37(3): 221–230.
13. Dial K, Hiemstra S, Thompson JE (2010) Simultaneous Measurement of Optical Scattering and Extinction on Dispersed Aerosol Samples. Anal. Chem. 82: 7885–7896.
14. Redmond H, Thompson JE (2011) Evaluation of a Quantitative Structure Property Relationship (QSPR) for Predicting Mid-Visible Refractive Index of Secondary Organic Aerosol (SOA). Phys. Chem. Chem. Phys. 13: 6872–6882.
15. Ma L and Thompson JE (2012) Optical Properties of Dispersed Aerosols in the Near UV (355 nm): Measurement Approach and Initial Data. Anal. Chem. 84: 5611–5617.
16. Halthore RN, Eck TF, Holben BN, Markham BL (1997) Sun photometric measurements of atmospheric water vapor column abundance in the 940–nm band. J. Geophys. Res. 102(D4): 4343–4352.
17. Staehelin J, Renaud A, Bader J, McPeters R, Viatte P, et al. (1998) Total ozone series at Arosa (Switzerland): Homogenization and Data Comparison. J. Geophys. Res. 103(D5): 5827–5841.
18. Alexandrov MD, Schmid B, Turner DD, Cairns B, Oinas V, et al. (2009) Columnar water vapor retrievals from multifilter rotating shadowband radiometer data. J. Geophys. Res. 114: D02306.
19. Pérez-Ramírez D, Navas-Guzmán F, Lyamani H, Fernández-Gálvez J, Olmo FJ, et al. (2012) Retrievals of precipitable water vapor using star photometry: Assessment with Raman lidar and link to sun photometry. J. Geophys. Res. 117: D05202.
20. Dubovik O, Smirnov A, Holben BN, King MD, Kaufman YJ, et al. (2000) Accuracy assessments of aerosol optical properties retrieved from Aerosol Robotic Network (AERONET) Sun and sky radiance measurements. J. Geophys. Res. 105(D8): 9791–9806.
21. Holben BN, Eck TF, Slutsker I, Tanré D, Buis JP, et al. (1998) AERONET—A Federated Instrument Network and Data Archive for Aerosol Characterization. Remote Sensing of Environment 66(1): 1–16.
22. Earth System Research Laboratory. Available: http://www.esrl.noaa.gov/gmd/grad/surfrad/aod/. Accessed on 26 August 2012.
23. Aeronet System. Available: http://aeronet.gsfc.nasa.gov/new_web/system_descriptions_calibration.html. Accessed 13 Sep 2012.
24. Shaw GE (1983) Sun Photometry. Bull. Amer. Meteor. Soc. 64: 4–10.
25. Shaw GE, Reagan JA, Herman BM (1973) Investigations of atmospheric extinction using solar radiation measurements made with a multiple wavelength radiometer. J. Appl. Meteorol. 12: 374–380.

26. Angstrom A (1970) Apparent solar constant variations and their relation to variability of atmospheric transmission. Tellus. 22: 205–218.

27. Ma L, Hsieh D, Holder D, Zobeck T, Morgan C, et al. (2011) Optical Properties of Aeolian Dusts Common to West Texas. Aeolian Research 3: 235–242.

28. Redmond H, Dial K, Thompson JE (2010) Light Scattering & Absorption by Wind Blown Dust: Theory, Measurement and Recent Data. Aeolian Research 2: 5–26.

8

Mixed Layer Depth Trends in the Bay of Biscay over the Period 1975–2010

Xurxo Costoya*, Maite deCastro, Moncho Gómez-Gesteira, Fran Santos

EPHYSLAB, Environmental PHYsics LABoratory, Facultad de Ciencias, Universidad de Vigo, Ourense, Spain

Abstract

Wintertime trends in mixed layer depth (MLD) were calculated in the Bay of Biscay over the period 1975–2010 using the Simple Ocean Data Assimilation (SODA) package. The reliability of the SODA database was confirmed correlating its results with those obtained from the experimental Argo database over the period 2003–2010. An iso-thermal layer depth (TLD) and an iso-pycnal layer depth (PLD) were defined using the threshold difference method with $\Delta T = 0.5°C$ and $\Delta \sigma_\theta = 0.125$ kg/m^3. Wintertime trends of the MLD were calculated using winter extended (December-March) anomalies and annual maxima. Trends calculated for the whole Bay of Biscay using both parameters (TLD and PLD) showed to be dependent on the area. Thus, MLD became deeper in the southeastern corner and shallower in the rest of the area. Air temperature was shown to play a key role in regulating the different spatial behavior of the MLD. Negative air temperature trends localized in the southeastern corner coincide with MLD deepening in this area, while, positive air temperature trends are associated to MLD shoaling in the rest of the bay. Additionally, the temperature trend calculated along the first 700 m of the water column is in good agreement with the different spatial behavior revealed for the MLD trend.

Editor: João Miguel Dias, University of Aveiro, Portugal

Funding: X. Costoya is supported by the Xunta de Galicia through the Plan Galego de Investigación, Innovación e Crecemento 2011-2015 (Plan I2C) in collaboration with the International Campus do Mar. The funders had no role in study design, data collection and analysis, decision to publish, or preparation of the manuscript.

* E-mail: xurxocostoya@uvigo.es

Introduction

Upper Ocean is characterized for a quasi-homogeneous layer where temperature, salinity and density scarcely vary with increasing depth [24]. This homogeneity layer is caused by turbulent vertical mixing that is driven by heat loss from the ocean to the atmosphere, as well as by wind stress. The deepest layer affected by this turbulent mixing is called mixed layer depth (MLD), which marks the width of the upper ocean that interacts with the atmosphere. MLD presents a high spatial and temporal variability. Thus, it varies at different temporal scales, from diurnal [3] to decadal scales [7] and from mesoscale to large spatial scale [8], [36]. The deepest MLDs in the North Atlantic occur in winter and early spring. During spring the onset of surface heating produces a restratification of the upper ocean, giving as result shallower MLDs, especially during summer. These variations of the MLD, that reach hundreds of meters over the year, have important biological implications [13], [47].

MLD plays an important role in influencing the ocean's role in air-sea interaction. Thus, it is fundamental in the exchange of heat and freshwater between the ocean and the atmosphere and influences other processes such as the formation of water masses or the activity of phytoplankton. [11] analyzed the influence of the winter MLD in the concentrations of nutrients and dissolved inorganic carbon in the Bay of Biscay. They confirmed that a deeper MLD as a consequence of a warmer winter produced higher surface winter concentrations of nutrients and dissolved inorganic carbon. Moreover, they related this fact with an

enhancement of the spring bloom in the Bay of Biscay. Similar results have been reported in the North Atlantic Ocean [2], [20].

Long-term MLD variability has been analyzed at some fixed moorings in the North Atlantic [34], [37]. Several studies on MLD variability have been carried out using global databases. Thus, for example, [25] used the World Ocean Atlas and [7] used the World Ocean Database. In fact, the main problem to develop a MLD long-term variability study is the lack and scarcity of temperature and salinity profiles to analyze the subsurface ocean variables, especially for salinity profiles [8]. After the development of the International Argo Project [1] the coverage of temperature and salinity profiles has been increased strikingly all over the world. However, its recent deployment prevents the realization of long term studies.

Taking in account the limitations commented above, it is necessary to find alternative databases to study subsurface ocean variables such as MLD at a regional scale. Some authors [4], [5], [6] have developed a project called Simple Ocean Data Assimilation (SODA). SODA is based on a method that uses an ocean model in conjunction with data assimilation providing a useful background of information beneath the sea surface. This fact allows to obtain a complete view of the different hydrographic processes that take place in the area, both at global [6], [51] and at basin scale [14], [18], [39], [52] which is the aim of the present study focused in the Bay of Biscay.

The Bay of Biscay is a semi-enclosed area located in the northeastern Atlantic Ocean between 0–10°W and 43–48°N. The oceanic part of the Bay of Biscay is influenced by subpolar and subtropical Atlantic gyres and it is characterized by following an anti-cyclonic, weak (1–2 cm/s) and variable circulation [26], [44]. A distinctive feature is the wide of the continental shelf along the French coast, especially the northern area, ranging from 60 to over 200 km and with a very gentle slope of 0.12% [38]. In contrast, the southern coast is characterized by a narrower continental shelf (7–20 km wide). Various studies have analyzed sea surface temperature trends for different periods and locations in the Bay of Biscay [10], [15], [16], [27], [40]. Furthermore, salinity and temperature variability beneath the sea surface were studied, focusing over the Eastern North Atlantic Central Water that is the main water observed at the upper layers in the Bay of Biscay [18], [19], [32]. In addition, [46] analyzed MLD variability using a one-dimensional water column model (GOTM) with data from NCEP/NCAR reanalysis over the period 1948–2008. They found that during 1970s and 1980s, MLDs were strikingly shallower than from 1995 onwards.

The aim of the present study is to analyze the trend of the MLD during wintertime for the whole Bay of Biscay over the period 1975–2010. The relevance of this study is the possibility to know whether MLD trends are similar at any location within the Bay of Biscay or there exist important spatial differences. This analysis provides complementary information respect to previous studies, which have analyzed the long-term MLD variability taking data from a particular area of the Bay of Biscay [46]. For this purpose two complementary datasets have been used. Namely, Argo is based on in situ measurements and SODA combines simulation and assimilated data. A comparison between them was carried out in order to confirm the reliability of SODA dataset. The used datasets and methods are described in section 2. Two different approaches were considered to calculate the wintertime MLD: the temperature criterion to define the isothermal layer depth (TLD) and the density criterion to define the iso-pycnal layer depth (PLD). This fact, together with the calculation of two different wintertime parameters (winter anomalies and annual maxima) gives as result a complete analysis of the wintertime MLD trends. Section 3 shows the results and the discussion. This includes an analysis of the atmospheric forcing carried out correlating air temperature, wind stress and precipitation minus evaporation trends with MLD results. Moreover, temperature trends along the

water column (0–700 m) were also considered to be linked with MLD. Finally, section 4 compiles the main conclusions.

Datasets and Methods

Two databases were used to obtain temperature and salinity data of the upper ocean in the Bay of Biscay. On the one hand, the "Simple Ocean Data Assimilation" (SODA) package (http://dsrs.atmos.umd.edu/), which was built following an ocean model based on Geophysical Fluid Dynamics Laboratory MOM2 physics was used [4], [5]. Assimilated data includes temperature and salinity profiles from the World Ocean Atlas-94 [28], as well as additional hydrography, sea surface temperature [42], and altimeter sea level from the ERS-1, Geosat, and TOPEX/Poseidon satellites. Reanalysis of world ocean climate variability are available from 1958 to 2010 at monthly scale, with a horizontal spatial resolution $0.5° \times 0.5°$ and a vertical resolution of 40 levels, ranging from 5 m to more than 5000 m decreasing the resolution with depth. The long coverage period of this dataset allows focusing the study on the period 1975–2010, when the higher temperature increasing was detected in the North Atlantic region [10], [12], [16], [17], [23], [27], [35].

On the other hand, the database derived from the International Argo Project was used [1]. Argo data were extracted from the US Global Ocean Data Assimilation Experiment (USGODAE) server (http://www.usgodae.org/argo/argo.html), which is one of the two Argo Global Data Assembly Center (GDAC) that stored Argo float data. Under this project, floats started to be deployed in 2000, increasing deployments year on year, until reaching more than 3500 active floats all over the world in 2013. This number of floats is enough to attain part of the goals that the Argo Project had when it was designed, such as a global resolution of $3°\times3°$. This amount of floats produces more than 100.000 profiles each year because they are designed to generate a profile each 10 days. Overall, profiles reach a maximum depth of 1500–2000 meters with higher resolution at the uppermost layers. The accuracy of pressure, temperature and salinity measurements is ±2.4dbar, ±0.005°C and ±0.01psu, respectively [1]. The recent onset of the Argo Program and the progressive deployment of Argo floats since 2002 involve that the number of profiles for the first years of the Argo Program in the Bay of Biscay were limited. In fact, there were not valid profiles for some months in 2003 and 2004.

Apart from these two datasets, meteorological large-scale reanalysis data of temperature, precipitation, latent heat and wind speed above sea surface were extracted from the National Center for Environmental Prediction/National Center for Atmospheric Research (NCEP/NCAR) website (www.cdc.noaa.gov). Data are supplied on a T62 Gaussian grid corresponding to a 1.9° both in latitude and longitude and monthly time resolution from 1948 on. Air temperature was considered at 2 m from surface, while wind was regarded at 10 m. Evaporation was calculated dividing the heat flux loss by the latent heat of water (2.5×10^6 J kg^{-1}) following [19]. Moreover, wind speed was used to calculate wind stress according to the formula: $\tau_{wind} = \rho_{air} C_D U_h^2$, where C_D is the wind-drag coefficient (0.0011), ρ_{air} is the air density (1.2250 kg m^{-3} at 15°C) and U_h is the wind speed above sea surface (10 m).

The Spearman rank correlation coefficient was used to analyze the significance of trends and correlations due to its robustness to deviations from linearity and its resistance to the influence of outliers [49].

The method used to determine the MLD was the threshold difference method that is based in the choice of a temperature or density threshold value. MLD is determined as the depth where the temperature or density exceeds the threshold value with

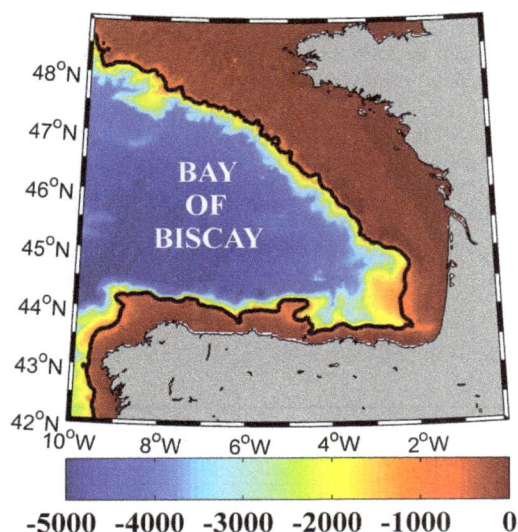

Figure 1. Bay of Biscay bathymetry (m). The isobath corresponds to -1000 m.

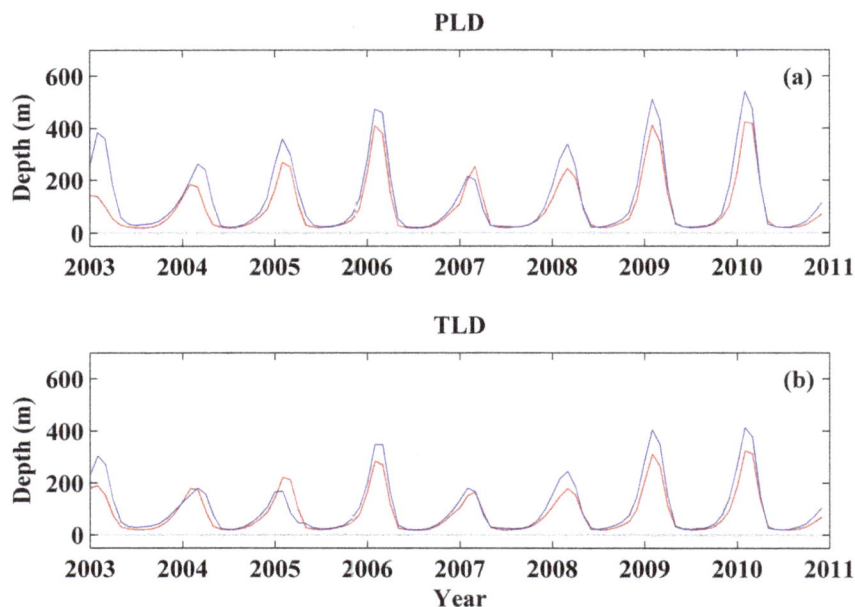

Figure 2. Comparison between Argo and SODA databases. (a) Isopycnal layer depth and (b) isothermal layer depth variability using Argo database (blue line) and SODA database (red line). A 1-2-1 filter was used to smooth signals only for visualization.

respect to a reference depth located near the sea surface. The threshold difference method was shown to be more stable than the gradient method based on temperature or density gradients [5]. Hence, an important point is the choice of the threshold value. There is not a consensus about what are the best threshold values, which is related to the fact that MLD features change both in space and in time due to the different physical processes and hydrographic features that characterize the ocean all over the world (see [8], [21], [24], [33] for recent discussions). Taking in account these studies, the most commonly used threshold values ($\Delta T = 0.5°C$ and $\Delta\sigma_\theta = 0.125$ kg/m^3) were selected for potential temperature and density (e.g. [36]).

Argo and SODA datasets allow taking advantage of their salinity and temperature vertical resolution. A potential temperature threshold to define the iso-thermal layer depth (TLD) and a potential density threshold to define the iso-pycnal layer depth (PLD) were considered following the terminology applied by [22] who analyzed different procedures to calculate MLDs using Argo floats. In this way we can detect possible biases derived of the salinity effect on the MLD that can produce important differences between TLD and PLD (e.g., [8], [22]). Potential values were calculated to avoid pressure problems with increasing depth following the method describe in [48]. Besides, a reference depth of 15 m was chosen to prevent the diurnal oscillations of

temperature and the effect of precipitation and evaporation that takes place in the first meters beneath sea surface [41]. This choice was adopted because the first two levels in the SODA database are 5 m and 15 m.

As it was previously mentioned, the deepest MLDs occur during winter and early spring in the North Atlantic region (e.g. [25], [36]). For this reason, the present study is focused on the analysis of winter MLD. Thus, annual maxima MLD and winter extended (December-March) anomalies were calculated.

First of all, both databases were compared for the period 2003–2010 in order to know the reliability of SODA database. To carry out this comparison, all available data among 2003–2010 in the Bay of Biscay (43.25–48.25°N//0.25–9.75°W) were extracted. In this case, the domain was slightly enlarged, with respect to the rest of the study, in order to include a larger number of Argo profiles, especially during the years 2003 and 2004 when the number of profiles was limited. Thus, a better sampling was obtained ensuring better results in the comparison. As we mentioned above, this area is characterized by its wide continental shelf, especially along the French coast (Figure 1). For this reason, to ensure that depth does not influence TLD/PLD calculations only profiles deeper than 1000 m were selected. A total of 101 points carried out the requirements in the SODA database, while a total of 3362 profiles, which corresponds to 58 floats, were extracted

Table 1. Correlations between annual maxima and winter anomalies of PLD/TLD calculated from ARGO and SODA databases for the period 2003–2010.

	Annual Maxima	Winter anomaly
PLD	0.66 (0.074)	0.74 (0.037)
TLD	0.64 (0.089)	0.71 (0.046)

Threshold values of $\Delta T = 0.5°C$ and $\Delta\sigma_\theta = 0.125$ kg/m^3 were used for potential temperature and for potential density, respectively. Statistical significance (p value) is showed in brackets.

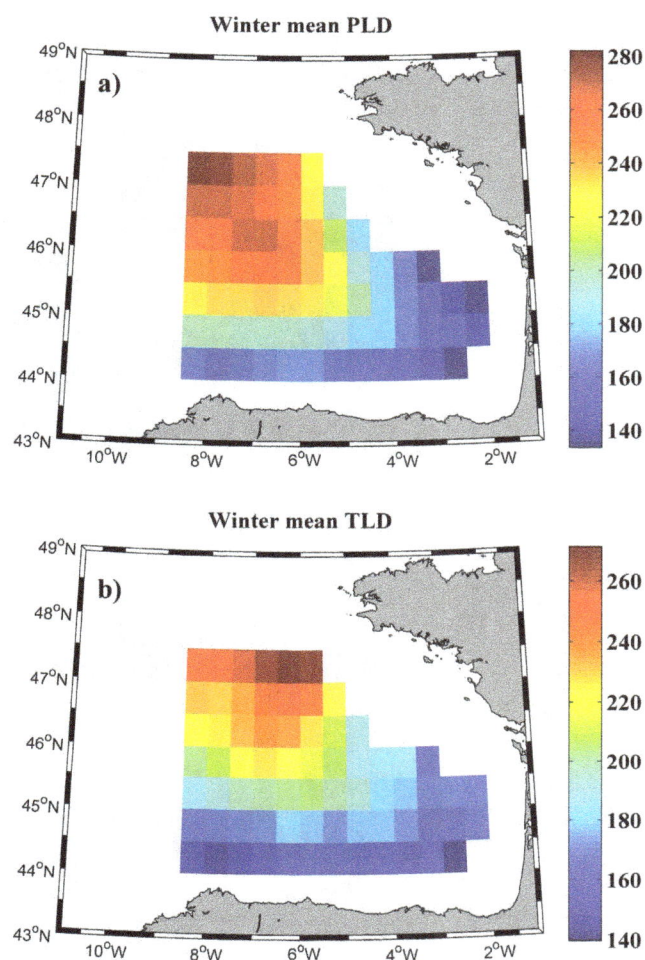

Figure 3. Mean values of winter mixed layer depth. (a) Winter mean isopycnal layer depth (m) and (b) winter mean isothermal layer depth (m) over the period 1975–2010.

from the ARGO database. These 3362 profiles were submitted to a quality control selecting only profiles with a control flag of 1, following the Argo quality control manual v2.8 [50] (http://www.argodatamgt.org/Documentation) what is enough to guarantee the quality of data. Argo and SODA profiles were interpolated for each meter following a linear interpolation in order to get the most accurate TLD/PLD value.

Data from the 101 points of SODA database and good quality profiles from ARGO database were horizontally averaged for the whole region and for each month over the period 2003–2010. Thus, an average profile was obtained for each month and for each database. Then, TLDs and PLDs were calculated based on each monthly averaged profile. Finally, annual maxima and winter anomalies were determined and correlated for both datasets. An annual maximum corresponds to the winter month when TLD/PLD attains the highest depth. TLP and PLD winter anomalies were calculated considering the available profiles for the December–March months and subtracting to each month, the mean value of that month for the whole period of time.

The area between 43.75–47.25°N and 2.25–8.25°W was selected to analyze trends in wintertime MLD over the Bay of Biscay for the period 1975–2010. Only grid points deeper than 1000 m were considered, which includes 70 points from SODA

database. The methodology was described above but TLDs/PLDs were calculated in this case for each of the 70 points in order to know if trends are similar in the whole bay. Trends were assumed to be linear and obtained by fitting time series (TLDs and PLDs) to a straight line in a least-squares sense. Finally, annual maxima and winter anomalies trends of TLD/PLD were correlated with atmospheric variables (air temperature, P-E balance and wind stress).

Results and Discussion

The correlation between Argo and SODA using annual maxima and winter anomalies is shown in Table 1. Correlation coefficients are slightly higher for winter anomalies (0.74 for PLD and 0.71 for TLD) than for annual maxima (0.66 for PLD and 0.64 for TLD). All these correlation coefficients are significant at a level higher than 90%. In addition, monthly averaged PLDs and TLDs for both databases are shown in Figure 2 over the period 2003–2010. Overall, a good agreement was observed between both databases for the whole period. Nevertheless, SODA (red line) underestimates the MLD depth both for TLD (2003, 2006, 2008, 2009 and 2010) and PLD (all years except 2007). We should note that PLD depends both on temperature and salinity and so the sources on uncertainty are higher compared to TLD. The largest differences were observed in winter 2003. This fact seems to be related to the few Argo profiles available during that year. The Argo program was launched in 2002, which caused a less uniform sampling in the bay over the first years.

Once the reliability of SODA database was confirmed using in-situ measurements, it was used to characterize the MLD in the bay over the period 1975–2010. Winter PLD/TLD means were calculated for each of the 70 grid points that constitute the bay (Figure 3). A remarkable southeast-northwest MLD gradient can be observed in both frames. In this way, the deepest mean MLDs, around 270 m, were found near the northern boundary while the shallowest mean MLDs, around 140 m, were noticed at the southeastern corner. Overall, not significant differences were found between winter mean PLD and winter mean TLD. The observed differences are related to haline forcing, which is a phenomenon produced by different oceanic processes such as surface freshwater fluxes or freshwater advection. These changes can produce that the halocline is located above the thermocline or vice versa. This effect on MLD has been previously described all over the world by different authors [9], [31].

Long-term trends of annual maxima and winter anomalies of TLD and PLD for each grid point in the Bay of Biscay for the period 1975–2010 are shown in Figure 4. Black dots represent points with statistical significance higher than 90%. Negative (positive) trends mean that MLD tends to be shallower (deeper). In all cases, the most important feature is that points located at the southeastern corner of the Bay of Biscay show a significant positive trend. MLD deepens in this zone and shallows in the rest of the area, especially at the central part.

As it was previously mentioned, annual maxima correspond to the month when TLD/PLD attain their maximum, which corresponds to winter or early spring (commonly March) in the Bay of Biscay. First, trends in annual maxima of MLD (Figure 4a,b) will be analyzed. Trends at grid points with significant positive values are observed to range from 20 to 60 m per decade, being higher for PLD (Figure 4a) than for TLD (Figure 4b). In a similar way, trends at grid points with significant negative values range from -10 m to -60 m per decade for TLD and are around -25 m per decade for PLD. The spatial distribution of significant positive and negative trends is similar for winter anomalies (Figure 4c,d). In

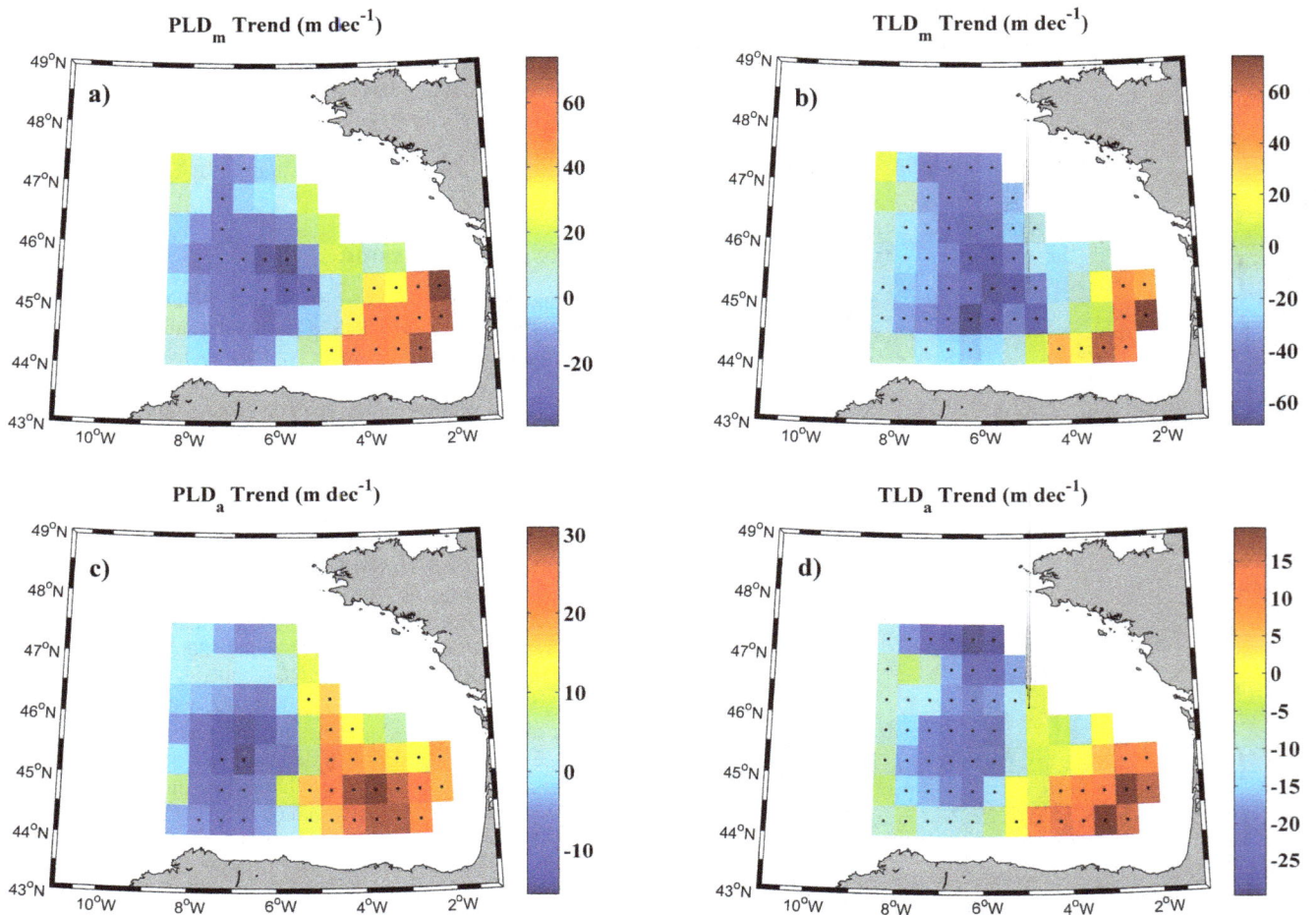

Figure 4. Trend values of mixed layer depth. (a) (b) Annual maxima trends and (c) (d) winter anomalies trends (m dec^{-1}) of PLD (left) and TLD (right) in the Bay of Biscay for the period 1975–2010. Black dots represent points with a significance level higher than 90%.

this case, positive MLD trends range from 5 to 25 m per decade and negative trends from -5 to -25 m. Differences between TLD and PLD can also be observed for winter anomalies as previously noticed for annual maxima. The positive MLD trend described above for the southeasthern of the bay is in good agreement with results obtained by [46]. They found that MLD was shallower during the 1970s and 1980s, becoming deeper from 1995 onwards.

In order to know the influence of the atmospheric forcing over the MLD trend, long-term trends calculated using the temperature and density criterion were correlated with different atmospheric variables (air temperature, P-E balance and wind stress) for the period 1975–2010. Trends, which were assumed to be linear, were calculated for these atmospheric variables following the same procedure used for PLDs and TLDs (Figure 5).

Only the air temperature trend shows a different sign depending on the area of the Bay of Biscay (Figure 5a), being positive over most of the bay, with values around 0.1°C per decade. However, negative trends, with maximum values of −0.3°C per decade, were detected at the southeastern corner. Dots represent trends with significance higher than 90%. Hence, the air temperature trend seems to be spatially consistent with annual maxima and winter anomalies TLD/PLD trends shown in Figure 2. P-E balance (Figure 5b) shows a significant negative trend, with

differences between -0.5 mm per decade in the northeast boundary and around -0.15 mm per decade in central and southeastern boundary. Regarding wind stress (Figure 5c), grid points show negative trends for the whole area, being higher in the north (-0.004 N m^{-2} per decade) than in the south (-0.001 N m^{-2} per decade) of the bay.

As we mentioned above, trends in air temperature seem to be correlated with trends in MLD (TLD and PLD). To analyze the significance of this resemblance between patterns we have proceeded as follows. Considering, for example, TLD as an ocean variable, it was spatially averaged for all grid points where the variable was observed to have significant positive trends (>90%). This resulted in a one-dimensional signal (TLD^{+}), where only a value is stored per month. The same protocol was followed for the grid points with significant decreasing trends (TLD^{-} signal). On the other hand, a similar signal was created for air temperature, in such a way that Tair^{+} (Tair^{-}) was obtained by averaging the grid points with significant positive (negative) trends. The protocol previously described for TLD was also followed for PLD.

Table 2 shows the correlation between ocean variables (TLD and PLD) and Tair. As it previously mentioned the areas where air temperature decreases (increases) coincide macroscopically with the areas where MLD increases (decreases). Thus, only the comparisons among ocean and atmospheric variables with

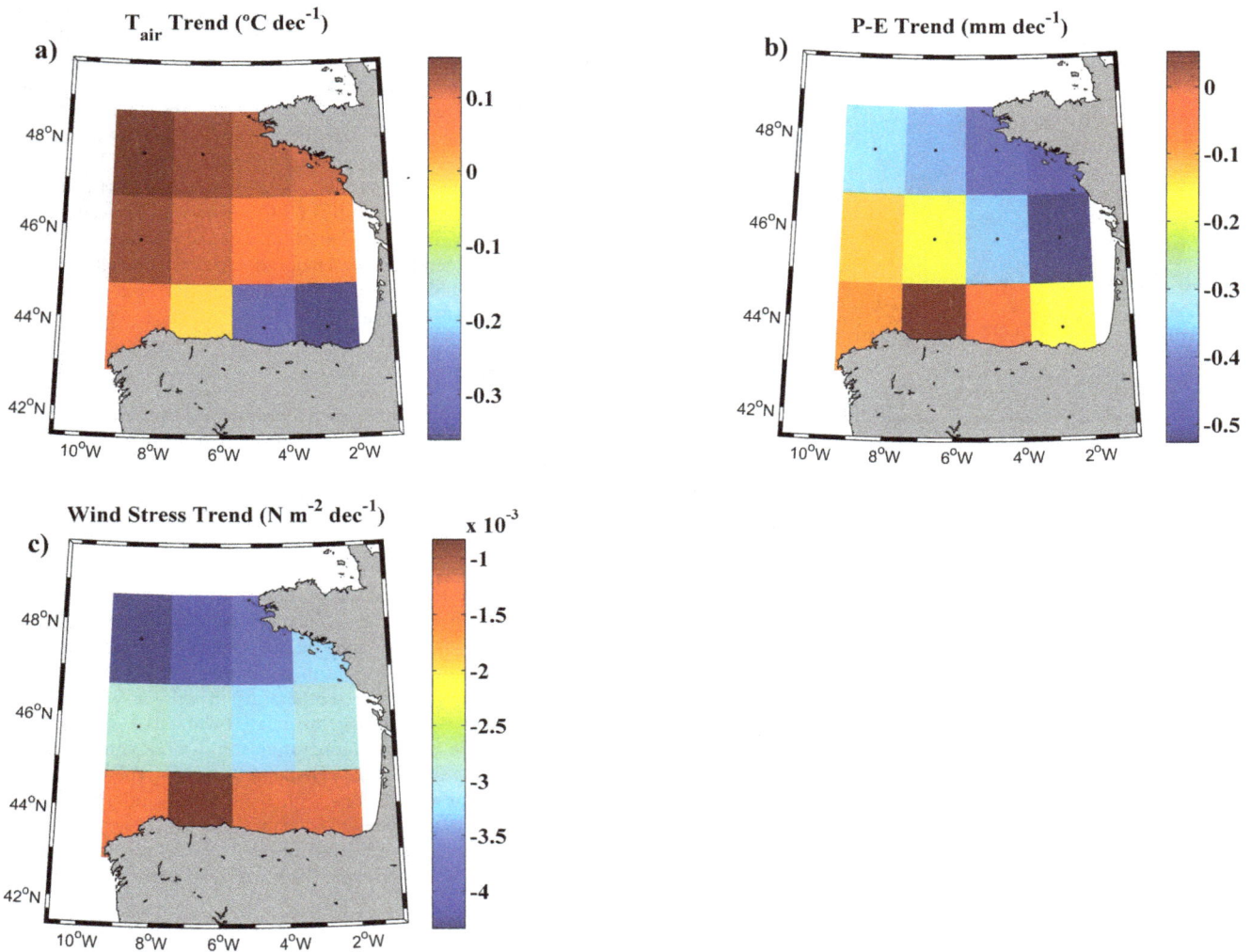

Figure 5. Trends of atmospheric variables. (a) Air temperature trend ($^{\circ}$C dec^{-1}), (b) P-E (precipitation minus evaporation) balance trend (mm dec^{-1}) and (c) wind stress trend (N m^{-2}dec^{-1}) calculated for the whole Bay of Biscay over the period 1975–2010. Black dots represent points with a significance level higher than 90%.

different sign were considered. MLD^{+} signals were only compared with Tair^{-} signals and vice-versa. Significant negative correlations were obtained between air temperature and TLD/PLD. This means that increasing air temperature results in the decrease of TLDs/PLDs depths. All correlations show a significance level higher than 95%. In particular, correlations between the Tair^{-}

and PLD^{+}/TLD^{+} (both correspond to the southeast corner of the Bay of Biscay) range from -0.35 for winter anomalies of TLD to -0.48 for annual maxima. Higher correlation coefficients were found between Tair^{+} and PLD^{-}/TLD^{-}, with values of -0.69 for winter anomalies of PLD and -0.83 for annual maxima. The significance of correlations proves the fact that could be foreseen

Table 2. Left, correlations between points of negative air temperature trends (T$_{air}^{-}$) and points of positive annual maxima (PLD$_m$/TLD$_m$) and winter anomalies (PLD$_a$/TLD$_a$) trends.

	T$^{-}_{air}$		T$^{+}_{air}$
PLD$^{+}_m$	-0.45 (0.007)	PLD$^{-}_m$	-0.7 (0.001)
TLD$^{+}_m$	-0.48 (0.004)	TLD$^{-}_m$	-0.83 (0)
PLD$^{+}_a$	-0.39 (0.019)	PLD$^{-}_a$	-0.69 (0.001)
TLD$^{+}_a$	-0.35 (0.035)	TLD$^{-}_a$	-0.8 (0)

Right, correlations between points of positive air temperature trends (T$_{air}^{+}$) and points of negative annual maxima (PLD$_m$/TLD$_m$) and winter anomalies (PLD$_a$/TLD$_a$) trends. Statistical significance (p value) is showed in brackets.

T_{Trend} (°C dec^{-1})

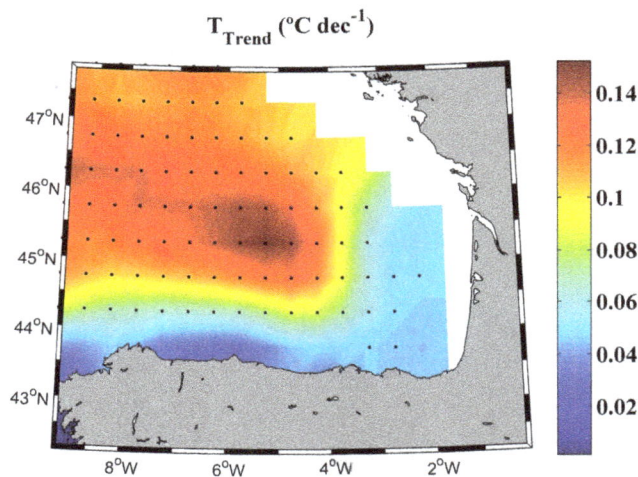

Figure 6. Winter ocean temperature trend for to the upper 700 m of the Bay of Biscay over the period 1975–2010. Black dots represent grid points with a significance level higher than 95%.

by visual inspection, namely, decreasing MLD trends are related to increasing in air temperature and vice-versa. On the other hand, correlations with P-E and wind stress did not show significant results for any of the possible combinations.

The influence of air temperature on MLD was inferred previously at the southeastern Bay of Biscay by [45]. They analyzed an extreme mixing event, characterized for a deeper than normal MLD in 2005. They found that winter 2005 was characterized by a relatively high sensible heat flux (the highest since 1965) which is a variable where air temperature plays a crucial role. Air temperature decrease causes an increase in net heat loss and this is directly associated with MLD sinking. Moreover, [46] indicated that winter net heat losses are also related to storminess activity. This fact agrees with results presented in Figures 5b since the balance between precipitation and evaporation (P-E) has a negative trend in most of the bay. On the other hand, wind stress does not play an important role in winter mixed layer formation in the Bay of Biscay according to these authors. Therefore, mixed layer development in the bay shows to be mostly conducted by convection processes than by wind stress.

Considering a global scale, this study shows that a relative small and semi-enclosed sea such as the Bay of Biscay may show opposite trends related to MLD variability. This fact is due to the high dependence of MLD on local atmospheric forcing and hydrographic conditions. For this reason, results shown in this study cannot be extrapolated to other regions. However, it seems that atmospheric forcing can play a similar role to that described for the Bay of Biscay in neighboring areas. Thus, [7] found that MLD variability is not closely related to variation in local wind speed in the North Atlantic (60W–30W, 35N–45N) for the period 1960–2004. However, these authors found that wind speed plays an important role to explain MLD variability in the North Pacific (180W–150W, 35N–45N) for the same period.

Winter temperature trend for the upper 700 m of the water column was also analyzed over the period 1975–2010 using SODA database (Figure 6). According to [29], [30] the most important variations in the heat content during the last decades occurred in the upper 700 m. This is also shown for the Iberian Peninsula [43] and the Bay of Biscay [18]. The whole area presents a positive trend but with different intensity. Warming is higher at the northwest corner and in the central part of the bay with values ranging from 0.1–0.15°C per decade. However trends are almost negligible at the south and at the southeastern corner with values around 0.04°C per decade. Macroscopically, the area of higher (negligible) warming coincides with the area where MLD has become shallower (deeper).

Conclusions

The present study analyzes wintertime MLD trends in the Bay of Biscay over the period 1975–2010, which coincides with the highest warming period ever detected in the North Atlantic. MLD was calculated in two different ways, namely, by using a potential temperature criterion to define the TLD and a potential density criterion to define the PLD. Time evolution of MLD was analyzed in terms of trends in winter anomalies and in annual maxima.

MLD tended to deepen at the southeastern corner and shallow at the rest of the bay. Correlations carried out between atmospheric variables and MLD trends evidence that air temperature plays a key role in regulating the MLD variability. In this way, negative air temperature trends, localized in the southeastern part of the bay, are related with MLD deepening at this area and positive air temperature trends observed for the rest of the bay are related to MLD shoaling. MLD evolution did not show significant correlations with wind stress and precipitation minus evaporation. Finally, warming calculated for the upper 700 m is observed to be more intense at the area where MLD became shallower and negligible at the area where MLD deepened.

The MLD analysis carried out in this study highlights that mesoscale features can play an important role even in small semi-enclosed areas like the Bay of Biscay, where the observed behavior is far from being spatially homogeneous. Most of the research previously conducted in the area was focused on local events which do not necessarily represent the overall behavior of the Bay.

The complex bathymetry of the area and the fact that the bay is surrounded by land in more than 50% of its perimeter give rise to important local variations in the oceanic and atmospheric variables. Thus, the inner part of the Bay, which is much shallower than the central one, is highly dependent on continental inputs. This is especially important for temperature, which is influenced by land masses in winter and summer and by the presence of mountain ranges like *Picos de Europa* that can modify the passage of fronts in winter.

Author Contributions

Conceived and designed the experiments: XC MdC MGG FS. Performed the experiments: XC MdC MGG FS. Analyzed the data: XC MdC MGG FS. Contributed reagents/materials/analysis tools: XC MdC MGG FS. Wrote the paper: XC Mdc MGG.

References

1. Argo Science Team (2001) Argo: The global array of profiling floats. In: Koblinsky CJ, Smith, NR, editors. Observing the Oceans in the 21st Century: GODAE Project Office, Bureau of Meteorology, Melbourne: 248–257.

2. Bates NR (2001) Interannual variability of oceanic CO_2 and biogeochemical properties in the Western North Atlantic subtropical gyre. Deep-Sea Research, Part II, 48: 1507–1528.

3. Brainerd KE, Gregg MC (1995) Surface mixed and mixing layer depths. Deep-Sea Research, Part I, 9: 1521–1543.

4. Carton JA, Chepurin G, Cao X, Giese B (2000a) A simple ocean data assimilation analysis of the global upper ocean 1950–95. Part I: Methodology. Journal of Physical Oceanography 30: 294–309.

5. Carton JA, Chepurin G, Cao X (2000b) A simple ocean data assimilation analysis of the global upper ocean 1950–95. Part II: Results. Journal of Physical Oceanography 30: 311–326.

6. Carton JA, Giese BS (2008) A reanalysis of ocean climate using Simple Ocean Data Assimilation (SODA). Monthly Weather Review 13: 2999–3017.

7. Carton JA, Semyon AG, Hailong L (2008) Variability of the oceanic mixed layer, 1960–2004. Journal of Climate 21: 1029–1047, doi:10.1175/2007JCLI1798.1.

8. de Boyer Montégut C, Madec G, Fischer AS, Lazar A, Iudicone D (2004) Mixed layer depth over the global ocean: An examination of profile data and a profile-based climatology. Journal of Geophysical Research 109: C12003, doi:10.1029/2004JC002378.

9. de Boyer Montégut C, Mignot J, Lazar A, Cravatte S (2007) Control of salinity on the mixed layer depth in the world ocean: 1. General description. Journal of Geophysical Research 112: C06011, doi:10.1029/2006JC003953.

10. deCastro M, Gómez-Gesteira M, Álvarez I, Gesteira JLG (2009) Present warming within the context of cooling–warming cycles observed since 1854 in the Bay of Biscay. Continental Shelf Research 29: 1053–1059.

11. Dumousseaud C, Achterberg EP, Tyrrel T, Charalampopoulou A, Schuster U, et al. (2010) Constrasting effects of temperature and winter mixing on the seasonal and inter-annual variability of the carbonate system in the Northeast Atlantic Ocean. Biogeosciences 7(5): 1481–1492, doi: 10.5194/bg-7-1481-2010.

12. García-Soto C, Pingree RD, Valdés L (2002) Navidad Development in the Southern Bay of Biscay: climate change and Swoddy structure from remote sensing and in situ measurements. Journal of Geophysical Research 107: C83118.

13. García-Soto C, Pingree RD (2009) Spring and summer blooms of phytoplankton (SeaWiFS/MODIS) along a ferry line in the Bay of Biscay and western English Channel. Continental Shelf Research 29: 1111–1122.

14. Giese BS, Ray S (2011) El niño variability in simple ocean data assimilation (SODA), 1871–2008. Journal of Geophysical Research 116: C02024.

15. Goikoetxea N, Borja A, Fontán A, González M, Valencia V (2009) Trends and anomalies in sea surface temperature, observed over the last 60 years within the southeastern Bay of Biscay. Continental Shelf Research 29: 1060–1069.

16. Gómez-Gesteira M, deCastro M, Alvarez I, Gesteira JLG (2008) Coastal SST warming trend along the continental part of the Atlantic Arc (1985–2005). Journal of Geophysical Research 113: C04010.

17. Gómez-Gesteira M, Gimeno L, deCastro M, Álvarez I, Nieto R, et al. (2011). The state of clima in NW Iberia. Climate Research 48: 109–144.

18. Gómez-Gesteira M, deCastro M, Santos F, Álvarez I, Costoya X (2013) Changes in ENACW observed in the Bay of Biscay over the period 1975–2010. Continental Shelf Research 65C: 73–80, doi:10.1016/j.csr.2013.06.014.

19. González-Pola C, Lavín A, Vargas-Yánez M (2005) Intense warming and salinity modification of intermediate water masses in the southeastern corner of the Bay of Biscay for the period 1992–2003. Journal of Geophysical Research 10: C05020, doi:10.1029/2004JC002367.

20. Gruber N, Keeling CD, Bates NR (2002) Interannual variability in the North Atlantic Ocean carbon sink. Science 298: 2374–2378.

21. Holte J, Talley L (2009) A new algorithm for finding mixed layer depths with applications to Argo data and Subantartic Mode Water formation. Journal of Atmospheric and Oceanic Technology 26: 1920–1939.

22. Hosoda S, Ohira T, Sato K, Suga T (2010) Improved description of global mixed-layer depth using Argo profiling floats. Journal of Oceanography 66: 773–787.

23. Intergovernmental Panel on Climate Change (IPCC) (2007) Climate change 2007: the physical science basis. Contribution of Working Group 1 to the Fourth Assessment Report of the Intergovernmental Panel on Climate Change. Cambridge University Press, Cambridge, UK.

24. Kara AB, Rochford PA, Hurlburt HE (2000) An optimal definition for ocean mixed layer depth, Journal of Geophysical Research 105: 16803–16821.

25. Kara AB, Rochford PA, Hurlburt HE (2003) Mixed layer depth variability over the global ocean. Journal of Geophysical Research 108: 3079, doi:10.1029/2000JC000736.

26. Koutsikopoulos C, Le Cann B (1996) Physical processes and hydrological structures related to the Bay of Biscay Anchovy. Scientia Marina 60: 9–19.

27. Koutsikopoulos C, Beillois P, Leroy C, Taillefer F (1998) Temporal trends and spatial structures of the sea surface temperature in the Bay of Biscay. Oceanologica Acta 21: 335–344.

28. Levitus S, Boyer T (1994) World Ocean Atlas 1994. vol. 4: Temperature, NESDIS Atlas series, NOAA, Washington, DC 117 pp.

29. Levitus S, Antonov J, Boyer T (2005) Warming of the world ocean, 1955–2003. Geophysical Research Letters 32: L02604.

30. Levitus S, Antonov JI, Boyer TP, Locarnini RA, Garcia HE, et al. (2009) Global ocean heat content 1955–2008 in light of recently revealed instrumentation problems. Geophysical Research Letters 36: L07608, doi:10.1029/2008GL037155.

31. Liu H, Grodsky SA, Carton JA (2009) Observed subseasonal variability of oceanic barrier and compensated layers. Journal of Climate 22: 6104–6119, doi:10.1175/2009JCLI2974.1.

32. Llope M, Anadon R, Viesca L, Quevedo M, Gonzalez-Quiros R, et al. (2006) Hydrography of the Southern Bay of Biscay shelf-break region: Interacting the multiscale physical variability over the period 1993–2003. Journal of Geophysical Research 111: C09021.

33. Lorbacher K, Dommenget D, Niiler PP, Kohl A (2006) Ocean mixed layer depth: A subsurface proxy of ocean-atmosphere variability. Journal of Geophysical Research 111: C07010, doi:10.1029/2003JC002157.

34. Michaels A, Knap A (1996) Overview of the US JGOFS Bermuda Atlantic Time-series Study and the Hydrostation S program. Deep Sea Research II 43(2–3): 157–198.

35. Michel S, Vandermeirsch F, Lorance P (2009) Evolution of upper layer temperature in the Bay of Biscay during the las 40 years. Aquatic Living Resources, 22: 447–461, doi:10.1051/alr/29009054.

36. Monterey G, Levitus S (1997) Seasonal Variability of Mixed Layer Depth for the World Ocean. NOAA Atlas NESDIS 14: 96 pp.

37. Nilsen JEO, Falck E (2006) Variations of mixed layer properties in the Norwegian Sea for the period 1948–1999. Progress in Oceanography 70: 58–90, doi:10.1016/j.pocean.2006.03.014.

38. Pascual A, Cearreta A, Rodríguez-Lázaro J, Uriarte A (2004) Geology and palaeoceanography. In: Borja A, Collins M, editors. Oceanography and Marine Environment of the Basque Country. Elsevier Oceanography Series no. 70, Elsevier, Amsterdam. pp.53–73.

39. Patti B, Guisande C, Riveiro I, Thejll P, Cuttitta A, et al. (2010) Effect of atmospheric CO2 and solar activity on wind regime and water column stability in the major global upwelling areas. Estuarine, Coastal and Shelf Science 88: 42–45.

40. Planque B, Beillois P, Jegou AM, Lazure P, Petitgas P, et al. (2003) Large- scale Hydroclimatic Variability in the Bay of Biscay: The 1900s in the Context of Interdecadal Changes. ICES Marine Science. Symposium 219: 61–70.

41. Price JF, Weller RA, Pinkel R (1986) Diurnal cycling: Observations and models of the upper ocean response to diurnal heating, cooling and wind mixing. Journal of Geophysical Research 91: 8411–8427.

42. Reynolds RW, Smith TM (1994) Improved global sea surface temperature analysis using optimum interpolation. Journal of Climate 7: 929–948.

43. Santos F, Gómez-Gesteira M, deCastro M, Álvarez I (2012) Variability of coastal and ocean water temperature in the upper 700 m along the western Iberian Peninsula from 1975 to 2006. PloS ONE 7, doi: 10.1371/journal.pone.0050666.

44. Saunders PM (1982) Circulation in the eastern North Atlantic. Journal of Marine Research 40 (Suppl.): 641–657.

45. Somavilla R, González-Pola C, Rodríguez C, Josey SA, Sánchez R, et al. (2009) Large changes in the hydrographic structure of the Bay of Biscay after the extreme mixing of winter 2005. Journal of Geophysical Research c01001, doi:10.1029/2008JC004974.

46. Somavilla R, González-Pola C, Ruiz Villarreal M,, Lavín A (2011) Mixed layer depth (MLD) variability in the southern Bay of Biscay. Deepening of winter MLDs concurrent to generalized upper water warming trends? Ocean Dynamics 61: 1215–1235.

47. Sverdrup H (1953) On conditions for the vernal blooming of phytoplankton. Journal du Conseil International pour l'Exploration de la Mer, 18: 287–295.

48. UNESCO (1983) Algorithms for computation of fundamental properties of seawater. UNESCO Technical Paper in Marine Science 44:53 pp.

49. Wilks DS (1995) Statistical Methods in the Atmospheric Sciences. Academic Press, San Diego: 467 pp.

50. Wong A, Keeley R, Carval T, and the Argo Data Management Team (2013) Argo Quality control manual, version 2.8. http://www.argodatamgt.org/content/download/15699/102401/file/argo-quality-control-manual-version2.8.pdf.

51. Zheng Y, Giese BS (2009) Ocean heat transport in Simple Ocean Data Assimilation: structure and mechanisms. Journal of Geophysical Research 114: C11009.

52. Zheng Y, Shinoda T, Kiladis GN, Lin J, Metzger EJ, et al. (2010) Upper-Ocean Processes under the Stratus Cloud Deck in the Southeast Pacific Ocean. American Meteorological Society, doi:10.1175/2009JPO4213.1.

9

The Potential Wind Power Resource in Australia: A New Perspective

Willow Hallgren*, Udaya Bhaskar Gunturu, Adam Schlosser

The MIT Joint Program on the Science and Policy of Global Change, Massachusetts Institute of Technology, Cambridge, Massachusetts, United States of America

Abstract

Australia's wind resource is considered to be very good, and the utilization of this renewable energy resource is increasing rapidly: wind power installed capacity increased by 35% from 2006 to 2011 and is predicted to account for over 12% of Australia's electricity generation in 2030. Due to this growth in the utilization of the wind resource and the increasing importance of wind power in Australia's energy mix, this study sets out to analyze and interpret the nature of Australia's wind resources using robust metrics of the abundance, variability and intermittency of wind power density, and analyzes the variation of these characteristics with current and potential wind turbine hub heights. We also assess the extent to which wind intermittency, on hourly or greater timescales, can potentially be mitigated by the aggregation of geographically dispersed wind farms, and in so doing, lessen the severe impact on wind power economic viability of long lulls in wind and power generated. Our results suggest that over much of Australia, areas that have high wind intermittency coincide with large expanses in which the aggregation of turbine output does not mitigate variability. These areas are also geographically remote, some are disconnected from the east coast's electricity grid and large population centers, which are factors that could decrease the potential economic viability of wind farms in these locations. However, on the eastern seaboard, even though the wind resource is weaker, it is less variable, much closer to large population centers, and there exists more potential to mitigate it's intermittency through aggregation. This study forms a necessary precursor to the analysis of the impact of large-scale circulations and oscillations on the wind resource at the mesoscale.

Editor: Francois G. Schmitt, CNRS, France

Funding: The authors gratefully acknowledge the financial support for this work provided by the MIT Joint Program on the Science and Policy of Global Change through a number of federal agencies and industrial sponsors (for the complete list, see http://globalchange.mit.edu/sponsors/current.html). The funders of the Joint Program had no role in study design, data collection and analysis, decision to publish, or preparation of the manuscript.

Competing Interests: The authors have declared that no competing interests exist.

* Email: hallgren@mit.edu

Introduction

The general climatology of the winds in Australia has been documented on a national basis [1,2,3] and at the state level [4,5,6,7], using a variety of methodologies [8]. Such climatologies indicate that Australia has wind resources that are in places comparable to those in northern Europe, and indicate that the location of the strongest winds is in western, southwestern, and southern Australia, and southeastern coastal regions [8].

The physical quantity conventionally used to describe the wind energy potential in Australia is wind speed in m/s, whereas in the USA, wind atlases show maps of wind power density (WPD) to describe the quality of the wind resource. Most previous published studies use the mean to characterize the central tendency of the wind resource, however histograms of the wind resource measured using wind power density are characteristically skewed with long-tailed distributions [9] (Figure S1). Therefore, wind power studies based only on the total mean WPD do not give a representative picture of the central tendency of the wind power potential and also omit valuable information in terms of wind intermittency, variability and the temporal distribution of power generation [10], which would affect estimates of power production and required backup [11].

Variability in the wind resource has major ramifications for the economics and therefore the feasibility of wind power generation and distribution, and hence measures of variability are useful for wind energy policy makers. Yet, very few atlases show maps of wind variability [9], and when they do it is typically in terms of the standard deviation of the wind speed or WPD. However, the economic viability of wind power as an alternative energy source strongly depends on how reliable the resource is, in terms of its availability and persistence, as well as other factors such as proximity to high-capacity power transmission lines, and how remote it is from population centers and the electricity grid. The reliability of wind power can in theory be increased by mitigating the natural intermittency of the wind resource, by aggregating power from wind farms that are geographically dispersed, with the aim of achieving a more continuous wind resource over large areas, and there have been several studies trying to address this issue [12,13].

Wind power production doubled in the 5 years to 2012, and has grown 340% since 1997, to meet 3.4% of Australia's total electricity demand and 26% of total renewable energy generated, which is a bit less than half that generated by hydropower [14]. Wind power is likely to become economically competitive in the coming decades, and is projected to grow by 350% when wind power projects currently in development come online in the next few years [15]. This projected expansion of wind energy conforms to national policies that were designed to lower carbon emissions, including legislation that was introduced to put a price on carbon, and the Renewable Energy Target of 20% by 2020 [15]. In light of this policy directive, there is a need to increase the accuracy and practical relevance of the assessment of Australia's wind power resource.

We assess Australia's potential wind power resource with alternative metrics of abundance, variability and intermittency that provide deeper insights about the stability of the wind resource at a widespread deployment scale [9,11] over long time periods, using a robust, multi-decadal dataset.

Several authors explore the variability and intermittency of the wind resource at many scales [16,17]. There are fewer studies at the mesoscale scale range than at smaller scale ranges, despite the fact that knowledge of variability at this scale is important to the management and control of wind power generation [16]. Our study focuses on variability and intermittency at the hourly scale and above -the mesoscale- and addresses the type of scenario, to take just one example, in which long wind lulls spanning weeks, during sustained periods of high pressure, have been known to occur in countries such as the UK and Germany (Oswald et al 2008 [18], telegraph article [19]). These instances have implications for the reliability of power generated, as well as the potential backup and storage required to sustain power delivery. The goal of the present paper is to characterize the wind resource in Australia and its inherent variability, as a necessary precursor to studies of the impact of large-scale climate oscillations on the variability of the wind resource at different scales.

Questions our study asks include: (1) What is the geographical distribution of the abundance, variability, availability, and persistence of wind power density (WPD), and do these differ with higher turbine hub heights? (2) Where can wind intermittency be mitigated by the aggregation of geographically dispersed wind farms?

Methods

2.1. Data

We have sought to address some of the limitations of previous wind resource studies that used data that had a coarse spatial and temporal resolution, a relatively short record length, and sparse and uneven coverage [20,11]. We used 31 years of hourly $1/2° \times 2/3°$ resolution MERRA (Modern Era Retrospective Analysis for Research and Applications [21] data (from 0030 on January 1st, 1979 to 2330 on 31st December, 2009) to reconstruct the wind field at several turbine hub heights 50 m, 80 m, and 150 m, since the MERRA dataset does not provide wind speeds at different hub heights. These heights were chosen to represent the recent 1990's (US) 50 m standard wind turbine hub height [22,23], and the 80 m hub height, which has become more common as technology develops, and the potentially much higher hub heights in the future.

Wind speed and then wind power density were computed at these different heights using boundary layer flux data (consisting of such parameters as surface roughness, displacement height and friction velocity) and similarity theory of the atmospheric boundary layer [9]. By doing this, we sought to improve on previous wind resource constructions that used a constant scaling exponent (irrespective of surface roughness) to scale the wind speed from a lower altitude (usually 10 m) to that of the turbine hub height. We use WPD ($W\ m^{-2}$) to describe the wind resource as it is a function of not only wind speed but also density, which also varies in space and time. It indicates how much wind energy can be harvested at a location by a wind turbine but is independent of wind turbine characteristics. In a recent study, Farkas [24] found that non-consideration of air density causes an root mean square (RMS) error of 16% in wind potential, which is a considerable difference, and therefore air density should be an important consideration in estimating the wind resource potential. The domain considered for our study spans the entire Australian continent plus Tasmania, between 10°S and 45°S latitudes and 110°E and 155°E longitudes.

While the resolution of the data used in this study is lower than the mesoscale, there have also been many studies that establish the utility of data at the GCM resolution (e.g. Schwartz and George, 1999) [25] for understanding the variability and impact of large-scale circulations at a regional scale. Several studies have used a similar dataset, although with a shorter record length, to estimate the potential wind resource in China [26] and also globally [27]. However, for studying inter-decadal variability, we argue that the longer record length of the data is as essential an attribute. This is because, according to sampling theorem, a dataset has to have at least 20 years of data for understanding inter-decadal variability. Hence this construction was designed, and is most appropriate, for such studies.

All other constructions that span only a few years fail to represent such variability. Moreover, studies such as those by Pryor, Barthelmie and Schoof (2006) [28], Chadee and Clarke (2013) [29], use data with a lower resolution than that used here, to study similar issues (inter-annual variability of wind indices across Europe, large-scale wind energy potential of the Caribbean, etc.). This would indicate that our data resolution is suitable for the purpose of our research, and represents an improvement to the resolution of a number of prior studies [22,26,27,28,29,30,31,32].

Since Gunturu and Schlosser (2012) [9] have already done a thorough evaluation of the lowest model layer wind speed data taken directly from MERRA, and since this study uses the same data, as it is a continuation of theirs, it is unnecessary to reproduce this validation of the MERRA data here. As the original MERRA wind data that this study employs is in the public domain, the description of the methodology will enable others to construct the wind power density dataset that is used in this study.

2.2. Comparison with Existing Wind Climatologies in Australia

Here, the wind resource is constructed for a hub height of 80 m, and was compared with a publicly available map of 80 m wind speed, since a publicly accessible wind power density map was not available. This was done in order to understand the ability of this constructed dataset to reproduce large-scale spatial features. This map was originally published by the Australian government and also used by various state governments [15]. We also use publicly available maps of the location of wind farms in South Australia and New South Wales to validate our wind resource construction [33,34].

2.3. Wind resource metrics

The metrics we use in our study are wind abundance, variability and intermittency in the form of availability and persistence [9,11]. Most previously published studies use the mean to characterize the central tendency of the wind resource. Since the mean is not a robust measure of the central tendency for distributions with long tails, we use the median, which is immune to the extreme values in the distribution, as a robust measure of the central tendency of the wind resource, and provides a better evaluation of it's abundance. As Pryor and Barthelmie (2011) [20] point out "there is a need for accurate data pertaining to metrics of the wind climate beyond the central tendency, and trends in annual mean wind speeds have little bearing on the viability of wind energy."

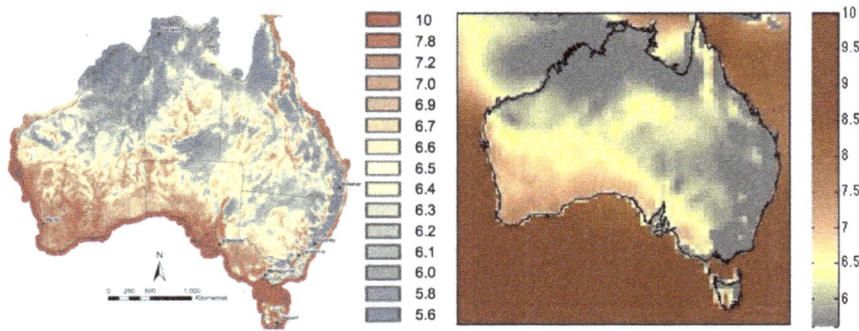

Figure 1. Comparison of mean wind speed (m/s) at an 80 m turbine hub height across Australia. (Left) Map developed by the Australian Government Department of the Environment, Water, Heritage and the Arts in 2008, and (right) the map constructed from MERRA data.

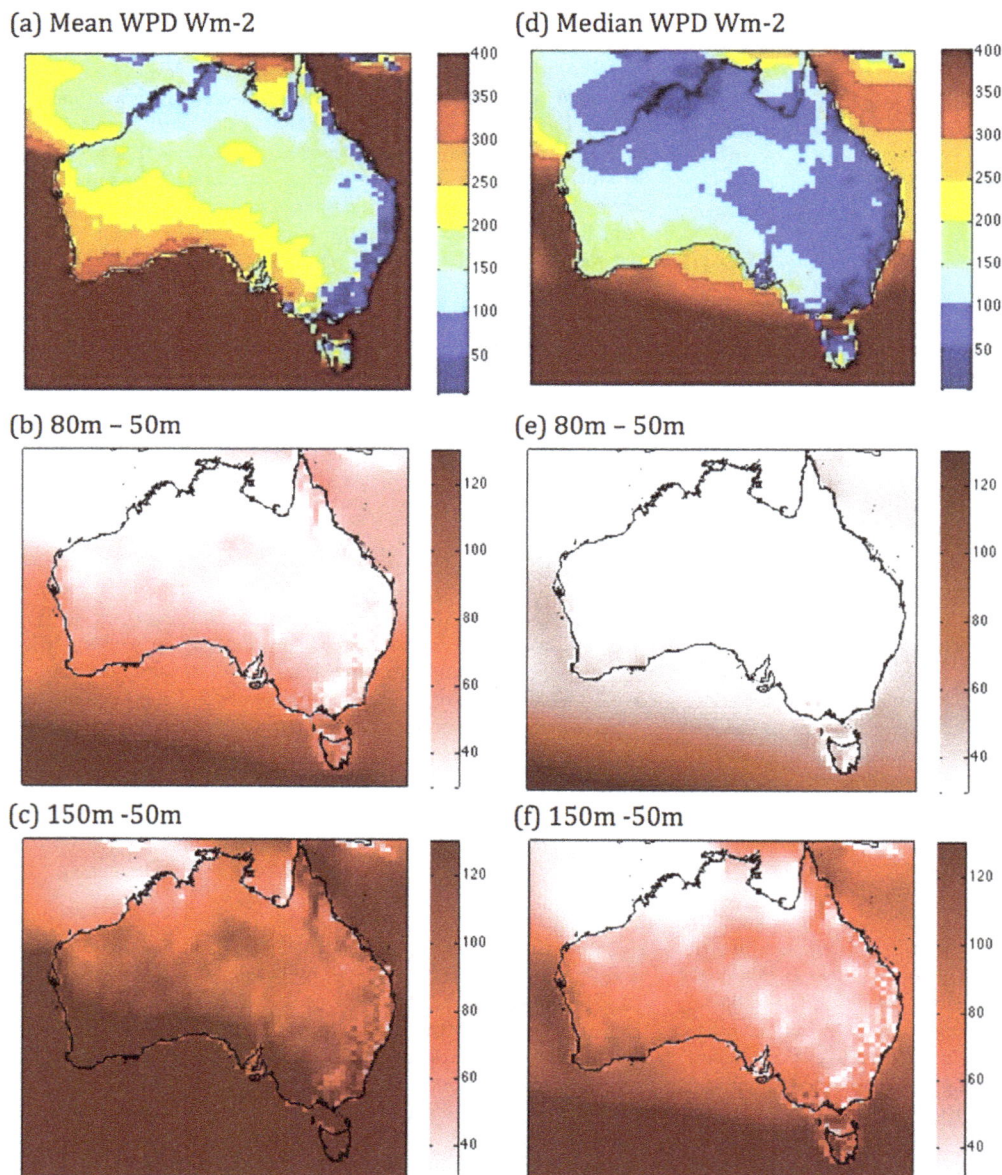

Figure 2. Measures of abundance. (a) The mean WPD at 50 m, (b) the change in the mean from 50 m to 80 m and from (c) 50 m to 150 m, (d) median wind power density at 50 m, (e) the change in the median from 50 m to 80 m and from (f) 50 m to 150 m. All units are W m^{-2}.

Instead of using the standard deviation to represent the variability of the wind resource, we argue that the variability of the wind resource is better captured in terms of the robust coefficient of variation (RCoV), since it is calculated using the median, which we argue is a more accurate representation of the wind power at a given site than the mean. We also use the Inter-quartile range (IQR) as a measure of the statistical dispersion, higher values of which can indicate the greater possibility of 'swings' of the WPD at a location, and therefore the amount of backup power that needs to be maintained.

In addition to these measures of variability, we also look at two measures of the intermittency of the wind - availability (or lack of) and persistence - since these are important indicators of intermittency, which is recognized as one of the key limitations to large-scale installation of wind power. We apply the reliability theory concept of availability to wind power, as a measure of the temporal distribution of the wind resource, and therefore of the reliability of a wind power generation system. We calculate the percentage of hours in our time series where WPD$>$200 W m^{-2}, and use the inverse of this - unavailability - of non-useful WPD (i.e. proportion of hours where $<$200 W m^{-2}), to characterize the geographic distribution of the reliability of the wind resource [11], and as one measure of intermittency. Our rationale for choosing the 200 W m^{-2} cutoff is the same as Gunturu and Schlosser (2011) [11], and incorporates a number of contributing arguments, which are detailed in Text S1. Mean episode length (i.e. number of hours of WPD above 200 W m^{-2}) was calculated as a measure of the persistence of the WPD, which is important in the planning and development of a robust deployment strategy for harvesting wind power.

We use Gunturu and Schlosser's (2011) [11] technique to analyze the potential value of aggregating the power generated by geographically dispersed wind farms in a roughly 1000$\times$1000 km box (19$\times$19 grid cells), in order to mitigate intermittency in the wind resource. Values of anticoincidence [35], and null-antic-oincidence were calculated for each grid cell (see Fig. 5 in Gunturu and Schlosser, 2011) by converting the time series of WPD at each grid point into a binary sequence of 1 s and 0 s depending on if the WPD is greater or less than the 200 W m^{-2} we use as the cutoff useful for viable commercial generation. We base our analysis of anticoincidence on these binary sequences. Two grid points are said to be anticoincident when the hourly time series of WPD is greater than 200 W m^{-2} at one of the two points, but not both, for 50% of the total length of the time series. We also calculate the null-anticoincidence, which offers a somewhat more relaxed criterion. Null-anticoincidence refers to the number of grid points in a roughly 1000$\times$1000 km area surrounding a central point which have usable wind power ($>$200 W m^{-2}), when the central point does not, for at least 50% of the time when there's no wind at the central point [11]. If the region within this analysis area shows higher values of anticoincidence then this means that there will be fewer coincident lulls in the wind resource across the region, and that aggregating power from geographically dispersed wind farms will be more likely to mitigate the intermittency of the wind resource across the region as a whole.

Our choice for using a box this size for the anticoincidence analysis was based on the fact we are looking at the wind resource at a regional scale, hence, this is the scale at which we studied anticoincidence: the mesoscale. For more information on the rationale for the box size, refer to the Supplementary Information. In terms of the temporal scale used in this study, while there have been several methods and technologies to mitigate intermittency at the operational scale, such intermittency for the grid operations occurs at micro-to-hundreds of seconds. But for the scale that this

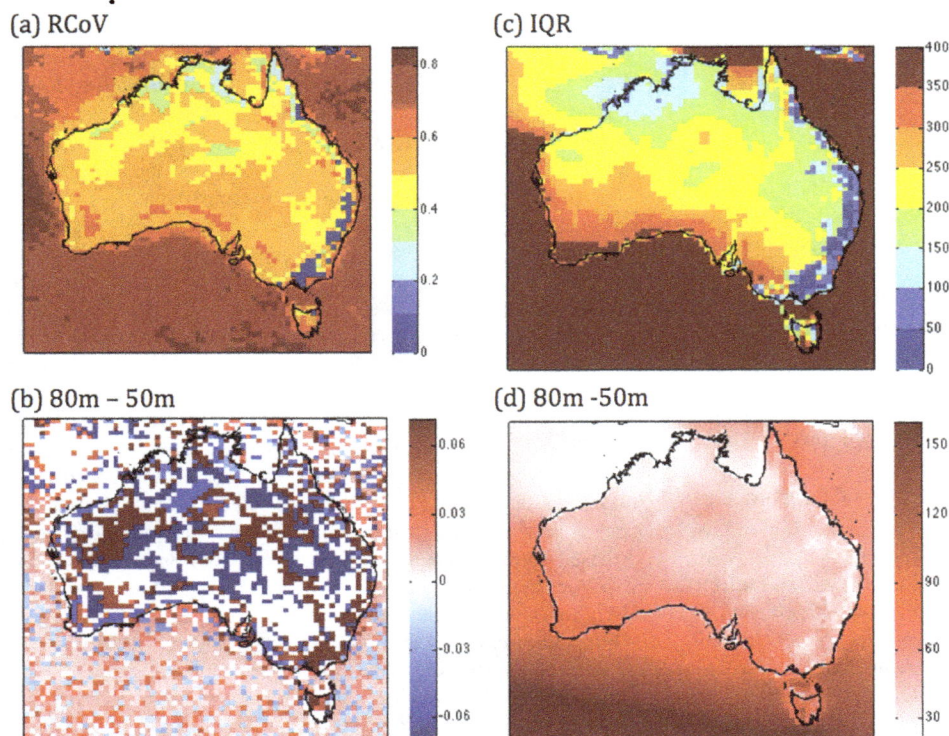

Figure 3. Measures of variation. (a) The robust coefficient of variation (RCoV -unitless) of WPD at 50 m, (b) the change in the RCoV from 50 m to 80 m, (c) inter-quartile range (IQR, W m^{-2}) at 50 m, (d) the change in the IQR from 50 m to 80 m.

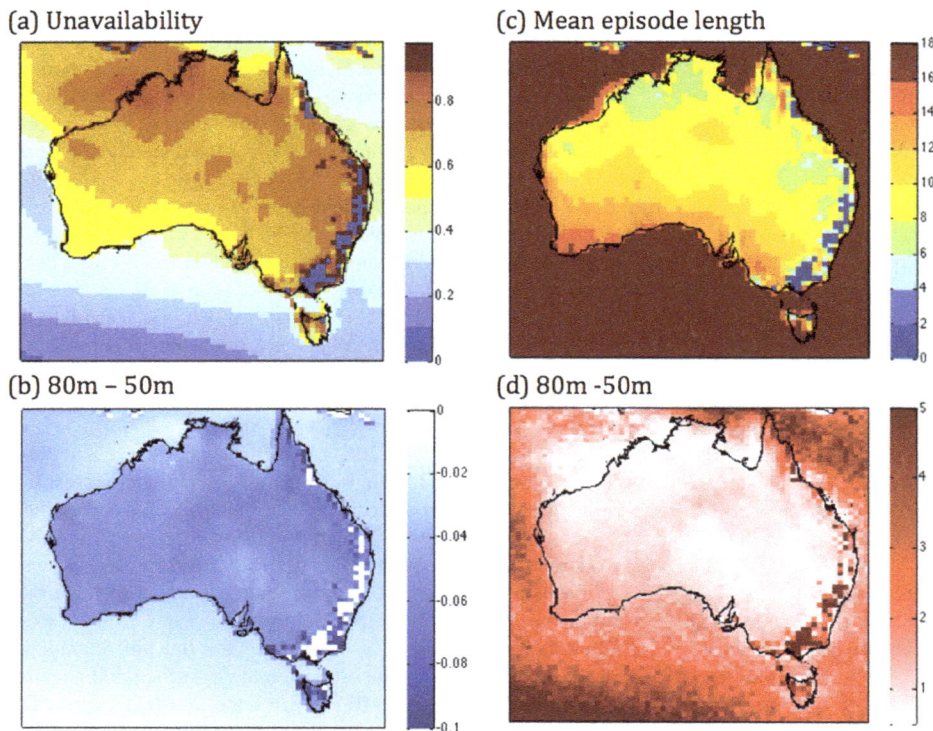

Figure 4. Measures of intermittency. (a) The unavailability of WPD at 50 m (fraction of time), (b) the change in the unavailability from 50 m to 80 m, (c) the mean episode length at 50 m (hours) (d) the change in the mean episode length from 50 m to 80 m.

study pertains to, no methods or technologies have yet been developed to deal with intermittency, to the knowledge of the authors, at the scale of one hour or more, in which case, the issues of back up and resource adequacy become important.

Results and Discussion

Wind speed and wind power density were computed at several wind turbine hub heights using boundary layer flux data from the Modern Era Retrospective-analysis for Research and Applications (MERRA) [21] and similarity theory of the atmospheric boundary layer [9]. We use wind speed to compare our results to existing wind atlases (as the reference atlas for Australia uses wind speed instead of wind power density to measure wind power potential), as well as a range of metrics to analyze wind power density, including wind abundance, variability, and intermittency in the form of availability and persistence [11,9]. Detailed descriptions of the data and methodology are described in the Methods section.

3.1. Comparison of MERRA and Australian Government maps of wind speed at 80 m

Our approximately 50 km×67 km (½ degree×⅔ degree) map of 80 m above ground level wind speed (Fig. 1) is quantitatively and geographically similar to the 9 km×9 km resolution map of wind speed at the same height produced by the Australian Government Department of the Environment, Water, Heritage and the Arts (hereafter referred to as AGD) [15]. This map was created by WindLab (www.windlab.com) for the AGD and is derived from observed weather station data taken from Bureau of Meteorology weather stations for the years 1995–2005, for the entire continent, and supplemented with commercially produced meteorological datasets, which are then assimilated into a high

resolution broad-area wind mapping model called WindScape [36]. WindScape uses a regional scale weather model (The Air Pollution Model (TAPM [37]) to improve the resolution of the observed data, and also a fine scale computational fluid dynamics model Raptor and/or Raptor-NL to create fine scale resolution maps of the wind resource over broad areas. The maps created are validated and adjusted to achieve consistency with observational data at ground level [38].

While our construction of the wind resource matches qualitatively and quantitatively very well with that of the AGD map overall, there are differences between the two maps in some regions. Our results mostly show slightly lower values for most areas compared to corresponding areas on the AGD map (Table 1). For example, a comparison of our map with the maps of NSW [33] and South Australian [34] wind farms, indicates regions where areas of better wind resources, as shown on our constructed map, coincide with existing wind farm deployments on [33] and [34], particularly in NSW, even though wind speed values in these areas might be slightly lower on our constructed map, than on the AGD map, as shown in Fig. 1. On that map, this is not always the case - it shows even better wind resources outside of these regions of wind farm deployment. This indicates that our map does actually capture areas of good wind resource in areas where there are existing wind farms.

Furthermore, our coarser-resolution map of the wind resource shows fewer orographic effects of the Great Dividing Range than the AGD map. Nevertheless, our map captures precisely the areas where there are existing wind farms on the New South Wales Southern Highlands and Blue Mountains [33,34]. So although our map has a resolution which does not capture as much topographical detail as the government map, it captures precisely the areas where there are existing wind farms, for instance, our

Table 1. Comparison of the range of values (m/s) in many areas of the 80 m wind speed map constructed from MERRA data to the one produced by the Australian Government.

Regions of similarity	MERRA data map	Australian government map
East coast, Tasmania	5.6–7.0	6.5–7.8
Western Victoria	6.5–7.0	Mostly >7.0
SE South Australia	6.4–7.2	Up to 7.8
Central Australia	5.6–7.0	5.8–6.6

The first region encompasses much of the East coast, and includes southeast and northeast QLD, and Tasmania.

map shows two small regions on the eastern seaboard of good wind resource, which is where all but one of the existing wind farms are currently located.

Reasons for the differences seen in these two maps could be due to the lower spatial resolution of our constructed map and the lower temporal record length of the AGD map. Since the AGD wind resource map has been constructed by running a mesoscale model (TAPM) for 11 years (and all other constructions also span only a few years), the record length of the construction is short compared to the record length of our construction, which represents an average over 31 years, that includes many years of low and high wind. Short record lengths do not represent interannual variability and climate scale (i.e. more than a few years) oscillations like the El Nino Southern Oscillation (ENSO) robustly.

3.2. Measures of abundance and variability

Reflecting the wind speed patterns of previous Australian wind atlases, our constructed map of mean WPD at 50 m (Fig. 2a) shows that the strongest wind resources occur in southwest Western Australia, southern South Australia, and Tasmania, and south-western Victoria. It is lowest in mountainous areas along the Great Dividing Range in eastern Australia, in northwest Australia, and northwest QLD. Most of the continent has mean WPD values below 300, and most of the populated east coast of the country has values below 200 W m^{-2} at this resolution, which is the cutoff for the production of usable power that turbines can produce, the rationale for which is detailed in Text S1. As turbine hub height increases to 80 (Fig. 2b) and 150 m (Fig. 2c), there is an increase in mean WPD of up to about 40 and 100 W m^{-2} in the northern two-thirds of Australia and 80 and 160 W m^{-2} (and higher in

Tasmania) in the south respectively. While the mean WPD construction reflects the other known datasets that illustrate wind speed, we extend the analysis that has historically been done, and look at other metrics of the resource that could be useful for assessing the economics of wind power generation and also for operational stability.

The map of median WPD at 50 m (Fig. 2d) indicates that a greater part of the continent has WPD below the 200 W m^{-2} value. Compared with the mean WPD in Fig. 2a, the median values are almost half of the mean values throughout much of the country. This implies that the distribution is very skewed, and hence we argue that the median is a much more robust measure of central tendency and therefore a more appropriate metric to represent WPD. As turbine hub height increases to 80 m (Fig. 2e) and then 150 m (Fig. 2f), there is less of an increase in median WPD compared to mean; up to about 30 and 80 W m^{-2} in the northern half of Australia, and up to about 50 and 120 W m^{-2} (and higher in Tasmania) along the southern part of the country. This scenario implies that the number of hours which show an increase in WPD are about the same as those which show a decrease, however the increase of WPD in those hours which show an increase, is greater than the decrease of WPD in the hours which show a decrease. We infer from this that variability and intermittency of the resource are increasing while the median resource is increasing.

Most maps of the variability of the wind resource use the standard deviation. We do not use the normal standard deviation. In line with our argument that the median is a better metric, being non-parametric, we use the 'robust coefficient of variation' (RCoV) that is the ratio of median deviation about the median to the median. Our results show that the highest RCoV values occur in southwest Tasmania and WA, and in southern South

Anti-coincidence   Null-anti-coincidence

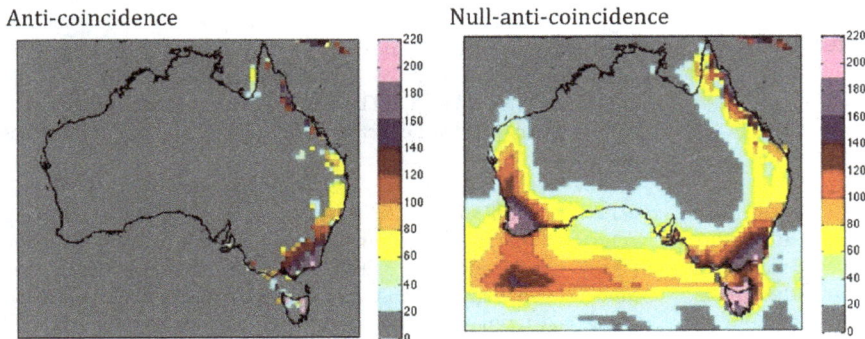

Figure 5. Anticoincidence (left) and Null-anticoincidence (right) of wind power density, at 50 m. Units indicate the number of grid points in a ~1000×1000 km box surrounding the gridpoint in question which are anticoincident to the central gridpoint, which is when the hourly time series of WPD is greater than 200 W m-2 at one of the two points, but not both, for 50% of the total length of the time series.

Australia, but inland from the coastline, which indicates these areas have relatively higher variability compared to the abundance in terms of the median (Fig. 3a). The lowest values, indicating a less variable, more reliable wind resource, occur along the southeastern seaboard and in parts of northern Australia near the coast.

RCoV increases with hub height in some areas (e.g. southeastern Australia), and decreases in others (much of inland SA) (Figs. 3b, S2 (a)). This is because although the median increases with height everywhere, variability decreases in some regions and increases in others. When the median increases with the hub height, and the variability also increases as much or more, RCoV (which is the ratio of deviation to central tendency) also increases. The RCoV decreases when the median increases but the variation does not increase so much (i.e. the ratio decreases). A scenario where RCoV decreases with height indicates that raising the hub height would better harvest the greater wind resources at higher hub heights, with lowered variability and intermittency. With greater surface friction, the standard deviation of the wind in the boundary layer increases [39]. Therefore, the boundary layer roughness predominantly determines the impact that raising the hub height has on the RCoV of the wind resource.

The interquartile range (Fig. 3c) is a measure of an important measure of dispersion in the wind resource since it is immune from the effect of outlying extreme values. Thus it is one of the robust measures of dispersion. As such, it can provide an insight as to the possibility of swings in the wind resource and therefore the amount of backup power that needs to be maintained. At 50 m, the areas that show high IQR (Fig. 3c) tend to coincide with areas that have the highest mean and median WPD (Figs. 2a and 2d, southwest and southern parts of the continent), and increases more with turbine hub height in these areas (Fig. 3d, S2 (b)). The regions that have low mean WPD also have the lowest IQR (e.g. east coast). IQR increases with turbine hub height across the country (Fig. 3d, S2 (b)).

If we consider just abundance and variability, regions that have high WPD and low variability (as shown by IQR) are areas where the wind resource could potentially be harnessed economically. Unfortunately, in Australia, our analysis indicates that at the resolution of this study, the areas which have mean WPD$>$ 200 W m^{-2} also have an IQR of at least the same magnitude if not greater, though undoubtedly there are isolated areas where this would not be the case – but our relatively coarse dataset is unable to show this. However, an additional, very important consideration for harnessing wind power economically at a widespread deployment scale is the extent of its episodic nature - or intermittency.

3.3. Measures of intermittency and the potential for its mitigation

To explicitly gauge the intermittency of WPD, we first consider a metric of unavailability (given as fraction of time WPD is less than a minimum threshold - see Data section). We find that unavailability, which decreases with height, is generally highest in the areas where mean (or median) WPD is low (far northwest Australia, northern Tasmania, and just west of the Great Dividing Range on the eastern seaboard). The lowest values are seen along the eastern seaboard, indicating more reliable winds in these areas. Large areas scattered throughout northern and eastern Australia exhibit relatively high values (above 0.65), with the southwestern third of the country exhibiting moderate values (Fig. 4a). Unavailability decreases with height, as might be expected (WPD increases, so given the 200 W m^{-2} threshold of availability, it also increases), except for the areas which have the lowest mean

WPD values – higher altitude areas along the eastern seaboard - which show a negligible change in unavailability with a change in height (Figs. 4b, S3 (a)).

The availability of WPD as a continuous resource over time is also considered. The spatial pattern of mean episode length (defined as the average time that WPD is continuously above the same threshold) closely resembles that of the mean WPD. We found that the mean episode length at 50 m hub height (c) is lowest in parts of the Great Dividing Range in the east of the country, where WPD is low, and highest in the southern Australia, south-west Western Australia, and Tasmania, where WPD is highest. Mean episode length increases with height most where the mean WPD is lowest, along the Great dividing range in the east (Figs 4d, S3 (b)). Conversely, areas where mean episode length is highest show only small increases ($<$2 hours) with increasing hub height to 80 m (Fig. 4d), and raising the hub height to 150 m results in a near linear response in terms of additional episode length (Fig. S3(b)).

The coincidence (or lack thereof - see Methods) of intermittent wind power in different places sets the scope of installed backup generation capacity required to maintain a steady power supply, as well as the benefits of the aggregation of wind resources. The areas with the lowest unavailability (suggesting low wind intermittency, or more reliable, steady winds) coincide with areas of moderate to high anticoincidence at 50 m, such as along the eastern seaboard. 'Anticoincidence' denotes the occurrence of one event without the simultaneous occurrence of another [35]. The greatest intensity of anticoincident points is in the southeast of the continent, including northeast Tasmania (Fig. 5). However, these areas also have a small episode length (suggesting less persistent winds), which suggests that the aggregation of wind farms may indeed help mitigate wind intermittency in the more densely populated southeast of Australia.

Davy and Coppin (2003) [40] found that the variability in the total wind power output in south east Australia can be reduced to some extent by wider distribution of numerous wind farms, but remains substantial, thus their analysis suggests some degree of anticoincidence of southeastern Australia's wind resource. Their analysis spanned 4 years from March 1999 to March 2003, and also used hourly automatic weather station data from nine sites located on the SE Australian coast. It is useful to note that this period includes a marked La Nina episode and so the wind record may contain some anomalies, and may not be suitable for more general inferences. The record length used in our study, by contrast, is much longer and spans many ENSO cycles, and therefore can be used to infer the mean picture more robustly.

There are areas in Australia with relatively high intermittency - high unavailability and quite low mean episode length - such as northern and northwest Australia, that overlap a vast swathe of the continent west of the Great Dividing Range that shows little anticoincidence of WPD. These are the areas where aggregating turbines would be least effective, at the spatial and temporal scales analyzed.

However, an analysis of the null-anticoincidence (Figure 5) across Australia suggests that there may be some merit in linking wind farms across large areas to increase the reliability of the power supply in areas which show low anticoincidence and moderate to high intermittency, such as parts of the northern QLD coast, inland NSW, and parts of western Victoria and Tasmania, all of which show high values of null-anticoincidence. This may improve the reliability of wind power in these areas. These results agree well with previous research that has shown the coexistence of higher values of anticoincidence with regions that have high topographical inhomogeneity (i.e. mountain ranges) and proximity to the sea. This research has also co-located low anticoincidence areas to low surface roughness (flat terrain), semi-arid climate and terrains, with climate characterized by anticyclones which occur over large areas, leading to a large coincidence of low wind states across these high pressure systems [11].

Summary and Conclusions

Our study suggests that many areas with the strongest widespread wind resource, in terms of both mean and median WPD (SW Western Australia, southern South Australia and Tasmania, and SW Victoria) also score relatively highly on measures of variability (IQR, RCoV) and exhibit moderate levels of intermittency, in terms of reliability (i.e. unavailability) and persistence (mean episode length). Much of the areas which have moderate to high wind intermittency also have very low anticoincidence, as defined in the Methods section, suggesting that there are large expanses of the continent in which aggregating turbines would be less effective, based on our study, at the spatial and temporal scales analyzed (keeping in mind the limitations of this study, described below). These areas also tend to be geographically remote from the bulk of the Australian population on the east coast (certainly in Western Australia, Northern Territory and South Australia), disconnected from the east coast's electricity grid (Western Australia, Northern Territory), and often are not connected or located near enough high capacity electricity infrastructure (parts of South Australia) [41], all of which would decrease the potential economic viability of wind farms in these locations.

However, in eastern Australia (along the Great Dividing Range and the eastern seaboard), many areas exhibit a comparatively poorer wind resource (in terms of the mean and median), and the broad scale mean WPD is below the 200 W m^{-2} cutoff. However, the variability is also lower in these areas, the reliability is better, and the potential to mitigate intermittency (in the form of relatively low persistence) by the aggregation of wind farms, is larger; these areas tend to have higher values of anticoincidence, and null-anticoincidence. Our results broadly agree with those of Davy and Coppin (2003) [40] who demonstrated that variability in the total wind power output in south east Australia can be reduced to some extent by wider distribution of numerous wind farms.

There are several assumptions and limitations of our study which require articulating, the most important being the mapping scale issues that this study raises, whereby coarser resolution maps can overestimate the area available at a given wind speed, and will also potentially fail to depict many areas with good resources which occur at a scale smaller than the resolution our study employs (1/2×2/3 degree, or about 55×73 km square) [8]. Therefore, we acknowledge that our results are at least partly scale and resolution dependent. That being said, the continuous assimilation of observations to run the model enhances the efficacy of the MERRA data, i.e. if there are many sites that have good subgrid scale wind resources, this will be taken into consideration because the observations at these point locations are fed into the data assimilation cycle.

We assumed a neutral boundary layer, as do most of the wind resource assessments, including that by National Renewable Energy Laboratory (NREL) [22]. The wind energy atlas of the United States [22] justifies the neutral boundary condition as a first approximation, because the wind speeds (4–25 m/s) at which much of the power is produced in turbines occur at neutral stability. Parameterization of boundary layer stability into wind

resource estimation is still a much researched area and we are working towards one such improvement.

The temporal resolution of the MERRA dataset is one hour, and as such, sub-hourly wind intermittency cannot be studied, even though this type of shorter scale intermittency can impact the voltage and frequency stability of a power grid [11]. Also, the MERRA data is created from the assimilation of observational data and satellite remote sensed data into a global model, and will reflect any imperfections of the model and the assimilation procedure, and will have an influence on the results presented here.

These limitations notwithstanding, we note that our data and results are not meant to be used for assessments of the deployability of wind farms at individual sites. Our wind resource construction is a tool to understand the geophysical nature of the resource at a regional scale and its variability, and the impact of large-scale atmospheric circulations and phenomena on the resource and its variability.

For this purpose, the multi-decade span of the MERRA data provides a more robust assessment of the temporal characteristics (i.e. mean, median, availability, intermittency, etc.) of wind power than that used in other studies. As described previously, while the data sets that exist have high spatial and temporal resolution, they do not have the record length required to assess the variability of the resource at the regional scale over longer time scales.

On the other hand, the constructed wind resource data described here uses a much longer record length, and this will allow future studies to utilize it to analyze the variability of the resource at different time scales (like the intra-seasonal and ENSO cycle time scales) and in response to different atmospheric oscillations like the El Nino Southern Oscillation and the Madden Julian Oscillation. This data will also be useful for analyzing the economic viability and the levelized costs of wind power compared to other energy sources, as well as for developing strategies for deployment such as the best pattern for aggregation. Studies such as this can conceivably delineate how far intermittency can be mitigated by aggregation and could play a role in the faster deployment of wind farms.

Supporting Information

Figure S1 An example of a histogram of wind power density that shows a typical skewed distribution.

Figure S2 Measures of variation. (a) the change in the RCoV from 50 m to 150 m, (b) the change in the IQR from 50 m to 150 m.

Figure S3 Measures of intermittency. (a) the change in the unavailability from 50 m to 150 m, (b) the change in the mean episode length from 50 m to 150 m.

Text S1 Rational for the cut-off employed to calculate the intermittency metrics.

Author Contributions

Conceived and designed the experiments: UBG AS. Performed the experiments: WH. Analyzed the data: WH UBG. Contributed reagents/materials/analysis tools: WH UBG. Wrote the paper: WH UBG.

References

1. Gentilli J (1971) The World Survey of Climatology. Volume 13: Climates of Australia and New Zealand. Amsterdam: Elsevier. 405 p.

2. Parkinson G (1986) Atlas of Australian Resources: Third series. Volume 4: Climate. Canberra: Division of National Mapping.

3. Mills D (2001) Assessing the potential contribution of renewable energy to electricity supply in Australia. PhD Thesis. Department of Geographical Sciences and Planning, University of Queensland.

4. Dear SJ (1991) Victorian Coastal Wind Atlas. Renewable Energy Authority Victoria and State Electricity Commission of Victoria.

5. Dear SJ, Lyons TJ, Bell MF (1990) Western Australian Wind Atlas. Minerals and Energy Research Institute of Western Australia.

6. Electricity Trust of South Australia (1989) South Australian Wind Energy Program 1984–1988. Technical Report. A joint project of ETSA, SADME and SENRAC. 83 p.

7. Blakers A, Crawford T, Diesendork M, Hill G, Outhred H (1991) The role of wind energy in reducing greenhouse gas emissions. Report R9317, Australian Government Department of Arts, Sport and Environment, Tourism and Territories (DASETT).

8. Coppin PA, Ayotte KA, Steggel N (2003) Wind Resource Assessment in Australia - A Planners Guide. Report by the Wind Energy Research Unit, CSIRO Land and Water. 96 p.

9. Gunturu UB, Schlosser CA (2012) Characterization of wind power resource in the United States. Atmos Chem Phys 12: 9687–9702.

10. Hennessey J Jr (1997) Some Aspects of Wind Power Statistics. Journal of Applied Meteorology 16: 119–128.

11. Gunturu UB, Schlosser CA (2011) Characterization of Wind Power Resource in the United States and its Intermittency. MIT Joint Program Report No. 209. 65 p.

12. Kahn E (1979) The Reliability of Distributed Wind Generators. Electric Power Systems Research 2: 1–14.

13. Archer CL, Jacobson MZ (2007) Supplying baseload power and reducing transmission requirements by interconnecting wind farms. Journal of Applied Meteorology and Climatology 46: 1701–1717.

14. Clean Energy Council (2012) Clean Energy Australia Report. Online report. Available: http://www.cleanenergycouncil.org.au/dms/cec/reports/2013/CleanEnergyAustraliaReport2012_web_final300513/Clean%20Energy%20Australia%20Report%202012.pdf. Accessed 2012 Jul 22.

15. Sinclair Knight Merz (2010) Renewable Resourceful Victoria, The renewable energy potential of Victoria, Part 2 - Energy Resources. Online report. Available: http://www.dpi.vic.gov.au/_data/assets/pdf_file/0006/38841/SKM-DPI-Renewable-Energy-Part2-v5_Part1.pdf. Accessed 2013 Jul 20.

16. Calif R, Schmitt F (2014) Multiscaling and joint multiscaling description of the atmospheric wind speed and the aggregate power output from a wind farm. Nonlinear Processes in Geophysics 21: 379–392.

17. Calif R, Schmitt F, Huang Y (2013) Multifractal description of the wind power fluctuations using arbitrary order Hilbert spectral analysis. Physica A: Statistical Mechanics and Its Applications 392(18): 4106–4120.

18. Oswald J, Raine M, Ashraf-Ball H (2008) Will British weather provide reliable electricity? Energy Policy 36(8): 3212–3225.

19. Mason R (2010) Wind farms produced 'practically no electricity' during Britain's cold snap. The Telegraph. Available: http://www.telegraph.co.uk/finance/newsbysector/energy/6957501/Wind-farms-produced-practically-no-electricity-during-Britains-cold-snap.html. Accessed 2013 Jan 7.

20. Pryor SC, Barthelmie RJ (2011) Assessing climate change impacts on the near-term stability of the wind energy resource over the United States. Proceedings of the National Academy of Sciences 108: 8167–8171.

21. Rienecker MM, Suarez MJ, Gelaro R, Todling R, Bacmeister J, et al. (2011) MERRA – NASA's Modern-Era Retrospective Analysis for Research and Applications, Journal of Climate 24: 3524–3648.

22. Elliott DL, Holladay CG, Barchet WR, Foote HP, Sandusky WF (1987) Wind Energy Resource Atlas of the United States. NASA STI/Recon Technical Report N, 87: 24819.

23. Elliott DL, Wendell LL, Gower GL (1991) An Assessment of the Available Windy Land Area and Wind Energy Potential in the Contiguous United States, Technical Report, Pacific Northwest Laboratories, Richland, WA, USA. 7306–7321.

24. Farkas Z (2011) Considering Air Density in Wind Power Production. arXiv preprint arXiv: 1103.2198. Available: http://arxiv.org/pdf/1103.2198.pdf. Accessed 2014 Apr 21.

25. Schwartz M, George R (1999) On the use of reanalysis data for wind resource assessment. National Renewable Energy Laboratory, Golden, Colorado. Presented at the 11th Applied Climatology Conference, American Meteorological Society, Dallas, Texas. January 10–15, 1999. Available: http://www.nrel.gov/docs/fy99osti/25610.pdf.

26. McElroy MB, Lu X, Nielson CP, Wang Y (2009) Potential for wind-generated electricity in China. Science 325: 1378–1380.

27. Lu X, McElroy MB, Kiviluoma J (2009) Global potential for wind-generated electricity. Proceedings of the National Academy of Sciences 106(27): 10933–10938.

28. Pryor SC, Schoof JT, Barthelmie RJ (2006) Winds of Change? Projections of Near-Surface Winds Under Climate Change Scenarios. Geophysical Research Letters, 33, L11702, doi:10.1029/2006GL026000.

29. Chadee XT, Clarke RM (2014) Large-scale wind energy potential of the Caribbean region using near-surface reanalysis data. Renewable and Sustainable Energy Reviews 30: 45–58.

30. Archer CL, Jacobson MZ (2003) Spatial and temporal distributions of US winds and wind power at 80 m derived from measurements. Journal of Geophysical Research 108. D9: 4289.

31. Archer CL, Jacobson MZ (2005) Evaluation of global wind power. Journal of Geophysical Research 110(D12): D12110.

32. Archer CL, Jacobson MZ (2007) Supplying baseload power and reducing transmission requirements by interconnecting wind farms. Journal of Applied Meteorology and Climatology 46(11): 1701–1717.

33. Clarke D (2013) Wind farms in New South Wales. Available: https://maps.google.com/maps/ms?ie=UTF&msa=0&msid=201184947181960624999.0004b83097a9d77f0c0f4&dg=feature. Accessed 2013 Dec 20.

34. Clarke D (2013) Wind farms in South Australia. Available: http://maps.google.com/maps/ms?ie=UTF&msa=0&msid=201184947181960624999.0004b827f26385521d5b3&dg=feature. Accessed 2013 Dec 20.

35. Wiktionary (2013) Available: http://en.wiktionary.org/wiki/anticoincidence. Accessed 2013 Sept 1.

36. Steggle N, Ayotte K, Davy R, Coppin P (2002) Wind prospecting in Australia with WINDSCAPE. Paris: Proceedings of Global Wind Power Conference.

37. Hurley PJ, Blockley A, Rayner K (2001) Verification of a prognostic meteorological and air pollution model for year-long predictions in the Kwinana industrial region of Western Australia. Atmospheric Environment 35: 1871–1880.

38. International Renewable Energy Agency (2012) IRENA Case Study 2013: Renewable Energy resource Mapping in Australia, International Renewable Energy Agency Report. 4 p.

39. Panofsky HA, Dutton JA (1984) Atmospheric turbulence: Models and methods for engineering applications. Wiley. 397 p.

40. Davy R, Coppin P (2003) South East Australia Wind Power Study. Report by the Wind Energy Research Unit, CSIRO Atmospheric Research. 24 p.

41. Geoscience Australia, ABARE (2010) Canberra: Australian Energy Resource Assessment. 358 p.

10

Direct Observation of Unstained Biological Specimens in Water by the Frequency Transmission Electric-Field Method Using SEM

Toshihiko Ogura*

Biomedical Research Institute, National Institute of Advanced Industrial Science and Technology (AIST), Umezono, Tsukuba, Ibaraki, Japan

Abstract

Scanning electron microscopy (SEM) is a powerful tool for the direct visualization of biological specimens at nanometre-scale resolution. However, images of unstained specimens in water using an atmospheric holder exhibit very poor contrast and heavy radiation damage. Here, we present a new form of microscopy, the frequency transmission electric-field (FTE) method using SEM, that offers low radiation damage and high-contrast observation of unstained biological samples in water. The wet biological specimens are enclosed in two silicon nitride (SiN) films. The metal-coated SiN film is irradiated using a focused modulation electron beam (EB) at a low-accelerating voltage. A measurement terminal under the sample holder detects the electric-field frequency signal, which contains structural information relating to the biological specimens. Our results in very little radiation damage to the sample, and the observation image is similar to the transmission image, depending on the sample volume. Our developed method can easily be utilized for the observation of various biological specimens in water.

Editor: Laurent Kreplak, Dalhousie University, Canada

Funding: This study was supported by KAKENHI Grant-in-Aid for Scientific Research (B) and for Challenging Exploratory Research of the Japan Society for the Promotion of Science. The funders had no role in study design, data collection and analysis, decision to publish, or preparation of the manuscript.

Competing Interests: The author has declared that no competing interests exist.

* E-mail: t-ogura@aist.go.jp

Introduction

Scanning electron microscopy (SEM) is an important technique for producing high-resolution images of biological samples [1–3]. To allow observation under high vacuum conditions and to avoid electrical radiation damage by SEM, the biological specimens must be prepared using glutaraldehyde fixation, negative staining, a cryo technique, and/or a metal coating [4–6]. These preparations also have positive effects in terms of contrast enhancement and allowing the biological specimens to remain uncharged. Until now, atmospheric and/or wet biological specimens have been observed using atmospheric holders [7–10]. However, using these methods, the specimens undergo heavy radiation damage caused by the electron beam (EB) [11–13], while unstained samples give a very poor contrast [9,10]. Therefore, these systems further require glutaraldehyde fixation with negative staining or use of gold labels [9,10]. In transmission electron microscopy (TEM) based system, the biological molecules in a water layer were observed by the environmental sample holder equipped to phase-plate TEM [14]. This system is effective to analyse the biological molecules in water layer.

Recently, we reported SEM-based methods for the analysis of biological specimens with low radiation damage [15–18]. These methods enable high-contrast imaging of atmospheric and/or wet unstained biological samples [15,16]. The high contrast is accomplished through the secondary electron and/or soft X-ray generated from the metal-coated silicon nitride (SiN) thin film. The irradiated electrons are almost scattered and absorbed in the metal-coated thin film, which results in low radiation damage to the unstained biological samples. However, it is difficult to observe specimens in a water layer with a thickness greater than 10 μm using these methods.

Here, we present a newly developed imaging method resulting in no damage to unstained biological specimens in water, on the basis of a frequency transmission electric-field (FTE) system using SEM. Wet biological specimens are enclosed in two SiN films: the upper SiN film is coated with tungsten (W) and nickel (Ni) layers with a thickness of 20 and 5 nm, respectively. The W-Ni-coated SiN film is irradiated using a focused modulation EB at a frequency of 30−60 kHz and a low-accelerating voltage. Irradiated electrons are almost absorbed in the W layer; thus, the negative electric-field potential arises at this position. This negative potential is oscillated at the EB modulation frequency of 30−60 kHz. The electric-field oscillation is transmitted to the lower SiN film through the biological samples in water. Finally, a measurement terminal under the SiN film detects the electric frequency signal, providing structural information relating to the biological specimens. In our method, the biological samples are not directly irradiated via the EB; thus, electron radiation damage is completely avoided.

Results

A schematic of our FTE imaging method using SEM is shown in Figure 1A. The electrostatic deflection system is installed in a thermionic SEM. A control, square wave, signal of 30−60 kHz from a function generator is applied to the deflection plate, producing the chopped EB. The metal-coated SiN film on the sample holder is irradiated using an EB with a low-accelerating voltage of 3−4 kV. A measurement terminal under the sample holder detects the transmission of an electric frequency signal to the biological specimens in water. The detection signal is output to a lock-in amplifier after pre-amplification. After the experimental procedure, FTE images are constructed using the detection signal and scanning signal of the EB from a data recorder.

Unstained and unfixed biological specimens in water are deposited under a metal-coated SiN film and sealed using a further 50-nm-thick SiN film, which is installed in the sample holder (Figure 1B). The sample holder consists of an upper aluminium part and a lower acrylic resin part. The upper part is connected to the system ground (GND); consequently, the metal layer on the SiN film is conducted to the GND. The lower, acrylic resin, part of the holder has a high resistivity; hence, measurement at the lower terminal of the holder is insulated from the metal-coated SiN film. This prevents direct detection of the EB signal through the metal layer on the SiN film. The metal-coated film consists of 20-nm-thick W and 5-nm-thick Ni layers on a 50-nm-

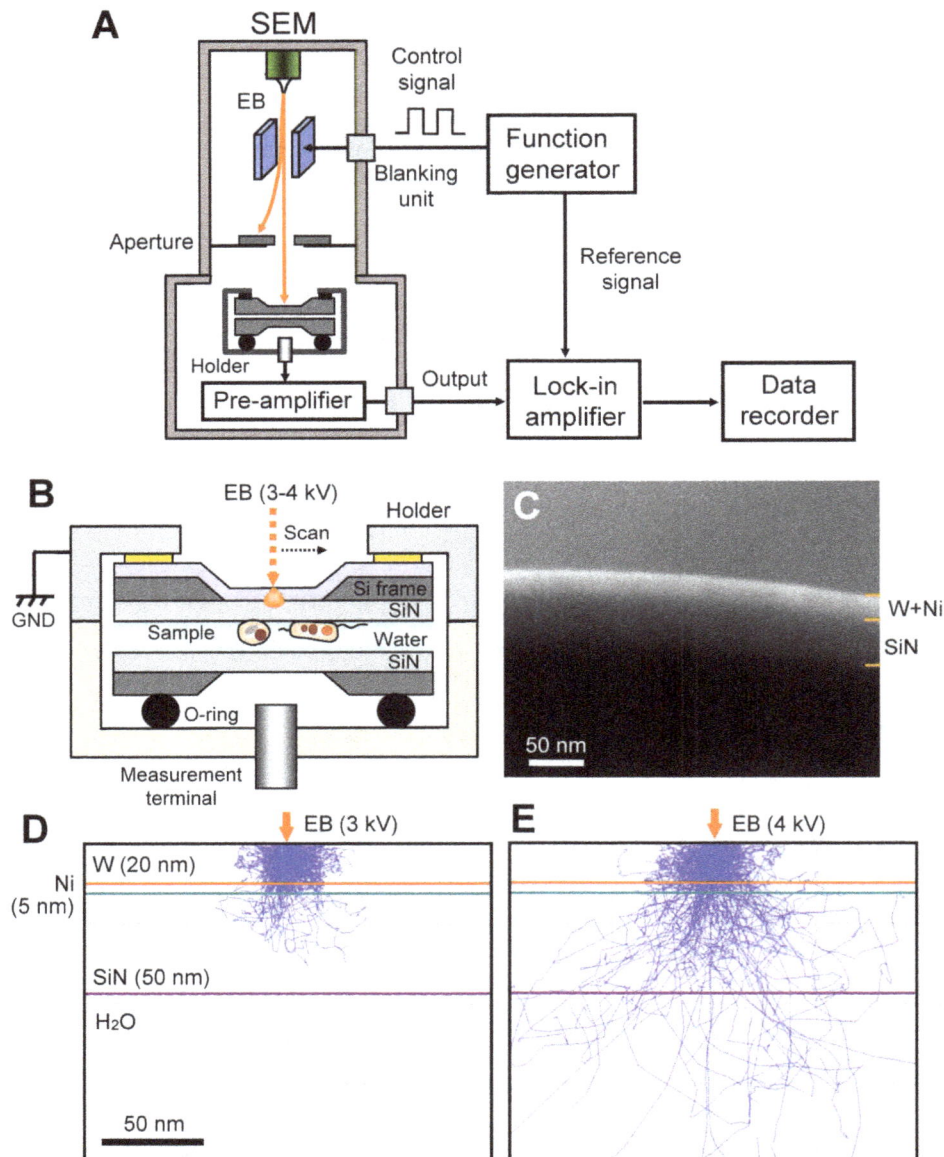

Figure 1. Experimental set-up and data acquisition system. (A) FTE imaging system. The scanning EB irradiates the upper side of the W−Ni-coated SiN film, which is modulated by the beam-blanking unit using a function generator at 30−60 kHz. The output signal from the lock-in amplifier is recorded by a data recorder. (B) Schematic of the atmospheric sample holder. The two SiN films with the liquid sample are sealed by the two sample-holding parts with double-sided tape, and the holding pieces are coupled using screws. The sample holder consists of an upper aluminium part and a lower acrylic resin. The wet biological specimens are enclosed in two SiN films: the upper side of the SiN film is coated with a W and Ni layer and connected to the system GND. (C) SEM cross-sectional image of a metal-coated SiN film consisting of a 20-nm-thick W layer, 5-nm Ni layer, and 50-nm SiN film. (D) MC simulation of the electron trajectories in the W−Ni-coated SiN film at 3-kV EB. The irradiated electrons are completely scattered and absorbed in the film. (E) MC simulation of 4-kV EB. A few electrons reach the water layer. Scale bars (C) and (D) 50 nm.

thick SiN film (Figure 1C). The 5-nm-thick Ni layer acts as adhesion between the W layer and the SiN film.

We calculate the scattered electron distribution in a W–Ni-coated SiN film using Monte Carlo (MC) simulation (Figures 1D and E). For 3-kV EB, the irradiated electrons are strongly scattered and completely absorbed in the 20-nm W layer (Figure 1D). At 4 kV, a few electrons reach the water layer (Figure 1E). Therefore, biological specimens under a W–Ni-coated SiN film are subject to very little damage from the irradiated electrons.

Our detection mechanism is shown in Figure 2. During EB irradiation, the electrons are scattered and absorbed by the W layer; hence, the negative electric potential arises in the irradiated position. Its negative potential influences the electric dipoles of the water molecules and other ions in the sample solution (Figure 2A). The negative potential at the irradiated position orientates the water molecules under the SiN film; thereby its negative potential is propagated to the bottom side of the SiN film. Finally, the potential is detected by a measurement terminal under the SiN film through a pre-amplifier. When the EB is turned off, the negative potential at the EB-irradiated position immediately disappears (Figure 2B). Therefore, water molecules under a SiN film adopt a random orientation, and the detection signal disappears. The detection state is iterated by the square wave of EB. When the biological specimens are irradiated with the EB, the detection signal is attenuated by its specimens (Figure 2C).

We first observe the FTE images of unstained yeasts in a water layer sealed in the gap between two SiN films (Figure 3). At an EB of 3 kV and 30-kHz frequency, the yeasts show a large ellipsoid with a diameter of 40 μm of black contrast (Figure 3A). Using 4-kV EB, the FTE image shows two yeasts with diameters of 10 and 20 μm, respectively, and clear contrast (Figure 3B). To further investigate the FTE image, the EB frequency is varied between 30 and 60 kHz (Figures 3C and D). The cluster yeast image is improved at the 60-kHz frequency.

Next, we image the bubble area in the specimens (Figure 3E). In the observation image, the left side is the air region and the right side is water. The yeast highlighted with the white arrow is located at the boundary of air and water. Interestingly, the edge of the

yeast at the air side is very sharp. On the other hand, the water side is broad. The difference in the edges is clearly identified from a line plot of the yeast centre (Figure 3F). The falling-edge width under air and water conditions is 0.25 μm and 1.53 μm, respectively, which is defined as the distance over which the normalized intensity decreases from 0.75 to 0.25 in the lineout. These differences may represent the different characteristics of the spread at transmission of the frequency electric signal between air and water. At the yeast of the right-lower side in Figure 3E, the edge between air and water is uneven. Because the electric dipoles of the water molecules are influenced by the negative potential oscillation in the chopped-EB-irradiated position, the water edge is probably pulsed according to its electric-field oscillation.

Finally, we measure the yeasts with environmental bacteria (Figure 4). At the 3-kV EB, the various bacteria of the spherical or cylinder shapes are scattered around the yeasts, with sizes of 1−3 μm (Figure 4A). At the 4-kV EB with 4000× magnifications, a tubular bacterium is clearly observed (Figure 4B). To further investigate the bacterium indicated by the black arrow, we create a pseudo-colour in an expanded image (Figure 4C). Some constrictions are visible in the bacterium. Moreover, the image intensity increases at the crossover point of the two bacteria (Figure 4C white arrow). This suggests that the FTE image is similar to the transmission image, depending on the sample volume.

Discussion

Our FTE imaging method enables observation of unstained biological specimens in water at high contrast. Furthermore, the irradiated electrons are scattered and absorbed in the W–Ni-coated SiN film on the biological specimens; hence, our method is capable of undamaged observation. The irradiated EB is modulated by 30−60 kHz chopping and 3−4 kV acceleration (Figure 1). Therefore, the negative potential oscillation arises at the EB-irradiated position in the W–Ni-coated SiN film, which is transmitted to the bottom SiN film through the biological samples in the liquid layer. Finally, the specimen images are observed by the electric frequency signal from the measurement terminal. A similar electric-field detection method was adopted for the medical

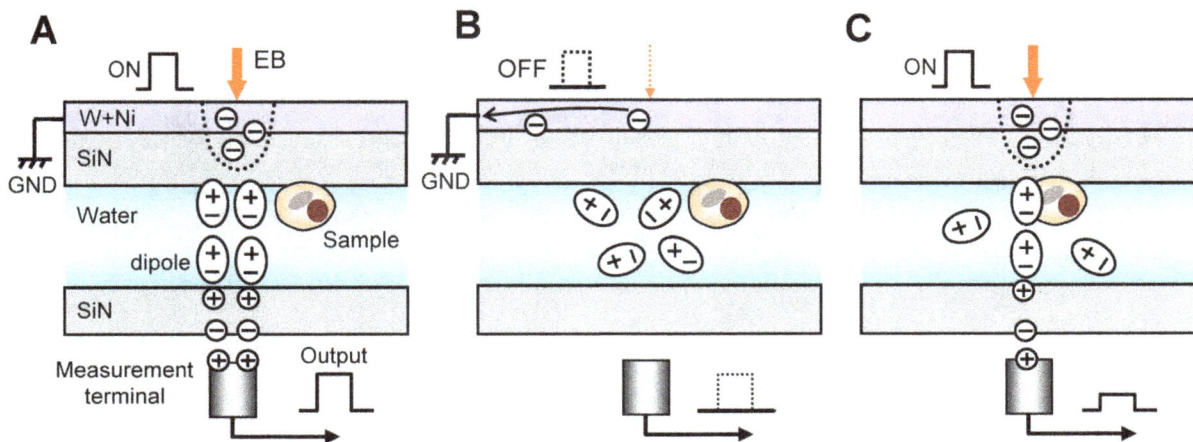

Figure 2. Schematic of our hypothesis of the detection mechanism of the FTE imaging system. (A) If EB irradiates the W–Ni-coated film, the electrons are scattered and absorbed in the film; hence, the negative potential arises in the irradiated position. The negative electric potential influences the electric dipoles of the water molecules and other ions in the sample solution. The negative potential at the irradiated position orientates the water molecules; thereby the negative potential is transmitted to the lower side of the SiN film. (B) When the EB is withdrawn, its negative potential immediately disappears. Therefore, water molecules under the SiN film resort to a random orientation; thereby the detection signal of a measuring terminal is 0 V. Its state is iterated by the square wave of EB. (C) When the EB is irradiated at the biological specimen position, the detection signal is attenuated by the specimens.

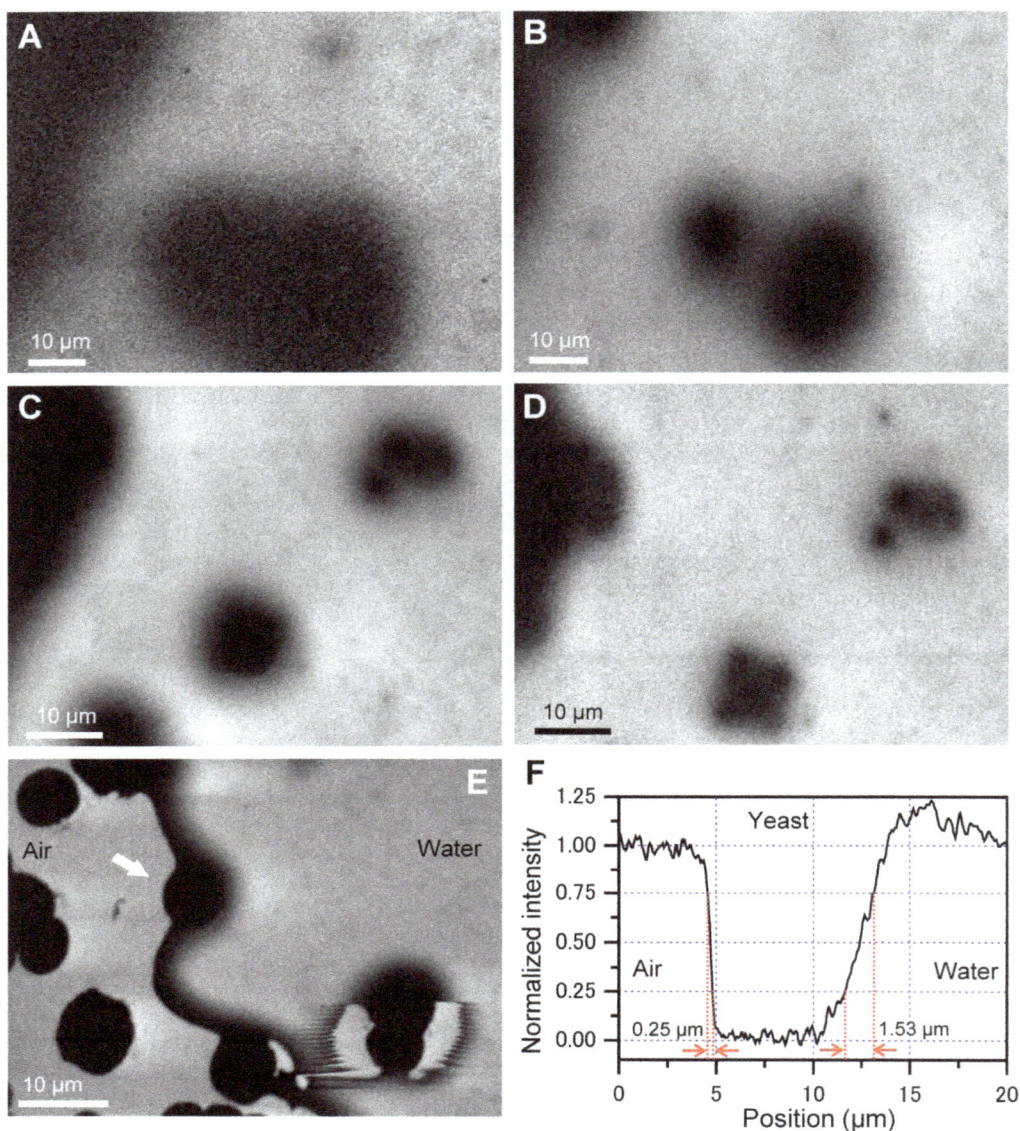

Figure 3. Images of unstained and unfixed yeast in the water. (A) FTE image of yeast obtained under 3-kV EB of 30-kHz chopped frequency and 1500× magnification. In the image, the yeasts show a large ellipsoid with a diameter of 40 μm in clear black contrast. (B) Image of 4-kV EB at 1500× of same area. The yeast image is clearer than that at 3-kV EB. (C) Image of 4-kV EB at 30-kHz frequency and 2000× at another detection area. The image shows two cluster yeasts. (D) Image of the same area of (C) taken at 4-kV and 60-kHz EB. The image clearly shows each yeast in both clusters. (E) A bubble area image, obtained from 4-kV EB of 30-kHz frequency at 2000×. In the image, the left side is the air region and the right side is water. The yeast at the position indicated by the arrow is between air and water. The yeast edge of the air side is sharp, while the water side is broad. (F) Line plot of the centre of the yeast at the white arrow in (E). The falling-edge width under air and water conditions are 0.25 μm and 1.53 μm, respectively. All scale bars are 10 μm.

imaging of the human body, which is called electrical impedance tomography (EIT) [19−23]. Furthermore, the ultrasonic imaging system by SEM (SEAM) was already developed from a focused modulated EB using the deflection plate [24,25]. In our FTE system, we combined both technologies of EIT and SEAM system using the atmospheric sample holder.

We assume that the mechanism of propagation of the electric frequency signal at the water layer is based on the electric dipoles of the water molecules and various ions (Figure 2). At the negative potential in the W–Ni-coated SiN film by the ON state of EB, the electric dipoles are aligned according to the electric potential. Water has a high electric permittivity; therefore, the electric-

potential oscillation in the metal-coated SiN film is propagated to the lower SiN film through the sample solution. On the other hand, the biological specimens consist of amino acids, organic matters, and lipids, with low electric permittivity. Therefore, these specimens decrease the transmission signal of the electric oscillation. The image contrast is influenced by the thickness of the biological specimens, which is similar to the transmission image (Figure 4C). In our method, yeasts with a diameter greater than 10 μm are sealed in the sample holder; therefore, the transmission length of the signal in water is approximately 10 μm. This length is greater than the high-voltage EB of 1 MeV [7,8]. Furthermore, our system is more convenient and inexpensive

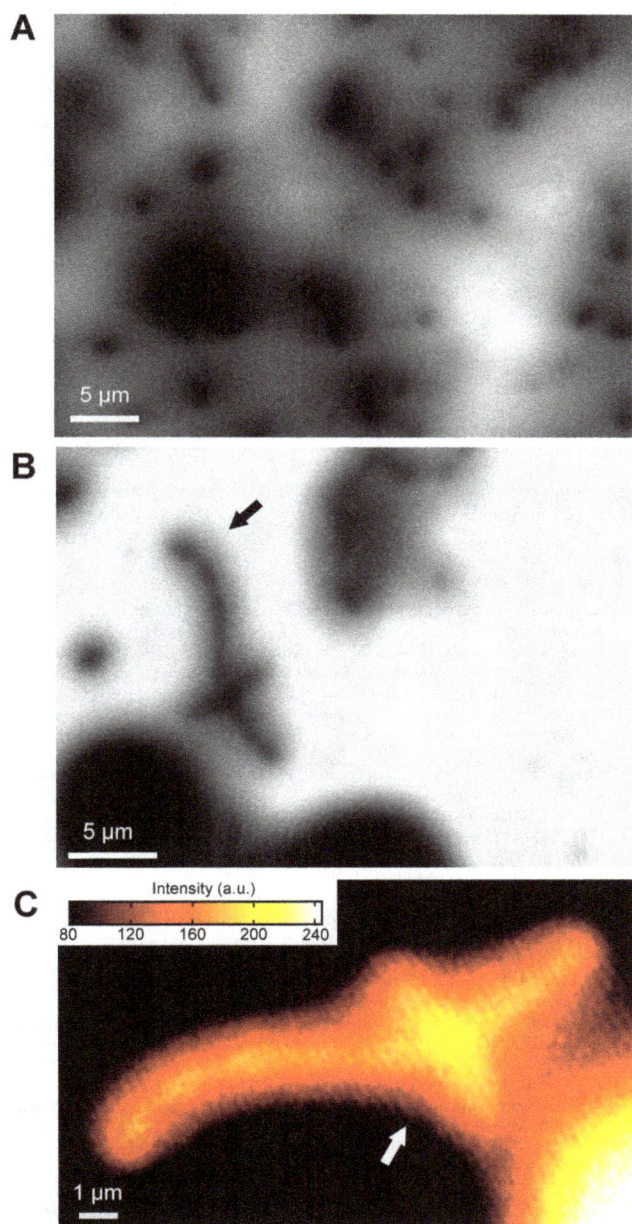

Figure 4. Images of unstained yeast with environmental bacteria in water. (A) At 3-kV EB of 30 kHz with 3000×, the various bacteria of the spherical or cylindrical shapes are scattered around the yeast, with sizes of 1−3 µm. (B) At 4-kV EB with 4000× magnifications, a tubular bacterium was clearly observed. (C) An expanded bacteria image at the reversed intensity of pseudo-colour indicated by the white arrow in (B). The image intensity is increased at the crossover point of two bacteria. Scale bars (A) and (B) 5 µm; (C) 1 µm.

improving the spatial resolution. To reach 10-nm resolution, we must investigate the physical process of the signal transmission mechanism. In the future, we plan to develop a high-resolution system using high-frequency EB chopping and a 1-GHz detection amplifier. Furthermore, this system will be introduced to high-resolution field-emission SEM. We expect our future FTE system to enable the imaging of unstained biological specimens in water with a resolution greater than 10 nm, allowing observation of living viruses and macromolecular proteins.

In conclusion, we have successfully imaged unstained specimens in water using a newly developed FTE system based on SEM. The specimens in the water layer are placed under the W–Ni-coated SiN film, which is sealed with another 50-nm SiN film using a sample holder. The resulting images clearly show unstained and unfixed yeasts and bacteria in water. Our method offers very low radiation damage to the sample, and the resulting image is similar to the transmission image, depending on the sample volume. Our developed method can easily be utilized for the observation of various biological specimens including living bacteria, viruses, and protein complexes in water. Furthermore, our FTE method can be used for diverse liquid samples across a broad range of scientific fields such as the nanoparticles, organic materials, and catalytic materials.

Materials and Methods

Metal coating on the SiN film

A 50-nm-thick SiN film supported by a 0.4 mm × 0.4 mm square window in a Si frame (4×4 mm^2, 0.38-mm thick; Silson Ltd., UK) was coated with W and Ni using a magnetron sputter machine (Model MSP-30T, Vacuum Device Inc., Japan). The Ni sputter conditions were 1.1-Pa Ar pressure, 200-mA current, and 5-sec sputter time. The W sputter conditions were 1.1 Pa, 200 mA, and 20 sec. The distance between the sputter target and SiN films were both 50 mm. The deposited W and Ni layers were 20-nm and 5-nm thick, respectively, as shown in the cross-section image obtained by a field-emission (FE) SEM (JSM-7000F, JEOL, Japan).

Sample preparation

A yeast sample of *Sacharomyces cerevisiae* was obtained from DCL Yeast Ltd. (UK). Active dried yeast (10 mg) was dissolved in 1 ml of solution containing 0.5% (w/v) trehalose (Hayashibara Inc., Japan) and 0.5% NaCl. To ascertain the presence of yeast, a sample solution was observed by optical microscope (400×, Carl Zeiss Axio Observer A1, Germany) before preparation of the sample holder. For the yeast solution with environmental bacteria, the fresh yeasts solution was preserved at room temperature (23 °C) for one week.

Atmospheric sample holder

The developed sample holder maintained the sample solution at atmospheric pressure in the sample space between a W–Ni-coated 50-nm SiN film and a 50-nm SiN film (Figure 1). The two SiN films incorporating the liquid sample were sealed using the two sample-holding parts with double-sided tape, and the holding pieces were coupled using two screws. The sample holder consisted of an upper aluminium part and a lower acrylic resin part. The upper part was connected to GND, allowing conduction to the metal layer on the SiN film. The lower part, made of resin, had a high resistivity; hence, measurement at the terminal underside of the holder was insulated from the metal-coated SiN film.

compared with high-voltage transmission electron microscopy. In our experiments, the biological specimens were attached to the SiN film. Therefore, its movements were not detected. However, when the specimens moved, its images probably are unclear in a present system under 160-sec scanning time. Therefore, our next system has to improve at faster scanning time to detect movements of the biological specimens.

The spatial resolution of our method is currently unsatisfactory at approximately 200 nm. Therefore, we are working on

SEM and FTE imaging system

A beam-blanking unit (Sanyu Electron Co., Japan) consisting of deflection plates was introduced into the thermionic emission SEM (JSM-6390, JEOL, Japan). The unit was controlled by a function generator (WF1974, NF Co., Japan) using a square wave between 0 and 10 V and $30-60$ kHz frequency. The atmospheric sample holder was fixed onto an aluminium stage on the upper side of the W–Ni-coated SiN film. The sample holder with the biological specimens in water was mounted onto the sample stage, and its measurement terminal was connected to the pre-amplifier (Figure 1A). The electric frequency signal from the pre-amplifier was entered into a lock-in amplifier (LI5640, NF Co., Japan). The XY scanning signal and the output of lock-in amplifier were recorded using a data recorder (EZ7510, NF Co., Japan). The data files were transferred to a personal computer (Intel Core i7, 2.8 GHz, Windows 7), and the FTE images were calculated using Matlab R2007b (Math Works Inc., USA). The observation conditions of SEM were captured under the following parameters: $1500-4000\times$ magnifications, 1280×960 pixels, 160-s scanning

time, 7-mm working distance, $3-4$ kV accelerating EB, and $300-700$ pA current.

Monte Carlo simulations

Electron trajectories in the W–Ni-coated SiN film were calculated by MC simulation using CASINO version 2.42 [26]. For the W and Ni layers and SiN film, the density was 19.3 g/cm^3, 8.9 g/cm^3, and 3.1 g/cm^3, and the thickness was 20 nm, 5 nm, and 50 nm, respectively. Physical models used for simulation were the same as in our previous study [15]. The simulation parameters were as follows: 1,000,000 electrons, EB accelerating voltages $3-4$ kV, and EB spot diameter 10 nm. The simulations were performed on a personal computer (Intel Core i7, 2.8 GHz, Windows 7).

Author Contributions

Conceived and designed the experiments: TO. Performed the experiments: TO. Analyzed the data: TO. Contributed reagents/materials/analysis tools: TO. Wrote the paper: TO.

References

1. Duckett JG, Ligrone R (1995) The formation of catenate foliar gemmae and the origin of oil bodies in the liverwort *odontoschisma denudatum* (Mart.) Dum. (Jungermanniales) : a light and electron microscope study. Annals of Botany 76: 405−419.
2. Motta PM, Makabe S, Naguro T, Correr S (1994) Oocyte follicle cells association during development of human ovarian follicle. A study by high resolution scanning and transmission electron microscopy. Arch histol cytol 57: 369−394.
3. Minoura N, Aiba S, Higuchi M, Gotoh Y, Tsukada M, et al. (1995) Attachment and growth of fibroblast cells on silk fibroin. Biochem Biophys Res Commun 208: 511−516.
4. Lamed R, Naimark J, Morgenstern E, Bayer EA (1987) Scanning electron microscopic delineation of bacterial surface topology using cationized ferritin. J Microbiol Meth 7: 233−240.
5. Allan-Wojtas P, Truelstrup Hansen L, Paulson AT (2008) Microstructural studies of probiotic bacteria-loaded alginate microcapsules using standard electron microscopy techniques and anhydrous fixation. LWT Food Sci Technol 41: 101−108.
6. Richards SR, Turner RJ (1984) A Comparative study of techniques for the examination of biofilms by scanning electron microscopy. Water Res 18: 767−773.
7. Nagata F, Ishikawa I (1972) Observation of wet biological materials in a high voltage electron microscope. Jpn J Appl Phys 11: 1239−1244.
8. Parsons DF (1974) Structure of wet specimens in electron microscopy. Science 186: 407−414.
9. Thiberge S, Nechushtan A, Sprinzak D, Gileadi O, Behar V, et al. (2004) Scanning electron microscopy of cells and tissues under fully hydrated conditions. Proc Natl Acad Sci USA 101: 3346−3351.
10. de Jonge N, Peckys DB, Kremers GJ, Piston DW (2009) Electron microscopy of whole cells in liquid with nanometer resolution. Proc Natl Acad Sci USA 106: 2159−2164.
11. Glaeser RM (1971) Limitations to significant information in biological electron microscopy as a result of radiation damage. J Ultrastruct Res 36: 466−482.
12. Henderson R, Glaeser RM (1985) Quantitative analysis of image contrast in electron micrographs of beam-sensitive crystals. Ultramicroscopy 16: 139−150.

13. Egerton RF, Li P, Malac M (2004) Radiation damage in the TEM and SEM. Micron 35: 399−409.
14. Inayoshi Y, Minoda H, Arai Y, Nagayama K (2012) Direct observation of biological molecules in liquid by environmental phase-plate transmission electron microscopy. Micron 43: 1091−1098.
15. Ogura T (2010) Direct observation of unstained wet biological samples by scanning-electron generation X-ray microscopy. Biochem Biophys Res Commun 391: 198−202.
16. Ogura T (2012) Direct observation of the inner structure of unstained atmospheric cells by low-energy electrons. Meas Sci Technol 23: 085402.
17. Ogura T (2008) A high contrast method of unstained biological samples under a thin carbon film by scanning electron microscopy. Biochem Biophys Res Commun 377: 79−84.
18. Ogura T (2012) High-contrast observation of unstained proteins and viruses by scanning electron microscopy. PLoS ONE 7: e46904.
19. Boone K, Lewis AM, Holder DS (1994) Imaging of cortical spreading depression by EIT: implications for localization of epileptic foci. Physiol Meas 15: A189−A198.
20. McEwan A, Cusick G, Holder DS (2007) A review of errors in multi-frequency EIT instrumentation. Physiol Meas 28: S197−S215.
21. Boverman G, Isaacson D, Saulnier GJ, Newell JC (2009) Methods for compensating for variable electrode contact in EIT. IEEE Trans Biomed Eng 56: 2762−2772.
22. Kulkarni R, Kao TJ, Boverman G, Isaacson D, Saulnier GJ, et al. (2009) A two-layered forward model of tissue for electrical impedance tomography. Physiol Meas 30: S19−S34.
23. Nguyen DT, Jin C, Thiagalingam A, McEwan AL (2012) A review on electrical impedance tomography for pulmonary perfusion imaging. Physiol Meas 33: 695−706.
24. Brandis E, Rosencwaig A (1980) Thermal-wave microscopy with electron beams. Appl Phys Lett 37: 98−100.
25. Cargill GS (1980) Ultrasonic imaging in scanning electron microscopy. Nature 286: 691−693.
26. Drouin D, Couture AR, Joly D, Tastet X, Aimez V, et al. (2007) CASINO V2.42 − A fast and easy-to-use modeling tool for scanning electron microscopy and microanalysis users. Scanning 29: 92−101.

11

The Path Integral Formulation of Climate Dynamics

Antonio Navarra[1]*, Joe Tribbia[2], Giovanni Conti[1]

1 Centro Euromediterraneo sui Cambiamenti Climatici, Bologna, Italy, **2** National Center for Atmospheric Research, Boulder, Colorado, United States of America

Abstract

The chaotic nature of the atmospheric dynamics has stimulated the applications of methods and ideas derived from statistical dynamics. For instance, ensemble systems are used to make weather predictions recently extensive, which are designed to sample the phase space around the initial condition. Such an approach has been shown to improve substantially the usefulness of the forecasts since it allows forecasters to issue probabilistic forecasts. These works have modified the dominant paradigm of the interpretation of the evolution of atmospheric flows (and oceanic motions to some extent) attributing more importance to the probability distribution of the variables of interest rather than to a single representation. The ensemble experiments can be considered as crude attempts to estimate the evolution of the probability distribution of the climate variables, which turn out to be the only physical quantity relevant to practice. However, little work has been done on a direct modeling of the probability evolution itself. In this paper it is shown that it is possible to write the evolution of the probability distribution as a functional integral of the same kind introduced by Feynman in quantum mechanics, using some of the methods and results developed in statistical physics. The approach allows obtaining a formal solution to the Fokker-Planck equation corresponding to the Langevin-like equation of motion with noise. The method is very general and provides a framework generalizable to red noise, as well as to delaying differential equations, and even field equations, i.e., partial differential equations with noise, for example, general circulation models with noise. These concepts will be applied to an example taken from a simple ENSO model.

Editor: Jürgen Kurths, Humboldt University, Germany

Funding: Centro Euro-Mediterraneo sui cambiamenti Climatici (CMCC). The funders had no role in study design, data collection and analysis, decision to publish, or preparation of the manuscript.

Competing Interests: The authors have declared that no competing interests exist.

* E-mail: antonio.navarra@cmcc.it

Introduction

The equations that govern the evolution of the atmosphere and the ocean have been known for a long time and have been extensively investigated. To investigate them, several numerical methods that exploit the first order time derivatives to obtain the time evolution, have been intensely developed. The equations showed a strong sensitivity to small perturbations, both in the initial conditions as well as in the parameters defining them, giving rise to the entire field of dynamical chaos [1].

The chaotic nature of the dynamics stimulated the application of methods and ideas derived from statistics and statistical dynamics. For instance, ensemble systems are used to make weather predictions which are designed to sample the phase space around the initial condition. Such an approach has been shown to substantially improve the usefulness of the forecasts since it allows forecasters to issue probabilistic forecasts. The implicit assumption is that the presence of various sources of errors, coupled with the intrinsic sensitivity of the evolution equations to small errors [1], makes a single forecast not so useful [2,3].

The concept has gained a large consensus because it has been shown to be relevant to various dynamical problems. Numerical experiments driven by external forcing, such as those used with prescribed SST (Sea Surface Temperature) or even prescribed concentrations of greenhouses gases in climate change experiments, have shown that the response to external forcing is still sensitive to errors, either because of uncertainties in the initial condition or in the model formulation. Ensemble experiments are now commonly used in these cases [4–6].

These works shifted the dominant paradigm of interpreting the evolution of atmospheric flows (and the ocean to some extent, see [7]) attributing an increasing importance to the probability distribution of the variables of interest rather than to a single representation. The ensemble experiments can be considered as crude attempts to estimate the evolution of the probability distribution of the climate variables, which is the relevant quantity for practice. Other interesting quantities, as variance and correlation functions, can be obtained from the Probability Distribution Function (PDF). The ensemble mean of temperature, for instance, cannot be considered simply as the average of the available ensemble members, but as the simplest estimation of the expectation value.

Finding an equation for the evolution of the PDF is far from trivial. Hasselmann [8] has shown that a stochastic component is consistent with the basic principles of the atmospheric/ocean dynamics and whereas other investigators [9–14] have shown that some aspects of the atmosphere dynamics can be described by simple models with a stochastic component. It is also possible to estimate the stochastic component from observations [15,16].

The addition of stochastic noise to the evolution equation results in a multidimensional Langevin-like equation that can be shown to support a Fokker-Planck equation for the evolution of the probability distribution of the state vector. This result is very interesting since the Fokker-Planck equation is linear, even if the corresponding evolution equation may be non-linear. However, the Fokker-Planck equation is obtained in a phase space with the dimensions corresponding to the number of degrees of freedom of the original equations. Even a very simple general circulation model can easily have hundreds of degrees of freedom and a numerical approach is not feasible.

This paper shows that it is possible to write the evolution of the probability distribution as a functional integral of the same kind introduced by Feynman [17] in quantum mechanics, using some of the methods and results developed in statistical physics [18,19]. The approach allows obtaining a formal solution to the Fokker-Planck equation corresponding to the Langevin-like equation of motion with noise. The method is very general and it provides a framework easily generalizable to red noise, as well as to delay differential equations, and even field equations, i.e. partial differential equations with noise. The approach has been proved useful in fields other than physics, such as polymer theory, chemistry and even financial markets [20–22]. There are also applications to other relevant problems in geosciences : turbulence fluids [23–25], Lyapunov exponents [26], data assimilation [27], or wave propagation in random media [28,29]. The first quantum field theory formalism describing additive noise was developed by Martin, Siggia and Rose [30], by using a different kind of approach, a method similar to the canonical quantization. The path integral technique, however, is relatively less known in the field of Climatology.

In this paper, the authors attempt to solve stochastic differential equations with the Path Integral technique. This method is applied to solve a linear simple model and a non-linear one, relevant to climatological problems, to demonstrate the power of this tool. Although the technique seems involuted, it could be very easily generalized and could also be the basis for applications to field equations arising in a field theory. This method has only been used with simple linear and non-linear ENSO models, which contain only time depending variables. The aim of this paper is to stimulate interest in the path integral technique for application in the investigation of the Global Climate System. The authors' hope is to use the formalism of the field variables to face, with this technique, more complicated models, by applying this method to study general circulation models with noise.

The remainder of this paper is organized as follows. Section 0 introduces and summarizes the general theoretical foundation and Section discusses the calculation of the integrals. Section introduces the concept of Green's matrix and functions. Section introduces a discussion of perturbation expansion applied to non-linear cases. In Section, these concepts are applied to an example taken from a simple ENSO model and Section concludes.

Methods

The Path Integral Formulation

Langevin equation and probability. The systems describing the atmosphere or the ocean can be written as coupled Langevin equations:

$$\dot{q}_\mu(t) = f_\mu(\mathbf{q}(t)) + \epsilon_\mu(t), \qquad (1)$$

where $\mathbf{q}(t) = (q_1(t), \ldots, q_d(t))$ represents a trajectory in $\mathbb{R}^d$ and $f_\mu(\mathbf{q})$ represents a differentiable function of $\mathbf{q}$. It is assumed that there are d degrees of freedom, and in what follows it will be considered a Gaussian white noise $\epsilon_\mu(t)$. This kind of noise is characterized by its 1-point correlation functions, the averages, that are equal to zero, and by the 2-points correlation functions:

$$\langle \epsilon_\mu(t)_\nu(t') \rangle = Q\delta_{\mu\nu}\delta(t - t'). \qquad (2)$$

In the equation above, $\delta_{\mu\nu}$ is the Kronecker delta and Q measures the strength of the correlation. For simplicity, Q is taken as a constant, and the variances of different $\epsilon_\mu(t)$ noise terms are equal. The equations above are not the most general stochastic first order differential equations. Time translation invariance has been explicitly assumed, and the same variance has been used for different variables, but those restrictions are not really limiting and it has been assumed for simplicity [31].

The Langevin equations (1) generate a time-dependent probability density function for a stochastic vector $\mathbf{q}(t)$, given the value of this vector at initial time, which can be written formally as:

$$P(\mathbf{q}, T | \mathbf{q}_0, t_0) = \langle \Pi_{\mu=1}^d \delta(q_\mu(T) - q_\mu) \rangle_\epsilon \qquad \text{with} \qquad T \geq t_0, \qquad (3)$$

in which $\mathbf{q}_0$ and t_0 are the initial conditions, and δ is the Dirac delta. This probability is just the ensemble average over the solutions of the Langevin equations (1); $\langle\rangle_\epsilon$ denotes an average with respect to the probability distribution of the realizations of the stochastic variables $\epsilon_\mu(t)$. $P(\mathbf{q}, T | \mathbf{q}_0, t_0)$ is the conditional probability to find the system in $\mathbf{q}$ at time T starting from the point $\mathbf{q}_0$ at the time t_0. $(q_\mu(T) - q_\mu)$ is the difference between a point of the trajectory obtained with the Langevin equation (1) at the time T, and a fixed point in the configurations space. The trajectory depends on the initial condition $\mathbf{q}_0$, at time t_0. Although $\mathbf{q}_0$ doesn't appear in the right-hand side of the equation, that expression implicitly depends on it by means $q_\mu(T)$.

Using the Gaussian nature of the noise, starting from the equation above, it is possible to write a Fokker-Planck equation for $P(\mathbf{q}, T | \mathbf{q}_0, t_0)$, see for example [31]:

$$\frac{dP(\mathbf{q}, T | \mathbf{q}_0, t_0)}{dT} = \sum_{\mu=1}^d \frac{\partial}{\partial q_\mu}\left(\frac{1}{2}Q\frac{\partial P(\mathbf{q}, T | \mathbf{q}_0, t_0)}{\partial q_\mu} - f_\mu(\mathbf{q})P(\mathbf{q}, T | \mathbf{q}_0, t_0)\right). \qquad (4)$$

The formal solution of this equation can be written as a path integral [19]

$$P(\mathbf{q}, T | \mathbf{q}_0, t_0) = \int_{\mathbf{q}_0 = \mathbf{q}(t_0)}^{\mathbf{q}_T = \mathbf{q}(T)} [\mathcal{D}\mathbf{q}(t)] \exp(-S(\mathbf{q})), \qquad (5)$$

where $[\mathcal{D}\mathbf{q}(t)]$ means that the integration is done over all paths $\mathbf{q}(t)$ that go from t_0 to T. The functional $S(\mathbf{q})$ is the continuous Onsager-Machlup action which in the white noise case

$$S(\mathbf{q}) = \frac{1}{2Q}\int_0^T \left((\dot{q}_\mu - f_\mu(\mathbf{q}))\delta_{\mu\nu}(\dot{q}_\nu - f_\nu(\mathbf{q})) + Q\frac{\partial f_\mu}{\partial q_\mu}\right)dt, \qquad (6)$$

for the last equation summed over repeated index it is used. The extra divergence term in the action is associated with the difficulty of defining the derivative of a stochastic process. These expressions are symbolic and, they have to be defined by a discretization rule. In fact, a functional integral is well-defined only if it is assigned a formal continuos expression and a discretization rule. The process paths, which are solutions of the Langevin equation, are continuous as $\Delta t \to 0$, but they are not differentiable, and the ordinary rules of calculus must be modified to come up with a consistent definition. In the case of a simple additive noise, the pathologies do not show up, but if there is multiplicative noise, it is absolutely necessary to choose an interpretation. In the following, the Stratonovich interpretation will be used as the discretization rule, which allows treating the fields as differentiable, and therefore to use them in the ordinary rules of calculus. In the case of weak additive noise, the divergence term drops simplifying the action:

$$S_{weak}(\mathbf{q}) = \frac{1}{2Q} \int_0^T \left((\dot{q}_\mu - f_\mu(\mathbf{q}))\delta_{\mu\nu}(\dot{q}_\nu - f_\nu(\mathbf{q})) \right) dt. \quad (7)$$

Expectation values for a generic quantity $F(\mathbf{q}(T))$ can be obtained by

$$< F(\mathbf{q}(T)) >$$
$$= \frac{\int \int_{\mathbf{q}_0 = \mathbf{q}(t_0)}^{\mathbf{q}_T = \mathbf{q}(T)} [\mathcal{D}\mathbf{q}(t)] F(\mathbf{q}(T)) \exp(-S(\mathbf{q})) P(\mathbf{q}_0, t_0) d\mathbf{q}_0}{\int \int_{\mathbf{q}_0 = \mathbf{q}(t_0)}^{\mathbf{q}_T = \mathbf{q}(T)} [\mathcal{D}\mathbf{q}(t)] \exp(-S(\mathbf{q})) P(\mathbf{q}_0, t_0) d\mathbf{q}_0}, \quad (8)$$

where $\langle \rangle$ here is just a time average, and $P(\mathbf{q}_0, t_0)$ is the distribution that describes the system at the initial time t_0. Integrating over the initial conditions using $P(\mathbf{q}_0, t_0)$, the average depends only on the point $\mathbf{q}(T)$. The correlation can be obtained by using a polynomial expressions of the $\mathbf{q}$ components on the functional F.

Stochastic equations and path integrals have a mathematical meaning only if it is a discretization is associated to them. One can apply a discretization, for instance denoting the initial and final times by t_0 and T, respectively,

$$\begin{aligned}
\Delta t &= (T - t_0)/N \\
t_n &= t_0 + n\tau \\
q_n &= q(t_n) \\
\epsilon_n &= \epsilon(t_n),
\end{aligned} \quad (9)$$

with $n = 1, \ldots, N$. The probability distribution of the discretized noise is given by

$$b(_n) = (2\pi Q)^{-1/2} \exp\left(-\frac{\epsilon_n^2}{2Q} \right).$$

If the Langevin equation (1) is integrated in an infinitesimal time interval Δt, the discretized equation becomes

$$\mathbf{q}_{n+1} - \mathbf{q}_n = \Delta t \mathbf{f}(\mathbf{q}_n) + \sqrt{\Delta t}\epsilon_n. \quad (10)$$

The conditional probability, that the system will be in the state $\mathbf{q}_{n+1}$ at time t_{n+1} given that it was in $\mathbf{q}_n$ at time t_n, could be defined with the following symbol,

$$p(\mathbf{q}_{n+1}, t_{n+1} | \mathbf{q}_n, t_n) = \int \delta(\mathbf{q}_{n+1} - \mathbf{q}_n - \Delta t \mathbf{f}(\mathbf{q}_n) - \sqrt{\Delta t}\epsilon_n) b(\epsilon_n) d\epsilon_\mathbf{n}$$
$$= \frac{1}{(2\pi Q \Delta t)^{-1/2}} \exp\left[-\frac{(\mathbf{q}_{n+1} - \mathbf{q}_n - \Delta t \mathbf{f}(\mathbf{q}_n))^2}{2Q\Delta t} \right],$$

where δ is the Dirac delta. On the right-hand side of the equation above, the only variables which appear are $\mathbf{q}_{n+1}$, $\mathbf{q}_n$, Δt, therefore it is necessary to always use a notation for which the transition probability depends explicitly on time, for instance t_{n+1} and t_n. This will be more explicit, as it will soon be shown, when the summation in the action is transformed into an integral with extremes depending on the initial and final time of the transition, see Eq. (6). This means that time is a variable, not an index, and it is coherent with the fact that the PDF, which satisfies the Fokker-Planck equation, is time-dependent. In order to obtain the unconditional probability $p(\mathbf{q}_{n+2}, t_{n+2})$, one would have to use the Kolmogorov-Chapman equation,

$$p(\mathbf{q}_{n+2}, t_{n+2} | \mathbf{q}_n, t_n)$$
$$= \int p(\mathbf{q}_{n+2}, t_{n+2} | \mathbf{q}_{n+1}, t_{n+1}) p(\mathbf{q}_{n+1}, t_{n+1} | \mathbf{q}_n, t_n) d\mathbf{q}_{n+1}.$$

the probability for the entire path can be obtained

$$p(\mathbf{q}_N, T | \mathbf{q}_0, t_0)$$
$$= \int_{\mathbf{q}_0 = \mathbf{q}(t_0)}^{\mathbf{q}_N = \mathbf{q}(T)} \frac{d\mathbf{q}_1 \cdots d\mathbf{q}_{N-1}}{(2\pi Q \Delta t)^{N/2}} \exp\left(-\frac{S_N(\mathbf{q}_0, \ldots, \mathbf{q}_N)}{2Q} \right), \quad (11)$$

where it has been defined

$$S_N(\mathbf{q}_0, \ldots, \mathbf{q}_N) = \sum_{n=0}^{N-1} \left(\frac{(\mathbf{q}_{n+1} - \mathbf{q}_n - \Delta t \mathbf{f}(\mathbf{q}_n))^2}{\Delta t} \right). \quad (12)$$

The S_N functional plays the role of the action as in classical mechanics and it is also known as the Onsager-Machlup functional. Probability cannot be exactly analytically computed for a non-linear $\mathbf{f}$, but with a linear $\mathbf{f}$ the integral is Gaussian and can be computed. From the Eq. (11) it is possible to see that $[\mathcal{D}\mathbf{q}(t)] \approx \frac{d\mathbf{q}_1 \cdots d\mathbf{q}_{N-1}}{(2\pi Q \Delta t)^{N/2}}$, these quantities always have to be considered in the limit approximation. There are $N-1$ integrations over the possible intermediate values of the path, and the end points $\mathbf{q}_0$, $\mathbf{q}_N$ are fixed. Note that there are N factors in the denominator of the Eq. (11), $\frac{1}{(2\pi Q \Delta t)^{N/2}}$, and so presumably a normalization factor will have to be introduced later, since they can be divergent when $N \to \infty$. The choice of the discretization is important because the term

$$(\mathbf{q}_n - \mathbf{q}_{n-1})\mathbf{f}(\mathbf{q}_{n-1})$$

is ill-defined when the small time step limit is studied, and it must be treated carefully. It turns out that Feynman's original choice of symmetrizing the term [17] as

$$(\mathbf{q}_n - \mathbf{q}_{n-1})\frac{\mathbf{f}(\mathbf{q}_{n-1}) + \mathbf{f}(\mathbf{q}_n)}{2}$$

is equivalent to choosing the Stratonovich interpretation. Different continuous formal expressions exist for the functional integral, which, with the appropriate discretization rule, define the same stochastic process. The Stratonovich mean point formulation is useful to analytically treat problems. In particular it is connected to the possibility of using the usual techniques of integral calculus. With this kind of discretization, it is possible to define all the terms that in the limits become the action seen before Eq. (6). The discretization, beyond giving meaning to the expressions above, gives a recipe to explicitly compute those quantities.

The propagator.   The probability of reaching $\mathbf{q}_N$ at T from any point $\mathbf{q}_0$ at t_0, obeying the initial distribution $P(\mathbf{q}_0,t_0)$, is then given by

$$
\begin{aligned}
&p(\mathbf{q}_N,T) \\
&= \int_{\mathbf{q}_0 = \mathbf{q}(t_0)}^{\mathbf{q}_N = \mathbf{q}(T)} \frac{d\mathbf{q}_1 \cdots d\mathbf{q}_{N-1}}{(2\pi Q \Delta t)^{N/2}} \\
&\quad \exp\left(-\frac{S_N(\mathbf{q}_0,\ldots,\mathbf{q}_N)}{2Q}\right) P(\mathbf{q}_0,t_0)d\mathbf{q}_0,
\end{aligned}
\tag{13}
$$

which describes the evolution of the probability distribution from time t_0 to time T. It is the solution to the Fokker-Planck equation. The final integration over $\mathbf{q}_0$ resolves the normalization issues previously mentioned and a final result is obtained. It is also possible to write Eq. (13) as

$$p(\mathbf{q}_N,T) = \int k(\mathbf{q}_N,T|\mathbf{q}_0,t_0)P(\mathbf{q}_0,t_0)d\mathbf{q}_0, \tag{14}$$

where a symbol for the kernel has been introduced

$$
\begin{aligned}
&k(\mathbf{q}_N,T|\mathbf{q}_0,t_0) \\
&= \int_{\mathbf{q}_0 = \mathbf{q}(t_0)}^{\mathbf{q}_N = \mathbf{q}(T)} \frac{d\mathbf{q}_1 \cdots d\mathbf{q}_{N-1}}{(2\pi Q \Delta t)^{N/2}} \exp\left(-\frac{S_N(\mathbf{q}_0,\ldots,\mathbf{q}_N)}{2Q}\right),
\end{aligned}
\tag{15}
$$

that propagates the solution from time t_0 to time $t_N = T$; this expression is also known as the propagator. This equation is the analogous of Eq. (5) discretized.

The concept of the path integrals recurring in these formulas is illustrated in Fig. 1. The probability of reaching $\mathbf{q}_N$ starting at $\mathbf{q}_0$ is composed by the sum of all paths that may take all possible intermediate values at intermediate times. Their contribution must be integrated for all possible values. For further details about the path integral, refer to [22].

Considering that:

$$\lim_{N \to +\infty} p(\mathbf{q}_N,T) = P(\mathbf{q},T) \qquad \text{and}$$

$$\lim_{N \to +\infty} k(\mathbf{q}_N,T|\mathbf{q}_0,t_0) = K(\mathbf{q},T|\mathbf{q}_0,t_0),$$

the expression for the probability in the continuous case is given by

$$P(\mathbf{q},T) = \int K(\mathbf{q},T|\mathbf{q}_0,t_0)P(\mathbf{q}_0,t_0)d\mathbf{q}_0, \tag{16}$$

and continuous time propagator from time t_0 to T is

$$K(\mathbf{q},T|\mathbf{q}_0,t_0) = \int_{\mathbf{q}(t_0)}^{\mathbf{q}(T)} [\mathcal{D}\mathbf{q}(t)] \exp\left(-S(\mathbf{q})\right). \tag{17}$$

Eq. (16) is the probability of finding the system in the state $\mathbf{q}$ at time T, given the initial distribution $P(\mathbf{q}_0,t_0)$ at time t_0.

Calculating the Path Integral

Practically, analytically computable path integrals are rare, and they are essentially limited to Gaussian integrals, which, as previously noticed, are obtained when $\mathbf{f}(\mathbf{q})$ is linear. They can be analytically calculated from the discretization previously introduced only in particular cases [22]. It is possible to consider an approximate method for the computation derived from the steepest descent method (or saddle point method) [31]. The path integral is dominated by the minima of the action, which are the trajectories that minimize the action functional. It can be approximated by a series of Gaussian integral, one for each minimum of the action, considering fluctuations around these

Discretized Path Integral

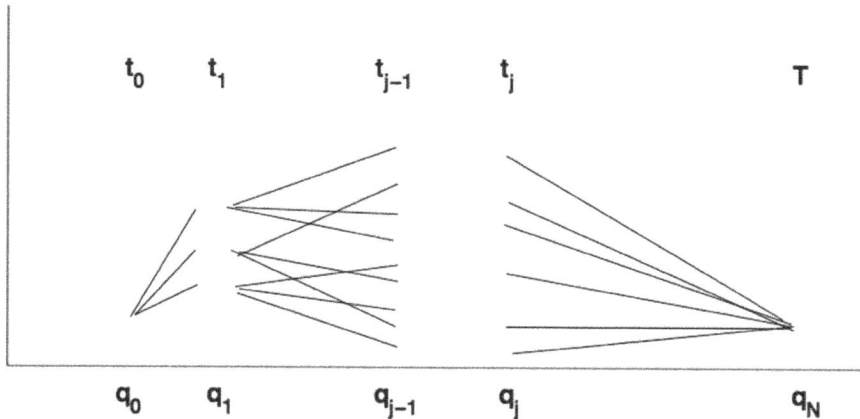

Figure 1. Discretization of the path integral. The initial $\mathbf{q}_0$ and final $\mathbf{q}_N$ variable are not integrated over.

trajectories and computing the approximate integral. In this way the path integral can be separated into two simpler factors; the first one contains the stationary conditions, and the second contains a term that can be transformed, using the projection on eigenfunctions, in Gaussian integrals. If $\mathbf{f}$ is considered as non-linear, this method results useful because it lets us use a simple perturbation expansion technique.

Let the function $\mathbf{f}(\mathbf{q}_n)$ be a linear operator $\mathbf{A}$. In this case the action can be written as

$$S(\mathbf{q}) = \frac{1}{2Q} \int_0^T \left(([\dot{\mathbf{q}} - \mathbf{A}\mathbf{q})^T (\dot{\mathbf{q}} - \mathbf{A}\mathbf{q})) + Q Tr(\mathbf{A}) dt \right) \quad (18)$$

and the path integrals become

$$K(\mathbf{q}, T; \mathbf{q}_0, t_0)$$
$$= \exp\left(-\frac{1}{2} Tr(\mathbf{A}) T \right) \int_{q(0)}^{q(T)} [\mathcal{D}\mathbf{q}(t)] \quad (19)$$
$$\exp\left(-\frac{1}{2Q} \int_0^T (\dot{\mathbf{q}} - \mathbf{A}\mathbf{q})^T (\dot{\mathbf{q}} - \mathbf{A}\mathbf{q}) dt \right).$$

Following the steepest descent method, trajectories that minimize the action must be found. However, there is a problem associated with the fact that, for a system of the present form, there are two solutions to the first order of variations, which correspond to the equation of motion without noise. The solutions correspond to the choice $\mathbf{q} = \mathbf{r}$, so that

$$\dot{\mathbf{r}} = \pm \mathbf{A}\mathbf{r} \qquad \mathbf{r}(0) = \mathbf{q}_0,$$

the unperturbed trajectory corresponds to the plus sign. Obviously it would be desirable to be able to investigate the perturbation around this solution, but this is complicated because the particular value of the action in this case is zero, making a traditional expansion impossible. However, as pointed out by [32], there is a method that allows the expansion along the correct solution and also satisfies both boundary conditions for the integration in the action. It is necessary to introduce a change of variables quantity $\mathbf{q} = \mathbf{r} + \mathbf{g}$, so that the action (18) can be written as

$$S = -\frac{1}{2Q} \int_0^T (\dot{\mathbf{r}} + \dot{\mathbf{g}} - \mathbf{A}(\mathbf{r} + \mathbf{g}))^T (\dot{\mathbf{r}} + \dot{\mathbf{g}} - \mathbf{A}(\mathbf{r} + \mathbf{g})) dt$$
$$= -\frac{1}{2Q} \int_0^T (\dot{\mathbf{g}} - \mathbf{A}\mathbf{g})^T (\dot{\mathbf{g}} - \mathbf{A}\mathbf{g}) dt \quad (20)$$

because $\mathbf{r}$ satisfies the equation of motion without noise. The boundary conditions on this expression are given by

$$\mathbf{g}(0) = 0 \qquad \mathbf{g}(T) = \mathbf{q}_T - \mathbf{r}(T).$$

The measure of the integral does not change, since it is a linear transformation, and $[\mathcal{D}\mathbf{q}(t)]$ is transformed in $[\mathcal{D}\mathbf{g}(t)]$ without the adjoint of a new factor in front of it. One can now substitute around an unperturbed trajectory $\mathbf{g}_c(t)$ so that deviations of order $\sqrt{Q}$ are introduced obeying the boundary conditions $\mathbf{y}(0) = \mathbf{y}(T) = 0$

$$\mathbf{g}(t) = \mathbf{g}_c(t) - \mathbf{y}(t)\sqrt{Q}. \quad (21)$$

Substituting Eq. (21) in the action (20), it is

$$S = \int_0^T (\dot{\mathbf{g}}_c - \mathbf{A}\mathbf{g}_c)^T (\dot{\mathbf{g}}_c - \mathbf{A}\mathbf{g}_c) + 2(\dot{\mathbf{y}} - \mathbf{A}\mathbf{y})^T (\dot{\mathbf{g}}_c - \mathbf{A}\mathbf{g}_c)$$
$$+ (\dot{\mathbf{y}} - \mathbf{A}\mathbf{y})^T (\dot{\mathbf{y}} - \mathbf{A}\mathbf{y}) dt, \quad (22)$$

and integrating by parts the various terms and using the boundary conditions, it is obtained

$$S = -\frac{1}{2Q} \left(\mathbf{g}_c^T \dot{\mathbf{g}}_c + \mathbf{g}_c^T \mathbf{A}\mathbf{g}_c \right)|_0^T -$$

$$\frac{1}{2Q} \int_0^T \mathbf{g}_c^T \left(-\ddot{\mathbf{g}}_c + (\mathbf{A}^T - \mathbf{A})\dot{\mathbf{g}}_c + \mathbf{A}\mathbf{A}^T \mathbf{g}_c \right) dt - \quad (23)$$

$$\frac{1}{2Q} \int_0^T \mathbf{y}^T \left(-\ddot{\mathbf{g}}_c + (\mathbf{A}^T - \mathbf{A})\dot{\mathbf{g}}_c + \mathbf{A}\mathbf{A}^T \mathbf{g}_c \right) dt -$$

$$\frac{1}{2Q} \int_0^T \mathbf{y}^T \left(-\ddot{\mathbf{y}} + (\mathbf{A}^T - \mathbf{A})\dot{\mathbf{y}} + \mathbf{A}\mathbf{A}^T \mathbf{y} \right) dt - .$$

Therefore, if a $\mathbf{g}_c$ is chosen, which satisfies the equation with the given boundary conditions

$$-\ddot{\mathbf{g}}_c + (\mathbf{A}^T - \mathbf{A})\dot{\mathbf{g}}_c + \mathbf{A}\mathbf{A}^T \mathbf{g}_c = 0, \quad (24)$$

the action can be divided into two parts: the explicit terms depending on the boundary conditions and implicitly on the unperturbed solution $\mathbf{r}$, and a term that depends only on the fluctuations $\mathbf{y}$,

$$S = -\frac{1}{2Q} \left(\mathbf{g}_c^T \dot{\mathbf{g}}_c + \mathbf{g}_c^T \mathbf{A}\mathbf{g}_c \right)|_0^T -$$

$$\frac{1}{2Q} \int_0^T \mathbf{y}^T \left(-\ddot{\mathbf{y}} + (\mathbf{A}^T - \mathbf{A})\dot{\mathbf{y}} + \mathbf{A}\mathbf{A}^T \mathbf{y} \right) dt = S_1 + S_2.$$

The term S_1 does not depend on the varying path $\mathbf{y}(t)$ and therefore can be taken out from the integration in Eq. (19), whereas the term S_2 will depend only on time T, which is often called the *prefactor*. The propagator (19) can then be written as

$$K(\mathbf{q}, T; \mathbf{q}_0, t_0) = \exp\left(-\frac{1}{2} Tr(\mathbf{A}) T \right)$$

$$\exp\left(-\frac{S_1}{2Q}\right)\int_{y(0)}^{y(T)}[\mathcal{D}\mathbf{y}(t)]$$

$$\exp\left(-\frac{1}{2Q}\int_0^T \mathbf{y}^T\left(-\ddot{\mathbf{y}}+(\mathbf{A}^T-\mathbf{A})\dot{\mathbf{y}}+\mathbf{A}\mathbf{A}^T\mathbf{y}\right)dt\right) \tag{25}$$

with boundary conditions $y(0)=y(T)=0$. The remaining calculation can be finished by observing that the action in the paths y is then equivalent to a Sturm-Liouville boundary problem for the differential operator Λ

$$\int_0^T \mathbf{y}^T\left(-\ddot{\mathbf{y}}+(\mathbf{A}^T-\mathbf{A})\dot{\mathbf{y}}+\mathbf{A}\mathbf{A}^T\mathbf{y}\right)dt=\int_0^T\left(\mathbf{y}^T\Lambda\mathbf{y}\right)dt \tag{26}$$

The operator Λ is self-adjoint and therefore has a complete orthonormal set of eigenfunctions ϕ_{l_i} with real eigenvalues μ_{l_i}, $l=1,\ldots,\infty$, $i=1,\ldots,d$. The eigenfunction and eigenvalues are d-multiple infinities as a consequence of the dimensionality d of the operator. The variables $\mathbf{y}$ can be expanded in a series of the complete orthonormal eigenfunctions and

$$\frac{1}{2Q}\int_0^T \mathbf{y}^T\Lambda\mathbf{y}\,dt$$

$$=\frac{1}{2Q}\int_0^T\sum_{l=1}^{\infty}\sum_{i=1}^d c_{l_i}\phi_{l_i}\sum_{l=1}^{\infty}\sum_{j=1}^d \mu_{l_j}c_{l_j}\vartheta_{l_j}=\frac{1}{2Q}\sum_{l=1}^{\infty}\sum_{i=1}^d c_{l_i}^2\mu_{l_i} \tag{27}$$

Using this approach, the path integral Eq. (17) can be written as an infinite set of Gaussian integrals over the coefficients of the expansion. A change of variables from the q_i's to the c_i's will allow the execution of the integral. The functional path integral becomes an integral for the coefficients $c_{l_i}^2$, because in varying them, all possible paths are obtained. Since Λ is self-adjoint, it can be diagonalized by a unitary transformation with a unit Jacobian for the change of variables, therefore the path integral measure remains the same, and the boundary conditions are satisfied by the eigenfunctions. The integral is then formed by an infinite number of Gaussian integrals, and it can be obtained that,

$$K(\mathbf{q},T;\mathbf{q}_0,t_0)$$

$$=\lim_{L\to\infty}\exp\left(-\frac{1}{2}Tr(\mathbf{A})T\right) \tag{28}$$

$$\exp\left(-\frac{1}{2Q}S_1\right)\frac{1}{(2Q\pi\Delta\tau)^{Ld/2}}\Pi_{l=1}^{L-1}\Pi_{i=1}^d\left(\frac{2\pi Q}{\mu_{l_i}}\right)^{1/2},$$

or

$$K(\mathbf{q},T;\mathbf{q}_0,t_0)$$

$$=\lim_{L\to\infty}\exp\left(-\frac{1}{2}Tr(\mathbf{A})T\right) \tag{29}$$

$$\exp\left(-\frac{1}{2Q}S_1\right)\frac{1}{2Q\pi(\Delta\tau)^{Ld/2}}\Pi_{l=1}^{L-1}\Pi_{i=1}^d\left(\frac{1}{\mu_{l_i}}\right)^{1/2}.$$

The product is reduced to the inverse root of the determinant of Λ.

This determinant and the constant, which contain the temporal step, are usually regularized considering the ratio between this propagator and the propagator for a free evolution.

Generating Functions

The calculation of the n-points correlation functions, that will be used to compute the correlations in the following examples, is complicated, but it can be simplified by introducing the moment generating functional

$$Z[\mathbf{J}]=\frac{\int[\mathcal{D}\mathbf{q}(t)]\exp\left(-S(\mathbf{q})+\int\mathbf{J}(t)\cdot\mathbf{q}(t)dt\right)}{\int[\mathcal{D}\mathbf{q}(t)]\exp\left(-S(\mathbf{q})\right)}. \tag{30}$$

The functional derivative of the expression above

$$\left(\frac{\delta}{\delta J_\mu(t)}Z[\mathbf{J}]\right)\Big|_{J=0}=\frac{\int[\mathcal{D}\mathbf{q}(t)]q_\mu(t)\exp\left(-S(\mathbf{q})\right)}{\int[\mathcal{D}\mathbf{q}(t)]\exp\left(-S(\mathbf{q})\right)}=<q_\mu(t)> \tag{31}$$

provides the expectation value for the mean. Remember that the functional derivative is defined as follows

$$\frac{\delta}{\delta J(x)}J(y)=\delta(x-y),$$

where δ on the right-hand side is a Dirac delta, while the notation on the left is the usual notation for the functional derivation. The higher order correlations can be obtained by repeating the process:

$$\left(\frac{\delta}{\delta J_\mu(t)}\frac{\delta}{\delta J_\nu(\tau)}Z[\mathbf{J}]\right)\Big|_{\mathbf{J}=0}$$

$$=\frac{\int[\mathcal{D}\mathbf{q}(t)]q_\mu(t)q_\nu(\tau)\exp\left(-S(\mathbf{q})\right)}{\int[\mathcal{D}\mathbf{q}(t)]\exp\left(-S(\mathbf{q})\right)}=<q_\mu(t)q_\nu(\tau)>, \tag{32}$$

and for a generic functional F, it is possible to prove that

$$\left(F[\frac{\delta}{\delta J_\mu(t)}]Z[\mathbf{J}]\right)\Big|_{\mathbf{J}=0}=\frac{\int[\mathcal{D}\mathbf{q}(t)]F[q_\mu(t)]\exp\left(-S(\mathbf{q})\right)}{\int[\mathcal{D}\mathbf{q}(t)]\exp\left(-S(\mathbf{q})\right)}$$

$$=<F[q_\mu(t)]>. \tag{33}$$

The formalism of the derivation operator, appearing within the scope of F, means that one has to substitute the functional derivatives in place of the usual variables on which the operator F is defined as in the Eq. (32), where $q_\mu(t)q_\nu(\tau)$ are substituted with the derivatives $\frac{\delta}{\delta J_\mu(t)}\frac{\delta}{\delta J_\nu(\tau)}$.

The following paragraph shows how it is possible to compute $Z[\mathbf{J}]$ for a general case with non-linear $\mathbf{f}$ in the Langevin equations.

Perturbation Expansions

Feynman diagrams. The path integral formulation adapts itself very naturally to the definition of perturbation corrections of various kinds, for example, it can be used to compute corrections to the probability distribution and to the correlation functions. In fact, because of the general complexity of the action, it will be difficult to know the exact distribution computing the integrals. Although the technique seems involuted, it can be very easily generalized and can be the basis for applications to field equations arising in a field theory.

Consider the propagator for a non-linear evolution, $\dot{\mathbf{q}} - \mathbf{A}\mathbf{q} - \mu\mathbf{f}(\mathbf{q}) = 0$, where μ is a parameter that measures the strength of the non-linear terms,

$$K(\mathbf{q}, T; \mathbf{q}_0, t_0)$$
$$= \int_{\mathbf{q}_0}^{\mathbf{q}T} [\mathcal{D}\mathbf{q}(t)] \qquad (34)$$
$$\exp\left(-\frac{1}{2Q}\int_0^T (\dot{\mathbf{q}} - \mathbf{A}\mathbf{q} - \mu\mathbf{f})^T(\dot{\mathbf{q}} - \mathbf{A}\mathbf{q} - \mu\mathbf{f}(\mathbf{q}))dt\right),$$

that is an extension of (19). The same coordinate transformation described in Sect. (), $\mathbf{q} = \mathbf{r} + \mathbf{g}$, can be introduced so that the action can be written as an extension of Eq. (20)

$$S = -\frac{1}{2Q}\int_0^T (\dot{\mathbf{g}} - \mathbf{A}\mathbf{g} - \mu\mathbf{f}(\mathbf{r}+\mathbf{g}))^T(\dot{\mathbf{g}} - \mathbf{A}\mathbf{g} - \mu\mathbf{f}(\mathbf{r}+\mathbf{g})). \qquad (35)$$

Clearly the new measure $[\mathcal{D}\mathbf{g}(t)]$ also has to be considered.

The quadratic nature of the action creates a potential problem because the expansion of the terms, according to powers of the coupling constant μ, generate terms of the form $\dot{\mathbf{g}}\mathbf{f}(\mathbf{g})$ that couples state variables with derivatives. It is possible to overcome this problem using the Hubbard-Stratonovich transformation [33,34] extended to the multidimensional case, that is a generalization of the identity

$$\exp\left(-\frac{x^2}{2a}\right) = \sqrt{\frac{a}{2\pi}}\int_{-\infty}^{\infty}\exp\left(-\frac{ay^2}{2} - ixy\right)dy,$$

for the functional integrals. If the propagator is considered in its discretized form Eq.(15), and for each integral that appears the identity above is used when the continuos limit is restored, the propagator becomes

$$K(\mathbf{g}, T; \mathbf{g}_0, 0) = \int \mathcal{D}[\mathbf{y}(t)]\int_{\mathbf{g}_0}^{\mathbf{g}T}\mathcal{D}[\mathbf{g}(t)]$$

$$\exp\left(-\int_0^T \frac{Q\mathbf{y}^T\mathbf{y}}{2} - i\mathbf{y}^T(\dot{\mathbf{g}} - \mathbf{A}\mathbf{g} - \mu\mathbf{f}(\mathbf{g}+\mathbf{r})\right) \qquad (36)$$
$$+ \frac{\mu}{2}\partial_i f_i(\mathbf{g}+\mathbf{r}) + Tr(\mathbf{A})dt.$$

The auxiliary functions $\mathbf{y}(t)$ are defined over the entire time axis. This transformation introduces new integrations that can be summarized as $\mathcal{D}[\mathbf{y}(t)]$. The field $\mathbf{f}(t) = -i\mathbf{y}(t)$ can be introduced and the trace of the linear part can be taken from the functional integrals, as it does not depend on the paths, yielding

$$K(\mathbf{g}, T; \mathbf{g}_0, 0) = \exp\left(-\int_0^T Tr(\mathbf{A})dt\right)$$

$$\int \mathcal{D}[\phi(t)]\int_{\mathbf{g}_0}^{\mathbf{g}T}\mathcal{D}[\mathbf{g}(t)]\exp\left[-\int_0^T \frac{Q\phi^*\phi}{2} + \phi^*(\dot{\mathbf{g}} - \mathbf{A}\mathbf{g})dt\right] \times$$

$$\exp\left[\int_0^T \mu\phi^*\mathbf{f}(\mathbf{g}+\mathbf{r}) - \frac{\mu}{2}\partial_i\mathbf{f}(\mathbf{g}+\mathbf{r})\right)\right]dt, \qquad (37)$$

or

$$K_V(\mathbf{g}, T; \mathbf{g}_0, 0)$$
$$= \int \mathcal{D}[\phi(t)]\int_{\mathbf{g}_0}^{\mathbf{g}T}\mathcal{D}[\mathbf{g}(t)]\exp(-S_0)\exp\left(\int_0^T V(\mathbf{g}(t), \phi(t))dt\right). \qquad (38)$$

The subscript V has been added to underscore the dependence of this propagator on the non-linear terms in the second exponential $\exp(V)$, whereas the quadratic terms are contained in S_0. The term $V(t)$ contains higher order terms in $q(t)$ (hence in $\mathbf{g}(t)$ and $\phi(t)$) that reflect the impact of the non-linear interactions. The propagator corresponding to the quadratic part describes the evolution of the system without interaction and therefore can be described as the free evolution of the system. Usually it can be computed exactly:

$$K_0(\mathbf{g}, T; \mathbf{g}_0, 0) = \int \mathcal{D}[\vartheta]\int_{\mathbf{g}_0}^{\mathbf{g}T}\mathcal{D}[\mathbf{g}(t)]\exp(-S_0), \qquad (39)$$

whereas, in the presence of interactions, it is

$$K_V(\mathbf{g}, T; \mathbf{g}_0, 0) =$$
$$\int \mathcal{D}[\phi]\int_{\mathbf{g}_0}^{\mathbf{g}T}\mathcal{D}[\mathbf{g}(t)]\exp(-S_0)\exp\left(\int_0^T V(\mathbf{g}, \phi)dt\right) =$$

$$< \exp\left(\frac{1}{2Q}\int_0^T V(\mathbf{g}, \phi)dt\right) >_0. \qquad (40)$$

In other words, the propagator for the problem is the expected value of the interaction with respect to the probability distribution of the unperturbed, usually linear, problem. In the presence of a small coupling constant μ, the exponential for the interaction can be expanded in series, yielding successive corrections to the free propagator

$$K_V(\mathbf{g}, T; \mathbf{g}_0, 0) = K_0\left(1 + \frac{1}{2Q} < \int_0^T V(\mathbf{g}, \phi)dt >_0 + \right.$$

$$\left. \frac{1}{4Q^2} < \frac{1}{2}\int_0^T\int_0^T V(\mathbf{g}(t), \phi(t))V(\mathbf{g}(t'), \phi(t'))dtdt' >_0 + \dots\right). \qquad (41)$$

These expectation values can be computed using the generating functional Eq.(33).

Perturbation expansion for the correlation functions. It is useful, for the following computations, to define a scalar product as $(\mathbf{x},\mathbf{y}) = \mathbf{x}^* \cdot \mathbf{y}$, where the asterisk indicates Hermitian conjugation, $(.^*) = (\bar{.})^T$. The generating function can also be written for the non-linear case using the transformed action (37). It is convenient to write it using the real vector $\mathbf{J} = (\mathbf{j},\mathbf{k}) = (j_1, j_2, k_3, k_4)$ as the source term, so that

$$Z(\mathbf{J}) =$$
$$\frac{\int \mathcal{D}[\phi(t)] \int_{\mathbf{g}_0}^{\mathbf{g}T} \mathcal{D}[\mathbf{g}(t)] \exp\left(-\int_0^T \frac{1}{2}Q\phi^*\phi + \phi^*(\dot{\mathbf{g}}-A\mathbf{g}) - \mathbf{g}^*\mathbf{j} - \phi^*\mathbf{k}\,dt\right) \exp\left(\int_0^T V(\mathbf{g},\mathbf{r},\phi)dt\right)}{\int \mathcal{D}[\phi(t)] \int_{\mathbf{g}_0}^{\mathbf{g}T} \mathcal{D}[\mathbf{g}(t)] \exp\left(-\int_0^T \frac{1}{2}Q\phi^*\phi + \phi^*(\dot{\mathbf{g}}-A\mathbf{g})dt\right)}, \quad (42)$$

where

$$\exp\left(\int_0^T V(\mathbf{g},\mathbf{r},\phi)dt\right) = \exp\left(\int_0^T \mu\phi^T \mathbf{f}(\mathbf{g}+\mathbf{r}) - \frac{\mu}{2}\partial_i \mathbf{f}(\mathbf{g}+\mathbf{r}))dt\right).$$

For a small coupling constant μ, the exponential in a Taylor series can be expanded to obtain

$$\exp\left(\int V(t)dt\right) = 1 + \mu \int V(t)dt + \frac{\mu^2}{2}\int\int V(t)V(t')dtdt'. \quad (43)$$

When the function of the path V is a polynomial, every term is the expectation value of the terms of the series expansion of the exponential, and each one can be obtained by differentiating the generating function of the free evolution. The series can be formally exponentiated and written for the generating function of the non-linear case

$$Z(\mathbf{J}) = \exp\left(V\left(\frac{\delta}{\delta \mathbf{J}}\right)\right)Z_0(\mathbf{J}), \quad (44)$$

analogously to Eq. (33), that must be normalized by $Z(0)$. The expression for the quadratic generating function can be written as

$$Z_0(\mathbf{J})$$
$$= \frac{\int \mathcal{D}[\phi] \int_{\mathbf{g}_0}^{\mathbf{g}T} \mathcal{D}[\mathbf{g}] \exp\left(-\int_0^T \frac{1}{2}Q\phi^*\phi + \phi^*(\dot{\mathbf{g}}-A\mathbf{g}) - \mathbf{g}^*\vartheta - \mathbf{f}^*\mathbf{k}\,dt\right)}{\int \mathcal{D}[\phi] \int_{\mathbf{g}_0}^{\mathbf{g}T} \mathcal{D}[\mathbf{g}(t)] \exp\left(-\int_0^T \frac{1}{2}Q\phi^*\phi + \phi^*(\dot{\mathbf{g}}-A\mathbf{g})dt\right)}, \quad (45)$$

where a zero subscript has been added to indicate that it is the generating function for the linear evolution. Introducing the vector $\mathbf{u} = (\mathbf{g},\phi)$, it is possible to write

$$Z_0(\mathbf{J}) = \frac{\int_{\mathbf{u}_0}^{\mathbf{u}T} \mathcal{D}[\mathbf{u}(t)] \exp\left(-\int_0^T \frac{1}{2}\mathbf{u}^* \Delta^{-1}\mathbf{u} - \mathbf{u}^*\mathbf{J}\,dt\right)}{\int_{\mathbf{u}_0}^{\mathbf{u}T} \mathcal{D}[\mathbf{u}(t)] \exp\left(-\frac{1}{2}\int_0^T \mathbf{u}^* \Delta^{-1}\mathbf{u}\,dt\right)}, \quad (46)$$

where Δ^{-1} is the Hermitian operator

$$\Delta^{-1} = \begin{pmatrix} 0 & -\partial_t + A^* \\ \partial_t + A & Q \end{pmatrix}. \quad (47)$$

It is possible to obtain an explicit form for $Z_0[J]$ by inserting $\mathbf{u} = \mathbf{u}_c + \mathbf{w}$, with which the numerator becomes:

$$Z_0[\mathbf{J}] = \int_{\mathbf{w}_0}^{\mathbf{w}T} \mathcal{D}[\mathbf{w}(t)] \exp\left(-\int_0^T \frac{1}{2}\mathbf{u}_c^* \Delta^{-1}\mathbf{u}_c + \frac{1}{2}\mathbf{w}^* \Delta^{-1}\mathbf{u}_c \right.$$
$$\left. + \frac{1}{2}\mathbf{u}_c^* \Delta^{-1}\mathbf{w} + \frac{1}{2}\mathbf{w}^* \Delta^{-1}\mathbf{w} - \mathbf{u}_c^*\mathbf{J} - \mathbf{w}^*\mathbf{J}\,dt\right). \quad (48)$$

We can find $\mathbf{u}_c$ so that $\Delta^{-1}\mathbf{u}_c - \mathbf{J} = 0$, and then

$$Z_0[\mathbf{J}] =$$
$$\int_{\mathbf{w}_0}^{\mathbf{w}T} \mathcal{D}[\mathbf{w}(t)] \exp\left(-\int_0^T \frac{1}{2}\mathbf{u}_c^*\mathbf{J} + \mathbf{w}^*\mathbf{J} + \frac{1}{2}\mathbf{w}^* \Delta^{-1}\mathbf{w} - \mathbf{u}_c^*\mathbf{J} - \mathbf{w}^*\mathbf{J}\,dt\right)$$
$$= \int_{\mathbf{u}_0}^{\mathbf{u}T} \mathcal{D}[\mathbf{w}(t)] \exp\left(\int_0^T \frac{1}{2}\mathbf{u}_c^*\mathbf{J}\,dt\right)\exp\left(-\frac{1}{2}\mathbf{w}^* \Delta^{-1}\mathbf{w}\,dt\right). \quad (49)$$

The remaining path integral over $\mathbf{w}(t)$ is eliminated by the normalization, therefore the generating function is given by

$$Z_0[\mathbf{J}] = \exp\left(\int_0^T \frac{1}{2}\mathbf{u}_c^*\mathbf{J}\,dt\right). \quad (50)$$

The solution $\mathbf{u}_c$ can be expressed in terms of the Green's function of the operator Δ^{-1},

$$\mathbf{u}_c(t) = \int_0^T \mathbf{G}(t,t')\mathbf{J}(t')\,dt' \quad (51)$$

and the final form of the generating function is

$$Z_0[\mathbf{J}] = \exp\left(\frac{1}{2}\int_0^T \int_0^T \mathbf{J}^*(t)\mathbf{G}^*(t,t')\mathbf{J}(t')\,dtdt'\right). \quad (52)$$

This is a general expression; in fact, in the linear case a relation formally identical to the one above is obtained.

Results and Discussion

The Case of the ENSO

A simple model of the ENSO system based on the recharge theory was proposed years ago [35]. Following this model, ENSO can be described by a simple linear system

$$\frac{dh}{dt} = -rh - \alpha\mu b_0 \theta$$

$$\frac{d\theta}{dt} = (\gamma\mu b_0 - c)\theta + \gamma h,$$

where θ is the SST anomaly (Sea Surface Temperature) in the West Pacific and h is the depth anomaly of the thermocline in the East Pacific. The parameter μ measures the strength of the

Critical $\mu = 2/3$

(a) Propagator

ENSO system at critical values

(b) Numerical Experiment

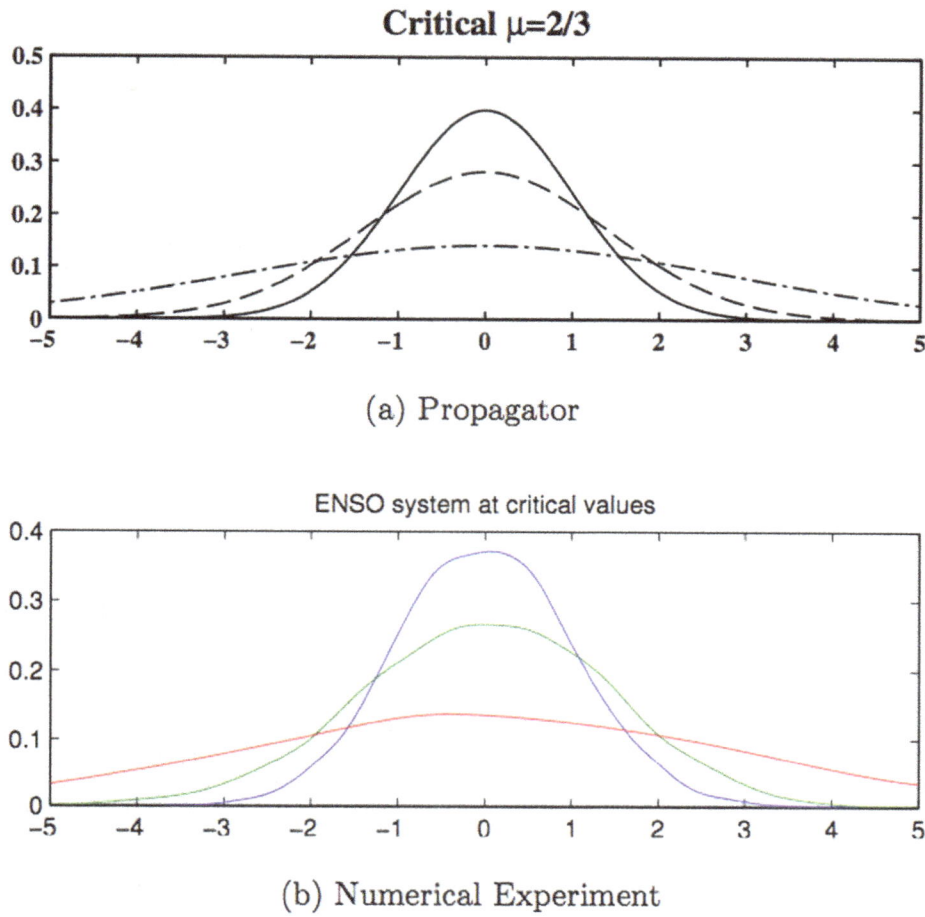

Figure 2. The probability distribution for the subcritical case ($\mu = 2/3$) from the propagator (a) and from 2000 numerical experiment (b). The solid line corresponds to T = 1, the dashed line to T = 2 and the dot-dashed line to T = 8.

interaction between the SST and the wind stress. Introducing the vector $\mathbf{q} = (h, \theta)$, it can be written as

$$\frac{d}{dt}\begin{pmatrix} h \\ \theta \end{pmatrix} = \begin{pmatrix} -r & -\alpha\mu b_0 \\ \gamma & \gamma\mu b_0 - c \end{pmatrix}\begin{pmatrix} h \\ \theta \end{pmatrix}$$

A coordinate transformation of the vector (h, θ) allows transforming the matrix to the standard form

$$\frac{d}{dt}\begin{pmatrix} z_1 \\ z_2 \end{pmatrix} = \begin{pmatrix} \beta & -w \\ w & \beta \end{pmatrix}\begin{pmatrix} z_1 \\ z_2 \end{pmatrix} \qquad (53)$$

The matrix in the equation above is indicated with L. The action for this system is given by

$$S = \frac{1}{2Q}\int_0^T (\dot{\mathbf{z}} - A\mathbf{z})^T (\dot{\mathbf{z}} - A\mathbf{z}) + Tr(A)dt.$$

The solution without noise

$$\mathbf{r}(t) = \begin{pmatrix} e^{\beta t}(z_1 0 \cos(tw) - z_{20}\sin(tw)) \\ e^{\beta t}(z_{20}\cos(tw) + z_1 0 \sin(tw)) \end{pmatrix},$$

around which the action must be expanded, is represented by an exponentially modulated oscillation. The period of the oscillation is w and the time scale of its exponential growth/decay is given by $1/\beta$. The oscillations are damped if $\beta < 0$, neutral oscillations occur for $\beta = 0$ and unbounded oscillations occur in case of $\beta > 0$. The solution of the stationarity equation (24), satisfying the boundary conditions that allow the calculation of the fluctuation prefactor, is given by the function

$$\mathbf{g}_c = \begin{pmatrix} \frac{\sinh(\beta t)\left(z_{20}\sin(tw - 2Tw)e^{\beta T} + z_1 0\cos(w(2T - t))e^{\beta T} - 2z_{1T}\cos(w(T - t)) + z_{2T}\sin(w(T - t))\right)}{\sinh(\beta T)} \\ -\frac{\sinh(\beta t)\left(2z_1 0\sin(tw - 2Tw)e^{\beta T} - z_{20}\cos(w(2T - t))e^{\beta T} + z_{2T}\cos(w(T - t)) + z_{1T}\sin(w(T - t))\right)}{\sinh(\beta T)} \end{pmatrix}$$

and therefore the propagator can be written, based on Eq. (25), as

$$K_0(z_{1T},z_{2T},T|z_{10},z_{20},0) = \frac{\beta e^{\beta T}}{2\pi Q \sinh(\beta T)}$$

$$\exp\left(-\frac{\beta}{2Q\sinh(\beta T)}\left[e^{-\beta T}(z_{10}^2+z_{20}^2)+e^{\beta T}(z_{1T}^2+z_{2T}^2)\right.\right.$$

$$\left.+2\sin(wT)(z_{1T}z_{20}-z_{10}z_{2T})-2\cos(wT)(z_{10}z_{1T}+z_{20}z_{2T})]\right). \quad (54)$$

With the choice of parameters proposed in [35], $c=1, \gamma=0.75, r=0.25, \alpha=0.125, b_0=2.5, \mu=2/3$, the system undergoes stable oscillations, and the entries of the corresponding matrix L are $\beta=0$ and $w=\sqrt{3/32}$. The corresponding propagator can be written as

$$K_c = \frac{1}{2\pi QT}\exp\left(-\frac{1}{2QT}\left(z_{10}^2+z_{1T}^2+z_{2T}^2+z_{20}^2\right.\right.$$

$$\left.-2\cos(Tw)(z_{10}z_{1T}+z_{20}z_{2T})+2\sin(wT)(z_{1T}z_{20}-z_{10}z_{2T}))\right) \quad (55)$$

Fig. 2 shows the probability distribution, obtained for a propagator for an initial probability distribution, that is, a delta function at the origin. It is a Gaussian (the figure shows only the section for $z_{2T}=0$), whose standard deviation increases with time. The system is analogous to a Brownian motion with the particle diffusing in the entire space. The period of the oscillation is close to 20 months and the separate members of the ensemble deviate rapidly as the system evolves. Fig. 3a shows the evolution of the individual members of the ensemble as the oscillation gains larger amplitude. The basic linear oscillation is neutral, so the stochastic fluctuations create the amplification effect, which later will result in the flattening of the probability distribution. For values of μ smaller than the critical value, the oscillation is damped, but the stochastic forcing can counterbalance it, permitting a statistical equilibrium. Fig. 3b shows the time evolution for the damped case, and it is possible to see how the divergence is considerably slowed down. Depending on the magnitude of the stochastic force Q, a different value of μ is necessary for equilibrium.

The probability distribution is correctly estimated by the propagator as it can be seen in Fig. 4. The zeroth order generating function can be obtained from the Green's function as in Eq. (55). The 2-point correlation function is given by the second functional derivative of $Z_0(\mathbf{J})$,

$$<z_1(\tau)z_1(t)> = \left(\frac{\delta}{\delta J_1(\tau)}\frac{\delta}{\delta J_1(t)}Z_0[\mathbf{J}]\right)\Big|_{\mathbf{J}=0} =$$

$$\frac{1}{2}\frac{\delta}{\delta J_1(\tau)}(\int_0^T\int_0^T \delta(t-\tau')G_{11}(\tau',\tau'')J_1(\tau'')d\tau'd\tau'' +$$

$$\int_0^T\int_0^T J_1(\tau')G_{11}(\tau',\tau'')\delta(t-\tau'')d\tau'd\tau'')Z_0[\mathbf{J}]\Big|_{\mathbf{J}=0} =$$

$$\frac{1}{2}\int_0^T \delta(\tau-\tau'')G_{11}(t,\tau'')d\tau''$$

$$+\frac{1}{2}\int_0^T G_{11}(\tau',t)\delta(\tau-\tau')d\tau' = \frac{1}{2}(G_{11}(t,\tau)+G_{11}(\tau,t)). \quad (56)$$

Considering more derivatives, one might also investigate higher order statistics such as the skewness. The Green's function G_{11} for the ENSO model in the transformed coordinates is given by

$$G_{11}(\tau,t) =$$

$$-\frac{Q\cos(tw-w\tau)\left(\frac{\sinh(\beta t)\sinh(\beta\tau)}{e^{\beta T}}-\frac{\sinh(\beta T)\sinh(\beta t)}{e^{\beta\tau}}\right)}{\beta\sinh(\beta T)}$$

$$\Theta_{(\tau-t)}$$

$$-\frac{Q\cos(tw-w\tau)\left(\frac{\sinh(\beta t)\sinh(\beta\tau)}{e^{\beta T}}-\frac{\sinh(\beta T)\sinh(\beta\tau)}{e^{\beta t}}\right)}{\beta\sinh(\beta T)} \quad (57)$$

$$\Theta_{(t-\tau)},$$

where Θ here is the sign function. In this way the standard deviation is given by equal time correlations ($\tau=t$)

$$<z_1(t)z_1(t)> = -\frac{Q\cosh(\beta T-2\beta t)-Q\cosh(\beta T)}{2\beta\sinh(\beta T)}.$$

Considering the evolution for a semi-infinite domain, when T becomes very large, it will be obtained

$$<z_1(t)z_1(t)> = -\frac{Q(e^{-2\beta t}-1)}{2\beta}$$

the equilibrium value

$$<z_1(t)z_1(t)>_{eq} = \frac{Q}{2\beta}.$$

It is interesting to note that the same time correlation does not depend on the oscillating part of the solution and the frequency w does not appear anywhere. The autocorrelation for positive lags $\tau'=\tau-t$ is given by

$$<z_1(t)z_1(t+\tau')> = \frac{Q\cos(\tau'w)(1-e^{-2\beta t})}{2\beta e^{\beta\tau'}},$$

and at the equilibrium value, when $t\to\infty$,

$$<z_1(t)z_1(t+\tau')>_{eq} = \frac{Qe^{-\beta\tau'}\cos(\tau'w)}{2\beta}.$$

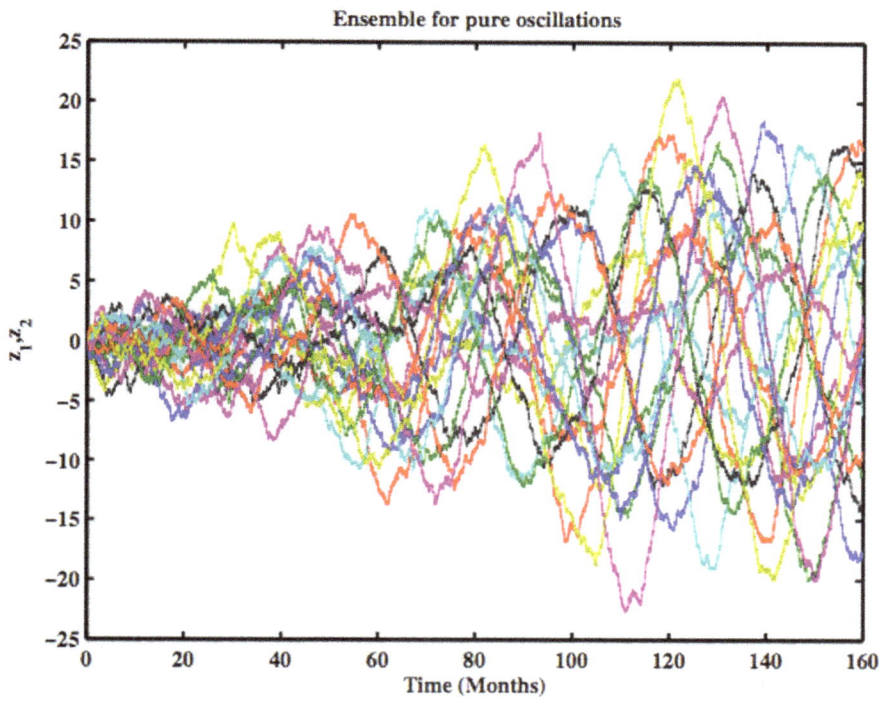

(a) Evolution for $\mu = 2/3$

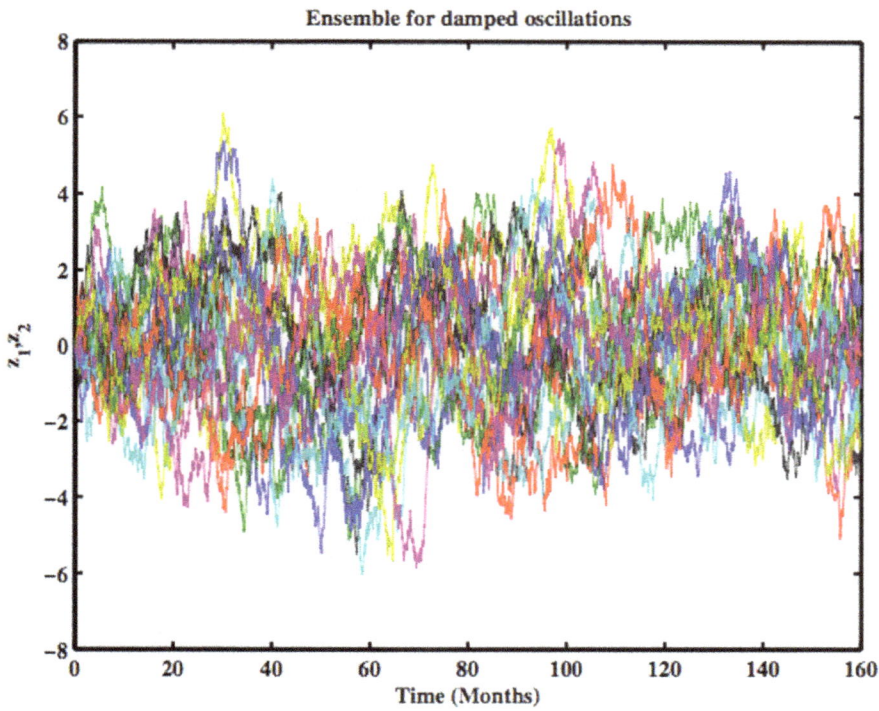

(b) Evolution for $\mu = 1/2$

Figure 3. The time evolution of 10 members for the critical case $\mu = 2/3$ (a) and the subcritical case $\mu = 1/2$ (b).

(a) Probability from Propagator

(b) Probability from Numerical Experiment

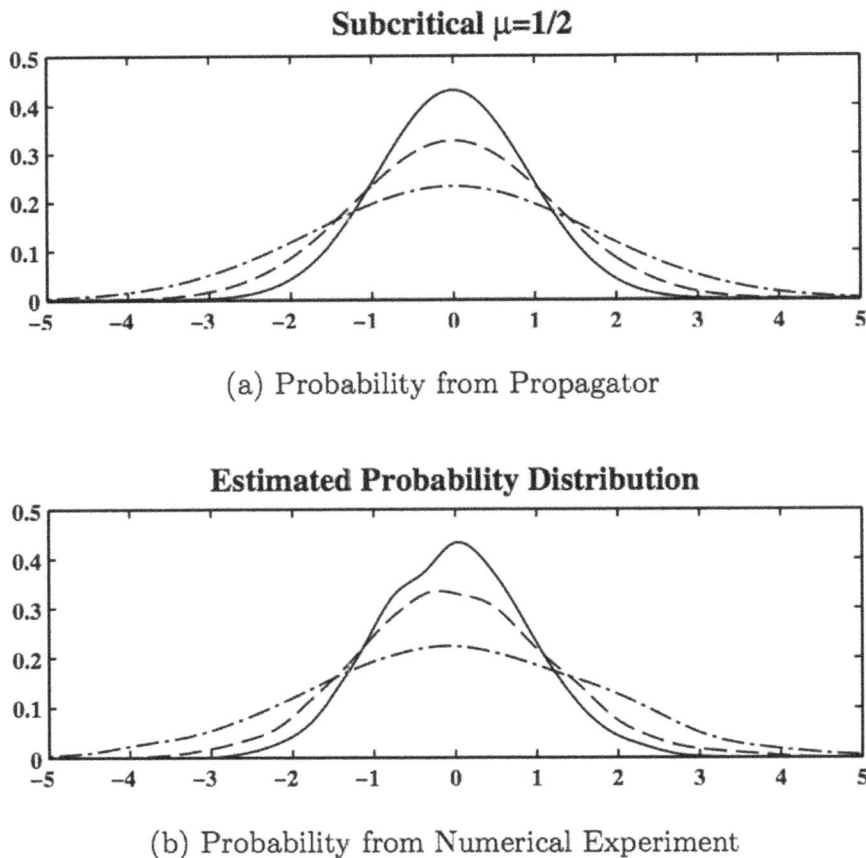

Figure 4. The probability distribution for the subcritical case ($\mu = 1/2$) from the propagator (a) and from 2000 numerical experiment (b). The solid line correspond to $T = 1$, the dashed line to $T = 2$ and the dot-dashed line to $T = 8$.

The cross-correlations in these coordinates are identically zero, but, going back to the (h, θ) coordinates, they will recover the correlations shown in [35].

In the same paper [35], a non-linear extension of the standard model is proposed. The non-linear terms represent the negative feedback of the thermocline, and involve the strength of the coupling between the wind stress and the SST; they are cubic in h and θ. The extra term appears only in the equation for the temperature as

$$-\epsilon (b\theta + h)^3$$

This expression can be used to get the non-linear terms in the action (36) to obtain the perturbation expansion in power of the interaction coefficient ϵ, which corrects the free (linear) propagator in the presence of non-linear terms. The expansion is rather

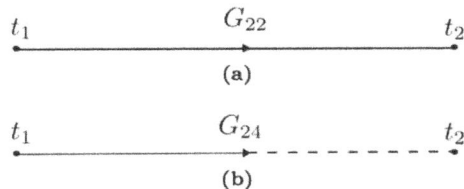

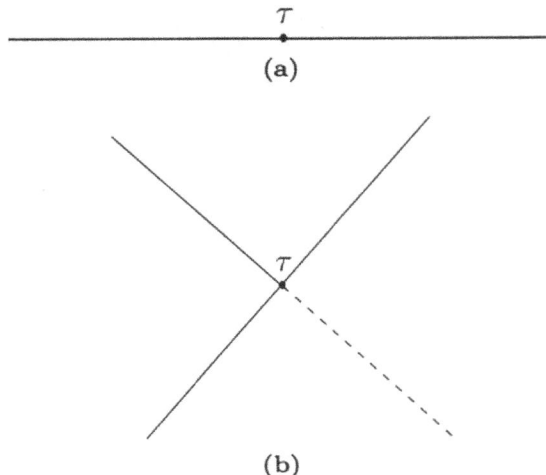

Figure 5. The propagators of the system: (a) the propagator for the variables (z_2, z_2) (b) the propagator for the variables (z_2, ϕ_2). A corresponding propagator can be obtained exchanging 2 and 4.

Figure 6. The internal vertex τ. (a) for the quadratic term z_2^2, (b) for the the quartic term $z_2^3 \phi_2$.

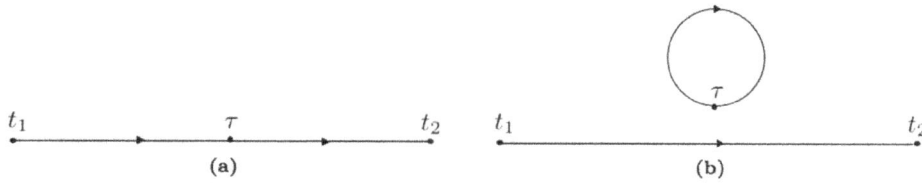

Figure 7. The graphical representation for the expressions (59) and (60).

tedious and, to illustrate the point, the system will be somewhat simplified reducing the non-linear term to a simple form, obtaining a simplified version of the cubic non-linear term in the system (53) that will result in

$$\frac{d}{dt}\begin{pmatrix} z_1 \\ z_2 \end{pmatrix} = \begin{pmatrix} \beta & -w \\ w & \beta \end{pmatrix}\begin{pmatrix} z_1 \\ z_2 \end{pmatrix} + \begin{pmatrix} 0 \\ -z_2^3 \end{pmatrix}, \qquad (58)$$

where ε, function of β, measures the strength of the non-linearity. The action for this system is given by Eq. (37), where $\mathbf{z}$ plays the role of $\mathbf{g}$. The relevant terms in the action are those deriving from $\phi^T \mathbf{f}(\mathbf{z}+\mathbf{r})$ which, in this case, reduce to the interaction terms between ϕ_2 and z_2, $-\varepsilon\beta\phi_2 g_2^3$. There are also terms deriving from the divergence in the action. The interaction terms are therefore given by:

$$V_I(\phi,z) = \varepsilon\beta\left(\frac{3z_2^2}{2} - \phi_2 z_2^3\right).$$

The generating function for these terms is then given by Eq. (44)

$$Z_V(\mathbf{J}) = \exp\left(V_I(\frac{\delta}{\delta\mathbf{J}})\right)Z_0(\mathbf{J}),$$

that can be expanded in power of ε,

where for convenience the numbering $\mathbf{j}=(j_1,j_2)$ and $\mathbf{k}=(k_3,k_4)$ have been introduced. As one can see from Eq. (33), the functional derivatives have to be evaluated at the same time point, τ, and they correspond to the powers of the dynamical variables.

As an example, the correction of the temporal covariance of z_1 will be computed to demonstrate the approach. This covariance is given by the 2-point correlation function, as in Sect. (4),

$$\langle z_1(t_1)z_1(t_2)\rangle = \left(\frac{1}{Z_V[\mathbf{J}]}\frac{\delta}{\delta J_1(t_1)}\frac{\delta}{\delta J_1(t_2)}Z_V[\mathbf{J}]\right)|_{\mathbf{J}=0}.$$

The basic rules of the functional derivation are given by

$$\frac{\delta f(t)}{\delta g(\tau)} = 0 \qquad \frac{\delta f(t)}{\delta f(\tau)} = \delta(t-\tau)$$

and therefore the two derivatives in $Z_V(\mathbf{J})$ will eliminate all terms with less than two j,k, whereas the terms with a larger number of $(\mathbf{j},\mathbf{k})$ will be eliminated by the evaluation at $\mathbf{J}=(\mathbf{j},\mathbf{k})=0$. Due to these mechanisms, the derivative only selects quadratic terms in the expansion of $Z_V(\mathbf{J})$. The other term in the first order expansion will be obtained by taking four derivatives, three with respect to j_2, and one with respect to k_4. There are two terms of this kind

$$j_2 G_{24}k_4 j_2 G_{22}j_2, \qquad k_4 G_{42}j_2 j_2 G_{22}j_2$$

$Z_V(\mathbf{J})$

$$= \left[1 + \int_0^T V_I\left(\frac{\delta}{\delta j_1(\tau)}, \frac{\delta}{\delta j_2(\tau)}, \frac{\delta}{\delta k_3(\tau)}, \frac{\delta}{\delta k_4(\tau)}\right)dt + \dots\right]Z_0(\mathbf{J})|_{\mathbf{J}=0},$$

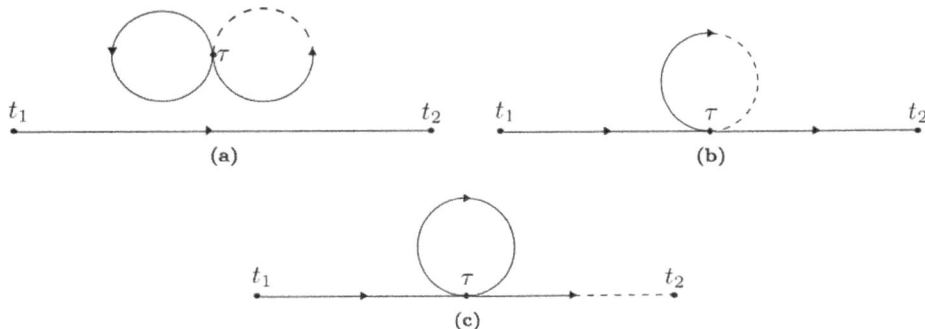

Figure 8. The terms of the perturbation expansion for the 2-point correlation, the variance. The full contribution can be obtained by using symmetry over all the vertices and adding the graphs obtained exchanging 2 with 4: (a) disconnected graph, corresponding to (61), (b) graph with G_{24} integrated over the internal vertex τ, corresponding to (62), (c) graph with G_{24} into an external point corresponding to (63).

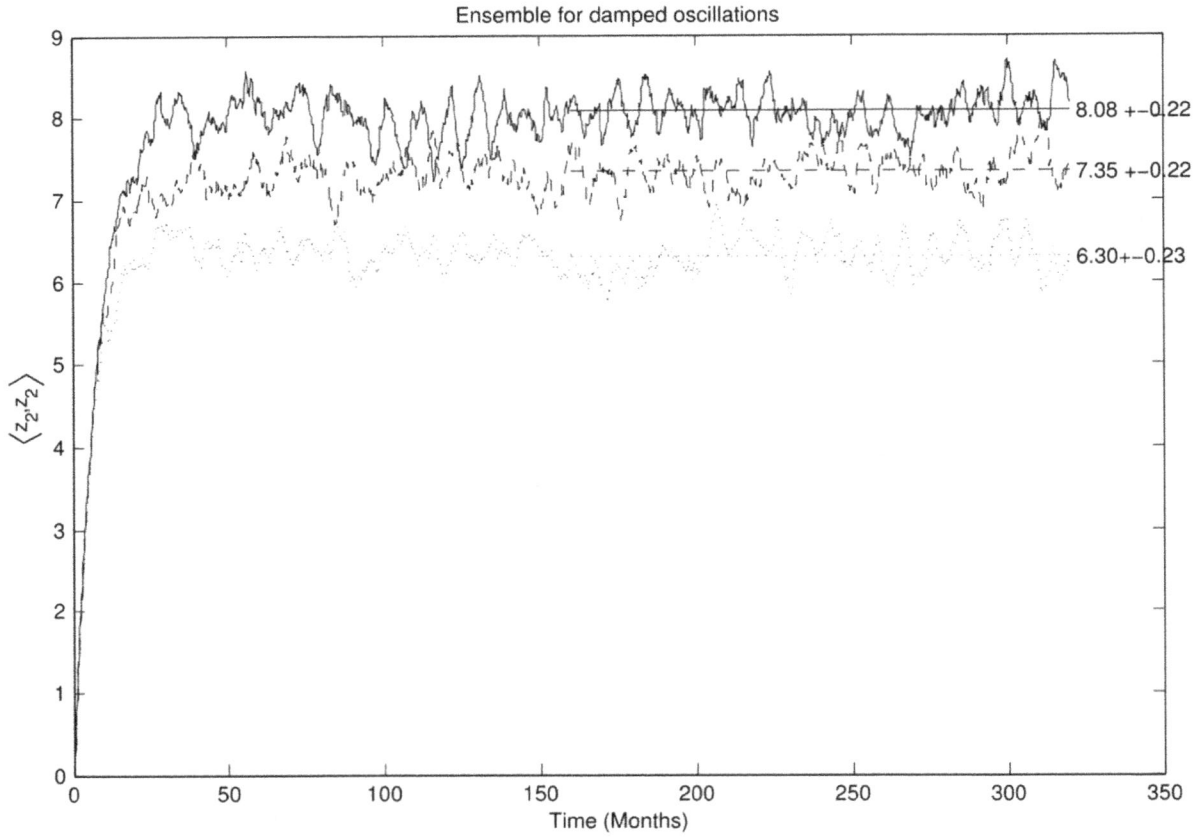

Figure 9. The evolution of the equal time variance $\langle z_2, z_2 \rangle$ for an ensemble of 2000 simulations for the test system. The averaged variance computed after equilibration and its standard deviation is shown to the right of the figure. The solid line represents the linear system, the dashed line is the non-linear system with $\varepsilon = 0.01$ and the dotted line is the non-linear system with $\varepsilon = 0.03$.

The denominator is given by the following expression

$$Z_V[\mathbf{J}]|_{\mathbf{J}=0} =$$

$$1 + \varepsilon \left(\frac{3}{2} \int_0^T G_{22}(\tau,\tau) d\tau + \frac{4!}{8} \int_0^T (G_{24}(\tau,\tau) + G_{42}(\tau,\tau)) G_{22}, (\tau,\tau) d\tau \right).$$

Propagators and interactions can be graphically seen in Fig. 5 and Fig. 6. The numerator is more complicated because now there are two more derivatives. The same arguments used before now lead to the conclusion that only the terms with three Green's functions will survive. The problem is combinatorial and is well known in quantum field theory. It is essentially the same as finding all possible combinations of six points in time: the "external" points, t_1, t_2, and the "internal" points τ that are going to be integrated over. Depending on which of the six j or k the derivatives will operate, different kinds of integrals will be generated. The zero order in ε is simply $G_{22}(t_1, t_2)$, but for the first order we need to count the contribution from V_I. The quadratic term in z_2 will result in

$$M_1 \frac{1}{8} \frac{3}{2} \int_0^T G_{22}(t_1, \tau) G_{22}(\tau, t_2) d\tau \tag{59}$$

$$M_2 \frac{1}{8} \frac{3}{2} \int_0^T G_{22}(t_1, t_2) G_{22}(\tau, \tau) d\tau. \tag{60}$$

The combinatorial analysis indicates that in all there are 4 4! terms given by the four time points; (t_1, t_2, τ, τ) are treated, organized in such a way that $M_1 = 16$ and $M_2 = 8$. More complicated expressions are obtained from the quartic terms. In this case there are three Green's functions involved: G_{22}, G_{24} and G_{42}. Firstly considering the combination with G_{24}, it can be seen that there are $5! * 3 = 360$ terms,

$$M_3 \frac{1}{3!} \frac{1}{8} \int_0^T G_{22}(t_1, t_2) G_{22}(\tau, \tau) G_{24}(\tau, \tau) d\tau \tag{61}$$

$$M_4 \frac{1}{3!} \frac{1}{8} \int_0^T G_{22}(t_1, \tau) G_{24}(\tau, \tau) G_{22}(\tau, t_2) d\tau \tag{62}$$

$$M_5 \frac{1}{3!} \frac{1}{8} \int_0^T G_{22}(t_1, \tau) G_{22}(\tau, \tau) G_{24}(\tau, t_2) d\tau. \tag{63}$$

with $M_3 = 144$, $M_4 = 144$, $M_5 = 72$. Another 360 terms will come from the symmetric terms containing G_{42}. However some simplifications can be obtained because the numerator can be factored to the first order in ε so that the normalization can be completely canceled at the denominator. The $G_{22}(t_1,t_2)$ can be collected to obtain for the numerator,

$$G_{22}(t_1,t_2)(1+\varepsilon\int_0^T \frac{3}{2} G_{22}(\tau,\tau)d\tau + 3\varepsilon\int_0^T G_{22}(\tau,\tau)G_{24}(\tau,\tau)d\tau$$

$$+3\varepsilon\int_0^T G_{22}(\tau,\tau)G_{42}(\tau,\tau)d\tau + other \quad terms \quad in \quad \varepsilon, \quad (64)$$

or at the first order in ε

$$\left(\begin{array}{c} 1+\varepsilon\int_0^T \frac{3}{2}G_{22}(\tau,\tau)d\tau + 3\varepsilon\int_0^T G_{22}(\tau,\tau)G_{24}(\tau,\tau)d\tau \\ +3\varepsilon\int_0^T G_{22}(\tau,\tau)G_{42}(\tau,\tau)d\tau \end{array}\right) \times$$

$$(G_{22}(t_1,t_2)+other \quad terms \quad in \quad \varepsilon). \quad (65)$$

The first parenthesis cancels with the numerator and the final expression for the variance is obtained

$$<z_1(t_1)z_1(t_2)> = G_{22}(t_1,t_2)+other \quad terms \quad in \quad \varepsilon.$$

This is the unperturbed variance corrected by the non-linear terms.

The terms in the perturbation expansion can be expressed with a graphical representation via Feynman diagrams like those in Fig. 5. In this problem there are three kinds of propagators, corresponding to the matrix entries of the Green's matrix. The diagonal entries generate the propagator of the state variable z, and the off diagonal terms, which turn out to be symmetric, generating the propagator connecting the state variable to the auxiliary variables ϕ. The Green's function $G_{22}(t_1,t_2)$ can be graphically expressed with a straight line. On the other hand, the G_{24} propagator can be seen as a dashed-continuos line. The points t_1 and t_2 are the external lines of the graph, the time point τ is recurring twice and is therefore special, because it has two lines that must be connected with the other point.

The quadratic terms Eq. (60) can be graphically written as in Fig. 7. The (b) graph in the figure represents the integral where the $G_{22}(t_1,t_2)$ propagator can be factored out. It is an example of the fact that these kinds of terms show up graphically since they are made up of separate parts. The so-called "disconnected" graph, in this example it is the product of $G_{22}(t_1,t_2)$ and $\int_0^T G_{22}(\tau,\tau)d\tau$.

The terms corresponding to $z_2^3\phi_2$ are more complicated. The internal vertex is of order four and has four lines, which must be connected with two external points. A four line vertex corresponds to the product of two Green's functions, in this case a G_{22} and a G_{24}, because there are only two external lines. The other two lines must be closed on themselves. The graphs are shown in Fig. 8, without showing all the possible symmetries and exchanges that produce all the 720 terms.

The disconnected graphs are the product of the component graphs, therefore the final correction to the variance or 2-point correlation can be written in the form

$$<z_1(t_1)z_1(t_2)> = G_{22}(t_1,t_2)+M_2\frac{\varepsilon}{8}\frac{3}{2}\int_0^T G_{22}(t_1,t_2)G_{22}(\tau,\tau)d\tau +$$

$$\varepsilon M_4\frac{1}{3!}\frac{1}{8}\int_0^T G_{22}(t_1,\tau)G_{24}(\tau,\tau)G_{22}(\tau,t_2)d\tau$$

$$+\varepsilon M_5\frac{1}{3!}\frac{1}{8}\int_0^T G_{22}(t_1,\tau)G_{22}(\tau,\tau)G_{24}(\tau,t_2)d\tau \quad (66)$$

The results are shown in Fig. 9. The figure shows the time evolution of the variance at equal times $t_1=t_2$ of an ensemble of 2000 numerical simulations. The solid line for the linear case concurs with the theoretical value at equilibrium, $Q/2\beta=8$, within the errors. The first order estimate of the non-linear equilibration gives 7.35 and 6.50 for $\varepsilon=0.1$ and $\varepsilon=0.3$ which are also in concordance with the results.

Conclusions

This paper has shown that the path integral formulation and functional methods can be used for stochastic equations derived from the type of equation of motion that are used to describe the atmosphere and the ocean. These equations pose special complications because the evolution equations are first order in time causing an action that introduces coupling terms between the velocity terms and the forcing function.

This problem prevents a straightforward application of the method as in quantum physics, however, it can be treated by a careful consideration of the boundary conditions. Complications in higher than one dimensions can be treated using the Stratonovich-Hubbard transformation. A perturbation expansion can then be designed for non-linear cases based on the calculation of the generating function for the n-points correlation functions and Feynman diagrams can be introduced.

In this paper the path integral technique is applied to solve a linear simple model and a non-linear one, related to the Climate System, to demonstrate of the power of this tool. Although the technique seems involuted, it could be very easily generalized and could also be the basis for applications to field equations arising in a field theory. This method has only been used with linear and non-linear simple ENSO models, which contain only depending on time variables. The aim of this paper is to stimulate interest in the path integral technique to study the Global Climate System. The authors' hope is to use the formalism of the field variables to face, with this technique, more complicated models, such as applying this method to study general circulation models with noise.

Acknowledgments

We would like to thank Gerardo Muñoz and Giorgio Turchetti for their numerous contributions.

Author Contributions

Conceived and designed the experiments: AN JT GC. Performed the experiments: AN JT GC. Analyzed the data: AN JT GC. Contributed reagents/materials/analysis tools: AN JT GC. Wrote the paper: AN JT GC.

References

1. Lorenz EN (1963) Deterministic nonperiodic ow. Journal of the Atmospheric Sciences 20: 130–141.
2. Epstein ES (1969) The role of initial uncertainties in prediction. Journal of Applied Meteorology 8: 190–198.
3. Leith CE (1974) Theoretical skill of monte carlo forecasts. Monthly Weather Review 102: 409–418.
4. Rodwell MJ, Doblas-Reyes FJ (2006) Medium-range, monthly, and seasonal prediction for europe and the use of forecast information. Journal of Climate 19: 6025–6046.
5. Toth Z, Kalnay E (1993) Ensemble forecasting at nmc: The generation of perturbations. Bulletin of the American Meteorological Society 74: 2317–2330.
6. Bodai T, Tel T (2012) Annual variability in a conceptual climate model: Snapshot attractors, hysteresis in extreme events, and climate sensitivity. Chaos 22.
7. Pinardi N, Bonazzi A, Scoccimarro E, Dobricic S, Navarra A, et al. (2008) Very large ensemble ocean forecasting experiment using the grid computing infrastructure. Bulletin of the American Meteorological Society 89: 799–804.
8. Hasselmann K (1976) Stochastic climate models. part i: Theory. Tellus 28: 473–485.
9. Farrell BF, Ioannou PJ (1995) Stochastic dynamics of the midlatitude atmospheric jet. Journal of the Atmospheric Sciences 52: 1642–1656.
10. DelSole T (2001) A simple model for transient eddy momentum uxes in the upper troposphere. Journal of the Atmospheric Sciences 58: 3019–3035.
11. Penland C (2003) Noise out of chaos and why it won't go away. Bulletin of the American Meteorological Society 84: 921–925.
12. Duane G, Tribbia J (2004) Weak Atlantic-Pacific teleconnections as synchronized chaos. Journal of the Atmospheric Sciences 61: 2149–2168.
13. Schneider EK, Fan M (2007) Weather noise forcing of surface climate variability. Journal of the Atmospheric Sciences 64: 3265–3280.
14. Sura P, Newman M (2008) The impact of rapid wind variability upon air–sea thermal coupling. Journal of Climate 21: 621–637.
15. Kravtsov S, Kondrashov D, Ghil M (2005) Multilevel regression modeling of nonlinear processes: Derivation and applications to climatic variability. Journal of Climate 18: 4404–4424.
16. Gritsun A, Branstator G (2007) Climate response using a three-dimensional operator based on the uctuation–dissipation theorem. Journal of the Atmospheric Sciences 64: 2558–2575.
17. Feynman RP, Hibbs A (1965) Quantum Mechanics and Path Integrals. McGraw Hill, 364pp pp.
18. Onsager L, Machlup S (1953) Fluctuations and Irreversible Processes. Physical Review 91: 1505–1512.
19. Haken H (1976) Generalized onsager-machlup function and classes of path integral of the fokkerplanck equation and the master equation. Z Physik 24: 321–326.
20. Risken H (1989) The Fokker-Planck equation. Springer, New York.
21. van Kampen NG (1992) Stochastic Processes in Physics and Chemistry. North-Holland, Amsterdam.
22. Kleinert H (2009) Path integrals in quantum mechanics, statistics, polymer physics and financial markets. World Scientific.
23. Gledzer EB, Monin AS (1974) The method of diagrams in perturbation theory. Russ Math Surv : 117–168.
24. Drummond IT (1982) Path integral methods for turbulent diffusion. J Fluid Mech : 59–68.
25. Thacker WI (1997) A path integral for turbulence in incompressible uids. J Math Phys : 300–320.
26. Gozzi E, Reuter M (1994) Lyapunov exponent, path-integrals and forms. Chaos, Solitons and Fractals : 1117–1139.
27. Quinn JC, Abarbanel HDI (2011) Data assimilation using a gpu accelerated path integral monte carlo approach. J Comput Phys : 8168–8178.
28. Dashen R, Munk WH, Watson KM, Zachariasen FSME F (1979) Transmission through a Fluctuating Ocean. Cambridge Univ. Press.
29. Klyatskin VI (2011) Lectures on Dynamics of Stochastic Systems. Elsevier.
30. Martin PC, Siggia ED, Rose HA (1973) Statistical dynamics of classical systems. Phys Rev A 8: 423–437.
31. Zinn-Justin J (1993) Quantum field theory and critical phenomena. Oxford University Press, 2nd ed. edition, xxiv, 996 p. : pp.
32. Muñoz G (1993) On path integrals and stationary probability distributions for stochastic systems. J Phys A: Math Gen 26: 4475–4478.
33. Lau AWC, Lubensky TC (2007) State-dependent diffusion: thermodynamic consistency and its path integral formulation. Phys Rev E 76: 11123–11140.
34. Arenas ZG, Barci DG (2010) Functional integral approach for multiplicative stochastic processes. Phys Rev E 81: 051113.
35. Jin FF (1997) An equatorial ocean recharge paradigm for enso. part i: Conceptual model. J Atmosph Sci 54: 811–829.

12

Long-Term Prediction of the Arctic Ionospheric TEC Based on Time-Varying Periodograms

Jingbin Liu[1]*, Ruizhi Chen[2], Zemin Wang[3], Jiachun An[3], Juha Hyyppä[1]

1 Department of remote sensing and photogrammetry, Finnish Geodetic Institute, Masala, Finland, 2 Conrad Blucher Institute for surveying & science, Texas A & M University Corpus Christi, Corpus Christi, United States of America, 3 Chinese Antarctic center of surveying and mapping, Wuhan University, Wuhan, China

Abstract

Knowledge of the polar ionospheric total electron content (TEC) and its future variations is of scientific and engineering relevance. In this study, a new method is developed to predict Arctic mean TEC on the scale of a solar cycle using previous data covering 14 years. The Arctic TEC is derived from global positioning system measurements using the spherical cap harmonic analysis mapping method. The study indicates that the variability of the Arctic TEC results in highly time-varying periodograms, which are utilized for prediction in the proposed method. The TEC time series is divided into two components of periodic oscillations and the average TEC. The newly developed method of TEC prediction is based on an extrapolation method that requires no input of physical observations of the time interval of prediction, and it is performed in both temporally backward and forward directions by summing the extrapolation of the two components. The backward prediction indicates that the Arctic TEC variability includes a 9 years period for the study duration, in addition to the well-established periods. The long-term prediction has an uncertainty of 4.8–5.6 TECU for different period sets.

Editor: Juan A. Añel, University of Oxford, United Kingdom

Funding: This work was supported in part by the Finnish Centre of Excellence in Laser Scanning Research, which is funded under project 272195 (JH) by the Academy of Finland (http://www.aka.fi). This work was also supported in part by the National Natural Science Foundation of China (grants 41174029(ZW), 41204028(JA), and 41231064(ZW))(http://www.nsfc.gov.cn). The funders had no role in study design, data collection and analysis, decision to publish, or preparation of the manuscript.

Competing Interests: The authors have declared that no competing interests exist.

* Email: jingbin.liu@fgi.fi

Introduction

As a result of climate change, the industrial and political importance of the Arctic area is growing significantly, and human activities are currently increasing in the Arctic region, including marine, terrestrial and space domains. In the Earth's ionosphere circulation, the polar ionosphere is located at the frontline of those areas responding to variations in the solar-terrestrial physical system because the polar ionosphere is directly connected to the interplanetary space and the Sun. The Total Electron Content (TEC) is an important parameter of the Earth's ionosphere. More detailed knowledge, modeling and predictions of Arctic TEC variability are of fundamental relevance in both engineering and science. Monitoring and predicting the Earth's ionosphere are among major tasks of the fields of solar-terrestrial physics and space weather [1–4]. In engineering fields, long-term predictions of the ionosphere over the scale of a decade can aid in evaluations of ionospheric effects on numerous radio navigation and communication systems as the ionosphere, which responses to solar activities, influences the technical systems in various ways: posing hazards to satellites, disrupting power-grids, causing blackouts in radio and telecommunication systems, even affecting the astronauts in space; predictions of ionosphere and other space weather are also required for guaranteeing effective operation, planning, and risk management of satellite and space exploration missions on time scales ranging from days to weeks to a solar cycle [5–6] because all of satellites and spacecraft are sensitive at some

level to ionosphere and solar cycle effects [3,7–8]. These kinds of space weather predictions will continue in research and operational settings in future, and the need for these predictions has moved from the science community to a global space weather user support system [8]. As an increasing number of countries are planning and implementing their satellite and space exploration missions, there is the need for any country with assets in space to monitor and predict space weather, including ionosphere condition and solar cycle, to protect their satellites and technology.

With the efforts of the International GNSS (Global Navigation Satellite Systems) Service (IGS) and geophysical research communities over the past two decades, the Global Positioning System (GPS) has become an endorsed ionosphere observation tool due to its ability to continuously observe the Earth's ionosphere over large spatial scales [9–13]. The IGS Ionosphere Working Group has constructed databases of GPS observables and TEC products derived from a continuously operating global network of ground-based GPS receivers [11,14]. Based on these long-term GPS TEC products, ionosphere climatology has recently been investigated on regional and global scales, as in the works [15–17]. In these studies, the time evolution of periodograms of regional and global TEC was reported, and empirical models of the ionospheric TEC were correspondingly constructed using input solar and geophysical indices, including Extreme Ultraviolet (EUV) irradiance, the 10.7 cm solar radio flux ($F10.7$) index and the geomagnetic activity index. Previously an empirical model can reconstruct past ionospheric TEC data using past solar and geophysical indices;

however, it cannot predict future TEC values without physical data input [16].

The state of the ionosphere can be predicted by either extrapolation methods or physical models [18–20]. Past studies have developed various empirical models to predict one or more physical parameters of the ionosphere on short time scales from days to weeks [21]. For example, the Advanced Stand Alone Prediction System allows for the prediction of radio communication conditions in the high-frequency and very-high-frequency radio spectrum, while the Ionospheric Communications Enhanced Profile Analysis and Circuit model predicts the maximum usable frequency parameter using the electron density profile. The widely acknowledged International Reference Ionosphere model utilizes predicted physical indices to provide expected ionospheric parameters and electron density profiles and can further predict the ionospheric TEC for given locations and dates in a short time scale [22]. An extrapolation method was developed to directly predict the global mean TEC for the next 2–7 years based on a stationary spectral analysis of GPS-derived TEC data for the past four years, which represented all of the data available at that time [14].

Based on the recent findings, this work develops a new extrapolation method of predicting the evolution of Arctic TEC parameters by utilizing time-varying periodograms of the variability of the Arctic TEC. Compared to the previous extrapolation method, this study uses data of the past 13.6 years, which allows us to analyze the periodograms of the TEC variability on the scale of a solar cycle [14]. The newly developed method in this study utilizes the time-varying periodograms of the Arctic TEC to perform long-term prediction over a solar cycle. The TEC time series is first divided into a component of periodic oscillations and a component of the average TEC, in addition to noise. This study investigates the variability of both components separately and forecasts the two components over the scale of a solar cycle. The prediction is conducted in both temporally backward and forward directions. The results of the backward prediction are compared with existing data to verify the performance of the prediction.

In this paper, Section 2 introduces the method of Spherical Cap Harmonic Analysis (SCHA) used to map the Arctic TEC and to estimate the Arctic mean TEC. The periodograms of the Arctic TEC variability are then investigated. Finally, the Arctic TEC values are predicted.

Methods and Materials of SCHA Mapping of the Regional Ionospheric TEC

The ionospheric TEC along GPS signal paths can be estimated using dual-frequency GPS observables, satellite orbit products, and the hardware delay parameters of receivers and satellites. The method of estimating the vertical TEC from GPS measurements has been presented in a number of works [2,23–25]. The previous work has developed a technique of carrier-phase smoothing to improve the accuracy of pseudorange-derived TEC estimates [26]. The estimated TEC data are used with the SCHA model to map the regional ionospheric TEC. In this work, a spherical cap is a regional part of a sphere, and it is defined by the geographical coordinate of a spherical cap pole (θ_P, λ_P), and the half angle that represents the size of the region in question (θ_0). The SCHA model consists of a set of spherical cap harmonics, which can constitute a convenient orthogonal basis over a specific spherical cap $(\theta_0 < \pi)$, and it is expressed as follows:

$$E_v(\theta_c, \lambda_c)$$
$$= \sum_{k=0}^{K_{\max}} \sum_{m=0}^{\min(k,M)} \tilde{P}_{n_k(m)}^m (\cos\theta_c)[\tilde{C}_k^m \cos(m\lambda_c) + \tilde{S}_k^m \sin(m\lambda_c)] \quad (1)$$

where (θ_c, λ_c) is the spherical cap coordinate of an Ionosphere Pierce Point (IPP), and is calculated using the geographical coordinate of the spherical cap pole and IPP;

$E_v(\beta_c, \lambda_c)$ is the vertical TEC at the IPP (θ_c, λ_c);

K_{MAX} and M are the maximum degree and order of the series, respectively;

$n_k(m)$ are non-integer degrees of the orders (m), k is the index of degrees;

$\tilde{P}(\cos\theta)$ is the normalized associated Legendre function; and $\tilde{C}_k^m$ and $\tilde{S}_k^m$ are normalized spherical cap harmonic coefficients.

The equation (1) has the similar formula expression of the global spherical harmonic function, while a major difference of them exists in the values of degrees [14]. In the case of spherical cap $(\theta_0 < \pi)$, the degrees $(n_k(m))$ are non-integer, and they are a function of the orders (m) given a specific half angle θ_0, whereas the degrees of spherical harmonic function $(\theta_0 = \pi)$ are simply natural numbers from 1 to N. The calculation method of non-integer degrees given a half angle θ_0 was given in the work [27].

The SCHA ionospheric model has been used by several research teams to map ionospheric TEC in different regions of the Earth, and it is suitable for large regions, particularly the polar areas, according to the comparisons between the SCHA method and the other regional models [27–34]. The zero-degree coefficient of the SCHA model represents the mean TEC of a specific region [14–16]. The mean TEC corresponds to an idealized ionosphere in which the TEC is uniformly distributed and, as a whole, has the same electron content as the actual ionosphere in the specific region; therefore, the mean TEC should represent the characteristics of the regional ionosphere [17]. The mean TEC has units of TECU

$$(1 TECU = 10^{16} electrons/m^2)$$

In the present study, the geographical North Pole is the spherical cap pole of the interested area, and the half angle is 30 degrees $(\theta_0 = 30°)$, the maximum degree is 8 $(K_{MAX} = 8)$ and the maximum order is 6 $(M = 6)$. The number of model parameters is 75 in total. The Arctic ionospheric TEC is estimated using GPS measurements from 44 IGS tracking stations located at high latitudes (above 55° North latitude, as shown in Figure 1) and related IGS products, including IGS precise orbit data, and differential code bias (DCB) products of receivers and satellites provided by the Center for Orbit Determination in Europe (CODE) (ftp://ftp.unibe.ch/aiub/CODE). Table I listed the used IGS stations with their geographical coordinates. One should note that some IGS stations located in the Arctic were not included in Table I because their DCB products of receivers are missing from the database. Some pairs of stations have very close coordinates because the two receivers share the observation facility. The measurement dataset of the study period from 2000 to 2013 is provided in RINEX (Receiver Independent Exchange) format by the IGS central bureau via ftp access (ftp://cddis.gsfc.nasa.gov/gps/data/daily/). Before the study period, GPS tracking stations in the Arctic region are not sufficient to map ionospheric TEC. The sample rate of the GPS measurements is 30 seconds, and the elevation cut-off threshold is 20 degrees in the data processing. The sp3 satellite orbit products are used to calculate the precise positions of satellites and further calculate the positions of the ionosphere pierce points and elevations of GPS signal paths. The Spline interpolation method is used to interpolate the satellite positions at the observation epochs. Based on the estimated Arctic TEC, the analysis and prediction are presented as follows. The reference index data such as solar and geomagnetic indices have been downloaded from the national geophysical data center (ftp://ftp.ngdc.noaa.gov/STP/).

Analysis on Time-varying Periodograms of the Arctic TEC

This section first analyzes periodograms of the Arctic TEC variability over the past 14 years (4961 days) from 2000 to 2013. For purposes of comparison, the Global Ionosphere Maps (GIM) product of CODE is used to calculate the Arctic mean TEC for the same region and the same time period [35]. The same dataset of GIM has been utilized in the studies of [16–17,36]. The GIM dataset represents global ionosphere TEC using a set of pre-defined grid points in the standard IONEX format [35]. As calculated by equation (2), GIM-derived regional mean TEC is the normalized weighted sum of TEC values of all IONEX grid points in the whole area of study.

$$\overline{TEC_{GIM}} = \frac{\sum \cos \varphi E_{\varphi,\gamma}}{\sum \cos \varphi} \qquad (2)$$

where $E_{\varphi,\gamma}$ is the TEC grid value associated with the geographic latitude and longitude (φ, γ) of the grid points, and $\cos \varphi$ is the weighting function for grid points of the geographic latitude φ. The sum of all weighting function $\sum \cos \varphi$ in the denominator is the normalization factor.

The top panel of Figure 2 shows the time series of the SCHA-derived Arctic mean TEC (black line) and the GIM-derived Arctic mean TEC (cyan line). The two time series have a high correlation coefficient of 0.9613. Over the whole time period, the mean difference between the two time series is 2.01 TECU with a standard deviation of 2.20 TECU, as shown in the middle panel of Figure 2. The SCHA-derived Arctic mean TEC is larger than the GIM-derived result under active ionosphere conditions (2000–2003 and 2012–2013), which is indicated by the $F10.7$ index showed in the bottom panel of Figure 2, while the two time series are comparable under calm ionosphere conditions (2008–2009). This observation indicates that the GIM-derived Arctic mean TEC is "averaged" by the global coverage, which is consistent with the conclusions regarding the hemisphere and latitude-band distribution of the mean TEC in [16–17,36].

In this study, the analysis and prediction are performed through the method of least-squares collocation, which is a generalization of least-squares adjustment, as presented in detail in [14]. The time series of the Arctic mean TEC over a given time interval is divided into the average TEC and a component of periodic oscillations with multiple periods, which is depicted mathematically using the harmonic expansion as follows [16–18].

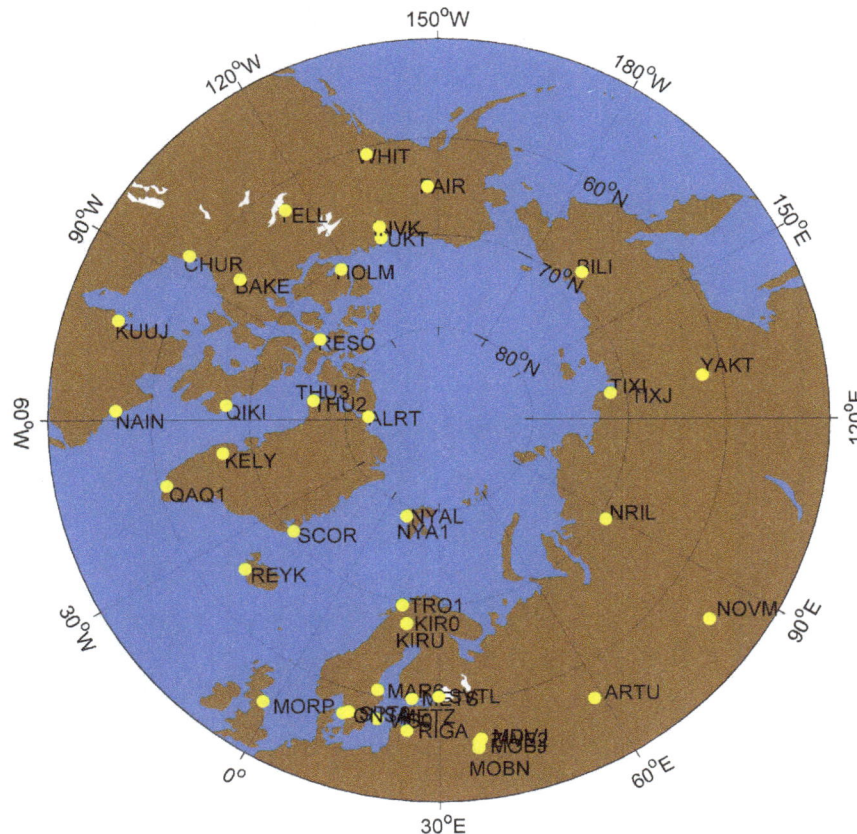

Figure 1. The geographical locations of the IGS tracking stations in the Arctic area. Land is indicated by brown, and sea/ocean is represented by blue. The yellow points indicate the locations of the IGS stations used in this study.

Table 1. The List and geographical locations of the IGS tracking stations in the Arctic area.

IGS name (4-char)	Latitude (degree)	Longitude (degree)	Elevation (meter)	IGS name (4-char)	Latitude (degree)	Longitude (degree)	Elevation (meter)
ALRT	297.6595	82.4943	78.11	MORP	358.3145	55.2128	144.40
ARTU	58.5605	56.4298	247.51	NAIN	298.3113	56.5370	33.48
BAKE	263.9977	64.3178	4.41	NRIL	88.3598	69.3618	47.89
BILI	166.4380	68.0761	456.24	NYA1	11.8653	78.9296	84.00
CHUR	265.9113	58.7591	−18.90	NYAL	11.8700	78.9300	82.00
FAIR	212.5008	64.9780	319.18	ONSA	11.9255	57.3953	45.50
HOLM	242.2391	70.7364	39.50	QAQ1	313.9522	60.7152	110.40
INVK	226.4730	68.3062	46.36	QIKI	295.9663	67.5593	13.30
KELY	309.0552	66.9874	229.81	RESO	265.1067	74.6908	34.90
WHIT	224.7779	60.7505	1427.00	REYK	338.0445	64.1388	93.10
VIS0	18.3673	57.6539	79.80	RIGA	24.0587	56.9486	34.70
YELL	245.5193	62.4809	181.00	SCOR	338.0497	70.4853	128.50
ZWE2	36.7601	55.7000	272.00	SPT0	12.8913	57.7150	219.90
KIR0	21.0602	67.8776	497.90	SVTL	29.7809	60.5329	77.10
KIRU	20.9684	67.8573	391.10	THU2	291.175	76.5370	36.10
KUUJ	282.2546	55.2784	−0.48	THU3	291.175	76.5370	36.10
MAR6	17.2585	60.5951	75.40	TIXI	128.8664	71.6345	46.98
MDVJ	37.2145	56.0215	257.40	TIXJ	128.8664	71.6345	47.05
METS	24.3953	60.2175	94.60	TRO1	18.9396	69.6627	138.00
METZ	24.3953	60.2175	94.50	TUKT	227.0057	69.4382	1.54
MOBJ	36.5697	55.1149	182.61	NOVM	82.9095	55.0305	149.98
MOBN	36.5695	55.1149	182.63	YAKT	129.6803	62.0310	103.37

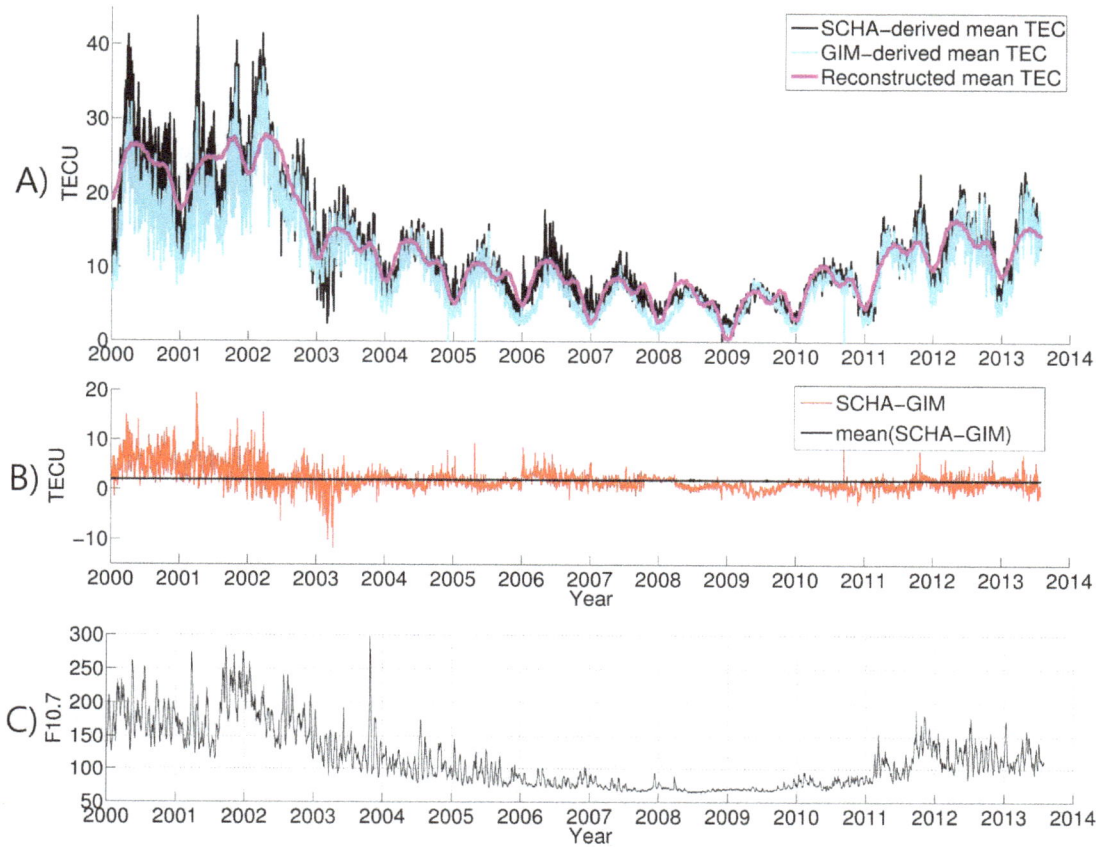

Figure 2. The Arctic TEC time series and corresponding solar activity condition of the time period. Top (A) Time series of the SCHA-derived Arctic mean TEC (black) and the GIM-derived Arctic mean TEC (cyan) and the reconstructed periodic oscillation component of the Arctic mean TEC based on the four periods (magenta). Middle (B) The difference between the SCHA- and GIM-derived mean TEC (cyan) and the mean of the difference (black). Bottom (C) The 10.7-cm radio flux for 2000–2013, indicating solar activity conditions.

$$\psi(t) = C_0 + \sum_{i=1}^{N} [C_i \cos(w_i t) + S_i \sin(w_i t)]$$

$$= C_0 + \sum_{i=1}^{N} [A_i \cos(w_i t - \phi_i)] \quad w_i = 2\pi/p_i \tag{3}$$

where $\psi(t)$ is the time series of the Arctic mean TEC for the sliding window, C_0 is the average TEC over a specific interval of the sliding window, N is the number of periods, w_i is the angular frequency with a period p_i, and C_i and S_i are parameters to be estimated, which define the phase and amplitude of the periodograms of the corresponding periodic oscillation components, $A_i = \sqrt{C_i^2 + S_i^2}$ is the amplitude of each periodic oscillation component, and ϕ_i is the corresponding phase.

In this study, a sliding window of one year is used, and the four selected periods ($N = 4$) include annual, semiannual, terannual and 27-day cycles. The sine and cosine coefficients (C_i and S_i) in equation (3) allow for a determination of both the phase and amplitude of each periodic oscillation component. As shown in Figure 3, the significant periodic variability in the phase and amplitude of the periodograms allows the oscillation components of the Arctic mean TEC to be predicted by simply extrapolating the periodograms.

In this study, the frequency spectra of the time-varying periodograms of the four oscillation components are analyzed separately, and the three most significant periods for the periodograms of each oscillation component are estimated. The three most significant periodic components represent more than 92% of the power in the frequency spectrum. Table II presents the significant periods and normalized powers of the corresponding periodograms. The spectral peaks near 27 days are due to sunspots co-rotating with the Sun's surface; these peaks spread over a certain range of periods because the angular velocity of sunspot rotation varies with solar latitude. In this study, we use a period of 27.8 days, which is the mean of the range associated with the spectral peaks. Compared to the SCHA-derived mean TEC, the time series of the TEC reconstructed using the time-varying periodograms has a Root Mean Square (RMS) error of 3 TECU under active ionosphere conditions and 1.2 TECU under calm ionosphere conditions, which correspond to approximately 10% of the ionosphere TEC, as shown in Figure. 2A).

In addition to the oscillation components, the average TEC estimated from the sliding window shows the variability over long-term periods. Figure 4 shows the time series of the average TEC (labeled as "Smoothing average TEC") and its periodic spectrum. In addition to the well-established spectral components, with periods of 11.22 years, 2 years, 1 year and 0.5 years, the spectrum also includes unexpected periods, e.g., 5.6 years and 9 years [16]. Figure 4 shows the time series of the average TEC reconstructed

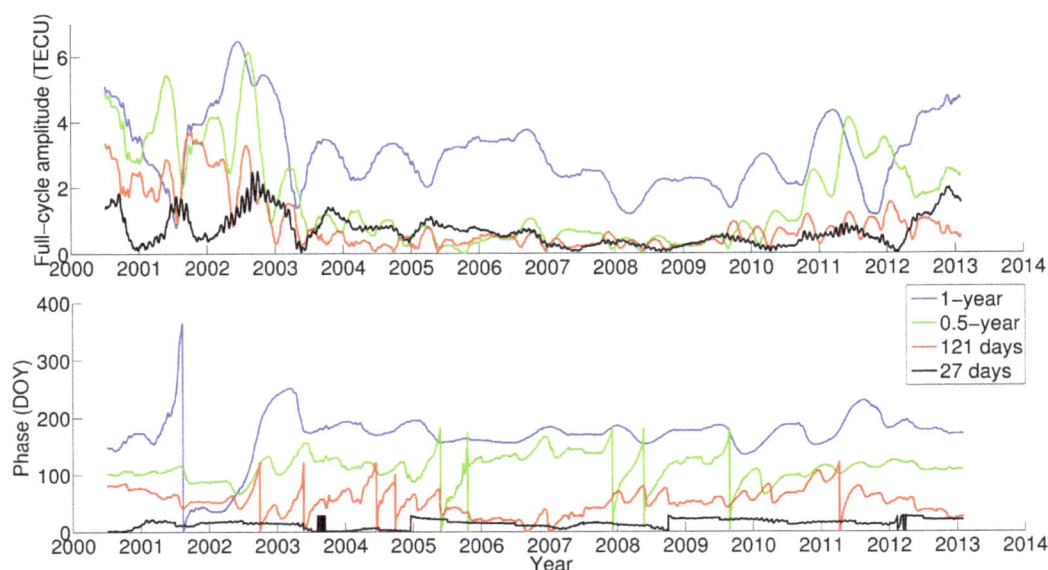

Figure 3. The time-varying periodograms of the selected periodicities of the Arctic TEC. Time evolution of the periodograms, including the full-cycle amplitude (top) and phase (bottom), for the four selected periods of the Arctic mean TEC.

using the well-established four periods alone and in combination with the two unexpected periods. Because the four periods of 11.22 years, 2 years, 1 year and 0.5 years are established according to general ionosphere physics and the average Arctic TEC of one-year-length sliding window has spectral components other than these periods, the reconstructed TEC does not fit exactly with the original TEC time series [16].

Some of these major periods are in accordance with well-established physical processes. The 11.22 years period is related to the cycle of solar sunspots. Both components of the annual and semi-annual periods have long been understood [16,37]. The 5.6 years period could be interpreted as a second Fourier harmonic of the 11.22 years solar activity cycle, which is related to the asymmetry of the TEC time series. The 9 years period could be an artifact of the recent unusually long solar activity minimum (solar cycle 23–24 minimum). The following section examines the results of backward prediction involving the 5.6 years and 9 years periods, respectively.

Results of Long-term Prediction of the Arctic TEC

Prediction of the Arctic mean TEC is accomplished by summing the predicted oscillation components and average TEC, which are calculated separately based on their respective periodograms. This section presents backward and forward

predictions of the Arctic mean TEC over the scale of a solar cycle. The backward prediction provides an opportunity to verify the reliability of the prediction method by comparing the predicted results with past observations, although the prediction process does not require any physical observations over the time interval of prediction.

The four oscillation components are extrapolated temporally based on their time-varying periodograms, as discussed in Section 3. For the average TEC prediction, this study compares the results of three sets of selected periods: A) the four well-established periods: 11.2 years, 2 years, 1 year and 0.5 years; B) the above four periods plus the 5.6 years period; C) the above four periods plus the 9 years period. Figure 5 shows the results of the backward prediction, which are the summation of the predicted oscillation components and the average TEC, for the preceding 11.2 years, from October 1988 to December 1999. The prediction results are set to zero when the values are negative. For purposes of comparison, Figure 5 also displays indices that are strongly correlated to the ionosphere for the same time period, including the solar sunspot number, the 10.7-cm solar radio flux and the geomagnetic index (Ap), which are used to examine the results of the backward prediction of the Arctic mean TEC.

The coefficients of correlation between the predicted mean TEC and the geophysical indices provide a measure of evaluation; as a baseline, we use the correlation coefficients between the

Table 2. Three Significant Periods and Normalized Powers of The Periodograms of The Oscillation Components of the Arctic Mean TEC.

Oscillation components	Period and power (in parentheses) of the periodograms		
Annual oscillation	1 year (0.622)	2 years (0.243)	11.22 years (0.135)
Semiannual oscillation	0.5 years (0.789)	1 year (0.059)	11.22 years (0.152)
Terannual oscillation	121.7 days (0.726)	1 year (0.046)	11.22 years (0.228)
27-day oscillation	27.7 days (0.806)	1 year (0.016)	11.22 years (0.178)

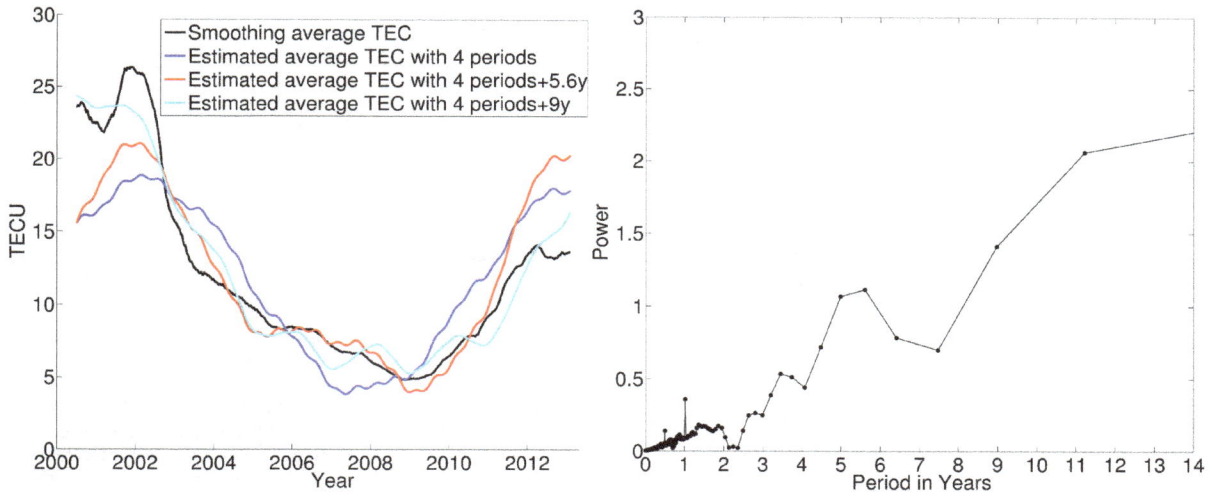

Figure 4. The time series and spectrum of average TEC related to different selections of periods. Left: Time series of the smoothed average TEC based on the sliding window (black), the average TEC reconstructed using the four well-established periods (blue) and with the inclusion of the periods of 5.6 years (red) and 9 years (cyan). Right: Periodic spectrum of the time series of the average TEC estimated with the sliding window.

SCHA-derived Arctic mean TEC and the geophysical indices for 2000–2013. The baseline correlation coefficients are 0.8427 for $F10.7$ and 0.8205 for the sunspot number. The reconstructed mean TEC for 2000–2013 based on period set C has the largest correlation coefficients, 0.8117 for $F10.7$ and 0.8057 for the sunspot number, while the mean TECs reconstructed using period sets A and B have lower correlation coefficients of approximately 0.7 for the both solar indices. For the predicted Arctic mean TEC

for 1988–1999, the results obtained for period sets A, B and C have comparable correlation coefficients of 0.7–0.75 for both solar indices over the same time interval. It should be noted that the prediction result of period set B has negative TEC values in the time 1994 to 1997, which is meaningless in physics and hence has been set to zero in Figure 5, although it is reasonable in mathematics that a prediction comes to a negative value due to prediction uncertainty when its true value is close to zero [38].

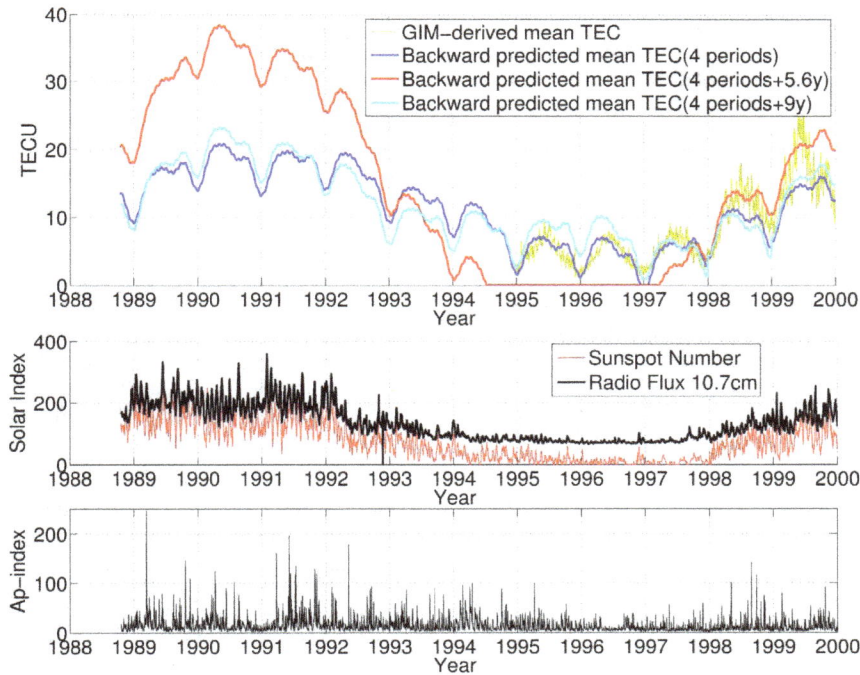

Figure 5. Backward predicted Arctic mean TEC for 1988–1999 and the corresponding geographical indices. Top: Backward predicted Arctic mean TEC for 1988–1999 based on different period sets of the average TEC, in addition with the GIM-derived Arctic mean TEC for 1995–1999. Negative TEC values are set to zero. Middle: Solar indices for 1988–1999, including the sunspot number and the 10.7-cm radio flux. Bottom: The geomagnetic index (Ap) for 1988–1999.

The Arctic ionospheric TEC has a low correlation coefficient of 0.1602 with the geomagnetic index (Ap). This result is consistent with the conclusions related to hemispherical and global scales, which are that the geomagnetic index is only related to the ionosphere over short-term periods, exhibiting a low correlation coefficient of approximately 0.2 with the long-term mean ionosphere [16–17].

This study calculates the GIM-derived Arctic mean TEC for 1995–1999 for comparison and accuracy evaluation. No data prior to this period are available. Table III compares the correlation coefficients for different time series of the SCHA-based and GIM-derived Arctic mean TEC. The results of backward prediction based on period sets A and C have correlation coefficients of more than 0.85 for the time series of the GIM-derived Arctic mean TEC, but the results for period set B have a significantly lower correlation coefficient of 0.4582. Compared to the GIM-derived Arctic mean TEC, the backward prediction results based on period sets A and C have RMS errors of 3.6 TECU and 3.1 TECU, respectively, for the entire duration of five years. The prediction results for period set B include negative TEC values for 1994 to 1997. For the entire duration of 11.2 years, the prediction accuracy for period sets A and C is estimated using the covariance of the parameters in the least-squares collocation [14,25]. The prediction uncertainty, which is depicted by the standard error of predicted values, is found to be 5.6 TECU for period set A and 4.8 TECU for period set C, which correspond to 25% of the average TEC under active ionosphere conditions. The discrepancy in the prediction results for the two period sets is within the prediction uncertainties.

The prediction results obtained for period set C, which includes the 9 years period, exhibited the maximum correlation with the observed solar indices and the GIM-derived mean TEC for 1995–1999; it also displayed the minimum RMS error with respect to the GIM-derived mean TEC for 1995–1999. The 9 years period may arise from the exceptionally prolonged solar cycle 23, lasting from 1996 to 2008 [39]. In fact, historical data of sunspot numbers since 1740 have shown that the length of a solar cycle varies from 9 years to 14 years [40]. It need be validated whether the 9 years period exists commonly in the Arctic TEC using a longer dataset, and the driving force of this variability should be physically interpreted.

Based on the two period sets A and C, the Arctic mean TEC is predicted for the following 11.2 years, from August 2013 to 2024, as shown in Figure 6. Both predictions display significant variability, such as the 11.2 years, annual, semi-annual and seasonal variations. The prediction uncertainty is the same as that of the backward prediction, and the prediction confidence is indicated by the shaded band, which becomes wider towards the next 11.2 years maximum. The both time series of predicted TEC based on period sets A and C show that the Arctic TEC will reach

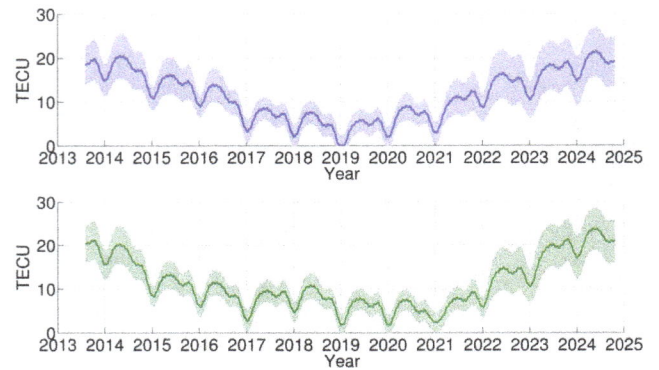

Figure 6. Forward predicted Arctic mean TEC and corresponding uncertainties for 2014–2024 using two period sets. Forward prediction and the corresponding uncertainty, indicated by the shaded band, based on different period sets of the average TEC: period set A (top) and period set C (bottom).

its next minimum in December 2018 and its next maximum in the summer of 2024, after the previous ionosphere maximum occurring in May 2013.

Conclusions

The dataset of the continuously operating GPS tracking stations in the Arctic region provides an unprecedented observation source of the Arctic ionospheric TEC in large spatiotemporal scales. In this study, a new method has been developed for predicting the Arctic TEC in the scale of a solar cycle using the past GPS dataset. The proposed method represents the time series of Arctic TEC with a component of periodic oscillations and a component of the average TEC. The periodograms of the variability of the Arctic TEC are then analyzed using a dataset for the past 13.6 years, from 2000 to 2013. The periodograms displayed time-varying evolution. Based on the time-varying periodograms, the newly developed method predicts the Arctic TEC over the scale of a solar cycle (11.2 years) using the technique of least-squares collocation. The prediction is performed in both temporally backward and forward directions. The backward predictions for the preceding solar cycle from 1988 to 1999 are compared to evaluate the performance with past physical data, including solar indices, the geomagnetic index and the GIM-derived Arctic TEC.

The proposed method of TEC prediction is based on the extrapolation approach that requires no input of physical observations of the time interval of prediction, and it is performed by summing the predicted periodic oscillation and average TEC components. The challenge in conducting long-term predictions of the ionosphere primarily arises in predicting the average TEC

Table 3. Correlation Coefficients for the SCHA-based TEC Time Series and the GIM-derived Arctic Mean TEC.

TEC time series	GIM-derived Arctic mean TEC	
	2000–2013	1995–1999
SCHA-derived daily mean TEC	0.9613	—
Reconstructed Arctic mean TEC for period set A	0.9044	0.8768
Reconstructed Arctic mean TEC for period set B	0.9125	0.4582
Reconstructed Arctic mean TEC for period set C	0.9183	0.8890

component. The proposed prediction method is verified using the backward prediction results, which show that TEC prediction result of involving the 9 years period is more consistent with the historical Arctic TEC dataset and geophysical data for the study duration. However, the 9 years period requires further confirmation and physical interpretation using a longer dataset because the duration of 14 years is a relatively short time, which represents all of the data available currently. Forward prediction for the future solar cycle from 2013 to 2024 is performed for two period sets of the average TEC. The prediction results for the two sets are consistent overall, with a standard error of 5.6 TECU and 4.8 TECU, respectively, which is equal to roughly 25% of the Arctic mean TEC under active ionosphere conditions.

Acknowledgments

The GPS observables used in this study have been downloaded from the IGS global data center (ftp://cddis.gsfc.nasa.gov/gps/data/daily/), the CODE products have been downloaded from CODE's data archive server (ftp://ftp.unibe.ch/aiub/CODE), and the solar and geomagnetic indices have been downloaded from the national geophysical data center (ftp://ftp.ngdc.noaa.gov/STP/).

Author Contributions

Conceived and designed the experiments: JL RC ZW. Performed the experiments: JL JA. Analyzed the data: JL RC ZW JA. Contributed reagents/materials/analysis tools: JL JH ZW. Wrote the paper: JL JH RC.

References

1. Garcia-Rigo A, Monte E, Hernandez-Pajares M, Juan JM, Sanz J, et al. (2011) Global prediction of the vertical total electron content of the ionosphere based on GPS data. Radio Science 46 (RS0D25). http://dx.doi.org/10.1029/2010RS004643.
2. Garner TW, Gaussiran T, Tolman BW, Harris RB, Calfas RS, et al. (2008) Total electron content measurements in ionospheric physics. Adv Space Res 42(4): 720–726. doi: 10.1016/j.asr.2008.02.025.
3. Dikpati M (2008) Predicting cycle 24 using various dynamo-based tools. Ann Geophys 26: 259–267, 2008. doi: 10.5194/angeo-26-259-2008.
4. Gulyaeva TL, Arikan F, Stanislawska I, Poustovalova LV (2012) Symmetry and asymmetry of ionospheric weather at magnetic conjugate points for two midlatitude observatories. Adv in Space Res 52(10): 1837–1844. DOI: 10.1016/j.asr.2012.09.038.
5. National Research Council (2003) The Sun to the Earth and beyond: A decadal research strategy in solar and space physics. Natl. Acad. Press, Washington, D.C.
6. National Space Weather Strategic Plan (1995) Official Fed. Coord. for Meteorol. Services, NOAA, Silver Spring, Md.
7. Hernández-Pajares M, Juan JM, Sanz J, Aragón-Àngel A (2012) Propagation of medium scale traveling ionospheric disturbances at different latitudes and solar cycle conditions. Radio Sci 47, RS0K05, doi: 10.1029/2011RS004951.
8. Pesnell WD (2012) Solar cycle predictions. Sol Phys 281(1): 507–532, doi: 10.1007/s11207-012-9997-5.
9. Sardon E, Rius A, Zarraoa N (1994) Estimation of the transmitter and receiver differential biases and the ionospheric total electron content from Global Positioning System observations, Radio Sci 29(3): 577–586.
10. Afraimovich EL, Astafyeva EI, Oinats AV, Yasukevich YV, Zhivetiev IV (2008) Global electron content: A new conception to track solar activity. Ann Geophys, 26: 335–344, doi: 10.5194/angeo-26-335-2008.
11. Hernández-Pajares M, Juan JM, Sanz J, Orus R, Garcia-Rigo A, et al. (2009) The IGS VTEC maps: a reliable source of ionospheric information since 1998, J Geod 83: 263–275, DOI 10.1007/s00190-008-0266-1.
12. Wu Y, Jin S, Wang Z, Liu J (2010) Cycle-slip detection using multi-frequency GPS carrier phase observations: a simulation study. Adv Space Res, 46: 144–149.
13. Scharroo R, Smith WHF (2010) A global positioning system-based climatology for the total electron content in the ionosphere, J Geophys Res 115, A10318, doi: 10.1029/2009JA014719.
14. Schaer S (1999) Mapping and predicting the earth's ionosphere using the global positioning system. Doctor Thesis, The University of Bern. Available: http://www.sgc.ethz.ch/sgc-volumes/sgk-59.pdf. Accessed 21 July 2014.
15. Liu J, Chen R, An J, Wang Z, Hyyppä J (2014) Spherical cap harmonic analysis of the Arctic ionospheric TEC for one solar cycle. J Geophys Res Space Physics, 119, doi: 10.1002/2013JA019501.
16. Lean JL, Meier RR, Picone JM, Emmert JT (2011) Ionospheric total electron content: Global and hemispheric climatology. J Geophys Res, 116, A10318, doi: 10.1029/2011JA016567.
17. Liu L, Wan W, Ning B, Zhang ML (2009) Climatology of the mean total electron content derived from GPS global ionospheric maps. J Geophys Res, 114, A06308, doi: 10.1029/2009JA014244.
18. Gulyaeva TL, Arikan F, Hernandez-Pajares M, Stanislawska I (2013) GIM-TEC adaptive ionospheric weather assessment and forecast system, Journal of Atmospheric and Solar-Terrestrial Physics, 102: 329–340, 2013. http://dx.doi.org/10.1016/j.jastp.2013.06.011.
19. Ivanov-Kholodny GS, Mikhailov AV (1986) The prediction of ionospheric conditions. ISBN 390-277-2143-2.
20. Zolesi B, Cander LR (2014) Ionospheric prediction and forecasting, Springer Geophysics Series XII.

21. Pietrella M, Perrone L, Fontana G, Romano V, Malagnini A, et al. (2009) Oblique-incidence ionospheric soundings over Central Europe and their application for testing now casting and long term prediction models. Adv Space Res43(11): 1611–1620.
22. Bilitza D, McKinnell LA, Reinisch B, Fuller-Rowell T (2011) The international reference ionosphere today and in the future. J Geod, 85: 909–920, DOI 10.1007/s00190-010-0427-x.
23. Jakowski N, Hoque M, Mayer MC (2011) A new global TEC model for estimating trans-ionospheric radio wave propagation errors, J Geod 85(12): 965–974, 2011. http://dx.doi.org/10.1007/s00190-011-0455-1.
24. Mannucci AJ, Wilson BD, Yuan DN, Ho CH, Lindqwister UJ, et al. (1998) A global mapping technique for GPS-derived ionospheric total electron content measurements. Radio Sci 33(3): 565–582, doi:10.1029/97RS02707.
25. Hernández-Pajares M, Aragón-Ángel À, Defraigne P, Bergeot N, Prieto-Cerdeira R, et al. (2014) Distribution and mitigation of higher-order ionospheric effects on precise GNSS processing. J Geophys Res Solid Earth, 119: 3823–3837, doi:10.1002/2013JB010568.
26. Liu J, Chen R, Wang Z, Zhang H (2011) Spherical cap harmonic model for mapping and predicting regional TEC. GPS Solut 15: 109–119.
27. Haines GV (1985) Spherical cap harmonic analysis. J Geophys Res 90(B3): 2583–2591.
28. Zahra B, Terkildsen M, Neudegg D (2010) Regional GPS-based ionospheric TEC model over Australia using Spherical Cap Harmonic Analysis. The 38th COSPAR Scientific Assembly, Bremen, Germany, 18–15 July 2010.
29. Ghoddousi-Fard R, Héroux P, Danskin D, Boteler D (2011) Developing a GPS TEC mapping service over Canada. Space Weather 9, S06D11, doi: 10.1029/2010SW000621.
30. Liu J, Wang Z, Wang H, Zhang H (2008) Modeling regional ionosphere using GPS measurements over China by spherical cap harmonic analysis methodology. Geomat and Inf Sci of Wuhan University 33(8): 792–795.
31. Liu J, Chen R, Kuusniemi H, Wang Z, Zhang H, et al. (2010) A preliminary study on mapping the regional ionospheric TEC using a spherical cap harmonic model in high latitudes and the Arctic region, Journal of Global Positioning Systems. 9(1): 22–32.
32. Wilson BD, Mannucci AJ, Edwards CD (1995) Subdaily northern hemisphere ionospheric maps using an extensive network of GPS receivers. Radio Sci 30: 639–648.
33. Komjathy A (1997) Global ionospheric total electron content mapping using the global positioning system, Doctor Thesis, The University of New Brunswick.
34. Liu J, Wang Z, Zhang H, Zhu W (2008) Comparison and consistency research of regional ionospheric TEC models based on GPS measurements. Geomat and Inf Sci of Wuhan University 33(5): 479–483.
35. Schaer S, Gurtner W, Feltens J, Feltens J (1998) IONEX: The IONosphere map Exchange format version 1. Available: ftp://cddis.gsfc.nasa.gov/reports/formats/ionex1.pdf. Accessed 21 July 2014.
36. Lean JL, Emmert JT, Picone JM, Meier RR (2011) Global and regional trends in ionospheric total electron content. J Geophys Res 116, A00H04, doi: 10.1029/2010JA016378.
37. Rishbeth H, Garriott OK (1969) Introduction to ionospheric physics. Volume 14 of International geophysics series, Academic Press, New York and London.
38. Omlin M, Reichert P (1999) A comparison of techniques for the estimation of model prediction uncertainty. Ecological Modelling 115: 45–59.
39. Strangeways HJ, Kutiev I, Cander LjR, Kouris S, Gherm V, et al. (2009) Near-earth space plasma modeling and forecasting. Ann Geophys, 52(3): 255–271.
40. Friis-Christensen E, Lassen K (1991) Length of the solar cycle: An indicator of solar activity closely associated with climate, Science, New Series, 254: 698–700.

13

Combining XCO$_2$ Measurements Derived from SCIAMACHY and GOSAT for Potentially Generating Global CO$_2$ Maps with High Spatiotemporal Resolution

Tianxing Wang*, Jiancheng Shi, Yingying Jing, Tianjie Zhao, Dabin Ji, Chuan Xiong

State Key Laboratory of Remote Sensing Science, Institute of Remote Sensing and Digital Earth, Chinese Academy of Sciences. Beijing, China

Abstract

Global warming induced by atmospheric CO$_2$ has attracted increasing attention of researchers all over the world. Although space-based technology provides the ability to map atmospheric CO$_2$ globally, the number of valid CO$_2$ measurements is generally limited for certain instruments owing to the presence of clouds, which in turn constrain the studies of global CO$_2$ sources and sinks. Thus, it is a potentially promising work to combine the currently available CO$_2$ measurements. In this study, a strategy for fusing SCIAMACHY and GOSAT CO$_2$ measurements is proposed by fully considering the CO$_2$ global bias, averaging kernel, and spatiotemporal variations as well as the CO$_2$ retrieval errors. Based on this method, a global CO$_2$ map with certain UTC time can also be generated by employing the pattern of the CO$_2$ daily cycle reflected by Carbon Tracker (CT) data. The results reveal that relative to GOSAT, the global spatial coverage of the combined CO$_2$ map increased by 41.3% and 47.7% on a daily and monthly scale, respectively, and even higher when compared with that relative to SCIAMACHY. The findings in this paper prove the effectiveness of the combination method in supporting the generation of global full-coverage XCO$_2$ maps with higher temporal and spatial sampling by jointly using these two space-based XCO$_2$ datasets.

Editor: Juan A. Añel, University of Oxford, United Kingdom

Funding: The work described in this paper has been jointly supported by project of "Climate Change: Carbon Budget and Related Issues" (Grant nr. XDA05040402) from Chinese Academy of Sciences (CAS), the CAS/SAFEA International Partnership Program for Creative Research Teams (Grant nr. KZZD-EW-TZ-09) and National Natural Science foundation of China (Grant nr. 41301177). The funders had no role in study design, data collection and analysis, decision to publish, or preparation of the manuscript.

Competing Interests: The authors have declared that no competing interests exist.

* Email: wangtx@radi.ac.cn

Introduction

In recent years, global warming caused by emission of CO$_2$ has attracted considerable attention from the public. During the past decade, although tremendous efforts have been made toward improving the understandings of the mechanism between CO$_2$ increase in the atmosphere and global warming, some uncertainties still exist in the spatiotemporal characteristics of CO$_2$ sinks/sources on regional and global scales due to the lack of high-density measurements of such variables with good accuracy [1,2]. To date, the estimates of CO$_2$ flux from inverse methods rely mainly on ground-based measurements [3,4]. Although providing highly accurate atmospheric CO$_2$ records, the traditional ground-based networks intrinsically suffer from sparse spatial coverage [2,5]. Satellite-based measurements with various spatial and temporal resolutions provide a unique opportunity to accurately map atmospheric CO$_2$ in both daytime and nighttime over large areas, thus having the potential to bridge this gap. As a result, various satellite-based platforms have been equipped in recent years for deriving the CO$_2$ concentrations.

Generally, methods for retrieving CO$_2$ from space can be grouped into two categories: (1) inferring CO$_2$ concentrations by measuring shortwave infrared (SWIR) reflected solar radiation around 1.6 and 2.0 µm with sufficient spectral resolution. This includes the Greenhouse gases Observing SATellite (GOSAT), operating since 2009 [6], the Scanning Imaging Absorption spectrometer for Atmospheric CartograpHY (SCIAMACHY), in orbit since 2002 [7], and the second Orbiting Carbon Observatory (OCO-2), which, as a rebuild of OCO [8,9], is planned to be launched in July 2014. In addition, CarbonSat will also be scheduled to be launched in 2018 (http://www.iup.uni-bremen.de/carbonsat/). These measurements have a nearly uniform sensitivity to CO$_2$ from the surface up through the middle troposphere, and thus are frequently used to derive the column-average dry air mole fraction of atmosphere CO$_2$ (XCO$_2$) during the daytime; (2) retrieving CO$_2$ concentrations by interpreting the recorded spectra of the Earth-atmosphere system in thermal infrared (TIR) bands (around 15 µm). Instruments that work in such a way include AIRS [10,11], IASI [12,13], and FTS (Band 4) of GOSAT [6]. These measurements bring the advantage that they can detect CO$_2$ during both day and night time, while the lack of sensitivity in the lower troposphere makes them inappropriate to estimate CO$_2$ near the surface where the largest signals of CO$_2$ sources and sinks occur [1]. The complementarities of these platforms allow us to combine the SWIR and TIR measurements for obtaining enhanced understanding of CO$_2$ spatiotemporal variations globally. Since XCO$_2$ is much less affected by vertical transport of CO$_2$, it is particularly useful for investigation of CO$_2$ sources and sinks using inversion modeling [14,15]. On the other hand, the spatial and temporal variations in XCO$_2$ are even smaller than that in the surface CO_2; therefore, unprecedented measurement precision and accuracy are highly required for such column measurements [16–19]. SCIAMACHY (operation stopped in April 2012) and GOSAT are two typical instruments that can be used to derive XCO_2 from space, and a variety of retrieval algorithms have been developed for SCIAMACHY [1,20–27] and GOSAT [2,4,5,28–30] with eyes on improving XCO_2 retrieval accuracy to a great extent. At present, a number of XCO_2 products have been released. These will definitely enhance our understanding of the global carbon cycle.

Unfortunately, almost all typical instruments currently used to derive atmospheric CO_2 concentration are working in the infrared spectral range (less than 16 μm). Thus, except for the instrument's observation mode (for example, GOSAT observes in lattice points), the spatial coverage of the derived CO_2 is severely restricted by the presence of clouds. In addition, the lower signal-to-noise level over ice/snow covered surfaces and ocean for SWIR instruments (e.g., SCIAMACHY) also contributes to the CO_2 sparse coverage. For instance, it has been pointed out that only about 10% of GOSAT data can be used for retrieval of XCO_2 due to the cloud contaminations [4]. The amount of CO_2 measurements will be even smaller if additional screening criteria such as quality of spectral fit, aerosol loadings, etc. are further applied. Although the amount of remaining CO_2 measurements from certain space-based instruments may largely surpass that of ground-based sites, it is still not sufficient enough for accurately quantifying the spatiotemporal distribution of CO_2 over the global scale. As a result, it is greatly desired to jointly use these available CO_2 measurements derived from various space-based data. Recently, a novel method has been proposed for combining CO_2 values from seven different algorithms, and a new Level-2 CO_2 database (EMMA) from one algorithm is composed according to the median of monthly average of seven CO_2 products in each $10° \times 10°$ latitude/longitude grid box [31]. In fact, this method cannot increase the number of CO_2 observations but chooses a product with moderate oscillation among the available products. Despite the usefulness of the XCO_2 measurements (Level 2) in their own right, further spatiotemporal analysis for interpreting their scientific merit is essentially necessary due to the retrieval uncertainties and sparse coverage of such Level-2 observations [32]. For this point, many works have attempted to generate global full-coverage (i.e., Level 3) maps from XCO_2 values derived from single satellite observations using a geospatial statistics approach [32–34]. However, as reflected in these studies (for instance, Fig. 1 in the work of [33]), a compromise has to be made between the interpolated accuracy and the spatiotemporal resolution of Level-3 product because of the limited amount of Level-2 XCO_2 observations being used. For this point, instead of using Level-2 XCO_2 from a single dataset (e.g., GOSAT or OCO-2) as performed in the existing literature, we attempt to explore the potential of combining two CO_2 datasets (GOSAT and SCIA-MACHY) in assisting in global Level-3 generation, aiming to: (1) propose a general strategy for combining (fusing) various CO_2 datasets with different instruments, algorithms, averaging kernels, etc.; and 2) increase the number of daily CO_2 points (utilized in Level-3 map interpolations) through the combination of two datasets, so that potentially improved Level-3 maps with higher accuracy and shorter time scale can be generated. The better the interpretation of the satellite-based CO_2 observations one can make, the higher the resolution (both temporal and spatial) of the generated global CO_2 maps.

Datasets

For GOSAT, the Fourier transform spectrometer (FTS) on GOSAT is the fundamental unit to retrieve atmospheric CO_2 and CH_4. It observes sunlight reflected from the earth's surface, and light emitted from the atmosphere and the surface. It is composed of three narrow bands in the SWIR region (0.76, 1.6, and 2.0 μm) and a wide TIR band (5.5–14.3 μm) at a spectral and spatial resolution of 0.2 cm^{-1} and 10.5 km, respectively [35]. Specifically, four CO_2 products from GOSAT have currently been released to the public: University of Leicester product [9,36], the RemoTeC product [28], NIES GOSAT product [35] and the product generated by NASA's Atmospheric CO_2 Observations from Space (ACOS) team (hereafter called ACOS product) [2,30]. The difference between some of the above mentioned products with various versions have been investigated in a recent study [37]. In the present paper, the ACOS product of 2009–2010 with version v2.9 has been employed.

SCIAMACHY was successfully launched on board Environmental Satellite (ENVISAT) in 2002 (unfortunately ceased in April 2012), which is a detector elements satellite spectrometer covering the spectral range 0.24–2.38 μm with a moderate spectral resolution of about 0.2–1.6 nm, and spatial resolution at nadir of 60×30 km [7]. It has eight spectral channels, with 1024 individual detector diodes for each band, observing the spectral regions 0.24–1.75 μm (band 1–6), 1.94–2.04 μm (band 7), and 2.26–2.38 μm (band 8) simultaneously in nadir and limb and solar and lunar occultation viewing geometries [22]. As mentioned in Section 1, till today, a number of CO_2-retrieval algorithms have been developed for SCIAMACHY. The IUP/IFE of University of Bremen has released two XCO_2 products, i.e., WFM-DOAS product [21,22] and the Bremen Optimal Estimation DOAS (BESD) product [1,26]. In this study, the BESD product with the versions of v02.00.08 for 2009–2010 is used.

In addition, CO_2 profiles of CT [38] are also collected here to allow the data mentioned above to be properly fused. CT is a NOAA data assimilation system, which provides the 3D profiles of CO_2 mole fractions in the atmosphere over the globe. For this study, CT data with version CT2011 is collected. This dataset provides global CO_2 profiles with $3° \times 2°$ latitude/longitude grid and 3 hours temporal resolution (a total 8 times from 01 to 22 in UTC) spanning the time period from January 2000 to December 2010. The CT dataset is used here mainly to assist in adjusting and time-shifting of the two CO_2 products being combined.

Methodologies

For combining the different space-based CO_2 measurements, three steps are adapted in this study. First, taking the global ground measurements of CO_2 as reference, remove the bias of the individual CO_2 retrievals for ensuring the accuracy of the fused CO_2 product; then make some adjustment for both the ACOS and BESD products, so that they can be physically comparable and thus combined; finally fuse the ACOS and BESD CO_2 products considering their retrieval uncertainties, spatial scales, differences in averaging kernels and overpass times, etc.

3.1 Global bias corrections

Removal of any global bias of the retrieved CO_2 when compared with the ground *in situ* measurements is essential before performing joint use. Many researches [4,25] frequently pointed out that CO_2 retrievals from GOSAT are low biased with different levels due to the uncertainties in pressure, radiometric calibration, line shape model, cloud and aerosol scattering, etc.

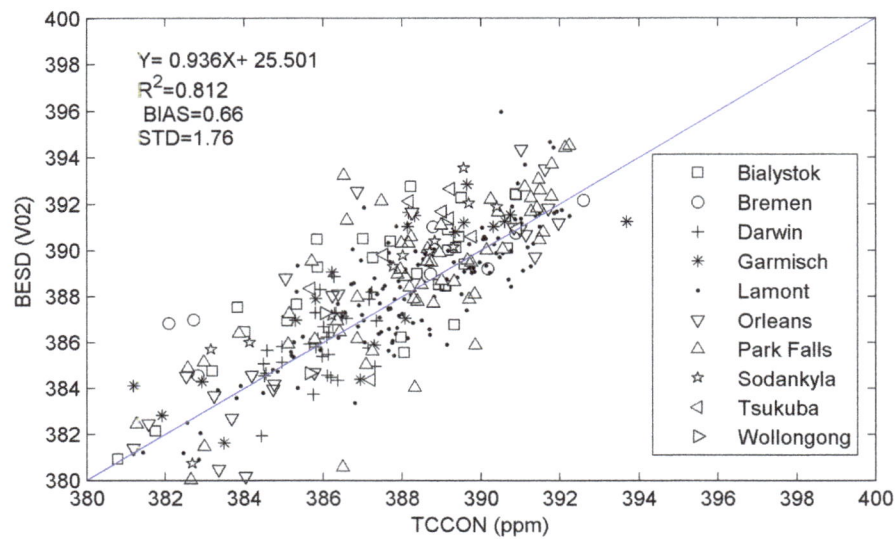

Figure 1. Validation of the BESD products against *in situ* **TCCON CO$_2$ measurements over globe for 2009–2010.**

Fortunately, a recent study has proposed a method for evaluating systematic errors in CO$_2$ and showed that the new version of ACOS product (v2.9) has a low global bias (<0.5 ppm) [39]. Thus, there is no global bias correction for the ACOS product being conducted here, but only the ACOS retrievals that pass the filter of table B1 in the work of [38] and marked as "good" in the quality flag are used. For the BESD product, we select Total Carbon Column Observing Network (TCCON) [15] measurements for 2009–2010 as the ground truth to determine its global bias. Specifically, BESD retrievals within ±2.5° and ±2.5° latitude/longitude box centered at each TCCON site and the mean FTS value (within ±1 h time window of satellite overpass time) are

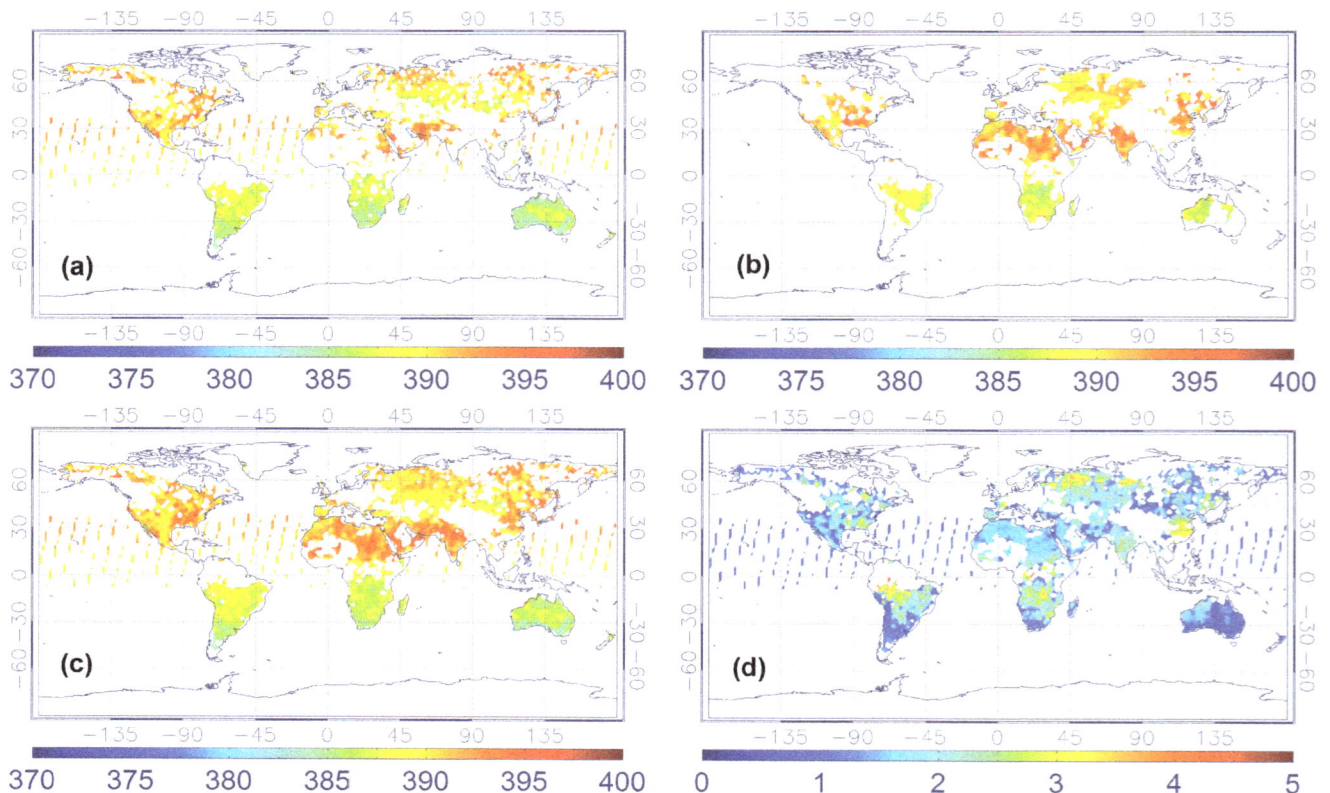

Figure 2. XCO$_2$ monthly mean maps in May of 2010. ((a) ACOS XCO$_2$, (b) BESD XCO$_2$, (c) combined product, and (d) XCO$_2$ uncertainties of the combined product).

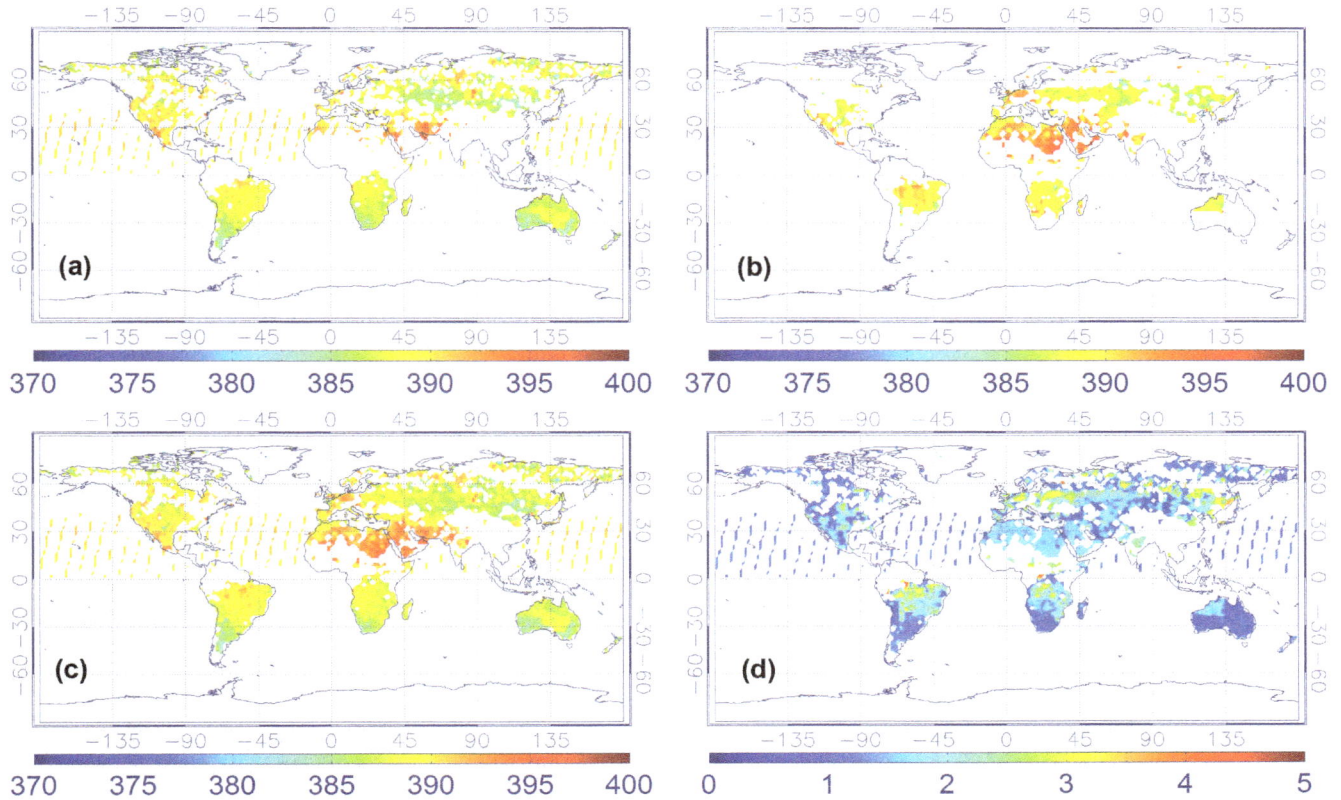

Figure 3. XCO$_2$ monthly mean maps in June of 2010. ((a) ACOS XCO$_2$, (b) BESD XCO$_2$, (c) combined product, and (d) XCO$_2$ uncertainties of the combined product).

extracted and compared (totally ten TCCON sites are utilized). The coincidence criteria mentioned above ultimately yield a total of 338 pairs of CO$_2$ measurements. The comparison result is shown in Fig. 1.

3.2 Retrieval adjustments

As pointed out by most researchers, it is not reasonable to directly compare or use two XCO$_2$ measurements. A suitable way to do that is to take the a priori profiles and variations in averaging kernel into account during the comparison [26,40]. To tackle the a priori issue, after correcting their global biases, both BESD and ACOS products are adjusted for a common a priori profile, which we assume to be the CT profile interpolated at the middle of the two overpass times (Equation (1)). Specifically, the a priori CO$_2$ profile of both the ACOS and BESD are first interpolated or extrapolated to the level of the CT CO$_2$ profile according to their pressure layers. After interpolation, the a priori profiles for both ACOS and BESD have the same dimension as the CT profile. Here the reason we take the CT profile at the middle of the two overpass times is that the time difference for GOSAT (1:00 pm) and SCIAMACHY (10:00 am) is relative large (3 hours), if we take one satellite time as reference, the induced error would be large for the other satellite measurements considering the CO$_2$ natural diurnal variation. So a middle time between these two satellite overpass times is selected for minimizing the CO$_2$ uncertainties during the adjustment.

$$XCO_2\_adj = XCO_2\_ret + (h^T I - a)(xCT - xa) \quad (1)$$

Here, XCO_2_adj is the adjusted XCO$_2$ for ACOS or BESD; XCO_2_ret corresponds to retrieved XCO$_2$ of ACOS or BESD; a is the column-averaging kernel (row vector) of ACOS or BESD; h is pressure-weighting function (column vector); I is an identity matrix; xCT and xa (column vectors) are the common CT CO$_2$ profile and the corresponding a priori CO$_2$ profile for ACOS or BESD, respectively.

While it is not trivial to accurately consider the smoothing error without an estimate of the true atmospheric variability which is generally not readily available for most cases [39]. Fortunately, some works revealed that the smoothing error is generally small [26,39]. Consequently, for the remainder of this paper, only the adjustment in Equation (1) is applied for both the ACOS and BESD CO$_2$ products (after bias corrections).

3.3 Combination and time shifting

Based on the processes described above, the world is divided into a number of $0.5° \times 0.5°$ latitude/longitude grid box (totally 720×360). For each grid cell, Equation (2) is used to combine the corresponding CO$_2$ measurements within that grid.

$$XCO_2\_Fued = \sum_{i=1}^{m} \left(XCO_2\_i \times \frac{1 - Uncert\_ratio^i}{\sum_{i=1}^{m}(1 - Uncert\_ratio^i)} \right) \quad (2)$$

where XCO_2_Fued is the combined XCO$_2$; m is the total number of space-based CO$_2$ retrievals (ACOS and/or BESD) within a certain grid; XCO_2_i is the ith XCO$_2$ retrieval in a grid for which

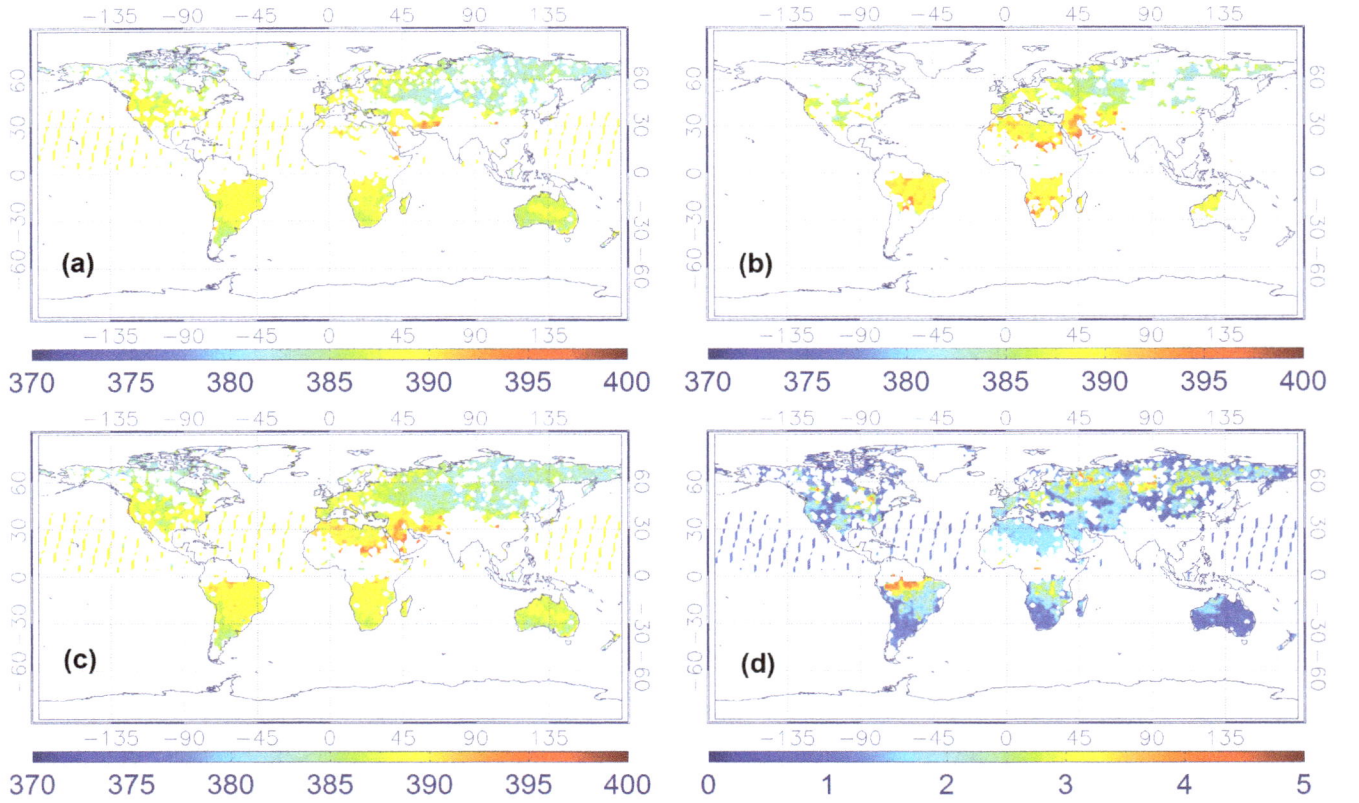

Figure 4. XCO_2 monthly mean maps in July of 2010. ((a) ACOS XCO_2, (b) BESD XCO_2, (c) combined product, and (d) XCO_2 uncertainties of the combined product).

the global bias and Equation (1) are supposed to be applied; $Uncert\_ratio^i$ is the ratio of uncertainty of the ith XCO_2 retrieval to its XCO_2 value.

Please note that since different CO_2 retrievals have distinct overpass times, it is necessary to unify them to avoid uncertainties induced from the time discrepancy before fusion. To this end, a method for considering the CO_2 shifting along time has been developed (Equation (3)). First, designate a specific time or select one overpass time as reference, then transfer CO_2 measurements at various overpass times to that of the reference time by interpolating the CT CO_2 at temporal scale. Here, it should be pointed out that despite the CO_2 absolute values of CT not being accurate enough, the daily cycle pattern of atmospheric CO_2 it reflects is assumed to be correct.

$$XCO_2\_ref = \frac{\omega^T X\_ref^{CT}}{\omega^T X\_t^{CT}} \times XCO_2\_t \quad (3)$$

Here, XCO_2_ref is the transformed XCO_2 (ACOS or BESD) at the reference time; XCO_2_t is the retrieved XCO_2 from ACOS or BESD at overpass time t; X_ref^{CT} and X_t^{CT} are CO_2 profiles of CT at times of reference and t, respectively; ω is the pressure-weighting vector (column vector).

Based on the time-shifting strategy proposed here, a global CO_2 map at any specific time can be theoretically produced by employing the pattern of the CO_2 daily cycle reflected by CT data. For instance, we can unify all XCO_2 retrievals being combined with various overpass times to that of UTC = 1.

Results

Evaluation analysis showed that the global bias for the BESD product is generally small. In this study, the bias of the BESD product is corrected by subtracting 0.6 ppm from all XCO_2 values according to the results in Fig. 1. Although the systematic bias of the XCO_2 retrievals is removed, it is supposed that the error characteristics (random error) within the data are still unchanged. The bias-corrected XCO_2 retrievals of both ACOS and BEDS are used as fundamental data for the combination algorithm.

By applying the series of processes shown in Section 3, daily, weekly, as well as monthly maps of combined XCO_2 for 2009 and 2010 are generated. Here, as an example, only four maps (from May to August) of monthly mean XCO_2 of 2010 are shown here (Fig. 2–Fig. 5). In addition, the total XCO_2 uncertainties of the combined product which mainly depend on the uncertainties of the original ACOS or BESD XCO_2 retrievals are also illustrated.

From Fig. 2–Fig. 5, it is not difficult to observe that the combined data realize the physical complementary of the two products in terms of spatial coverage. The number of valid CO_2 measurements in the fused product is the union of the CO_2 data from both the ACOS and BESD at the same geographical location. In addition, the combined XCO_2 demonstrates similar spatiotemporal characteristics with that of ACOS and BESD over the globe, which implies that all processes associated with the combination do not distort the essential information of the original XCO_2 products (ACOS or BESD). Similar findings can also be observed in the daily mean and weekly mean XCO_2 maps. To quantitatively investigate the improvement of fused XCO_2 in spatial coverage, the fractional coverage of all three variables

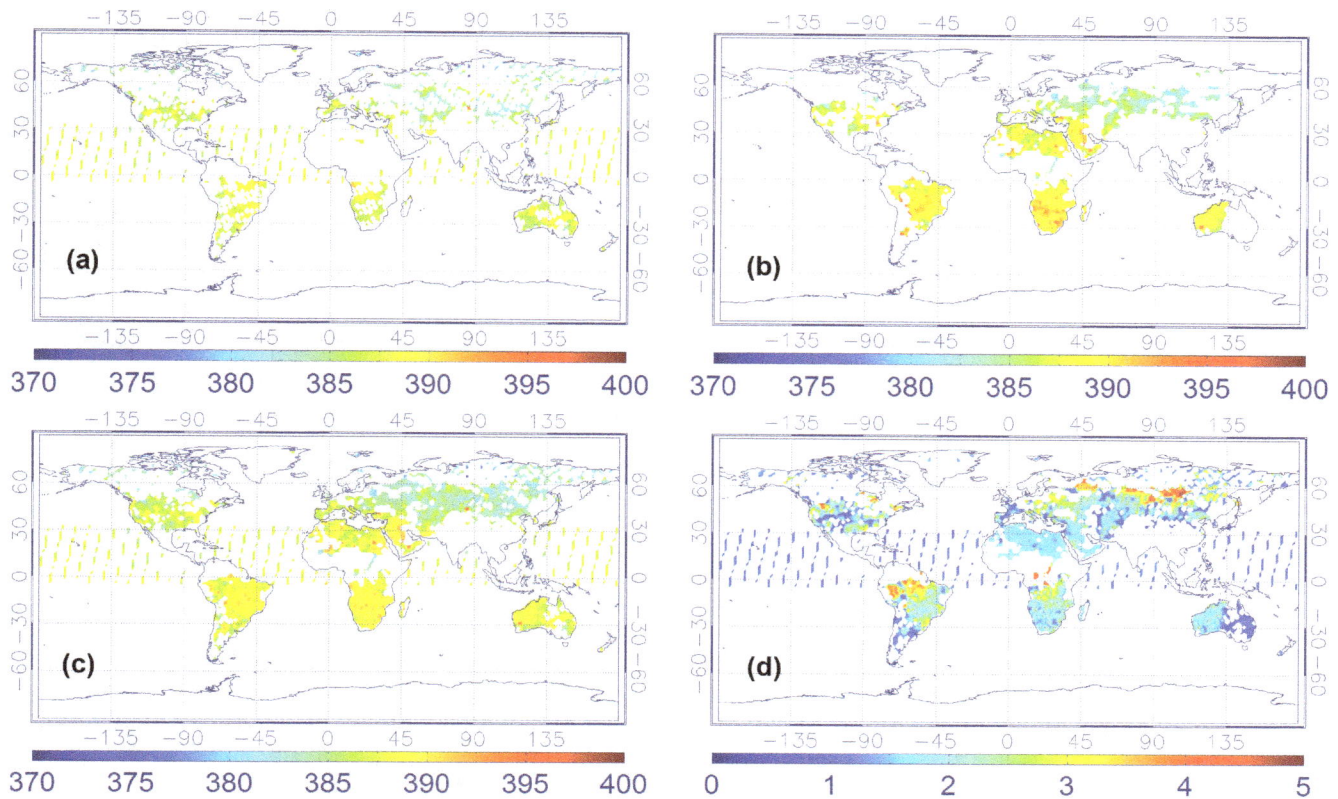

Figure 5. XCO$_2$ monthly mean maps in August of 2010. ((a) ACOS XCO$_2$, (b) BESD XCO$_2$, (c) combined product, and (d) XCO$_2$ uncertainties of the combined product).

(ACOS, BESD, and combined XCO$_2$) on both daily and monthly scales is calculated (Fig. 6). From Fig. 6, it can be seen that the average global coverage of ACOS and BESD is around 0.46% and 0.21%, respectively, on a daily scale. The monthly mean coverage of such products accounts for about 5.70% and 3.75%, respectively. While spatial coverage of combined XCO$_2$ can reach up to 0.65% and 8.42% on daily and monthly scales, respectively, it accounts for increments of 41.3% and 47.7% on the daily and

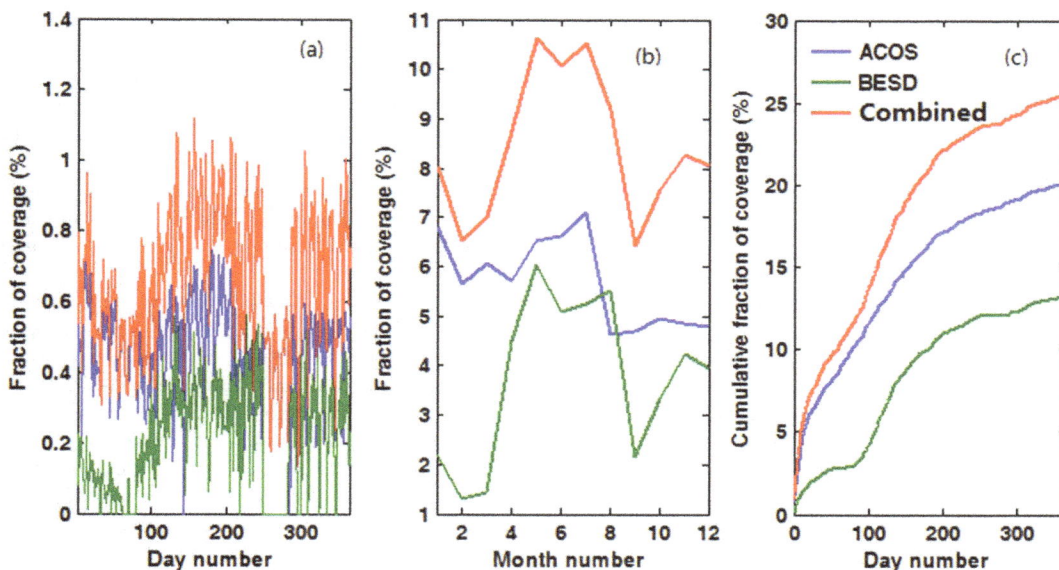

Figure 6. XCO$_2$ fraction of coverage of ACOS, BESD, and combined products. (a) Daily coverage. (b) Monthly coverage. (c) Cumulative coverage.

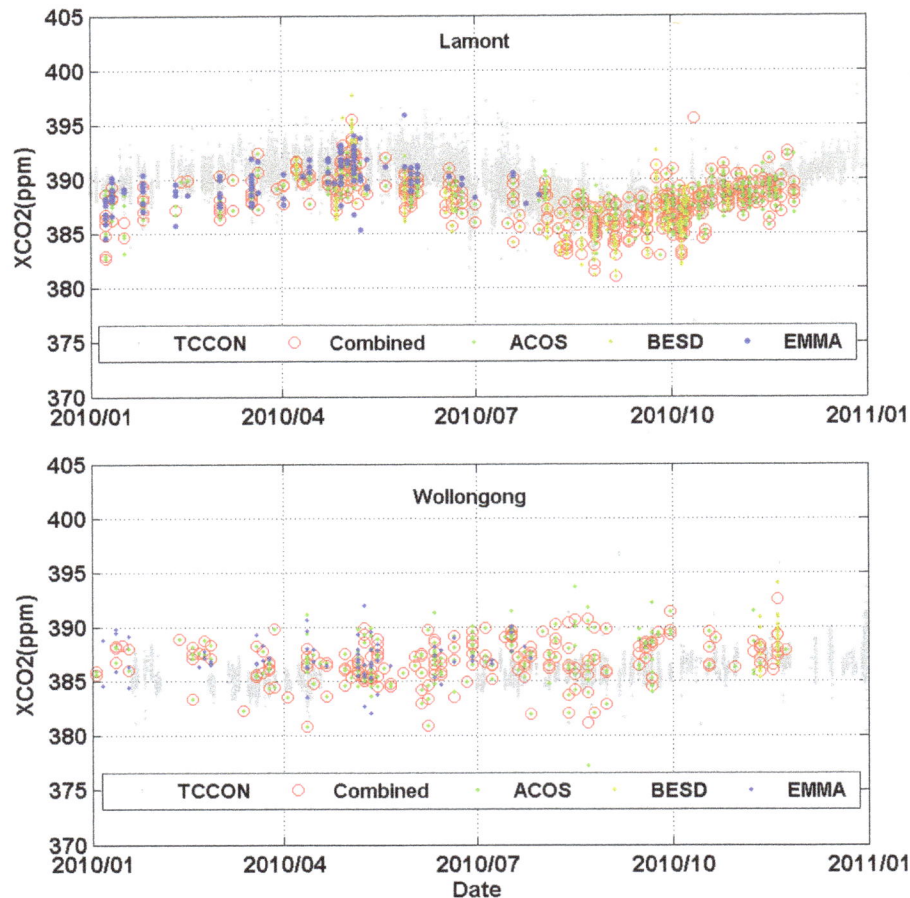

Figure 7. Comparison of XCO$_2$ measurements from TCCON, ACOS, BESD, EMMA, and our new combination method over Wollongong and Lamont sites (distance<0.25 degree, temporal difference<1 hour).

monthly scales with respect to that of GOSAT and it is even higher relative to the coverage of SCIAMACHY. Likewise, the cumulative fraction of coverage of the combined XCO$_2$ has risen to 25% when compared with 20% and 13% for ACOS and BESD, respectively. The increase in the XCO$_2$ spatial coverage indicates the potential advantage of the combined XCO$_2$ observations in generating global Level-3 XCO$_2$ maps when compared with any single dataset by providing more satellite-based XCO$_2$ retrievals used for optimal interpolating.

For evaluating the performance of our combination strategy, the combined XCO$_2$ values are compared with that retrieved from ACOS and BESD as well as XCO$_2$ in the EMMA database at two TCCON sites (Fig. 7). The results reveal that the XCO$_2$ values from the combination method show generally consistent variation in time with TCCON measurements except for a small overall bias (especially for the Lamont site). On the whole, the new combined XCO$_2$ product shows good consistency with the EMMA data, and they are comparable in terms of CO$_2$ magnitude, while the combined XCO$_2$ are shown with a longer time period, which is in line with the satellite observations, and possess more data points even over the same period.

Discussions and Conclusions

Despite the fact that space-based measurements can provide a unique opportunity to map atmospheric CO$_2$ over large areas, the number of valid CO$_2$ measurements from a single space-based instrument is generally limited for a certain day over a specific region due to the presence of clouds. In addition, although these Level-2 XCO$_2$ retrievals themselves are very important for inversion modeling of surface carbon sources/sinks, further comprehensive analysis by investigating the spatiotemporal full-coverage XCO$_2$ (Level 3) distribution is needed for interpreting their significant scientific merit [32]. While the limited satellite observations restrict the generation of Level-3 XCO$_2$ maps with high spatial and temporal resolutions when only a single satellite-based XCO$_2$ dataset is considered. This is our main motivation in this paper.

In this study, a strategy for combining SCIAMACHY and GOSAT CO$_2$ measurements has been proposed by fully accounting for the CO$_2$ global bias, differences in averaging kernels and overpass times, and the Level-2 retrieval errors of the CO$_2$ measurements being used. The results indicated that the average global coverage of both ACOS and BESD is less than 0.5% on a daily scale, and less than 6% on a monthly scale. While spatial coverage of combined XCO$_2$ can reach up to 0.65% and 8.42% on daily and monthly scales, respectively, the comparison analysis reveals that the combined XCO$_2$ product is consistent with TCCON and EMMA in both temporal variation and magnitude except for a small bias when compared with the TCCON measurements. All these findings herein prove the effectiveness of the combination method in supporting generation global full-coverage XCO_2 maps with higher temporal and spatial sampling by jointly using two space-based XCO_2 datasets. Similar to the existing studies (e.g. [32–34]), although these combined XCO_2 are not intended to be used in inverse modeling studies, they deliver a key complement for such research, and can be deemed as an independent dataset for comparison with model predictions. Similar to the existing study [31], an improved fusion approach (based on multiple XCO_2 datasets) to create Level-2 XCO_2 measurements that can be directly used for inverse modeling is also attempted and will be presented in another paper.

A last point that needs to be addressed is that although we employed CO_2 data of GOSAT and SCIAMACHY in this study, the proposed strategies are not restricted to such data. As a general strategy, it can be refined and adapted to further combine other XCO_2 products, such as OCO-2, CarbonSat, etc. in the future,

and even to be applied to the fusion of other trace gases, such as O_3, CH_4.

Acknowledgments

The authors would like to thank the SCIAMACHY team at University of Bremen IUP/IFE as well as the ACOS scientific teams for providing us the CO_2 products. The authors also thank the anonymous reviewers for their helpful and valuable comments to improve this work.

Author Contributions

Conceived and designed the experiments: TW JS YJ TZ DJ CX. Performed the experiments: TW JS YJ TZ DJ CX. Analyzed the data: TW JS YJ TZ DJ CX. Contributed reagents/materials/analysis tools: TW JS YJ TZ DJ CX. Wrote the paper: TW JS YJ TZ DJ CX.

References

1. Reuter M, Bovensmann H, Buchwitz M, Burrows JP (2010) A method for improved SCIAMACHY CO_2 retrieval in the presence of optically thin clouds. Atmos. Meas. Tech 3: 209–232.

2. O'Dell CW, Connor B, Bösch H, O'Brien D, Frankenberg C, et al. (2012) The ACOS CO_2 retrieval algorithm – Part 1: Description and validation against synthetic observations. Atmos Meas Tech 5(1): 99–121.

3. Baker DF, Law RM, Gurney KR, Rayner P, Peylin P, et al (2006) TransCom 3 inversion intercomparison: Impact of transport model errors on the interannual variability of regional CO_2 fluxes, 1988–2003. Global Biogeochem Cy 20: GB1002, doi:10.1029/2004GB002439.

4. Morino I, Uchino O, Inoue M, Yoshida Y, Yokota T, et al. (2011) Preliminary validation of column-averaged volume mixing ratios of carbon dioxide and methane retrieved from GOSAT short-wavelength infrared spectra. Atmos Meas Tech 4: 1061–1076.

5. Butz A, Hasekamp OP, Frankenberg C, Aben I (2009) Retrievals of atmospheric CO_2 from simulated space-borne measurements of backscattered near-infrared sunlight: accounting for aerosol effects. Appl Opt 18: 3322–3336.

6. Kuze A, Suto H, Nakajima M, Hamazaki T (2009) Thermal and near infrared sensor for carbon observation Fourier-transform spectrometer on the Greenhouse Gases Observing Satellite for greenhouse gases monitoring, Appl Opt 35: 6716–6733.

7. Bovensmann H, Burrows JP, Buchwitz M, Frerick J, Noel S, et al (1999) SCIAMACHY –Mission objectives and measurement modes. J Atmos Sci 56: 127–150.

8. Crisp D, Atlas RM, Breon FM, Brown LR, Burrows JP, et al (2004) The Orbiting Carbon Observatory (OCO) mission. Adv Space Res 34: 700–709.

9. Boesch H, Baker D, Connor B, Crisp D, Miller C (2011) Global Characterization of CO_2 Column Retrievals from Shortwave-Infrared Satellite Observations of the Orbiting Carbon Observatory-2 Mission. Remote Sens 3: 270–304.

10. Aumann HH, Chahine MT, Gautier C, Goldberg MD, Kalnay E, et al. (2003) AIRS/AMSU/HSB on the Aqua Mission: Design, science objectives, data products, and processing systems. IEEE Trans Geosci Remote Sens 41: 253–264.

11. Chahine M, Barnet C, Olsen ET, Chen L, Maddy E (2005) On the determination of atmospheric minor gases by the method of vanishing partial derivatives with application to CO_2. Geophys Res Lett 32: L22803, doi:10.1029/2005GL024165.

12. Phulpin T, Cayla F, Chalon G, Diebel D, Schlussel P (2002) IASI on board Metop: Project status and scientific preparation. Proceedings of the 12th International TOVS Study Conference. Lorne, Vi ctoria, Australia.

13. Turquety S, Hadji-Lazaro J, Clerbaux C, Hauglustaine DA, Clough SA, et al. (2004) Operational trace gas retrieval algorithm for the Infrared Atmospheric Sounding Interferometer. J Geophys Res 109: D21301, doi:10.1029/2004JD004821.

14. Yang Z, Washenfelder RA, Keppel-Aleks G, Krakauer NY, Randerson JT, et al. (2007) New constraints on Northern Hemisphere growing season net flux. Geophys Res Lett 34: L12807. doi:10.1029/2007GL029742.

15. Wunch D, Toon GC, Blavier JFL, Washenfelder R, Notholt J, et al. (2011) The Total Carbon Column Observing Network. Philos T Roy Soc A 369: 2087–2112.

16. Rayner PJ, O'Brien DM (2011) The utility of remotely sensed CO_2 concentration data in surface source inversions. Geophys Res Lett 28: 175–178.

17. Houweling S, Breon FM, Aben I, Rodenbeck C, Gloor M, et al. (2004) Inverse modeling of CO_2 sources and sinks using satellite data: a synthetic inter-comparison of measurement techniques and their performance as a function of space and time. Atmos Chem Phys 4: 523–538.

18. Olsen SC, Randerson JT (2004) Differences between surface and column atmospheric CO_2 and implications for carbon cycle research. J Geophys Res 109; D02301. doi:10.1029/2003JD003968.

19. Miller CE, Crisp D, DeCola PL, Olsen SC, Randerson JT, et al. (2007) Precision requirements for space-based XCO_2 data. J Geophys Res 112: D10314, doi:10.1029/2006JD007659.

20. Buchwitz M, Beek R, Burrows JP, Bovensmann H, Warneke T, et al. (2005) Atmospheric methane and carbon dioxide from SCIAMACHY satellite data: initial comparison with chemistry and transport models, Atmos Chem Phys 5: 941–962.

21. Buchwitz M, Beek R, Nol S, Burrows JP, Bovensmann H, et al. (2005) Carbon monoxide, methane and carbon dioxide columns retrieved from SCIAMACHY by WFM-DOAS: year 2003 initial data set. Atmos Chem Phys 5: 3313–3329.

22. Buchwitz M, Rozanov VV, Burrows JP (2000) A near-infrared optimized DOAS method for the fast global retrieval of atmospheric CH_4, CO, CO_2, H_2O, and N_2O total column amounts from SCIAMACHY Envisat-1 nadir radiances. J Geophys Res 105: 15231–15245, doi:10.1029/2000JD900191.

23. Houweling S, Hartmann W, Aben I, Schrijver H, Skidmore J, et al. (2005) Evidence of systematic errors in SCIAMACHY-observed CO_2 due to aerosols. Atmos Chem Phys 5: 3003–3013.

24. Barkley MP, Frieβ U, Monks PS (2006) Measuring atmospheric CO_2 from space using Full Spectral Initiation (FSI) WFM-DOAS, Atmos Chem Phys 6: 3517–3534.

25. Schneising O, Buchwitz M, Burrows JP, Bovensmann H, Reuter M, et al. (2008) Three years of greenhouse gas column-averaged dry air mole fractions retrieved from satellite – Part 1: Carbon dioxide. Atmos Chem Phys 8: 3827–3853.

26. Reuter M, Bovensmann H, Buchwitz M, Burrows JP, Connor BJ, et al. (2011) Retrieval of atmospheric CO_2 with enhanced accuracy and precision from SCIAMACHY: Validation with FTS measurements and com-parison with model results. J Geophys Res 116: D04301, doi:10.1029/2010JD015047.

27. Bösch H, Toon GC, Sen B, Washenfelder RA, Wennberg PO, et al. (2006) Space-based near-infrared CO_2 measurements: Testing the Orbiting Carbon Observatory retrieval algorithm and validation concept using SCIAMACHY observations over Park Falls, Wisconsin. J Geophys Res 111: D23302, doi:10.1029/2006JD007080.

28. Butz A, Guerlet S, Hasekamp O, Schepers D, Galli A, et al. (2011) Toward accurate CO_2 and CH_4 observations from GOSAT. Geophys Res Lett 14: L14812. DOI:10.1029/2011GL047888.

29. Yoshida Y, Ota Y, Eguchi N, Kikuchi N, Nobuta K, et al. (2011) Retrieval algorithm for CO_2 and CH_4 column abundances from short-wavelength infrared spectral observations by the Greenhouse Gases Observing Satellite. Atmos Meas Tech 4: 717–734.

30. Crisp D, Fisher B, O'Dell C, Frankenberg C, Basilio R, et al. (2012)The ACOS CO_2 retrieval algorithm - Part II: Global XCO_2 data characterization. Atmos Meas Tech 5: 687–707.

31. Reuter M, Bösch H, Bovensmann H, Bril A, Buchwitz M, et al. (2013) A joint effort to deliver satellite retrieved atmospheric CO_2 concentrations for surface flux inversions: the ensemble median algorithm EMMA. Atmos Chem Phys 13: 1771–1780.

32. Hammerling DM, Michalak AM, O'Dell C, Kawa SR (2012) Global CO_2 distributions over land from the Greenhouse Gases Observing Satellite(GOSAT). Geophys Res Lett 39: L08804, doi:10.1029/2012GL051203.

33. Hammerling DM, Michalak AM, Kawa SR (2012) Mapping of CO_2 at high spatiotemporal resolution using satellite observations: Global distributions from OCO-2. J Geophys Res 117: D06306, doi:10.1029/2011JD017015.

34. Zeng Z, Lei L, Hou S, Ru F, Guan X, et al. (2014) A Regional Gap-Filling Method Based on Spatiotemporal Variogram Model of CO_2 Columns. IEEE Trans Geosci Remote Sens 6: 3594–3603.

35. Yokota T, Yoshida Y, Eguchi N, Ota Y, Tanaka T, et al. (2009) Global Concentrations of CO_2 and CH_4 Retrieved from GOSAT: First Preliminary Results. SOLA 5: 160–163.

36. Parker R, Boesch H, Cogan A, Fraser A, Feng L, et al. (2011) Methane observations from the Greenhouse Gases Observing SATellite: Comparison to

ground-based TCCON data and model calculations. Geophys Res Lett 15: L15807, doi:10.1029/2011GL047871.

37. Wang TX, Shi JC, Jing YY, Xie YH (2013) Investigation of the consistency of atmospheric CO_2 retrievals from different space-based sensors: Intercomparison and spatio-temporal analysis, Chin Sci Bull, 33: 4161–4170.

38. Peters W, Jacobson AR, Sweeney C, Andrews AE, Conway TJ, et al. (2007) An atmospheric perspective on North American carbon dioxide exchange: Carbon Tracker. Proc Natl Acad Sci U.S.A. 48: 18925–18930.

39. Wunch D, Wennberg PO, Toon GC, Connor BJ, Fisher B, et al. (2011) A method for evaluating bias in global measurements of CO_2 total columns from space. Atmos Chem Phys 11: 12317–12337.

40. Rodgers CD (2000) Inverse Methods for Atmospheric Sounding: Theory and Practice. World Scientific Publishing Co. Ltd.193 p.

14

The Deep Atmospheric Boundary Layer and Its Significance to the Stratosphere and Troposphere Exchange over the Tibetan Plateau

Xuelong Chen[1,2], Juan A. Añel[3,4], Zhongbo Su[2], Laura de la Torre[4], Hennie Kelder[5], Jacob van Peet[6], Yaoming Ma[1]*

1 Key Laboratory of Tibetan Environment Changes and Land Surface Processes, Institute of Tibetan Plateau Research, Chinese Academy of Sciences, Beijing, China, **2** Faculty of Geo-Information Science and Earth Observation, University of Twente, Enschede, The Netherlands, **3** Smith School of Enterprise and the Environment, University of Oxford, Oxford, United Kingdom, **4** EPhysLab, Fac. of Sciences, Universidade de Vigo, Ourense, Spain, **5** Eindhoven University, Eindhoven, The Netherlands, **6** Royal Netherlands Meteorological Institute (KNMI), Utricht, The Netherlands

Abstract

In this study the depth of the atmospheric boundary layer (ABL) over the Tibetan Plateau was measured during a regional radiosonde observation campaign in 2008 and found to be deeper than indicated by previously measurements. Results indicate that during fair weather conditions on winter days, the top of the mixed layers can be up to 5 km above the ground (9.4 km above sea level). Measurements also show that the depth of the ABL is quite distinct for three different periods (winter, monsoon-onset, and monsoon seasons). Turbulence at the top of a deep mixing layer can rise up to the upper troposphere. As a consequence, as confirmed by trajectory analysis, interaction occurs between deep ABLs and the low tropopause during winter over the Tibetan Plateau.

Editor: Vanesa Magar, Plymouth University, United Kingdom

Funding: This research was funded by the Chinese National Key Programme for Developing Basic Sciences (2010CB951701), and the Chinese National Natural Science Foundation (40825015 and 40810059006). Xuelong Chen acknowledges the support of KNAW. The authors thank all the participants from China and Japan for their work during the JICA Tibetan Plateau meteorological observations. Laura de la Torre was partially supported by the European Regional Development Fund. The funders played no part in study design, data collection and analysis, the decision to publish, or the preparation of the manuscript. No additional external funding was received for this study.

* E-mail: ymma@itpcas.ac.cn

Introduction

The Tibetan Plateau, or the "Roof of the World" [1,2], is the world's largest and highest plateau. It exerts a profound thermal and dynamical influence on the local atmosphere [3], and atmospheric processes above the plateau are crucial for regional studies of climate and weather. Research indicates that the evolution of the atmospheric boundary layer (ABL) over elevated terrain can influence its development in its near and downstream regions [4]. In particular, the plateau's ABL influences its Eastern downstream area [5,6], and some synoptic systems over East China originate in the ABL over the plateau [7,8]. Convections over the Tibetan Plateau can influence the hydration of the global stratosphere [9]. A clear understanding of the ABL in this region is key to attempts to obtain a better picture of the thermodynamic influence of the plateau on weather and climate in Asia [10,11].

Over the past thirty years, several comprehensive observational experiments have revealed important characteristics of the plateau's land surface processes and of the structure of the ABL over the plateau [3,12–17]. It is now known, for example, that the ABL over the plateau is deeper than that over the lowlands [3,15,18,19]. However, few studies have investigated the connection of the high ABL with the upper troposphere and lower stratosphere (UTLS), even though the Tibetan Plateau is regarded to be a pathway of mass exchange between the troposphere and stratosphere [20].

The elevation of the Tibetan Plateau varies between 3000 and 8848 m above sea level (ASL). The top of the ABL may be as high as 9000 m ASL, which is close to the location of the tropopause: this is a result of both the plateau's elevation and the depth of the ABL. This can result in a stronger interaction between the UTLS and the ABL than in lowland areas. Chen et al. [21] concluded that the Tibetan Plateau is one of three key source regions for transport from the boundary layer to the tropopause in the Asian monsoon region. Studies have shown that multi-tropopause events, which are closely related to tropopause folds, frequently occur over the Tibetan Plateau [22–24]. Tropopause folds can cause stratospheric air to be transported downwards to the ABL through a number of different processes [25]. A previous study indicated that dynamic transport was the main factor that influences the vertical distribution of ozone over the Tibetan Plateau [26]. Surface ozone over the Tibetan Plateau is sensitive to ozone perturbation in the upper layers [27], suggesting that ozone above the planetary boundary layer may strongly influence surface ozone. It is clear that the interaction between the plateau's ABL and the UTLS is also important for troposphere-stratosphere exchanges.

Table 1. Observation period for each IOP and its representative season.

	IOP1	IOP2	IOP3
Observation date	25, Feb-19, Mar	13, May-12, Jun	07, Jul-16, Jul
Representative season	Winter	Monsoon-onset	Monsoon

The depth of the ABL changes in both space and time, and varies from less than one hundred to several thousand meters [28,29]. To present our investigation of this behavior, this paper is organized as follows. The next section presents the data and methods used to derive the depth of the ABL. Next, the depth is analyzed as a function of season and tropopause height, based on intensive observations during the winter, monsoon-onset and monsoon seasons. The next section describes simultaneous variations in both tropopause folds and the ABL, and is followed by a Lagrangian analysis to determine the possibility of air mass exchange between a deep ABL and the UTLS. Finally, the discussion and conclusions are presented.

Data and Methods

In a joint Sino–Japanese project, radiosonde observations were carried out over the Tibetan Plateau in 2008 [2]. The sounding dataset of the Gerze station (32.17°N, 84.03°E, 4415 m above mean sea level) was chosen to compare the ABL depth and tropopause height over the plateau, because of the absence of high mountains in the vicinity. In 2008, three intensive observation periods (IOPs) were used. The detail information about each IOP is listed in table 1. Vaisala RS-92 radiosondes were released every day at 01:00, 07:00, 13:00 and 19:00 BST (Beijing Standard Time, UTC+8) during these IOPs.

The height of the tropopause was determined using the World Meteorological Organization (WMO) lapse rate tropopause (LRT) definition [30], which defines the tropopause to be the lowest level at which the lapse rate decreases to 2 K/km or less, provided that the average lapse rate between this level and all higher levels within 2 km does not exceed 2 K/km.

The top of the convective boundary layer (CBL) was determined using the simple parcel method, which is a reliable method in unstable conditions [31–33]. This method equates the top of the CBL with the intersection of the actual potential temperature profile with the dry-adiabatic ascent, starting at 'near-surface temperature'.

The ECMWF ERA-Interim data [34] was used to analyze the UTLS structures around the plateau during the radiosonde ascents. Trajectory models are often used to trace the movement of air particles (e.g. Ladstätter-Weißenmayer et al. [35], Bergman et al. [36]). In order to check possible exchanges between the high CBL and the stratosphere, a trajectory model forced with ERA-Interim data was used to simulate whether the air mass can be transported from troposphere to stratosphere, and vice versa. The

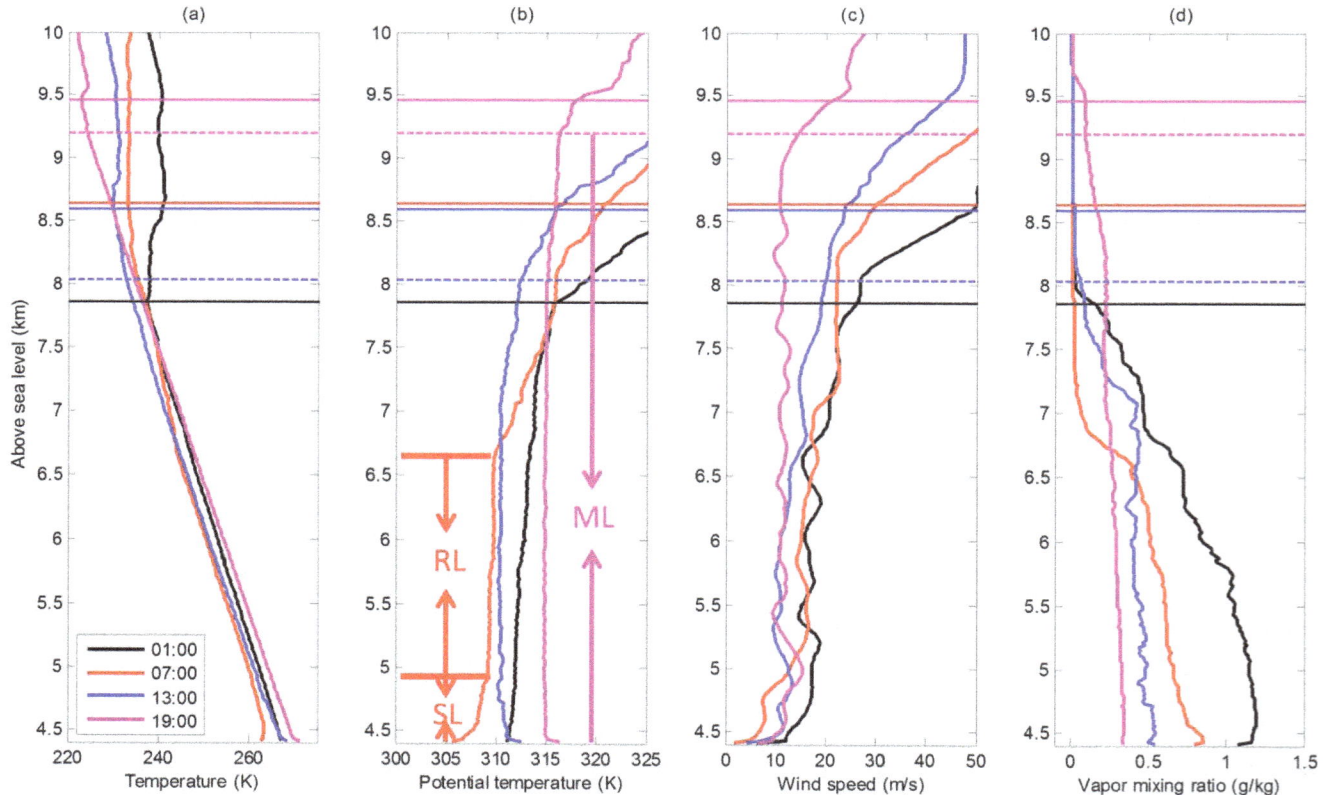

Figure 1. Profiles of (a) temperature, (b) potential temperature, (c) wind speed, (d) water vapor content on 25 Feb 2008. Profiles were recorded at 01:00 (dark line), 07:00 (red line), 13:00 (blue line), and 19:00 (magenta line) BST. The horizontal dashed lines show the corresponding tops of the CBL, and horizontal solid lines show the positions of the tropopause. The stable layer (SL), residual layer (RL), and mixed layer (ML) are also marked.

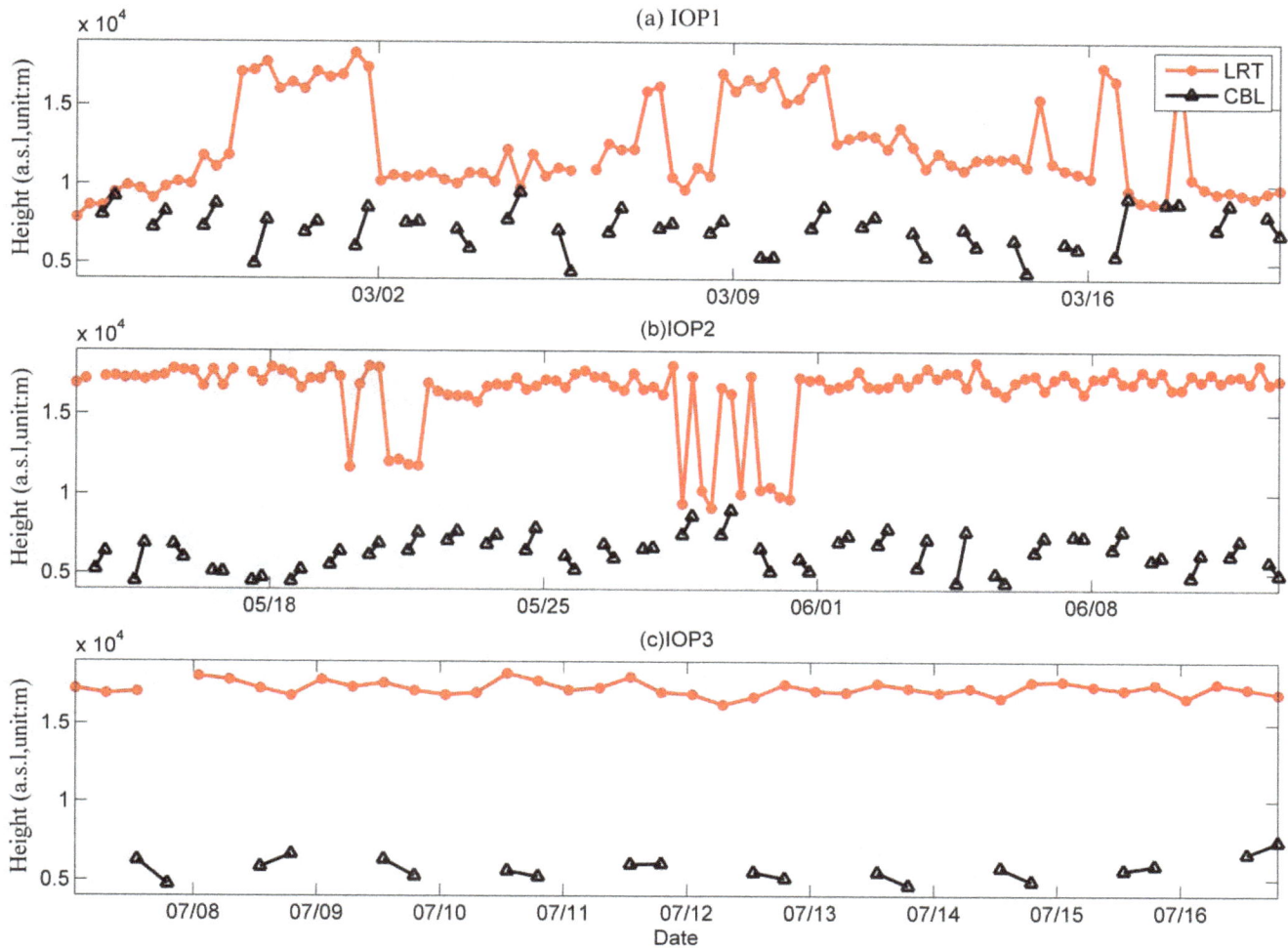

Figure 2. Variation of CBL depth and height of LRT during the three IOPs.

FLEXPART v8.1 model [37,38] was used to perform the trajectory analysis.

Ozone profiles from the GOME-2 instrument [39] were also used to study the structure of the UTLS from space. GOME-2 measures backscattered solar radiation from the Earth's atmosphere between 250 and 790 nm in four channels, with a relatively high spectral resolution (0.2–0.4 nm). In its normal mode, the instrument has an almost global daily coverage with a cross-track swath width of 1920 km, split into ground pixels of size 640×40 km. Ozone profiles were retrieved using the optimal estimation method described in Rodgers [40], with a pressure grid of 41 levels between the surface and the top of the atmosphere.

Results

The Variations of ABL Depth and Tropopause Height

According to meteorological measurements, IOP1, IOP2 and IOP3 represent winter time, monsoon-onset, and monsoon time respectively. From IOP1, 25 February was chosen as a typical sunny day, in order to analyze the detailed structure of a very deep ABL (Fig. 1). A surface-based inversion formed during the early morning, which was eliminated by 13:00 (Fig. 1a). The 07:00 sounding revealed a cool, stable layer (SL) 500 m deep

near the surface. Above this was a residual layer (RL) (Fig. 1b) of 1.7 km depth, which had a nearly constant potential temperature. After sunrise, the net radiation quickly increased, causing the surface to warm rapidly. By 13:00, this surface heating had caused sufficient growth of the mixed layer (ML) to replace the stable surface layer and residual layer. By 19:00, a mixed layer 4780 m high had been formed (equal to 9195 m ASL). Despite the fast development of the mixed layer, the water vapor and wind speed were also well mixed by turbulence, as generated by surface heating (Fig. 1c and 1d). ABL growth also entrained relatively dry air from above, which reduced the water content of the mixed layer between 01:00 and 19:00. The water vapor content in the mixed layer decreased slightly with increasing height, due to surface evaporation. All these phenomena suggest a mixed layer dominated by convection.

The wind shear in the troposphere on 25 February was weaker than on other days. At around 9.3 km ASL the wind speed increased very quickly, and the water vapor content dropped quickly. The layers of high gradient around 9.3 km ASL can be regarded as the top of the CBL, because of the well mixed layers below this height, or as an indication of the UTLS. Using the definitions mentioned previously, the top of the CBL and the tropopause height are indicated by horizontal dashed lines and

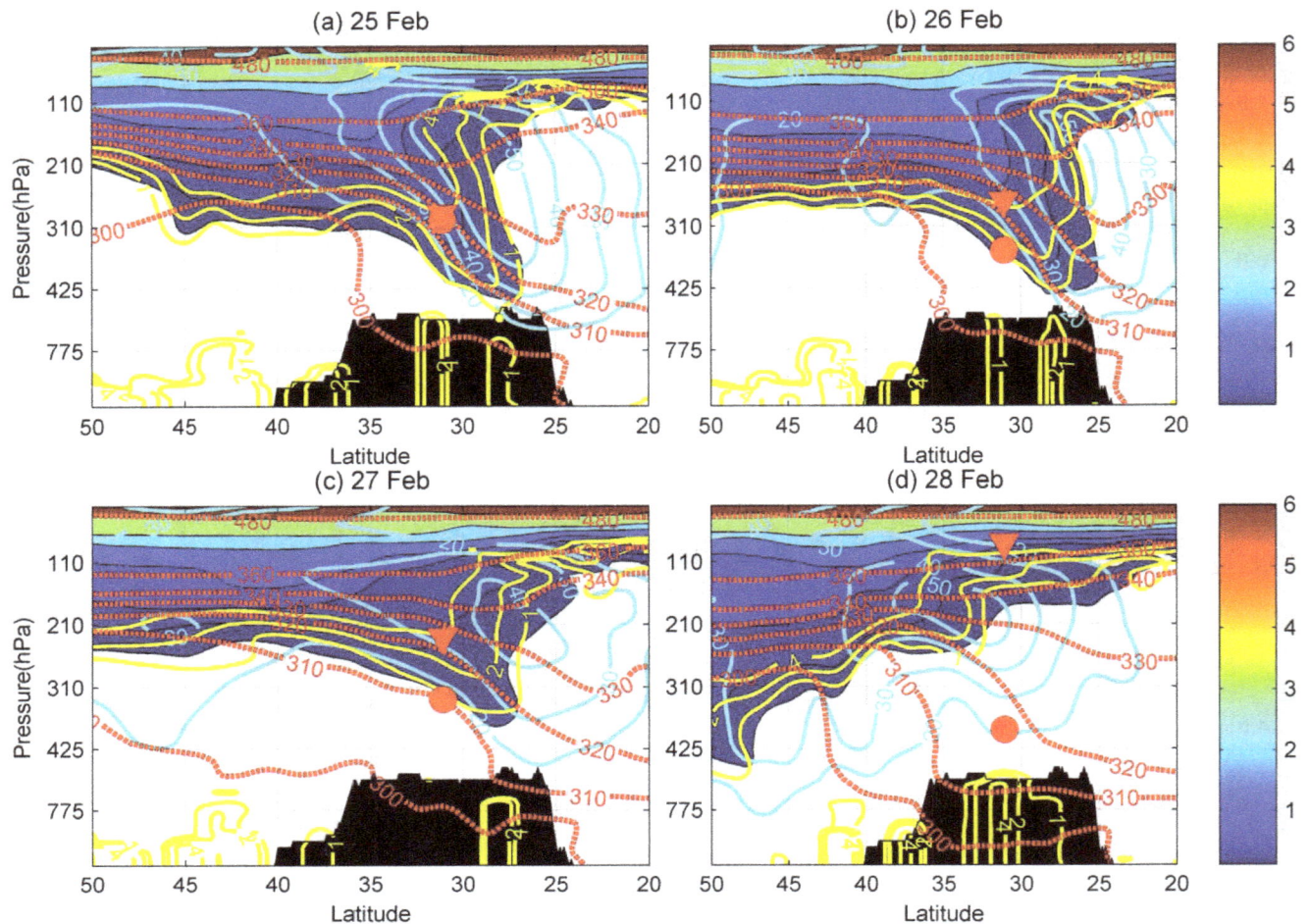

Figure 3. Meridional cross-section at 84.25° E (over Gerze site) at 20:00 BST, for the period 25 Feb. to 28 Feb. 2008, derived from ERA interim data, including zonal winds (cyan contours, m/s), potential vorticity (yellow lines, contours of 1, 2, 4 PV units), ozone (solid color, ×10⁶ kg/kg) and potential temperature (red contours, K). The color bar is the scale of ozone concentration. The area in black shows the cross section of the Tibetan Plateau terrain. The red triangles and circles show the position of the LRT and the top of CBL.

solid lines respectively. From 01:00 to 19:00, the height of the tropopause and of the top of the CBL both increased. At 19:00, these two characteristic layers were very close together. The turbulent mixing in the CBL developed by the surface heating may have intersected with the downward intrusion of stratospheric air caused by the tropopause fold (shown later).

Figure 2 shows variations of the tropopause height and the CBL depth during each IOP. The results indicate that the mixing layer can be up to 5 km deep, which is greater than other previous measured values over the plateau [3,18,41] and deeper than predicted [42]. The tropopause and top of the CBL were very close during IOP1 especially on 25 and 26 February, and on 4, 16, 17 and 18 March. However, these two characteristic layers can still be differentiated, and the measurements place the top of the CBL below the tropopause in each case.

During IOP2 and IOP3, the potential temperature in the mixing layer was also well mixed by turbulence. The height of the CBL was lower than that in IOP1. On most days during IOP1, the height of the tropopause was below 12 km, while during IOP2 and IOP3 the tropopause height was usually around 17 km. In the afternoons, the top of the CBL ranged from 5–10 km ASL during IOP1, from 4.5–9 km ASL during IOP2, and from 5–7.5 km ASL during IOP3.

These results indicate that the tropopause was lowest and the CBL was deepest during IOP1, while during IOP2 and IOP3 the tropopause was higher and the CBL was shallower. Chen et al. [24] demonstrated that variations in the height of the tropopause were related to tropopause folds. In the next section, the displacement of tropopause folds were analyzed with synchronous CBL variations. For winter days with a low tropopause and a deep CBL, we use trajectory analysis to test the extent of interactions between the CBL and the UTLS over the Plateau.

The Correspondence between Variations of Tropopause Folds and the CBL

The UTLS structure around the plateau was analyzed using ERA-Interim reanalysis data, which are plotted as meridional cross sections of the atmosphere at 12:00 GMT for 25–28 February during IOP1 (Fig. 3). The ABL is usually fully developed in the afternoon, so the ERA-Interim data at 12:00 GMT was used here. The observed heights of lapse-rate tropopause (LRT) and top of CBL, derived from radiosonde measurements at 19:00 BST, are labeled with red triangles and circles in the two figures. Potential vorticity (PV) isolines, zonal wind, and potential temperature derived from the reanalysis data are also shown on these graphs.

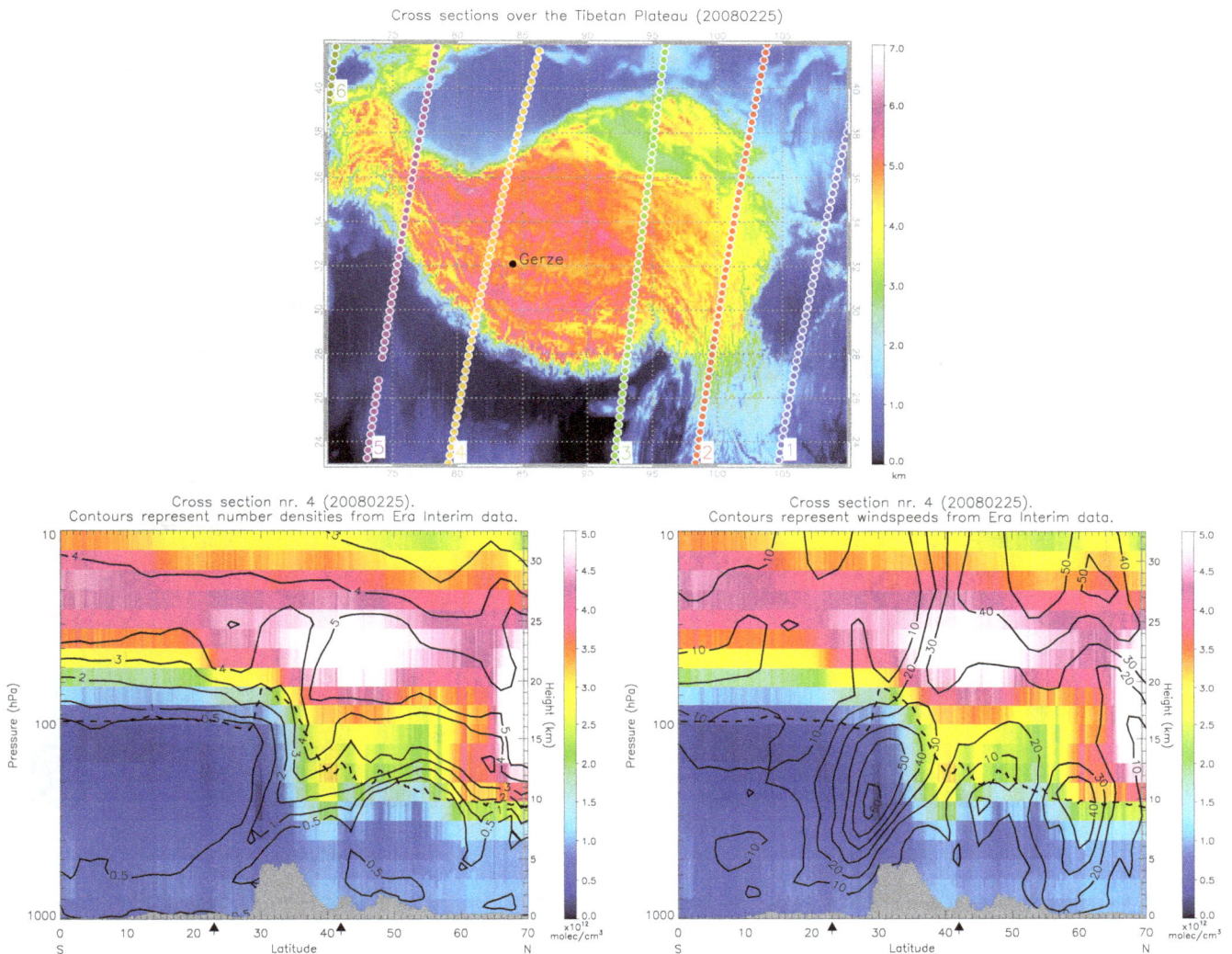

Figure 4. GOME-2 ozone results. Top: location of GOME-2 orbits over the Tibetan Plateau, with the altitude scale in km. Bottom left: GOME-2 ozone profiles in units of 10^{12} molecules/cm^3, with ERA-Interim number density contours superimposed. Bottom right: GOME-2 ozone profiles with ERA-Interim horizontal wind speed contours superimposed. The arrows below the x-axis in the bottom two plots show the extent of the Tibetan Plateau as illustrated in the top plot. The dashed line is the position of the thermal tropopause, according to the WMO definition.

The ozone content acts as a marker to indicate the tropopause fold events.

During IOP1 the westerly jet stream runs along the 28° N parallel, directly above the southern Tibetan Plateau (Fig. 3a–3d). The dynamical tropopause, identified by the isolines of 1 and 2 potential vorticity unit (PVU), exhibits a folded structure over the plateau. Several studies have revealed that wintertime tropopause folding beneath the subtropical jet stream can lead to a downward transport of stratospheric ozone into the middle troposphere, the lower troposphere or even the ABL [25,43–45]. A southern downward motion on the poleward side of the jet transports stratospheric air downwards isentropically. On 25 February, the layer between 2 PVU and 4 PVU, which can be considered to represent the UTLS, reached down to 425 hPa over the Plateau and gradually decayed between 26 and 28 February. Figs. 3a and 3b indicated that the top of the CBL on 25, and 26 February was inside the tongue of the stratospheric intrusions.

To validate the modeled ozone distribution derived from ERA-Interim, it was compared with data from the GOME-2 orbit closest to the Gerze station (see Fig. 4) on 25 February 2008. The observed ozone concentrations agree very well with the ERA-Interim data, and the tropopause folding event is clearly linked to the location of the jet stream over the Tibetan Plateau.

During IOP3, the jet core moved north of the plateau to 42° N, as did the slightly folded UTLS structure (Fig. 5). The quasi-vertical lines of potential temperature over the Tibetan Plateau moved to the north, outside of the plateau domain. The potential temperature was more horizontally stratified over the plateau, and the troposphere was more stable during the monsoon season than during the winter time. Griffiths et al. [46] showed that the tropopause folds increase the generation of potential instability in its vicinity. When the ABL developed sufficiently to reach the height of the unstable intrusion area, it grew even higher. During the deep intrusions of 25 and 26 February the ABL was deeper than during the shallow intrusion of 28 February. During IOP3, when there were no tropopause folds, the ABL was much shallower than during IOP1.

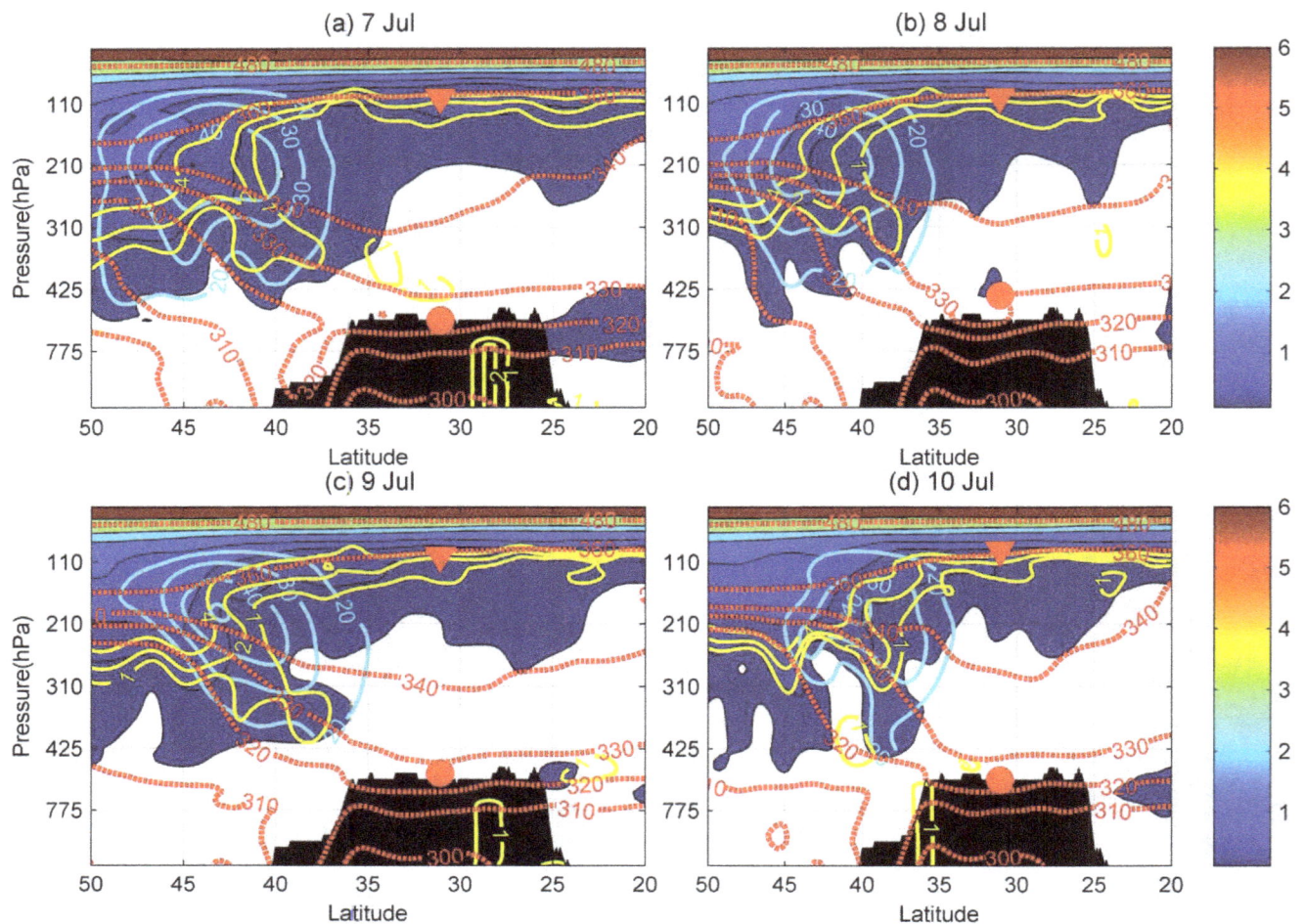

Figure 5. Meridional cross-section at 84.25°E (over Gerze site) at 20:00 BST, for period 7 Jul. 2008 to 10 Jul. 2008 BST. As Figure 3, but for a period in IOP2.

Fig. 6 gives a simple illustration of tropopause folds, westerly jet displacement and convective boundary layer variations during winter and summer time. The strongest westerly jet is situated above the plateau, with a tropopause fold below it. The frequency of stratospheric intrusions caused by tropopause folds is high during winter, which makes the upper tropospheric air unstable and facilitates ABL development. When the jet moves to the north of the Plateau and becomes weaker in summer time, the stratospheric intrusions rarely reach the upper tropospheric air above the plateau. However, whether these variations generally happen still needs more climatological study.

Lagrangian Analysis

To gain a better insight into the relationship between a deep ABL and stratosphere and troposphere exchange (STE), trajectories of air masses for the IOP1 were computed, using the Lagrangian particle dispersion model FLEXPART v8.1 forced with ERA-Interim reanalysis data. This approach is useful for studying the origin of air masses in the UTLS [47]. The analysis was focused on IOP1, because the deep ABL and tropopause folds create a higher potential for STE.

The spatial domain used for FLEXPART was 0–180°E×0–60°N, with a vertical extent of up to 20 km, well above the top of the ABL and the tropopause. To perform the cluster analysis only the particles within a 10°×10° box around the station of Gerze at 12:00 UTC on 25 February 2008 were used.

The analysis was split by height within the reference box for the central date into three layers: 5–9 km, 9–10 km and 10–14 km ASL, which roughly represent the ABL-upper troposphere, tropopause, and lowermost stratosphere layers. The results are also clustered according to the main patterns of movement of particles determined by the model. The light grey part represents later simulated times and the dark part earlier times (see Fig. 7). The small square shows the central point (12:00 UTC on 25 Feb 2008). Each point/diamond represents one time step (±3 hours).

The positions of most of the clusters in the three vertical layers show that the air masses stayed within the same layer for the analyzed period. We do not include the corresponding figures here. However, it is noteworthy that two clusters, accounting for a higher percentage of particles for the 10–14 km layer, show a substantial air mass from lower levels (around 8 km) and that the PV for these air masses decreased after the central date. This result indicates the possibility that air at UTLS levels is irreversibly mixed with air from the upper ABL, which can contribute to the development of a high ABL.

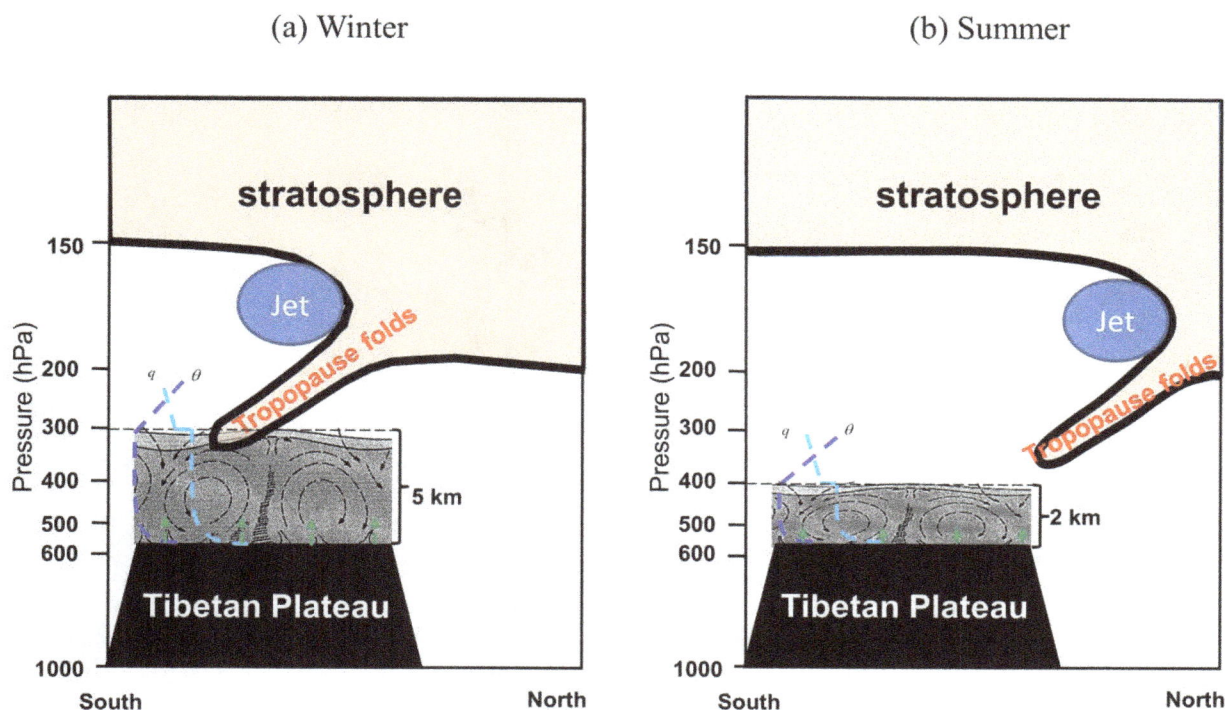

Figure 6. A schematic illustration of the correspondence between tropopause folds, westerly jet displacement and convective boundary layer variations during winter and summer time. The profile of water vapor (q) and potential temperature (θ) in the CBL and surface sensible heating (up green arrow) were included.

Discussion and Conclusions

The structure of the troposphere over the high-altitude terrain of the Tibetan Plateau is still poorly understood, despite its impacts on regional synoptic and atmospheric circulation. Based on high-resolution radiosonde observations, we have shown measurements of a deeper boundary layer than in any previous research. The radiosonde sounding dataset of three different periods in one year demonstrates a significant seasonal contrast in ABL height. Following the suggestion of Santanello et al. [48] that atmospheric stability is the most influential variable controlling ABL development, we compared the atmospheric stability in each observation period and found that in winter time the stratification of the troposphere was related to tropopause folding events. Due to the high frequency of tropopause folds co-existing with the westerly jet situated above the plateau in wintertime [49], the isentropic surfaces in the middle and lower troposphere intersect with the Tibetan Plateau. This distribution of potential temperature lines demonstrates that the troposphere over the plateau is fairly stable during winter time, which makes it easier for a dry and warm eddy to be transported upwards. The instability associated with tropopause folds also provides a potential interpretation of the high ABL. Due to convection and vigorous vertical mixing caused by dry heating at the plateau's surface, a dry-adiabatic lapse rate is established in the high CBL. By the afternoon, the dry thermal convection originating from the heated surface can reach the upper layers of the troposphere. A well-mixed potential temperature and water vapor layer can clearly be identified, when westerly wind dominates over the plateau. The mechanical turbulence caused by shear of the westerly wind can also contribute to the mixing of potential temperature. All these factors make the ABL much deeper, allowing the formation of

larger thermals and eddies [50]. It should be pointed out that radiosonde observations are single points in time and space. We have taken radiosonde data from one station to be representative of the desired area. The mixing height is also influenced by advection, radiation and ABL cloud, which have been ignored here. Indeed, due to the low air density and intense solar radiation, the ABL of the plateau has the potential to grow much deeper than the ABL over lowlands [3]. Thermodynamic sounding profiles suggest that the direct heating of the Tibetan Plateau during winter daytime can rise to as high as 5 km above ground. These results call for a reinterpretation of the response of the ABL over the plateau to uplifted surface heating and the regional meteorological situation.

By comparison with deserts and other arid regions that also have a high ABL [51–53], the elevation of the Tibetan Plateau makes it easy to demonstrate how transport processes in the UTLS over the plateau affect the ABL. A close surface-troposphere-stratospheric coupled system may exist over the plateau. It has been already pointed out that the ozone flux from stratosphere to troposphere to ABL in spring is greatest over the Tibetan Plateau [54]. Johnson and Viezee [25] identified four mechanisms governing the fate of stratospheric air injected into the lower troposphere. Their mechanism 2 is a stratospheric intrusion down to the ABL, and the lower portion of the intrusion is mixed down to the ground by turbulent eddies and convection at the top of the boundary layer. Our observations show that the top of the CBL on 25 and 26 February was connected with the stratospheric intrusions, which is a typical illustration of mechanism 2. In this case, air from the lower troposphere can be mixed by turbulence and transported upwards to the tongues of the intrusions. The intrusions can also transport ozone downwards into the ABL, even down to the ground level. It has been pointed out that the strong

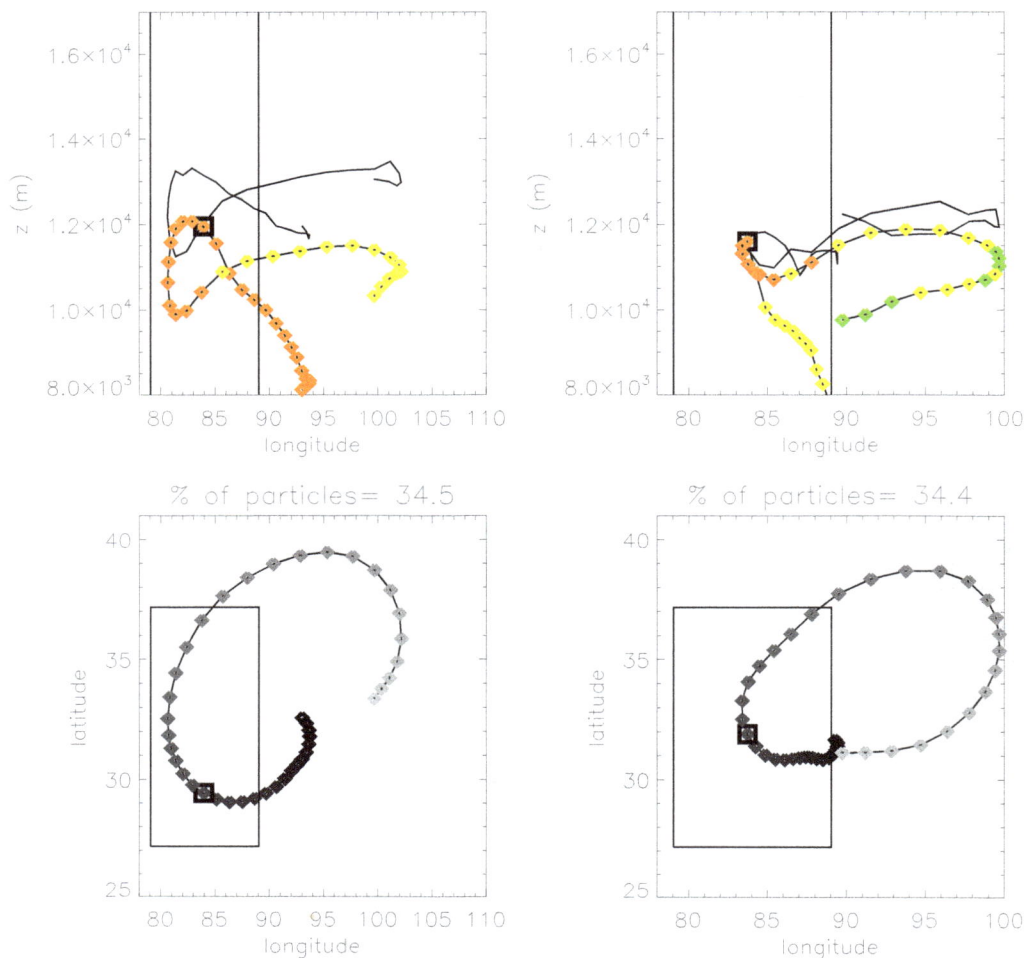

Figure 7. Clusters of trajectories for particles within a 10°×10° box around the station of Gerze and within the 10–14 km layer at 12:00 UTC on 25 February 2008. The smallest square shows the central date (12:00 on 25 Feb 2008 GMT). Each point/diamond represents one time step (3 hours). The upper plots show the height-longitude representation: the solid black line represents the height of the tropopause for each time step as calculated by the FLEXPART model, and the diamond at the lowest level corresponds to the earliest time simulated. Colors correspond to the mean PV of the air mass: orange represents 2.5–3.0 PVU, yellow 2–2.5 PVU, green 1.5–2 PVU. The lower plots show the latitude-longitude representation: the dark grey part represents earlier times in the simulation, while the light grey part represents later times.

vertical mixing in the daytime CBL can bring upper-level ozone downward to augment surface ozone production [55,56]. Therefore, the connection between a deep CBL and the UTLS is important to the plateau's surface ozone pollution, which is already considered as a dominant factor in diurnal variation of surface ozone in America [57,58]. Wang et al. [59] suggested that the surface high-ozone events in the northeastern Tibetan Plateau were mostly caused by the downward transport of upper tropospheric air. The deep ABL and its coupling to the UTLS are also significant to the vertical distribution of ozone over the Tibetan Plateau.

Acknowledgments

The authors thank all the participants from China and Japan for their work during the JICA Tibetan Plateau meteorological observations. The authors would also like to thank Profs. Thomas Foken and Kun Yang for some fruitful discussions held during the work. They also thank Profs. Xiangde Xu and Toshio Koike for their coordination of the Sino-Japanese JICA project.

Author Contributions

Conceived and designed the experiments: YMM XLC. Performed the experiments: XLC. Analyzed the data: XLC JAA LT HK JVP. Contributed reagents/materials/analysis tools: ZBS. Wrote the paper: XLC JAA HK.

References

1. Ma Y, Su Z, Li Z, Koike T, Menenti M (2002) Determination of regional net radiation and soil heat flux over a heterogeneous landscape of the Tibetan Plateau. Hydrological Processes 16: 2963–2971.
2. Xu X, Zhang R, Koike T, Lu C, Shi X, et al. (2008) A New Integrated Observational System Over the Tibetan Plateau. Bulletin of the American Meteorological Society 89: 1492–1496.

3. Yang K, Koike T, Fujii H, Tamura T, Xu X, et al. (2004) The Daytime Evolution of the Atmospheric Boundary Layer and Convection over the Tibetan Plateau: Observations and Simulations. Journal of the Meteorological Society of Japan 82: 1777–1792.
4. Stensrud DJ (1993) Elevated Residual Layers and Their Influence on Surface Boundary-Layer Evolution. Journal of the Atmospheric Sciences 50: 2284–2293.

5. Li Y, Gao W (2007) Atmospheric boundary layer circulation on the eastern edge of the Tibetan Plateau, China, in summer. Arctic, Antarctic, and Alpine Research 39: 708–713.

6. Xu GR, Cui CG (2009) Observational Analysis on Atmospheric Boundary Layer Height over Eastern Qinghai-Tibet Plateau and its Downstream Key Area. Torrential Rain and Disasters (in Chinese) 28: 112–118.

7. Gao Y, Tang M, Luo S, Shen Z, Li C (1981) Some Aspects of Recent Research on the Qinghai-Xizang Plateau Meteorology. Bulletin of the American Meteorological Society 62: 31–35.

8. Tao S, Ding Y (1981) Observational evidence of the influence of the Qinghai-Xizang (Tibet) Plateau on the occurrence of heavy rain and severe convective storms in China. Bulletin of the American Meteorological Society 62: 23–30.

9. Fu R, Hu Y, Wright JS, Jiang JH, Dickinson RE, et al. (2006) Short circuit of water vapor and polluted air to the global stratosphere by convective transport over the Tibetan Plateau. Proceedings of the National Academy of Sciences 103: 5664–5669.

10. An Z, Kutzbach JE, Prell WL, Porter SC (2001) Evolution of Asian monsoons and phased uplift of the Himalaya-Tibetan plateau since Late Miocene times. Nature 411: 62–66.

11. Boos WR, Kuang Z (2010) Dominant control of the South Asian monsoon by orographic insulation versus plateau heating. Nature 463: 218–222.

12. Yanai M, Li C, Song Z (1992) Seasonal heating of the Tibetan Plateau and its effects on the evolution of the Asian summer monsoon. Journal of the Meteorological Society of Japan 70: 319–351.

13. Yanai M, Li CF (1994) Mechanism of Heating and the Boundary Layer over the Tibetan Plateau. Monthly Weather Review 122: 305–323.

14. Xu X, Mingyu Z, Jiayi C, Lingen B, Guangzhi Z, et al. (2002) A comprehensive physical pattern of land-air dynamic and thermal structure on the Qinghai-Xizang Plateau. Science in China Series D: Earth Sciences 45: 577–594.

15. Li MS, Dai YX, Ma YM, Zhong L, Lv SH (2006) Analysis on structure of atmospheric boundary layer and energy exchange of surface layer over Mount Qomolangma region (in Chinese). Plateau Meteorology 25: 807–813.

16. Sun F, Ma Y, Li M, Ma W, Tian H, et al. (2007) Boundary layer effects above a Himalayan valley near Mount Everest. Geophys Res Lett 34.

17. Ma Y, Wang Y, Wu R, Hu Z, Yang K, et al. (2009) Recent advances on the study of atmosphere-land interaction observations on the Tibetan Plateau. Hydrol Earth Syst Sci 13: 1103–1111.

18. Zuo H, Hu Y, Li D, Lü S, Ma Y (2005) Seasonal transition and its boundary layer characteristics in Anduo area of Tibetan Plateau. Progress in Natural Science 15: 239–245.

19. Fan SJ, Fan Q, Yu W, Luo XY, Wang BM, et al. (2011) Atmospheric boundary layer characteristics over the Pearl River Delta, China, during the summer of 2006: measurement and model results. Atmos Chem Phys 11: 6297–6310.

20. Zhou X, Li W, Chen L, Liu Y (2006) Study on ozone change over the Tibetan Plateau. Acta Meteorologica Sinica 20: 129.

21. Chen B, Xu XD, Yang S, Zhao TL (2012) Climatological perspectives of air transport from atmospheric boundary layer to tropopause layer over Asian monsoon regions during boreal summer inferred from Lagrangian approach. Atmos Chem Phys 12: 5827–5839.

22. Randel WJ, Seidel DJ, Pan LL (2007) Observational characteristics of double tropopauses. J Geophys Res 112: D07309.

23. Añel JA, Antuña JC, de la Torre L, Castanheira JM, Gimeno L, et al. (2008) Climatological features of global multiple tropopause events. Journal of Geophysical Research-Atmospheres 113: D00B08.

24. Chen X, Ma Y, Kelder H, Su Z, Yang K (2011) On the behaviour of the tropopause folding events over the Tibetan Plateau. Atmos Chem Phys 11: 5113–5122.

25. Johnson WB, Viezee W (1981) Stratospheric ozone in the lower troposphere –I. Presentation and interpretation of aircraft measurements. Atmospheric Environment 15: 1309–1323.

26. Chen C, Tian WS, Tian H-Y, Huo Y, Shu J (2012) Vertical Distribution of Ozone and Stratosphere-Troposphere Exchanges on the Northeastern Side of Tibetan Plateau. Plateau Meteorology 31 295–303.

27. Yang X, Wang M, Li X (1999) Numerical study of surface ozone in China during summer time. J Geophys Res 104: 30341–30349.

28. Ao CO, Waliser DE, Chan SK, Li J-L, Tian B, et al. (2012) Planetary boundary layer heights from GPS radio occultation refractivity and humidity profiles. J Geophys Res 117: D16117.

29. Seidel DJ, Zhang Y, Beljaars A, Golaz J-C, Jacobson AR, et al. (2012) Climatology of the planetary boundary layer over the continental United States and Europe. J Geophys Res 117: D17106.

30. WMO (1957) Meteorology: A three dimensional science. WMO Bull 6: 134–138.

31. Holzworth CG (1964) Estimates of mean maximum mixing depths in the contiguous United States. Monthly Weather Review 92: 235–242.

32. Seibert P, Beyrich F, Gryning S-E, Joffre S, Rasmussen A, et al. (2000) Review and intercomparison of operational methods for the determination of the mixing height. Atmospheric Environment 34: 1001–1027.

33. Hennemuth B, Lammert A (2006) Determination of the Atmospheric Boundary Layer Height from Radiosonde and Lidar Backscatter. Boundary-Layer Meteorology 120: 181–200.

34. Dee DP, Uppala SM, Simmons AJ, Berrisford P, Poli P, et al. (2011) The ERA-Interim reanalysis: configuration and performance of the data assimilation system. Quarterly Journal of the Royal Meteorological Society 137: 553–597.

35. Ladstätter-Weißenmayer A, Meyer-Arnek J, Schlemm A, Burrows JP (2004) Influence of stratospheric airmasses on tropospheric vertical O3 columns based on GOME (Global Ozone Monitoring Experiment) measurements and back-trajectory calculation over the Pacific. Atmos Chem Phys 4: 903–909.

36. Bergman JW, Jensen EJ, Pfister L, Yang Q (2012) Seasonal differences of vertical-transport efficiency in the tropical tropopause layer: On the interplay between tropical deep convection, large-scale vertical ascent, and horizontal circulations. J Geophys Res 117: D05302.

37. Stohl A, James P (2004) A lagrangian analysis of the atmospheric branch of the global water cycle. Part I: method description, validation, and demonstration for the August 2002 flooding in Central Europe. Journal of Hydrometeorology 5: 656–678.

38. Stohl A, James P (2005) A lagrangian analysis of the atmospheric branch of the global water cycle. Part II: moisture transports between Earth's ocean basins and river catchments. Journal of Hydrometeorology 6: 961–984.

39. Callies J, Corpaccioli E, Eisinger M, Hahne A, Lefebvre A (2000) GOME-2 metops second-generation sensor for oper- ational ozone monitoring. ESA bulletin 102: 28–36.

40. Rodgers CD (2000) Inverse methods for atmospheric sounding:theory and practice. London: World Scientific Publishing Press. 238p.

41. Zhang G, Xu X, Wang J (2003) A dynamic study of Ekman characteristics by using 1998 SCSMEX and TIPEX boundary layer data. Advances in Atmospheric Sciences 20: 349–356.

42. Engeln A, Teixeira J, Wickert J, Buehler S (2006) CHAMP Radio Occultation Detection of the Planetary Boundary Layer Top In: Foelsche U, Kirchengast G, Steiner A, editors. Atmosphere and Climate: Springer Berlin Heidelberg. 265–272.

43. Langford AO, Reid SJ (1998) Dissipation and mixing of a small-scale stratospheric intrusion in the upper troposphere. J Geophys Res 103: 31265–31276.

44. Betts AK, Gatti LV, Cordova AM, Silva Dias MAF, Fuentes JD (2002) Transport of ozone to the surface by convective downdrafts at night. J Geophys Res 107: 8046.

45. Cooper OR, Stohl A, Hubler G, Hsie EY, Parrish DD, et al. (2005) Direct transport of midlatitude stratospheric ozone into the lower troposphere and marine boundary layer of the tropical Pacific Ocean. Journal of Geophysical Research-Atmospheres 110: D23310. doi:23310.21029/22005JD005783.

46. Griffiths M, Thorpe AJ, Browning KA (2000) Convective destabilization by a tropopause fold diagnosed using potential-vorticity inversion. Quarterly Journal of the Royal Meteorological Society 126: 125–144.

47. Añel JC, de la Torre L, Gimeno L (2012) On the origin of the air between multiple tropopauses at midlatitudes. The Scientific World Journal 2012: 1–5.

48. Santanello JA, Friedl MA, Kustas WP (2005) An empirical investigation of convective planetary boundary layer evolution and its relationship with the land surface. Journal of Applied Meteorology 44: 917–932.

49. Sprenger M, Croci Maspoli M, Wernli H (2003) Tropopause folds and cross-tropopause exchange: A global investigation based upon ECMWF analyses for the time period March 2000 to February 2001. J Geophys Res 108: 8518.

50. Stull RB (1988) An introduction to boundary layer meteorology. Berlin: Springer. 670 p.

51. Gamo M (1996) Thickness of the dry convection and large-scale subsidence above deserts. Boundary-Layer Meteorology 79: 265–278.

52. Cuesta J, Edouart D, Mimouni M, Flamant PH, Loth C, et al. (2008) Multiplatform observations of the seasonal evolution of the Saharan atmospheric boundary layer in Tamanrasset, Algeria, in the framework of the African Monsoon Multidisciplinary Analysis field campaign conducted in 2006. J Geophys Res 113: D00C07.

53. Zhang Q, Zhang J, Qiao J, Wang S (2011) Relationship of atmospheric boundary layer depth with thermodynamic processes at the land surface in arid regions of China. SCIENCE CHINA Earth Sciences 54: 1586–1594.

54. Skerlak B, Sprenger M, Pfahl S, Sodemann H, Wernli H (2012) A global ERA-interim climatology of stratosphere-troposphere exchange and high-resolution model case studies; 2012 22–27 April, 2012 Vienna, Austria. p.1249.

55. Zhang J, Rao ST (1999) The role of vertical mixing in the temporal evolution of ground-level ozone concentrations. Journal of Applied Meteorology 38: 1674–1691.

56. Rao ST, Ku JY, Berman S, Zhang K, Mao H (2003) Summertime characteristics of the atmospheric boundary layer and relationships to ozone levels over the Eastern United States. Pure and Applied Geophysics 160: 21–55.

57. Lin J-T, Youn D, Liang X-Z, Wuebbles DJ (2008) Global model simulation of summertime U.S. ozone diurnal cycle and its sensitivity to PBL mixing, spatial resolution, and emissions. Atmospheric Environment 42: 8470–8483.

58. Hu X-M, Doughty DC, Sanchez KJ, Joseph E, Fuentes JD (2012) Ozone variability in the atmospheric boundary layer in Maryland and its implications for vertical transport model. Atmospheric Environment 46: 354–364.

59. Wang T, Wong HLA, Tang J, Ding A, Wu WS, et al. (2006) On the origin of surface ozone and reactive nitrogen observed at a remote mountain site in the northeastern Qinghai-Tibetan Plateau, western China. J Geophys Res 111: D08303.

15

Role of Megafauna and Frozen Soil in the Atmospheric CH$_4$ Dynamics

Sergey Zimov*, Nikita Zimov

Northeast Science Station, Pacific Institute for Geography, Russian Academy of Sciences, Cherskii, Russia

Abstract

Modern wetlands are the world's strongest methane source. But what was the role of this source in the past? An analysis of global ^{14}C data for basal peat combined with modelling of wetland succession allowed us to reconstruct the dynamics of global wetland methane emission through time. These data show that the rise of atmospheric methane concentrations during the Pleistocene-Holocene transition was not connected with wetland expansion, but rather started substantially later, only 9 thousand years ago. Additionally, wetland expansion took place against the background of a decline in atmospheric methane concentration. The isotopic composition of methane varies according to source. Owing to ice sheet drilling programs past dynamics of atmospheric methane isotopic composition is now known. For example over the course of Pleistocene-Holocene transition atmospheric methane became depleted in the deuterium isotope, which indicated that the rise in methane concentrations was not connected with activation of the deuterium-rich gas clathrates. Modelling of the budget of the atmospheric methane and its isotopic composition allowed us to reconstruct the dynamics of all main methane sources. For the late Pleistocene, the largest methane source was megaherbivores, whose total biomass is estimated to have exceeded that of present-day humans and domestic animals. This corresponds with our independent estimates of herbivore density on the pastures of the late Pleistocene based on herbivore skeleton density in the permafrost. During deglaciation, the largest methane emissions originated from degrading frozen soils of the mammoth steppe biome. Methane from this source is unique, as it is depleted of all isotopes. We estimated that over the entire course of deglaciation (15,000 to 6,000 year before present), soils of the mammoth steppe released 300–550 Pg (10^{15} g) of methane. From current study we conclude that the Late Quaternary Extinction significantly affected the global methane cycle.

Editor: Ben Bond-Lamberty, DOE Pacific Northwest National Laboratory, United States of America

Funding: The authors have no support or funding to report.

Competing Interests: The authors have declared that no competing interests exist.

* E-mail: sazimov55@mail.ru

Introduction

Ice core analyses indicate that during the Pleistocene-Holocene transition (18,000 to 11,000 year before present (BP)), coincident with a rise in Greenland's temperatures, atmospheric methane content increased from ~1000 Tg (1 Tg = 10^{12} g) to ~2000 Tg (Fig. 1A) [1–5]. The modern atmospheric lifetime of methane is approximately 10 years [6] (atmospheric methane lifetime is the ratio between atmospheric methane content and global annual methane flux). If the hypothesis that oxidation rates of methane in the atmosphere did not vary substantially in the past is accepted, and the lifetime of methane in the atmosphere stayed stable, then global CH$_4$ emissions during Pleistocene-Holocene transition (18–11 ka BP) increased from ~100 Tg/yr to ~200 Tg/yr. In the Holocene, relative to the late Pleistocene, the climate was stable while global methane emission was not (Fig. 1B). From these data, it can be assumed that during 15–6 ka BP strong methane sources existed. By 6 ka BP, these sources vanished or substantially decreased in strength. This, in turn, caused a decrease in atmospheric concentrations of methane. Later, in the second part of the Holocene, other sources appeared (or were activated) and global emissions of methane rose by approximately 50 Tg/yr.

A comparison of ice core data from Greenland and Antarctica indicated that in the Last Glacial Maximum (LGM), atmospheric methane concentrations were roughly equal in the southern and northern hemispheres (Fig. 1B) [5]. Which indicate approximately equal methane production in both northern and southern hemisphere. However, simultaneous with a CH$_4$ concentration rise, an interhemispheric CH$_4$ gradient appeared, indicating the arrival of a strong northern source of methane. Based on this gradient, a northern methane source of 40–50 Tg/yr for the Bølling-Allerod (~14.5–13 ka BP), 40 Tg/yr for the Younger Dryas (~13–11.8 ka BP), and 60–70 Tg/yr for the Preboreal (~11.8–9 ka BP) was determined [2–5,8]. In contrast to the atmospheric methane concentration, the interhemispheric gradient stayed relatively stable over the course of the Holocene.

The origins of the dynamics noted above are actively debated. The rise in atmospheric methane during Pleistocene-Holocene transition can be explained by the expansion of boreal or tropical wetlands [1], the destabilization of gas clathrates [7], or the production of CH$_4$ following the thawing of organic rich permafrost under anaerobic conditions [8].

Between 5 and 1 ka BP in a stable climate, methane emissions increased by at least 50 Tg/yr (Fig. 1B). The reason for this phenomenon is unclear to us. We are not convinced by the hypothesis that humans early farming have caused this rise [9]. Modern rice paddies and livestock produce 70 and 90 Tg CH$_4$/yr

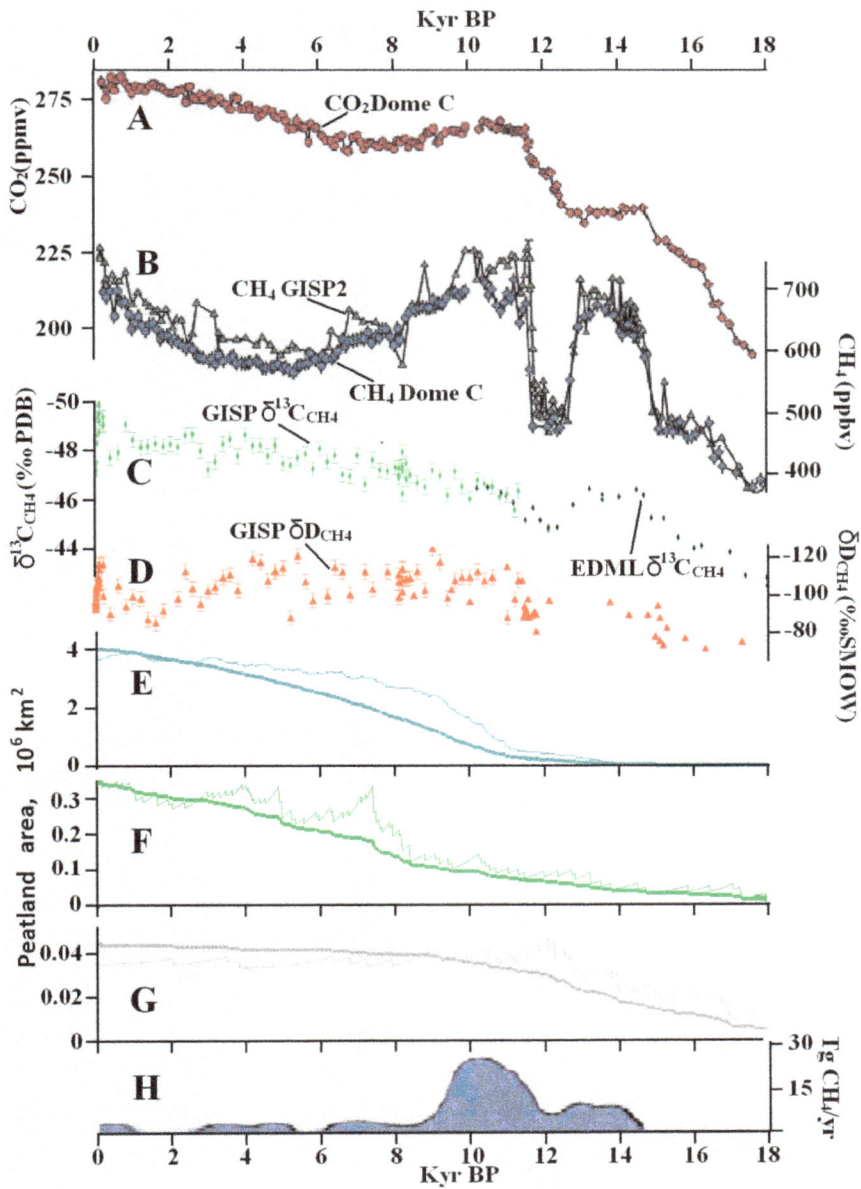

Figure 1. The dynamics of atmospheric: A) CO_2 [17]; B) CH_4 [17]; C) $\delta^{13}CH_4$ [12,13]; and D) δDCH_4 [11,13]. GISP1&2–Greenland Ice Sheet Projects. EDML - EPICA Dronning Maud Land Ice Core. The dynamics of peatland areas: E) boreal [17,18]; F) tropical [18]; and G) southern [18]. Thin lines represent emissions from these wetlands based on the model of wetland succession (arbitrary unit). H) Methane emissions from northern Siberian permafrost [8].

respectively [6].During 5 to 1 ka BP, the human population was two orders of magnitude lower than at present [10], therefore we think that increase in human population is not a sufficient explanation for the rise. Additionally, the first rice paddies were cultivated in areas of natural wetlands, and domestic animals replaced wild animals in pastures.

The isotopic composition of atmospheric methane obtained from ice core analyses allows us to understand the dynamics of primary methane sources [6,11,12]. Methane from different sources is different in isotopic composition (Fig. 2). The dynamic of isotopic methane signature from the LGM to the present is now known (Fig. 1C and 1D) [11–13]. This allows evaluation of the input to global methane emissions from different sources.

An attempt to estimate primary CH_4 sources during Pleistocene-Holocene transition was made by *Fischer et al.* [12]. The authors employed a model that describes the dynamics of atmospheric methane, its interhemispheric gradient, and the ^{13}C and deuterium budgets. However Monte Carlo approaches allowed the dynamics for only two sources to be calculated and according to their results, biomass burning emissions in the LGM for the 3.7 year atmospheric CH_4 lifetime were 65 Tg/yr, and for the 5.6 year lifetime equalled 41 Tg/yr (the same as the biomass burning emissions of today [6]). Boreal wetland emissions during the LGM were 4 Tg/yr and later increased to 54 Tg/yr in the Preboreal, regardless of lifetime duration [12]. The boreal wetland dynamic was predictable, since in the model the presence of an interhemispheric gradient was fully dependent on the presence of a

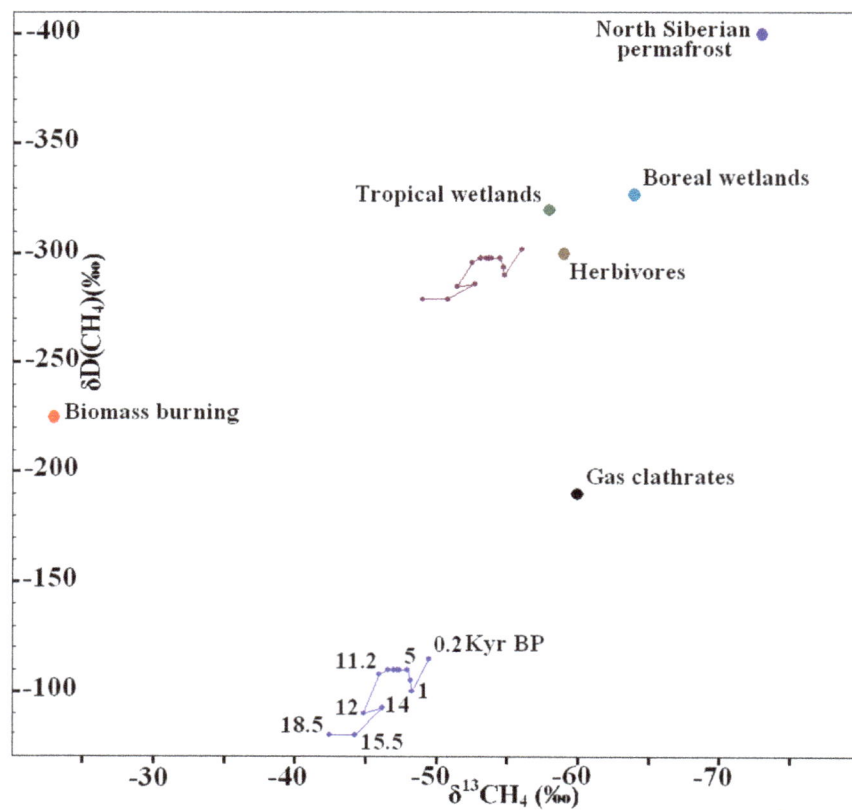

Figure 2. The isotopic signatures (¹³C and deuterium) of primary natural sources [12]. The blue line represents the dynamics of the isotopic content of methane in the atmosphere above Greenland from the LGM to the pre-industrial [11–13]; and purple line represents the dynamic of the isotopic content of the weighted average source calculated for conditions as in the scenario in Fig. 3A.

single northern source – boreal wetlands (permafrost was not considered). The Fischer model assumed that interhemispheric gradient in the LGM was small and, therefore the model showed that boreal wetland emissions were small, as well. Over the course of Pleistocene-Holocene transition interhemispheric gradient substantially increased and therefore, the model was supposed to show and did show a strong rise in boreal wetland emissions regardless of the dynamic of the isotopic content of atmospheric methane and its lifetime.

Below, we present our isotopic model. The general idea and equations of the model are similar to the Fischer model and most input parameters are the same. However, the following differences between our and Fischer's models allowed us to constrain or estimate all main methane sources from the LGM to the present: 1) we obtained measurements not only for the Pleistocene-Holocene transition but also for the Holocene; 2) we added into the model an additional northern source – permafrost; 3) we did not add restrictions connected with the interhemispheric gradient; interhemispheric methane gradient dynamics data was utilized only to test the results of our modelling: we checked by modelling whether the strengths of the northern sources corresponded with the values of the interhemispheric gradient; 4) we set emissions dynamics for boreal and tropical wetlands using a model based on a large global database of basal peat initiation dates. These allowed us to reduce the number of unknowns in the isotopic model and, in contrast to Fischer's results, allowed us to calculate our system of equations, not with the Monte Carlo approach, but analytically.

Background

Before analysing the isotopic model, we want to discuss current knowledge about the glacial–interglacial dynamic of the main methane sources. Below we compare the results obtained from the modelling with these initial estimates.

Gas Clathrates

The storage of gas clathrates in the bottom sediments of seas and oceans is approximately 10000–20000 Pg [14]. But we doubt that this source seriously influenced the methane concentration in the atmosphere. Our observations of methane bubbling in different bottom sediments [15], have indicated that if CH_4 production in bottom sediments is low, bubbling occurs only at low pressure. In lakes, bubbling occurs at record low atmospheric pressure. In rivers, channels, and on the shelf of the East-Siberian Sea it occurs at low water levels. Record low pressure in the sediments occurs rarely and during that short time period releases the methane which had accumulated in the sediments for several months or even years. As soon as the pressure (water level) increases, bubbling immediately stops [15]. We assume that the same processes are correct for gas clathrates on the continental slope and the ocean floor. During deep glacial ocean regression the pressure in the bottom sediments declined much more strongly than in lakes and rivers. Gas clathrates emissions during the regression should have increased, with the maximum emissions being at the deepest regression, and after the LGM peak, the emissions should have immediately stopped as soon as ocean levels began to rise. However, for atmospheric methane dynamics we did not determine such an outcome. Gas clathrate methane has a very high concentration of deuterium isotope [11,12] and, therefore, even minor changes in the emissions of this source into the atmosphere, should substantially change the deuterium content in the atmospheric methane; however, before and after the LGM, the deuterium content in the atmosphere remained stable (Fig. 1C) [11].

Gas clathrate destabilization during ocean regression must have taken place [14]. But this methane did not penetrate to the atmosphere. Field experiments indicate that methane bubbles of a few centimetres in diameter lose ~20% of their methane content per 10 m of updrift (methane within the bubble is replaced by nitrogen and oxygen) [15]. Therefore, methane from the sea sediments (from hundreds of metres in depth and more) was dissolved in the ocean and oxidized to CO_2 [16]. During Pleistocene-Holocene transition the deuterium content of atmospheric CH_4 decreased (Fig. 1C). Amongst natural sources, the δDCH_4 of gas clathrates is the heaviest (Fig. 2). Therefore, the destabilization of gas clathrates could not be the cause of an atmospheric CH_4 rise [11], since a source with low deuterium content is required [11].

Wetlands

Wetland CH_4 emission dynamics can be reconstructed using the dynamics of peat accumulation [17,18]. The conditions required for methanogenesis and peat accumulation are similar – plant remains should appear under anaerobic conditions (i.e., isolated by a layer of water). Some wetlands become episodically covered with water. In such places peat does not accumulate, but methane is still occasionally emitted. The conditions for the existence of peatlands and episodic wetlands are similar – the more humid the climate and the less the drainage, the wetter the soils. If, in a specific region due to climate change, the area of peatlands increases, then the area of episodic wetlands will increase as well. Therefore, the dynamics of wetland areas as well as their CH_4 emissions should be correlated with the dynamics of peatland areas.

The dynamics of peatland areas for three climatic zones are shown in Figures 1E, 1F and 1G (thick lines). These graphs were built based on 1,700 ^{14}C basal peat initiation dates [17,18]. This is the world's biggest global database of peat initiation dates and it can allow reconstruction of the world wetland dynamics. From these graphs it can be concluded that during Pleistocene-Holocene transition, tropical and boreal peatlands were rare and also the rates of peat accumulation were an order of magnitude slower than during the stages of active peatland expansion [18]. Basal peat dates have not been determined for the LGM, but there are peatlands older than the LGM [17,18]. This means that in the LGM there were few peatlands.

There is no other evidences of abundancy of wetlands during the LGM [19]. An analysis of paleo-vegetation maps indicates that, during the LGM, forested areas were ten times smaller than during the Holocene [19,20]. Sphagnum spores (a reliable proxy for moist conditions) are abundant during the Holocene, but they are absent in pollen spectra during the LGM [20,21]. For northern territories with underlying permafrost, another reliable indicator of anaerobic conditions is soil methane [22]. In the frozen soils of the Holocene and other interglacial periods, methane is frequently present [22]. In soils that experience thawing, methane is always abundant, while in all of the frozen soils that accumulated during the LGM it is absent [22]. One of the additional reasons for glacial landscape dryness is a lowering of the erosion basis. At ocean regression in the Pleistocene, river mouths lowered, inclinations of the rivers increased and river beds deepened. Following the erosion basis, drainage of the territory increased and the area of

floodplains reduced. Most of the current vast moist plains in the glaciation were well drained uplands, and only when deglaciation finished and the water level (ground water level) reached the modern level, did landscape saturation occur. The Holocene climate was relatively stable but because of slow processes like land grading, silting of river beds and depressions, the area of peatlands increased over the course of Holocene. Accounting for all the above, we accept that wetlands during the LGM were rare. Now, knowing the dynamic of the wetlands area, we will try to reconstruct the methane emission dynamic from the wetlands.

The work of *McDonald et al.* [17] noted that, during the initial stage of peatland formation (high productive grass-herb formation), CH_4 emissions were higher than during later low productive stages. Therefore, for reconstructing peatland methane emissions, we utilized a model in order to take peatland succession into consideration. In the model we accept that maximum methane emission occurs at the time of peatland initiation, and declines afterwards. We assumed various scenarios of successions and in all cases the peatland succession model indicates that methane emission graph differ slightly from the dynamics graph of the peatland area. Only by accepting a very long stage of succession are the differences visible. In Figures 1E, 1F, and 1G the thin lines were built by assuming that, at the moment of initiation of every peatland, methane production was at a maximum, and very slowly declined. In 600 years, emissions were halved; in 1,600 years they were a quarter of their initial level, and six times lower during late succession. Even in the cases of long and deep succession, differences in graph shapes were not that pronounced.

Peat is a very good proxy for anaerobic conditions [17,18]. It is likely that the graphs in Figures 1D–F are the most precise data reflecting the dynamics of methane from wetlands. These graphs reflect the dynamic of methane emission in arbitrary units, but if we know the modern (pre-industrial) emission of boreal and tropical wetlands then, based on these graphs, we can estimate wetland methane emissions for any other time slice (up to the LGM). Utilizing these data we can determine whether wetlands could be the reason for the methane rise during Pleistocene-Holocene transition.

Area of southern wetlands (mostly Patagonian) smoothly grew from 18 to 12 ka BP (Fig. 1G), but their area was ~1% of the modern boreal wetland area. Therefore, it is unlikely they had an effect on the global dynamic of atmospheric methane. One can clearly see that the strong expansion of boreal wetlands only began around 11 ka BP, three thousand years after the methane atmospheric concentration reached Holocene levels and a strong interhemispheric gradient appeared (Fig. 1B). Tropical wetland expansion took place at an even later date. From these dynamics we can conclude that both tropical and boreal wetlands had little influence on the methane concentration rise during Pleistocene-Holocene transition. Between 11 and 7 ka BP numerous young wetlands appeared worldwide (Figs. 1D–F), so maximum wetland emissions should be expected during this period of time. However, atmospheric methane concentrations did not rise and instead declined (Fig. 1B), indicating that Pleistocene-Holocene transition other stronger sources existed that sharply declined in strength from 11 to 6 ka BP.

Permafrost

As shown during Pleistocene-Holocene transition few wetlands existed and the main source of methane could have been permafrost [8].

In northern Siberia during glaciation, mammoth steppe ecosystem existed [8,23,24]. In this ecosystem, water was often a limiting resource. Grass roots looking for water pierced the entire soil layer (down to the permafrost) where the temperature was never substantially higher than 0°C. Due to low decomposition, organic carbon accumulated not only on the surface but also in the deep active layer [24]. During glaciation on the planes of Siberia, massive loess strata (yedoma) accumulated. As dust accumulated on the soil surface, the lower soil horizons, together with organic remains, became incorporated into the permafrost. The carbon content in yedoma, despite low humus, is tens of kg/m^3 and sometimes reaches 1 ton/m^2 [25,26].

During deglaciation the depth of the summer thaw increased and ice wedges began to melt. In such places, water filled depressions, and migrating thermokarst lakes appeared. During the course of deglaciation these lakes eroded half of the yedoma. At the present time, yedoma is only preserved on slopes or in the form of hills where new lakes cannot form. On lake bottoms, all yedoma became thawed and turned into many metres of silt. Part of the organic material was then transformed by methanogens into CH_4. The diffusion of CH_4 from sediment is/was very slow, and the main source is/was bubbling [27,28].

Permafrost degradation is a very strong methane source (up to hundreds grams of CH_4 per square meter per year) [27,28]. On the one hand, the CH_4 emitted from permafrost can be considered as a fossil since it is strongly depleted in ^{14}C. (Note that the ^{14}C content of atmospheric methane decreased during Pleistocene-Holocene transition [29].) On the other hand, the ^{13}C and the deuterium content of methane from the Siberian permafrost is substantially lower than for all other sources ($\delta^{13}CH_4 = -73$‰ (−58 to −99‰), $\delta DCH_4 = -400$‰ (−338 to −479‰) [22,27,28]).

Permafrost methane emission is the only source that fulfils all of the following demands: it is the northern source; it is depleted in both deuterium and ^{14}C; and its dynamics (Fig. 1H) are correlated with methane atmospheric dynamics in Pleistocene-Holocene transition (Fig. 1B).

During the LGM, the mammoth steppe resting on permafrost was the largest biome. It expanded from France to Canada and from the Arctic islands to China [3,20]. In many regions of this biome, sediment deposition took place. As these sediments were accumulating they became incorporated into the permafrost. Such vast territories (besides the yedoma territory in the north of Siberia and Alaska) could be found in Europe, America, Siberia and China. Deposition of frozen sediments usually occurred with the accumulation of loess (area of $3*10^6$ km^2, see Fig. 1 in ref. [8]) but also at the foot of slopes and in river valleys. On all of these territories thick strata, similar to yedoma, accumulated; similar in carbon content, but with fewer ice wedges [8,24,25]. In places where there was no notable sedimentation, soils were thin. At the beginning of the glaciation, when permafrost appeared, it was ~1.6 m below the surface [24]. During glaciation, active layers decreased to ~0.6–0.8. Therefore, the bottom portion of the soils became incorporated into the permafrost, even without sedimentation. Since these thin soils had developed for many thousands of years, they accumulated humus and the C concentration within them was two times higher than in yedoma [24]. During deglaciation, frozen soils of the mammoth steppe and "southern yedoma" thawed, not only underneath lakes (as in the north of Siberia), but ubiquitously; they lost their ice and passed the stage of anaerobic decomposition, which was followed by the drainage and aerobic decomposition stages – resulting in the loss of Pleistocene carbon and the transition to Holocene soils and loess [24,25].

Methane emissions from northern Siberian permafrost have a very light isotopic signature. The light isotopic content of this CH_4 is due to the following factors: (i) methane is released quickly and does not become fractionated by methanotrophs; (ii) methane production occurs very slowly at very low temperatures [22]; and (iii) methane production occurs in deep sediments through the decomposition of Pleistocene organic matter in Pleistocene water [30,31]. The content of heavy isotopes declines in precipitation water as clouds move from south to north and from oceans to the interior of continents. Therefore, in northern Siberia the isotopic content of Pleistocene precipitation was very low, as indicated in the yedoma ice [30,31]. Since the hydrogen used in photosynthesis and methanogenesis originates from precipitating water, the deuterium content in CH_4 obtained from yedoma was also low. Our experimental methane isotope data was obtained for the coldest permafrost and for the most remote from the warm oceans sites in northern Siberia. The isotopic signature of methane from other "warmer" regions should be substantially heavier with isotopes. During the beginning of deglaciation, southern permafrost was largely degrading. For example, in Europe and China the signature of this source must have been substantially closer to the signature of boreal wetlands. With deglaciation, as permafrost degraded further on the north, the isotopic signature of the permafrost source must have turned lighter. The possible dynamics of this signature are shown in Table 1.

Unlike wetlands where annual emissions of carbon into the atmosphere are roughly equilibrated with carbon consumption with photosynthesis, degrading permafrost is only a source of carbon. The permafrost carbon reservoir was filled for tens of thousands of years within glacial steppes. Decomposition of this organic matter occurred in different conditions. The type of surface (thermokars lake, glacial lake, sea, wetland, or dry land) is not important for methanogenesis to occur during permafrost thawing. Organic materials decompose year round in deep sediments.

Local permafrost thawing, only in northern Siberia, could have released up to 26 Tg CH_4/yr (Fig. 1H) [8]. We may expect that at least the same amounts were released from the total thawing of American, European, west and south Siberian, and Chinese permafrost, as well as permafrost from other regions, including Patagonian permafrost. However, in the southern hemisphere permafrost was rare, having mostly a northern source. Consequently, the permafrost source is enough to maintain the interhemispheric methane gradient [8]. The dynamics of permafrost emissions should reflect not the dynamics of temperatures in the northern hemisphere, but the dynamics of the above- and belowground deglaciation (lots of frozen soils were hidden under glaciers [32]). The dynamic of methane emissions from thawing European permafrost should be close to that of the north Siberian permafrost (Fig. 1H). After 6 ka BP, this source should have quickly depleted.

Modern permafrost emissions, connected with permafrost degradation under lakes and shallow seas, contribute approximately 10 Tg CH_4/yr [8,33]. It is unlikely that such a high emission could be recorded for the entire Holocene. Most likely, modern high permafrost emissions are connected with the climate warming and the active exploration of northern Siberia in the 20th century. Before that, northern Siberia was poorly populated. In most ecosystems, the soil surface was covered with a thick layer of moss. Moss is a good heat insulator, and the depth of summer thaw under the moss in the yedoma region is only 20–40 cm. But if the moss layer is burned then the depth of the summer thaw (active layer) increases several fold. This often causes erosion and permafrost degradation [34]. This in turn leads to the creation of numerous ponds and thermokarst lakes [28]. In the 20th century active exploration, mining operations and city and town construction took place. Annual precipitation on the yedoma territories is approximately 200 mm/yr (~100 mm in summer) – it is a very dry region. According to our estimates, around half of

Table 1. The parameters values assumed in the model for A and B scenarios for all investigated time slices.

ka BP	18	15.5	14	12	11.2	10.6	9.2	8.4	7.4	5	3	1	0.2
ATM, ppb	370	470	600	500	730	730	730	660	650	580	610	675	730
ATM δC13,‰	−42.5	−44.3	−46.25	−44.9	−46	−46.6	−47	−47.25	−47.4	−48	−48.2	−48.3	−49.5
ATM δD, ‰	−80	−80	−93	−90	−108	−110	−110	−110	−110	−110	−105	−100	−115
WT δC13, ‰	−58	−58	−58	−58	−58	−58	−58	−58	−58	−58	−58	−58	−58
WT δD, ‰	−320	−320	−320	−320	−320	−320	−320	−320	−320	−320	−320	−320	−320
WB δC13, ‰	−64	−64	−64	−64	−64	−64	−64	−64	−64	−64	−64	−64	−64
WB δD, ‰	−327	−327	−327	−327	−327	−327	−327	−327	−327	−327	−327	−327	−327
HV δC13, ‰	−59	−59 (−61.5)	−59 (−63)	−59 (−63)	−59 (−65)	−59 (−65)	−59 (−65)	−59 (−65)	−59 (−65)	−59 (−65)	−59 (−65)	−59 (−65)	−59 (−65)
HV δD, ‰	−300	−300	−300	−300	−300	−300	−300	−300	−300	−300	−300	−300	−300
BB δC13, ‰	23	−23	−23	−23	−23	−23	−23	−23	−23	−23	−23	−23	−23
BB δD, ‰	−225	−225	−225	−225	−225	−225	−225	−225	−225	−225	−225	−225	−225
GC δC13, ‰	−60	−60	−60	−60	−60	−60	−60	−60	−60	−60	−60	−60	−60
GC δD, ‰	−190	−190	−190	−190	−190	−190	−190	−190	−190	−190	−190	−190	−190
PF δC13, ‰	−65.7	−65.7	−65.7	−65.7	−69.3	−69.8	−70.6	−71.3	−72	−72	−72	−72	−72
PF δD, ‰	−340	−340	−340	−340	−370	−374	−380	−387	−393	−393	−393	−393	−393

ka BP – thousands of years before present; ATM, ppb - atmospheric methane concentration in parts per billion units; δC13 and δD isotopic signatures of atmospheric methane or methane from specific sources. WT – tropical wetlands; WB – boreal wetlands; HV – herbivores methane source; BB –biomass burning methane source; GC – geological methane source; PF – permafrost methane source. If values differed between the A and B scenarios, the value for the second scenario is presented in brackets.

all forests in the yedoma regions burned in the 20th century. This disturbance caused increased active layer depth, rise of autumn-winter CO_2 emissions from the soils and rise of seasonal amplitudes of atmospheric CO_2 [35]. We suppose that must have increased the methane flux from the permafrost region as well.

Herbivores

The LGM is a period of permafrost zone expansion and the accumulation of carbon in permafrost [25,36]. ^{14}C dating of thermokarst lake initiation dates indicated stable permafrost in the LGM (Fig. 1H). Therefore, we can infer that there was no methane emission from degrading permafrost during the LGM. As noted above, wetland and gas clathrates emissions were also minor. Therefore, some other sources must have constituted the main contribution during the LGM. From the list of well-known methane sources [12] only one big source is left – herbivores.

Lets make some investigation and rough estimates on whether could this source have played important role the late Pleistocene? Modern methane emissions from wild herbivores are approximately 2–6 Tg/yr [37], but herbivores were more abundant in the past. During the last deglaciation, tens of megafauna species became extinct, and a hypothesis even exists which states that during the Younger Dryas CH_4 decline was caused by a decline in megafauna in North America [38].

Current animal populations in grasslands are mostly limited by human regulations, while in the Pleistocene, before man invented reliable hunting tools and learned how to hunt all animal species, herbivore biomass was limited by the forage available for animals. During the winter/dry season, under high animal diversity, everything that grew during the summer/rainy season was eaten [23,34,39,40]. Everything uneaten by bulls and horses was eaten by omnivorous goats. If available food resources were left unused on the pastures in the spring, mortality related to starvation was rare and the animal population grew in the following summer, and by the following year all the available forage was eaten. For all wild ruminant animals (e.g., bulls, antelope, deer, goats, etc.), methane production is 28.3 kg per ton of forage, or ~100 kg/yr per ton of animal weight [37]. For non-ruminant animals (e.g., proboscideans, horses, pigs, etc.), CH_4 production is ~4.5 times smaller [37].

Currently, the area of pasture and cropland on Earth is 35.3×10^6 km². Harvests from these territories maintain a "megafauna" population of ~1.2 Pg (34 tons per km² of pastures), one-third of which is human biomass and two-thirds of which comprises domestic animals, mainly ruminants, that produce ~90 Tg CH_4/yr [37]. During the LGM, forested areas were ten times smaller than during the Holocene, and the area of grass–herb dominated ecosystems, reached 70×10^6 km² [19]. I.e. pasture area was twice of modern pastures and agricultural fields.

Vegetative periods on wild pastures are longer than on tilled fields. On pastures the entire biomass is utilized, and not only seeds and fruits. In the LGM there were no "technical" plant species, like cotton. On pasture, nutrients are not removed from the fields with the harvest but quickly and uniformly returned to the bio cycle. Taking all this into account, we can suppose that herbivore methane emissions in the LGM could be no less than the modern 90 Tg/yr.

We obtained more precise estimates of animal densities and their methane emissions for northern Siberia, where, because of high rates of yedoma accumulation, high numbers of late Pleistocene animal skeletons and bones are preserved [34,40]. We did estimates with different methods and for different regions. As a result, it was ascertained that in the late Pleistocene (over the 40 ka period) on the plains of the Siberian north, on each square kilometre approximately 1,000 of mammoths, 20,000 bison,

30,000 horses and 80,000 reindeer had died. From that it was calculated that, on average, for every moment of time on each square kilometre one mammoth, five bison, 7.5 horses and 15 reindeer roamed. Accounting for other more rare animals, the biomass of herbivores equalled 10.5 ton/km² [34,40]. Half of that biomass belonged to ruminant bison and reindeer with the rest being non-ruminant mammoths and horses [34,40].To maintain this biomass, approximately 100 tons of dry biomass per annum was supposed to be consumed from each square kilometre.

Even today such a high animal density in high- and mid-lattitudes can be observed. The modern density of semi-wild Yakutian horses on the highly productive meadows of northern Siberia –30 horses/km² [34,40], which roughly corresponds with our calculations. The same density can be observed in the pastures of the" Pleistocene Park" in the Kolyma river lowland [34,40]. In the "Oostvaardersplassen" park in the northern Netherlands, animal density is not regulated by man, but determined only by pasture productivity. There animal density is stable around 100 hoofed animals per square kilometre [41].

In the LGM, animal density in the north was substantially lower than on average for the period of the late Pleistocene [34,40]. Therefore, we can accept that on each square kilometre of plane land (yedoma) pasture in LGM only 50–70 tons of forage were consumed. North of Siberia is the coldest and driest part of the mammoth steppe biome and this biome in turn comprised the most severe steppes in the world. In warmer and wetter climates, productivity (and consequently animal density) was probably higher. If it is assumed that, on average, on the planet herbivores consumed 100–140 tons/km², the area of pastures was 70×10^6 km² and ruminant animals constituted half of all the big herbivores, then global methane herbivore emissions could potentially reach 120–170 Tg/yr.

One of the features of the herbivore source is a quick and strong response to changes in precipitation. If in arid climates precipitation increased, then the productivity of grasses and herbs would increase in the same year, and in 1–2 years the number of ungulate and herbivore methane emission would increase respectively. This most likely occurred in the beginning of the Bølling warming (similarly, increase in herbivore density and herbivore methane productions should have been observed during all events of increased precipitation over the course of the Pleistocene). Activation of other sources, such as wetlands, or permafrost thawing appeared later. Eventually forest expansion caused pasture areas to decline and the herbivore methane flux have probably decreases. Additionally, the warming favoured the expansion of man to the north and to America. Due to increased hunting pressure, a number of herbivores declined at that time [34,40].

Biomass Burning

A direct proxy of fire intensity dynamics is the amount of charcoal in stratigraphic columns, but such data are rather sparse. Relatively reliable data are published only for the territory of the USA [42]. They showed a strong rise in fire activity in the Bølling-Allerod and Preboreal [42]. For north-eastern USA, data are obtained simultaneously with data on the abundance of dung fungus spores, which characterize the amounts of dung in the sediments and consequently the animal density. These data indicated that during the LGM, in north-eastern USA fires were rare, herbivores consumed everything, and there was nothing to burn [43]. After herbivore disappearance, fires became "persistent" [43]. Similar dynamics have been recorded for various regions ([39] and references therein).

Methods. A Methane Isotope Model

During methane oxidation, isotopic fractionation causes the isotopic content of atmospheric CH_4 to differ from the isotopic content of the weight averaged CH_4 source through the values of KIE (kinetic isotopic effects).

The KIE coefficients of $\delta^{13}CH_4$ and δDCH_4 are only moderately dependent on temperature and methane lifetime, and can be considered to be constants, allowing one to reconstruct the dynamics of the weighted average CH_4 source's isotopic content by knowing the dynamics of the isotopic content of atmospheric methane [6,11,12].

To reconstruct the dynamics of the main sources of CH_4, for every time slice we investigated we used the following set of equations [6,11,12]:

$$WB_i + WT_i + BB_i + GC_i + PF_i + HV_i = 1 \qquad (1)$$

$$WB_i \cdot \delta_{WB} + WT_i \delta_{WT} + BB_i \cdot \delta_{BB} + GC_i \cdot \delta_{GC} + PF_i \cdot \delta_{PF}$$
$$+ HV_i \cdot \delta_{HV} = \delta_{ATMi} \cdot (KIE_{13C}/1000 + 1) + KIE_{13C} \qquad (2)$$

$$WB_i \cdot D_{WB} + WT_i \cdot D_{WT} + BB_i \cdot D_{BB} + GC_i \cdot D_{GC} + PF_i \cdot D_{PF}$$
$$+ HV_i \cdot D_{HV} = D_{ATMi} \cdot (KIE_D/1000 + 1) + KIE_D \qquad (3)$$

where WB, WT, BB, GC, PF, HV are CH_4 sources from respectively boreal wetlands, tropical wetlands, biomass burning, gas clathrates, permafrost and herbivores, units are the ratio of total global emission; D and δ are the isotopic signatures of all of the corresponding sources, in ‰ (presented in Table 1 and fig. 2 [12]); KIE_{13C} and KIE_D are the coefficients of atmospheric isotopic fractionation, in ‰.i, number of time slice (see column labels in the Table 1 for details on investigated time slices).

These formulas show the relationship between the isotopic signatures of methane sources and the global atmosphere. As input parameters we have the isotopic record from the Greenland ice cores [11,12,13] (blue line in Fig. 2 and Table 1)(i.e., we know the isotopic signatures of the atmosphere above Greenland), but for our calculation globally averaged values are necessary. Therefore, as a first step we need to find a connection between the global atmosphere and the atmosphere above Greenland. In the southern hemisphere, non-tropical sources are minor. Therefore, the situation where the entire methane atmosphere becomes isotopically lighter, while becoming heavier above Greenland during the same period of time, is not possible. The dynamics of Greenland methane should be roughly parallel with the global methane dynamics. The $\delta^{13}CH_4$ from Greenland is different, by less than 1‰, from the $\delta^{13}CH_4$ of Antarctica and the averaged atmosphere, and remained the same for the period of time spanning from deglaciation to 1 ka BP [13]. In Figure 1, one can see that the values of $\delta^{13}CH_4$, obtained for the Younger Dryas through the Preboreal periods in Greenland (GISP 2) and Antarctica (EDML), are similar. We assumed that during the LGM the difference also remained small.

The δDCH_4 of the atmosphere above Greenland 1 ka BP, differed from Antarctic values by ~12‰, and from averaged atmospheric values by ~8‰ [13]. LGM emissions of isotopically light methane from boreal wetlands and permafrost were near zero and the interhemispheric methane gradient was low. Herbivores and biomass burning doesn't influence inter-hemispheric methane gradient. Therefore, the LGM interhemispheric gradient of methane isotopes should also be low. As a first approximation, we can assume that interhemishperic gradients of atmospheric methane concentrations and the isotopic contents are proportional to each other. We assumed that the δDCH_4 from Greenland differed by 8‰ from the averaged atmospheric values for the Holocene, by 4‰ for the Younger Dryas (12 ka BP), by 6‰ for the Bølling-Allerod period (14 ka BP), and by 2‰ for the LGM and 15.5 ka BP.

As a result, we ascertained the dynamic of isotopic signatures of global atmospheric methane. During atmospheric methane oxidation isotopic fractionation takes place, therefore methane in the atmosphere is much heavier than the source methane. This fractionation is described by parameters of KIE for both deuterium and ^{13}C. The *Fischer et al.* [12] model utilized KIE values of -6.8‰ for $\delta^{13}CH_4$ and -218‰ for deuterium. We utilized this pair of KIEs in our first scenario (Fig. 3A).

Inclusion of both isotopic dynamic of atmospheric CH_4 and KIEs allows the calculation of the right hand sides of Equations 2 and 3. It is the dynamic of the isotopic content of the weighted average global source (purple line in Fig. 2).

We have three equations and six sources (strengths of six sources of methane) for each time slice. But for the LGM time slice we know that the emissions from permafrost, tropical and boreal wetlands are close to zero. Therefore, for the LGM we can solve the equations directly: herbivore emissions constituted 69% of global emissions, biomass burning 39% and gas clathrates emission 2%. Without knowing the lifetime of methane in the atmosphere in the LGM it is impossible to reconstruct the size of the total global methane emission. But if we were to accept that total herbivore methane emissions in the LGM were 100 Tg/yr, then global emissions were 145 Tg/yr, and accounting for the methane content in the atmosphere in the LGM (~1000 Tg CH4), the life of methane was ~6.9 years.

The third step is to set values of tropical and boreal wetland methane emissions for all time slices. The wetland succession model we created based on peatland initiation dates (thin lines in Figs. 1D and 1E) allow the detection of the relative dynamic of these sources. Recent estimates for modern tropical and boreal wetland emissions are 110 and 55 Tg CH_4/yr [44]. Accepting these values for the pre-industrial time slice allows the reconstruction of the wetland methane source from the LGM to the present for all time slices. In Equation 1 all source units are the ratios of total global methane flux. To estimate the strength of wetlands sources not in Tg/yr units but in the ratios of total, for every time slice the methane lifetime should be considered.

Lifetime is an unknown and poorly constrained parameter in our model [6]. But solving the system of equations for numerous time slices, from the LGM to the present, and having reasonable (or any) solutions is possible only in narrow range of input parameters, including lifetime.

Accepting tropical and boreal wetland emissions of 110 and 55 Tg CH_4/yr [44] for the pre-industrial time slice, we automatically set the upper limit of methane lifetime values. If we were to accept a lifetime of more than 8.6 years then late Holocene solutions of Equations 1–3 would be lost. In turn, if the lifetime is shorter than 7.6 years, herbivore emissions during the late Holocene would exceed reasonable values (25 Tg/yr). At that time, wild herbivores were already rare and domestic animals were still few. Therefore, for the first scenario, we assumed 8 years for all time slices. It is likely that this value is correct for the entire

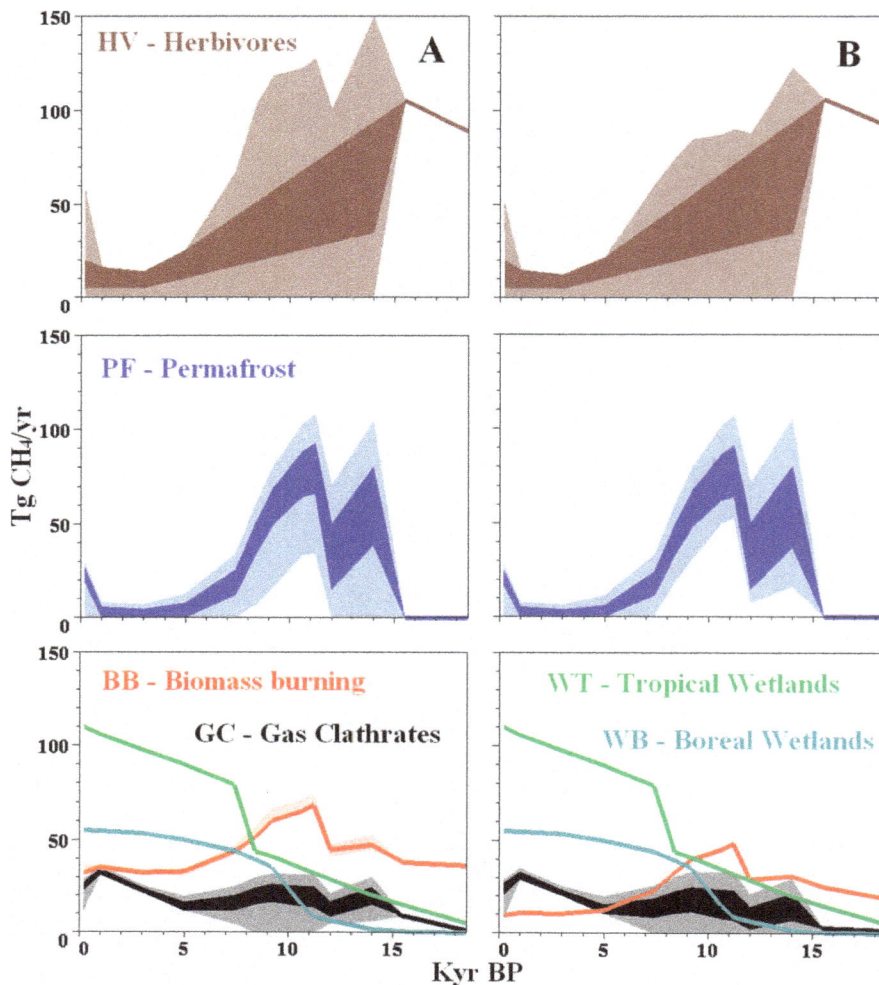

Figure 3. Model estimated ranges of the possible values for methane emissions for herbivore, permafrost, biomass burning and gas clathrates for time slices from the LGM to the pre-industrial. Dark shaded areas represent calculated ranges obtained following the additional restriction placed on herbivore and permafrost emissions. Wetland dynamics are provided as generalized graphs in Figs. 1F and 1G. A – scenario calculated for KIEs −6.8‰ and −218‰ [12]; B- scenario calculated for KIEs −10.8‰ and −227‰ [6].

Holocene. For Pleistocene-Holocene transition and the LGM it could be different. However wetland emissions were minor at that time; therefore changes in lifetime have little effect on the relative strength of other sources. If required, we can change the lifetime of methane during Pleistocene-Holocene transition and in the LGM, which would consequently change the strength of all sources in Figure 3 at that time. If we change the lifetime of methane in the atmosphere, it would change the strength of all sources in Tg/yr units, but emissions in units of the ratio from global emissions would stay intact.

After accepting boreal and tropical wetland emissions values for every time slice as ratios of global methane emission we obtain a system with three equations (Eq. 1–3) and four unknowns (biomass burning, gas clathrates, permafrost and herbivores). This does not allow us to calculate the dynamic of the source, but it allows us to estimate possible ranges of strengths of these sources for every time slice (Fig. 3). True estimates could only be made for the LGM and 15.5 ka BP time slices, where permafrost source was taken as 0.

The model's solution can be presented visually. For each time slice, in Figure 2, global emissions should be divided between individual sources, in the way "centre of gravity" would coincide

with the position of the weighted average source (knots in the purple line). Model reactions to different changes can also be evaluated visually. "Remote" sources – biomass burning, gas clathrates and permafrost – have a "long lever", therefore the shift in strength by a few Tg/yr of these sources, noticeably shifts the "centre of gravity". On the other hand, an error in the isotopic signature of these sources does not influence solutions very much. For example, a few per mil shift in the biomass burning source signature in Figure 2 would still leave this source far to the left from the "centre of gravity". Tropical wetlands and herbivores have a "short lever" therefore the change in strength by a few Tg/yr of these sources is not noticeable in the solutions, while an error in the isotopic signatures (position of these sources in Fig. 2), can substantially bias our solutions.

Modelling results are the most sensitive to changes in KIEs. These two parameters define the position of the purple line on Figure 2. If this line were slightly shifted towards the biomass burning, then the ratio of this source in global emissions would increase (and the ratio of other sources would consequently decrease). If the purple line were shifted to the bottom right, then the ratio of gas clathrates would increase. But if the purple line were further shifted towards the gas clathrates, then for fixed wetland emissions and lifetimes, the solutions in our model would be lost – the dynamic of methane isotopes in the atmosphere (Figs. 1B and 1C) does not allow substantial gas clathrate sources, regardless of the assumed KIE parameters.

In the available literature, KIE coefficients are broadly varied [6]. For $\delta^{13}CH_4$ an estimate of $-6.8‰$ exists, but a more recent value of $-10.8‰$ also exists [6]. The KIE for deuterium is $-218\pm50‰$ [5]. The *Fischer et al.* [12] model utilized KIE values of $-6.8‰$ and $-218‰$. We utilized this pair of KIEs in our first scenario as well (Fig. 3A). In this scenario, we had very high biomass burning methane emissions in the LGM (Fig. 3A). But this contradicts the high density of animals at that time – actively grazed pastures do not supply much organic material for fires [43].

For the second scenario (Fig. 3B) we used KIEs of $-10.8‰$ and $-227‰$, shifting the purple line in Figure 2 to the upper right. For these KIEs, solutions existed for all time slices. Moreover, we obtained solutions for the modern methane budget as well, but in order to do so had to make "permafrost corrections" in the modern budget. The ratio of fossil sources (coal and gas) in the modern budget is calculated through the $^{14}CH_4$ budget [6]. The ^{14}C age of methane in the atmosphere allows the calculation of the ratio of all sources that are fossil sources. Permafrost is also a fossil source (strongly depleted in ^{14}C) but it was not considered in reconstructions of the modern methane budget. Unlike gas and coal heavy with isotopes [6], permafrost is very light. For the modern budget it was necessary to add permafrost, and correspondingly reduce the ratio of other fossil sources. The current permafrost source is relatively small (~10 Tg/yr) but it has very distinct deuterium and ^{13}C signals compared to coal and gas; therefore taking even a small weight from one long "lever" and placing it with the opposite long "lever" (Fig. 2) substantially changes the isotopic content of global emissions. In the modern budget several additional anthropogenic sources are considered. By adding 10 Tg/yr emissions from permafrost to it, for KIEs of $-10.8‰$ and $-227‰$ we have managed to obtain values of all modern sources in the frames of their known ranges. Therefore, we saw no reason not to use KIE values of $-10.8‰$ and $-227‰$ for paleo reconstructions.

Since herbivores are a strong source with an isotopic signature close to the weighted average source signature, the parameter with the second strongest impact on our model results was the isotopic signature of this source. In the first scenario we used $\delta^{13}C$ for herbivores, as in the Fischer model, for all time slices. *Fischer et al.* used a $\delta^{13}C$ for HV of $-59‰$, as for modern animals, with a high proportion of C_4 plants (e.g. corn etc.) in the diet [6,45]. This value is likely to be true for the LGM also, since, during low atmospheric CO_2, the ratio of C_4 plants increased and the $\delta^{13}C$ in C_3 plants decreased by 2‰ [46]. However, for other time slices, CO_2 concentrations were higher. For Pleistocene-Holocene transition we used the herbivore signature in compliance with atmospheric CO_2 concentrations [46]. So, in the second scenario for the LGM, we assumed $-59‰$ (as in Fischer model); for the time slice of 15.5 ka BP we used $-61.5‰$, for the Bølling-Allerod and Younger Dryas we used $-63‰$, and for the Preboreal and Holocene we used 65‰. In calculating the modern budget we have utilized the same value as Fischer, $-59‰$. Detailed data of all assumed isotopic signatures of all sources can be found in Table 1.

For the second scenario, using the same restrictions as for the first, the lifetime was determined to be 9 years.

Results

Modelling results for both scenarios are presented in Fig. 3, we see that for the herbivore and permafrost ranges in our model an additional correction was required. The model returned clearly overestimated maximum values for these parameters. During deglaciation, permafrost emissions could not be substantially greater than the northern flux obtained by the interhemispheric gradient. During the Holocene, forested areas increased and pasture areas decreased. Additionally, numerous herbivore species became extinct. If current herbivore emissions are 90 Tg/yr, then 200 years ago, when human and domestic animal populations were much lower, emissions could not be as high as 50 Tg/yr. Any corrections for any ranges can easily be made using Figure 3. Since the equations in our model are linear, the values for all of the calculated sources are interconnected. The herbivore source is located inside the triangle of permafrost, gas clathrates and biomass burning (Fig. 2). Therefore, the herbivore source maximum corresponds to the minimum for permafrost, gas clathrates and biomass burning, and vice versa. If, for any time slice in Figure 3, we, for example, restrict the range of permafrost by one-third from the top, then the gas clathrates and biomass burning ranges would be restricted by one-third from the top as well, and the herbivore source range would also be restricted by one-third, but from the bottom. If we accept a middle value from the range of one specific source then the values of the other three sources would also be located in the middle of their ranges.

For our work, we only made the most obvious and least controversial correction (see paragraph above) – we restricted the ranges of herbivores and permafrost from the top. However, reducing the range of herbivores from the top, restricted the permafrost, gas clathrates and biomass burning source ranges from the bottom; while restricting the range of permafrost sources from the top led to restricted range of biomass burning and gas clathrates sources from the top and restricted herbivore source ranges from the bottom. As a result we obtained relatively narrow ranges for all sources (the dark shaded sector in Fig. 3).

Testing our model we changed all of the initial parameters over a wide range, and implemented additional sources. In all cases, if solutions existed, the dynamics of the sources were similar to the results in Figure 3. We are not confident that 110 and 55 Tg CH_4/yr are correct estimates, and tested our model for other values of modern wetlands emissions. This required accepting different lifetimes of methane. However, the main results of our work stayed intact.

All modelling was done on Maple v.10.0 software.

Biomass burning emissions in all scenarios had a very narrow range (Fig. 3), and responded strongly to variations in initial parameters but in all cases strongly increased from 15.5 to 12 ka BP. As was noted earlier, similar dynamics for fires were reconstructed using charcoal records [42].

Gas clathrate emissions, as we predicted, were low for all of the scenarios. For the end of the Holocene, for all scenarios, we determined a clear peak of gas clathrates (as suggested by Sowers, [13]).

Earlier we mentioned that methane from the bottoms of deep seas could not penetrate into the atmosphere therefore GC emissions during the Holocene (Fig. 3) could be connected with the land [47] or with shallow Siberian seas [33]. During the LGM, their bottoms were land and were armoured with a gas-proof thick frozen layer. Only when these layers thawed could trapped gas escape. Soils of the mammoth steppes are relatively thin (10–60 m) and thaw quickly. Gas clathrates are located deep, and, in order to thaw hundreds of metres of permafrost, thousands of years are required. Therefore, as hundreds of metres of permafrost degraded in the second part of the Holocene, gas clathrates emissions could grew (Fig. 3).

In all scenarios global **permafrost** emissions closely followed atmospheric methane dynamics and permafrost degradation in the Siberian north (Fig. 1H). If emissions from boreal wetlands and permafrost were combined then the dynamic of this united northern source would be compiled with the interhemispheric CH_4 gradient. By integrating the permafrost emission data (Fig. 3), we found that the source released ~300–550 Pg CH_4 (225–412 Pg C) to the atmosphere from 15 to 6 ka BP. During the Holocene, permafrost sources quickly diminished, while boreal wetlands emissions grew. This explains why the interhemispheric gradient was relatively stable in the Holocene.

In pre-industrial times (200–400 years ago), a sharp decline in the deuterium content in atmospheric methane took place (Fig. 1C). The model was sensitive to this decline and showed that this decline was caused by the activation of permafrost emissions (Fig. 3). Increased degradation of permafrost at that time made sense; at that time, active colonization (exploration) of Siberia and Alaska was underway and in the north many settlements and towns, connected with a transport network, appeared. Sable trapping became an activity not only for colonizers, but also for indigenous people for the first time [48]. With colonization, fire intensification must have taken place and permafrost degradation must have been activated.

Herbivore emissions during the Pleistocene, for a methane lifetime of 8–9 years, were 90–100 Tg/yr. If we were to accept a shorter lifetime, then herbivore emissions could be even higher. These estimates are close to our estimates obtained via global forage productivity (see Herbivore section in Background). Herbivore emissions until 15.5 ka BP were close to the capability of world pastures, decreased during the Holocene, and reached a minimum during the late Holocene. Wild animals were already rare, and domestic animals were still found in small numbers.

Conclusions

Today wetlands are main natural source of methane and it is natural and easiest to suppose that it was same in the past. However we haven't found any proofs of this hypothesis. The main conclusion of our work is that permafrost and herbivores in the past were not only important but also the main methane sources. The main fact from which these two conclusions follow is that in the LGM and during Pleistocene-Holocene transition there is little evidence of peat or anaerobic conditions in the soils. From that it follows that, at that time, wetland methane emissions were at least several times weaker then today. Therefore, some other sources were major at that time. Since, besides boreal wetlands, we know of only one northern methane source (permafrost), we can conclude that permafrost was responsible for the interhemispheric gradient dynamic in deglaciation. In the LGM, permafrost was stable and the main methane source (in the absence of wetlands) could only be herbivores.

These results were obtained independently from isotopic model. The model in turn also depends on data about the lack of wetlands in the LGM and during Pleistocene-Holocene transition, but it is independent of data about the interhemispheric gradient.

The isotopic model we used is simple and broadly accepted [6,11,12]. We have only added the unknown permafrost source, and utilized the most obvious restrictions – herbivore methane emissions in the Holocene could not be higher than in the LGM and in the end of the Holocene could not exceed 25 Tg/yr;

permafrost emissions could not be higher than indicated by the interhemispheric gradient.

The model has allowed the dynamic of all sources to be reconstructed in more detail. Additionally, noting that only the dynamic was reliably reconstructed, absolute values of the strength of each source have been estimated very approximately. Our estimates rely on modern estimates of wetland emissions. If true values differ from the assumed, then to solve the model equations different methane lifetimes should be taken. Our model estimates are rough, we did not consider some of the sources; for example, in the late Holocene some of the wetlands were transformed into rice paddies [9].

We know the values of the isotopic signatures of all sources and the KIE, possibly with substantial errors. These values can be corrected in the future. But if the position of each source in Figure 2 relative to each other will not change, then with new corrected values either solution of the model will not exist (wrong incoming values) or the model will show that herbivores were the main source in the LGM and permafrost during deglaciation.

Discussion

The cold and well aerated soils of mammoth steppe are unlikely to have produced methane in the Pleistocene, and are likely to have accumulated carbon [24,34,40]. This would certainly have affected concentrations of greenhouse gases in the atmosphere. Even modern permafrost soils are rich in carbon and are estimated to contain 1672 Pg C [49] globally. In the LGM, the mammoth steppe was the biggest biome and permafrost covered a much larger area. Estimating carbon storage in permafrost during the LGM has been a significant challenge [24], however, we can now make rough estimates using methane emissions from permafrost sources estimated in this study. Integrating the curve in figure 3, we estimate that permafrost emitted 400 Pg CH4 (300 Pg C) into the atmosphere during the deglaciation. What was the total carbon loss from the northern soils over that time period? Even in fully anaerobic conditions only part of the stored organic ($28 \pm 12\%$) can be transformed into methane [50]. When thick (tens of meters) frozen ice-rich soils (yedoma and its southern analogues) thaw they are initially anaerobic, however, as permafrost degradation continues, underlying gravel and sand thaws and water begins to drain from shallower soil layers. These soils become aerobic and carbon is transformed into the CO_2. However on most territories occupied by the mammoth steppe, soils (both including active layer and soils incorporated into the permafrost) were shallow (less than 2–3 meters deep), and carbon stored in these soils was mostly decomposed under aerobic conditions when the climate warmed. Therefore, if we accept that 15% of permafrost carbon loss was transformed into methane, we can estimate that permafrost soils lost 2000 Pg C during deglaciation (15–6 ka BP).

As current climate warming proceeds, permafrost temperatures are increasing and widespread permafrost degradation is projected. Currently, permafrost soils contain 1672 Pg of carbon [49]. What portion of this carbon will be transformed into methane as these soils begin to thaw? We assumed 15% during deglaciation. We could not take smaller value, since that would have returned an unrealistically big estimate of the permafrost carbon reservoir in the Pleistocene [24]. In modern permafrost, a higher proportion of carbon has already been thawed in the Holocene and then refrozen. It is unlikely that at future thawing such soils would produce lots of methane because much of the easily-decomposed forms of carbon have already been processed. We estimate that no more than 10% of organic carbon in modern permafrost is likely to be transformed into methane as the climate continues to warm. Thus, we estimate that modern permafrost pool has a potential to release 167 Pg C in the form of methane. If future permafrost degradation (in contrast to the Pleistocene-Holocene transition) is rapid and a quarter of this methane is released within 200 years, average methane emissions will be ~250 Tg/yr. This value will make degrading permafrost world biggest methane source.

If our calculations of animal densities in the Pleistocene are correct (or at least close), it appears that wild nature solely on recycled resources managed to sustainably maintain biomass higher than the entire biomass of modern civilization (humans and all domestic animals) which is supported only by burning enormous amounts of non-refundable resources. This results draws attention to the ineffectiveness of current human management of the precious, finite resources available on our planet.

Author Contributions

Conceived and designed the experiments: ZS ZN. Performed the experiments: ZN. Analyzed the data: ZS. Contributed reagents/materials/analysis tools: ZN. Wrote the paper: ZS ZN.

References

1. Chappellaz J, Blunier T, Raynaud D, Bamola JM, Schwander J, et al. (1993) Synchronous changes in atmospheric CH4 and Greenland climate between 40 and 8 kyr BP. Nature 366, 443.
2. Chappellaz J, Blunier T, Kints S, Dallenbach A, Bamola JM, et al. (1997) Changes in the atmospheric CH$_4$ gradient between Greenland and Antarctica during the Holocene. J.Geophys. Res. 102, 15987.
3. Severinghaus JP, Sowers T, Brook EJ, Alley RB, Bender ML (1998) Timing of abrupt climate change at the end of the Younger Dryas interval from thermally fractionated gases in polar ice. Nature 391, 141.
4. Brook EJ, Harder S, Severinghaus JP, Steig EJ, Sucher CM (2000) On the Origin and Timing of Rapid Changes in Atmospheric Methane During the Last Glacial Period. Global Biogeochem. Cylces 14, 559.
5. Dällenbach A, Blunier T, Flückiger J, Stauffer B, Chappellaz J, et al. (2000) Changes in the atmospheric CH$_4$ gradient between Greenland and Antarctica during the Last Glacial and the transition to the Holocene. Geophys. Res. Let. 27, 7, 1005, doi:10.1029/1999GL010873.
6. Quay P, Stutsman J, Wilbur D, Snover A, Dlugokencky E, et al. (1999) The isotopic composition of atmospheric methane. Global Biogeochem Cycles 13, 445.
7. Kennet JP, Cannariato KG, Hendy IL, Behl RJ (2003) Methane hydrates in Quaternary climate change AGU, Washington, D.C.
8. Walter KM, Edwards ME, Grosse G, Zimov SA, Chapin III FS (2007) Thermokarst lakes as a source of atmospheric CH4 during the last deglaciation. Science 318, 633.
9. Ruddiman WF, Guo Z, Zhou X, Wud H, Yu Y (2008) Early rice farming and methane emissions. Quaternary Science Reviews 27, 1291–1295 doi:10.1016/j.quascirev.2008.03.007.
10. Ferretti DF, Miller JB, White JWC, Etheridge DM, Lassey KR, et al. (2005) Unexpected Changes to the Global Methane Budget over the Past 2000 Years. Science 309, 1714–1717.
11. Sowers T (2006) Late quaternary atmospheric CH4 isotope record suggests marine clathrates are stable. Science 311, 838.
12. Fischer H, Behrens M, Bock M, Richter U, Schmitt J, et al. (2008) Changing boreal methane sources and constant biomass burning during the last termianation. Nature 452, 864.
13. Sowers T (2010) Atmospheric methane isotope records covering the Holocene period. Quat. Sci. Rev. 29, 213.
14. Kennett JP, Cannariato KG, Hendy IL, Behl RJ (2000) Carbon isotopic evidence for methane hydrate instability during quaternary interstadials. Science 288, 128.
15. Zimov SA, Voropaev YV, Davydov SP, Zimova GM, Davydova AI, et al. (2001). In Paepe R, Melnikov V editors. Permafrost Response on Economic Development, Environmental Security and Natural Resources, Kluwer Academic Publishers, 511–524.
16. Kessler JD, Valentine DL, Redmond MC, Du M, Chan EW, et al. (2011) A Persistent Oxygen Anomaly Reveals the Fate of Spilled Methane in the Deep Gulf of Mexico. Science 331, 312–315.
17. MacDonald GM, Beilman DW, Krementski KV, Sheng Y, Smith LC, et al. (2006) Rapid early development of circumarctic peatlands and atmospheric CH4 and CO2 variations. Science 314, 285.
18. Yu Z, Loisel J, Brosseau DP, Beilman DW, Hunt SJ (2010) Global peatlands dynamic since the Last Glacial Maximum. Geophys. Res. Lett 37, L13402.
19. Adams JM, Faure H (1998) A new estimate of changing carbon storage on land since the last glacial maximum, based on global land ecosystem reconstruction. Global and Plan. Change 16–17, 3.
20. Adams JM, Faure H, Faure-Denard L, McGlade JM, Woodward FI (1990) Increases in terrestrial carbon storage from the Last Glacial Maximum to the present. Nature 348, 711–714.
21. Gajewski K, Viau A, Sawada M, Atkinson D, Wilson S (2001) Sphagnum peatland distribution in North America and Eurasia during the past 21,000 years. Global Biogeochemical Cycles 15: 297–310.
22. Rivkina EM, Kraev GN, Krivushin KV, Laurinavichus KS, Fyodorov-Davydov DG, et al. (2006) Methane in permafrost of northern arctic. Earth Cryosphere 10, 23–41.
23. Zimov SA (2005) Pleistocene Park: return of mammoth's ecosystem. Science 308, 796–798.

24. Zimov NS, Zimov SA, Zimova AE, Zimova GM, Chuprynin VI, et al. (2009) Carbon storage in permafrost and soils of the mammoth tundra-steppe biome: role in the global carbon budget. Geophys. Res. Lett. 36, L02502, doi:10.1029/2008GL036332.
25. Zimov SA, Schuur EAG, Chapin, III FS (2006) Permafrost and the global carbon budget. Science 312, 1612.
26. Schirrmeister L, Siegert C, Kuznetsova T, Kuzmina S, Andreev A, et al. (2002) Paleoenvironmental and paleoclimatic records from permafrost deposits in the Arctic region of Northern Siberia. Quat. Int. 89, 97–118.
27. Zimov SA, Voropaev YV, Semiletov IP, Davidov SP, Prosiannikov SF, et al. (1997) North Siberian lakes: a methane source fueled by Pleistocene carbon. Science 277. 800–802.
28. Walter KM, Zimov SA, Chantom JP, Verbyla D, Chapin, III FS (2006) Methane bubbling from Siberian thaw lakes as a positive feedback to climate warming. Nature 443, 71.
29. Petrenko VV, Smith AM, Brook EJ, Lowe D, Riedel K, et al. (2009) ^{14}CH$_4$ measurements in Greenland ice: investigating Last Glacial Termination CH$_4$ sources. Science 324, 506.
30. Vasil'chuk YuK, Kotlyakov VM (2000) Principles of Isotope Geocryology and Glaciology. Moscow University Press. 616 p.
31. Brosius LS, Walter Anthony KM, Grosse G, Chanton JP, Farquharson LM, et al. (2012) Using the deuterium isotope composition of permafrost meltwater to constrain thermokarst lake contributions to atmospheric CH4 during the last deglaciation. J. Geoph. Res. 117, G01022, doi:10.1029/2011JG001810.
32. Zeng N (2003) Glacial-interglacial atmospheric CO2 change – the glacial burial hypothesis. Advances in Atmospheric Sciences 20, 677–693.
33. Shakhova N, Semiletov I, Salyuk A, Yusupov V, Kosmach D, et al.(2010) Extensive Methane Venting to the Atmosphere from Sediments of the East Siberian Arctic Shelf. Science 327, 1246–1250.
34. Zimov SA, Zimov NS, Chapin III FS (2012), The Past and Future of the Mammoth Steppe Ecosystem, In: Louys J editor. Paleontology in Ecology and Conservation. Springer-Verlag Berlin Heidelberg 2012. 193–224.
35. Zimov SA, Daviodov SP, Zimova GM, Davidova AI, Chapin III FS, et al. (1999) "Contribution of Disturbance to Increasing Seasonal Amplitude of Atmospheric CO2. Science 284, 5422.
36. Zech R, Huang Y, Zech M, Tarozo R, Zech W (2010) A permafrost glacial hypothesis to explain atmospheric CO2 and the ice ages during the Pleistocene. Clim. Past Discuss. 6, 2199–2221. 201.
37. Crutzen P, Aselmann I, Seiler W (1986) Methane Production By Domestic Animals, Wild Ruminants, Other Herbivorous Fauna, and Humans. Tellus 38B (3–4), 271–284.
38. Smith FA, Elliott SM, Lyons SK (2010) Methane emissions from extinct megafauna. Nature Geoscience 3, 374.
39. Johnson CN (2008) Ecological consequences of Late Quaternary extinctions of megafauna. Proc. R.Soc. B doi:10.1098/rsb.2008.1921.
40. Zimov SA, Zimov NS, Tikhonov AN, Chapin III FS (2012) Mammoth steppe: a high-productivity phenomenon. Quat. Sci. Rev. 57, 26–45.
41. Vera FVM (2009) Large-scale nature development – the Oostvaardersplassen. British Wildlife, June 2009.
42. Marlon JR, Bartlein PJ, Walsh MK, Harrison SP, Brown KJ, et al. (2009) Wildfire responses to abrupt climate change in North America. PNAS 106, 2519.
43. Gill JL, Williams JW, Jackson ST, Lininger KB, Robinson GS (2009) Pleistocene megafaunal collapse, novel plant communities, and enhanced fire regimes in North America. Science 326, 1100.
44. Bousquet P, Ringeval B, Pison I, Dlugokencky EJ, Brunke E-G, et al. (2011) Source attribution of the changes in atmospheric methane for 2006–2008 Atmos. Chem. Phys., 11, 3689.
45. Cicerone RJ, Oremland RS (1988) Biogeochemical aspects of atmospheric methane. Global Biogeochem. Cycles 2, 299.
46. Van de Water PK, Leavitt SW, Betancourt JL (1994) Trends in stomatal density and ^{13}C/^{12}C ratios of Pinus flexilis needles during last glacial-interglacial cycles. Science 264, 239.
47. Walter-Anthony KM, P Anthony, Grosse G, Chanton J (2012) Geologic methane seeps along boundaries of Arctic permafrost thaw and melting glaciers. Nature Geoscience, doi:10.1038/ngeo1480.

48. Syroechkovskii VE (1986) Severnii Olen'. Agropromizdat. Moscow (in Russian).

49. Tarnocai C, Canadell JG, Schuur EAG, Kuhry P, Mazhitova G, et al. (2009) Soil organic carbon pools in the northern circumpolar permafrost region. Global Biogeochem Cycles 23: 2023.

50. Walter Anthony KM, Zimov SA, Grosse G, Jones MC, Anthony P, et al. (2014) Permafrost thaw by deep lakes: from a methane source to a Holocene carbon sink. Nature. Submitted.

16

Hailstones: A Window into the Microbial and Chemical Inventory of a Storm Cloud

Tina Šantl-Temkiv[1,2,3], Kai Finster[2,3], Thorsten Dittmar[4], Bjarne Munk Hansen[1], Runar Thyrhaug[†5], Niels Woetmann Nielsen[6], Ulrich Gosewinkel Karlson[1]*

1 Department of Environmental Science, Aarhus University, Roskilde, Denmark, **2** Microbiology Section, Department of Bioscience, Aarhus University, Aarhus, Denmark, **3** Stellar Astrophysics Centre, Department of Physics and Astronomy, Aarhus University, Aarhus, Denmark, **4** Max Planck Research Group for Marine Geochemistry, Institute for Chemistry and Biology of the Marine Environment, University of Oldenburg, Oldenburg, Germany, **5** Department of Biology, University of Bergen, Bergen, Norway, **6** Danish Meteorological Institute, Copenhagen, Denmark

Abstract

Storm clouds frequently form in the summer period in temperate climate zones. Studies on these inaccessible and short-lived atmospheric habitats have been scarce. We report here on the first comprehensive biogeochemical investigation of a storm cloud using hailstones as a natural stochastic sampling tool. A detailed molecular analysis of the dissolved organic matter in individual hailstones via ultra-high resolution mass spectrometry revealed the molecular formulae of almost 3000 different compounds. Only a small fraction of these compounds were rapidly biodegradable carbohydrates and lipids, suitable for microbial consumption during the lifetime of cloud droplets. However, as the cloud environment was characterized by a low bacterial density (Me = 1973 cells/ml) as well as high concentrations of both dissolved organic carbon (Me = 179 μM) and total dissolved nitrogen (Me = 30 μM), already trace amounts of easily degradable organic compounds suffice to support bacterial growth. The molecular fingerprints revealed a mainly soil origin of dissolved organic matter and a minor contribution of plant-surface compounds. In contrast, both the total and the cultivable bacterial community were skewed by bacterial groups (γ-*Proteobacteria*, *Sphingobacteriales* and *Methylobacterium*) that indicated the dominance of plant-surface bacteria. The enrichment of plant-associated bacterial groups points at a selection process of microbial genera in the course of cloud formation, which could affect the long-distance transport and spatial distribution of bacteria on Earth. Based on our results we hypothesize that plant-associated bacteria were more likely than soil bacteria (i) to survive the airborne state due to adaptations to life in the phyllosphere, which in many respects matches the demands encountered in the atmosphere and (ii) to grow on the suitable fraction of dissolved organic matter in clouds due to their ecological strategy. We conclude that storm clouds are among the most extreme habitats on Earth, where microbial life exists.

Editor: Stefan Bertilsson, Uppsala University, Sweden

Funding: TST was supported by a Ph.D. fellowship granted by the Danish Agency for Science, Technology and Innovation (Forsknings- og Innovationsstyrelsen). Funding for the Stellar Astrophysics Centre is provided by The Danish National Research Foundation. The research is supported by the ASTERISK project (ASTERoseismic Investigations with SONG and Kepler) funded by the European Research Council (Grant agreement no.: 267864). The funders had no role in study design, data collection and analysis, decision to publish, or preparation of the manuscript.

Competing Interests: The authors have declared that no competing interests exist.

* E-mail: uka@dmu.dk

† Deceased.

Introduction

Airborne bacteria have lately generated a lot of interest, due to their ubiquitous presence and the accumulating evidence of their activity in the atmosphere [1]. Previous studies indicate that terrestrial habitats, in particular soils and plant leaf surfaces, are the major sources of airborne bacteria, whereas marine environments are a less prominent source [2]. By performing a meta-analysis of the composition of the airborne community and of their potential source environments, Bowers et al [3] identified bacterial taxa indicative for soil and plant-surface origin. Generally, they found that the airborne community was more similar to plant-surface than to soil communities. Depending on the land-use type, however, either soil or plant-surface bacteria were found to dominate the community. As the atmospheric bacterial community was distinct from its source communities, which was driven by the different relative abundances of bacterial taxa, the existence of a microbial community characteristic for the atmosphere was implied [3].

Diverse bacterial communities have been described in the atmosphere [4] and in clouds [5], [6]. However, bacterial communities in cloud water may be distinct from bacterial communities in the dry atmosphere, as the chances of airborne bacteria to enter into cloud droplets are increased for those that can act as cloud condensation nuclei [7]. After entering cloud droplets, bacteria are thought to influence physical and chemical processes in the atmosphere [1]. They may do this both by the means of their outer membrane structures as well as their metabolic activity. During their residence time in clouds, a group of mainly epiphytic Gram-negative bacteria could influence patterns of precipitation by facilitating the formation of ice crystals [8]. The so-called ice nucleation active (INA) bacteria are among the most efficient described ice nucleators. By forming large aggregates of INA proteins, which are anchored in their outer membrane [9], INA bacteria substantially elevate the freezing temperature of water. Thus, they may be important in mixed phase clouds, where subzero temperatures are often too high for water to freeze in the absence of ice nucleators.

There is also growing evidence that some cloudborne bacteria proliferate in cloud droplets. It was observed for two cloud events that the majority (72% and 95%) of cloud bacteria were viable [10]. Also, Hill et al [11] showed that on average 76% of cloudborne bacteria from two clouds were metabolically active. A couple of studies confirmed that the indigenous bacterial communities from rain- and cloud water could grow on either naturally present or supplemented organic compounds [12], [13]. Several isolates from clouds were shown capable of metabolizing nutrients present in cloud water [14] at rates that make them competitive with photooxidation [15]. However, it remains unclear whether cloud bacteria are in fact active *in situ*.

Inside storm clouds water droplets can coalesce into hailstones. During their formation, hailstones collect cloud and rain droplets in a non-selective way as they circulate inside the cloud, following unpredictable individual paths. We have recently shown that hailstones, which preserve the samples by freezing in real time, are useful sampling tools of storm cloud water and, indirectly, of air from the atmospheric boundary layer that has been sucked up by the storm cloud [6]. The storm cloud bacterial community was diverse with the estimated total bacterial richness of 1800 operational taxonomic unites (OTUs) at the species level and with a medium species evenness as estimated from Lorenz curves [6]. We also suggested that the highly diverse community encompasses strains with opportunistic ecologic strategy, which may grow despite the short residence times in clouds. Although some of the isolates have been characterized as opportunists [6], it remains unclear, whether the pool of organic chemicals can support the metabolism of these bacteria and if selective enrichment of some bacterial groups actually occurs in clouds. We report here on a comprehensive biogeochemical study, analyzing large hailstones from the same hail event [6]. By performing a detailed molecular characterization of water-soluble organic matter in hailstones and by aligning the potential substrates with the characteristic bacterial genera present in the cloud, we investigate the possibility of microbial growth in the storm cloud.

Materials and Methods

Ethics Statement

All sampling sites were public property and non-protected areas. In addition, in Slovenia there is no legal requirement for obtaining permits for taking precipitation samples. Thus, there were no specific permits required for the described field studies. Endangered or protected species were not in any way affected by or involved in the sampling activity.

Collection and cleaning of the hailstones

Forty two hailstones were collected after a thunderstorm discharged over Ljubljana, Slovenia in the late afternoon of May 25th, 2009. Hailstones were collected into sterile bags within 5 minutes after they fell on ground and stored at $-20°C$. For molecular characterization and analysis of dissolved organic carbon (DOC) as well as total dissolved nitrogen (TDN), the surface of 18 hailstones was cleaned by rinsing with deionized water. Ice cubes of deionised water, with their surface contaminated by soil and grass, were treated in the same way as a control for the rinsing procedure. All plastic and glass lab ware was acid washed; all metal equipment used was treated by dry-heat-

sterilization (160°C, over night). The cleaning of hailstones was done under conditions minimizing contamination by organic vapour.

For microbiological analysis, the surface of 24 hailstones was sterilized under sterile conditions as previously described [6]. For flow cytometry analysis 1.8 ml each of 12 hailstones was fixed in 2% glutaraldehyde, the remainder of these 12 hailstones was either refrozen, stored at $-20°C$, or used for the enumeration of colony forming units (CFU) using R2A plates [16].

Determination of dissolved organic carbon (DOC) and total dissolved nitrogen (TDN)

DOC and TDN were analyzed in 18 hailstones by low-volume manual injection and catalytic high-temperature combustion on a Shimadzu TOC-V analyzer with a total nitrogen module (TNM-1) [17]. Samples were acidified to pH = 2 with HCl (p.a.) and purged for 10 minutes with synthetic air prior to analysis to remove inorganic carbon. The accuracy of the analysis was confirmed with deep-sea reference water samples provided by the University of Miami. The accuracy with respect to deep-sea water was within 5% relative error and detection limits were 5 μM for DOC and 1 μM for TDN. Procedural blanks did not yield detectable amounts of DOC and TDN. Eight controls for the cleaning procedure were analyzed in the same way as hailstones and showed significantly lower values than the hailstones (Mann–Whitney U test, W = 144, p<0.0001 for both DOC and TDN). The negative controls were used for blank-correcting DOC and TDN concentrations. As the data were not normally distributed (Shapiro-Wilk normality test, W = 0.4, p<0.0001 for both DOC and TDN), we report the median (Me) together with the quartile 1–quartile 3 values (Q1–Q3) and use a nonparametric test for the analysis of correlation (Spearman's rank correlation coefficient).

Characterization of dissolved organic matter (DOM)

On three individual hailstones, covering the DOC concentration range, a detailed molecular characterization was performed using ultrahigh-resolution mass spectrometry on a 15 Tesla Bruker Solarix electrospray ionization Fourier-transform ion cyclotron resonance mass spectrometer (ESI-FT-ICR-MS). For FT-ICR-MS analysis, DOM was isolated from the hailstones via solid phase extraction [18]. DOM was directly infused into the mass spectrometer in methanol:water (1:1). The samples were ionized by electrospray ionization (ESI) in negative and positive mode. This ionization technique produces singly charged ions and keeps covalent bonds intact. 500 scans were accumulated in broad band mode for each sample. Procedural blanks did not contain detectable impurities. The mass spectra were internally calibrated. A mass error of <20 ppb was achieved for each detected mass. Based on this ultrahigh precision, molecular formulae were calculated for each peak. Programs used for data analysis and interpretation were Bruker Solarix Control, Bruker Data Analysis, Microsoft Access, and Ocean Data View. The difference between the three analyzed hailstones was insignificant compared to triplicate analysis of the same sample; therefore we discuss the average of the three samples.

Total bacterial abundance

Total bacterial abundance was determined using a FacsCalibur flow cytometer (Becton Dickinson, Franklin Lakes, NJ) equipped with an air-cooled laser providing 15 mW at 488 nm employing a standard filter set-up. The samples were stained with SYBRGreen I (final concentration 0.02% of the stock solution, Molecular Probes Inc., Eugene, OR) for 15 min in the dark, at room temperature [19]. Fluorescent microspheres (Molecular Probes Inc.) with a diameter of 0.95 µm were analyzed as a standard. Sterilized ice cubes of deionized water were fixed and analyzed in the same way as the samples. The densities in the negative controls were significantly lower than the densities in hailstones (Mann–Whitney U test, W = 47, p<0.005). As the data were not normally distributed (Shapiro-Wilk normality test, W = 0.5, p<0.0001), we report the median (Me) and the quartile 1–quartile 3 values (Q1–Q3).

The analysis of bacterial sources

The 16S rRNA gene sequences of the clones and the isolates, which have been previously reported under GenBank accession numbers JQ896628–JQ897350 [6], were analyzed for the community composition of individual hailstones. Operational taxonomic units (OTUs) were created by 99% similarity using the CD-HIT Suite: Biological Sequence Clustering and Comparison [20]. The Ribosomal Database Project (RDP) classifier was used for naive Bayesian classification of sequences [21]. The taxa that were independently sampled by at least 3 hailstones were considered characteristic and the presence of taxa only sampled by 1 or 2 hailstones was regarded as coincidental. The cultivable community was investigated on the genus level, whereas the total community was analyzed on the phylum, class or order level.

Results and Discussion

The composition of dissolved organic matter (DOM) was determined in terms of quantity and quality. Our bulk analysis of dissolved organic carbon (DOC) and total dissolved nitrogen (TDN) in 18 hailstones revealed high concentrations of both DOC and TDN. Concentrations of DOC ranged between 90 and 1569 µM, with a median DOC concentration of 179 µM (Q1–Q3 = 132–220 µM). TDN concentrations ranged between 23 and 228 µM, with a median of 30 µM (Q1–Q3 = 27–35 µM). Similar DOC concentrations were previously reported for cloud water from orographic clouds [22] and rain [23]. The concentration of TDN was in the range of values reported for TDN in precipitation [24]. On average, more than two thirds of TDN was present as dissolved inorganic nitrogen (DIN) in the form of nitrate and ammonium [6]. It has previously been reported for precipitation in both rural and urban areas world-wide that inorganic nitrogen accounts for the major fraction of dissolved nitrogen [24]. Considering that the concentrations of DOC and TDN in storm clouds are within the same range as the concentrations measured in rivers, lakes and oceans [25], storm clouds can be classified as eutrophic environments. There was a significant correlation (Spearman's rank correlation coefficient, rho = 0.749, p<0.001, n = 18) between DOC and TDN concentrations (Figure 1), which suggests that carbon and nitrogen were derived from the same organic source, which either served as a condensation nucleus or got dissolved in cloud water. Subsequently, the source was diluted by deposition of water vapour or coalescence of other cloud droplets, causing the range of concentrations that we observed between individual hailstones. As most of the TDN in hailstones was inorganic, we assert that a mineralization process, involving photochemistry or biodegradation, took place after dissolution of the source organic compound into cloud droplets.

The ability of heterotrophic microorganisms to metabolize DOM is not only dependent on the quantity of DOM that is available, but also on the molecular composition of the DOM pool. Not all compounds may be equally degradable by the microorganisms that are co-occurring in the cloud droplets. Using ultrahigh-resolution mass spectrometry we characterized the

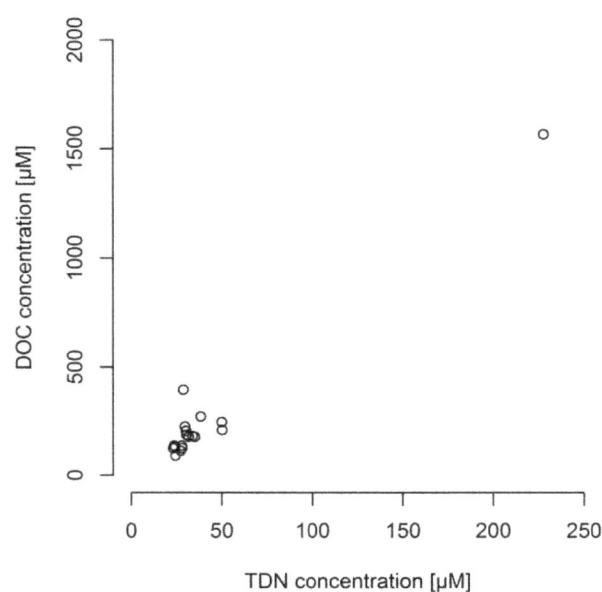

Figure 1. The correlation between DOC and TDN. Dissolved organic carbon (DOC) is presented as a function of dissolved total nitrogen (TDN). The DOC and TDN concentrations were significantly correlated (Spearman's rank correlation coefficient, rho = 0.749, p<0.001, n = 18).

molecular composition of DOM in three individual hailstones. FT-ICR-MS is the only method that allows obtaining molecular information on individual compounds in complex DOM mixtures. The method has been successfully used to get insights into the molecular composition of DOM in marine and in freshwater systems in hitherto unprecedented detail [26]. Here, we applied this advanced analytical technique for the first time on hailstones. Very small volatile organic compounds (<150 Da) escaped our analytical window. Their ubiquitous presence in the atmosphere has already been described elsewhere [22], thus we focused on molecules of the higher molecular mass range. The molecular formulae of 2839 compounds were identified. More than 99% of them were in the molecular mass range of 150–1000 Da. The median mass of all compounds was 354 Da. The median molecular formula was $C_{16}H_{22}O_7$, i.e. half of the detected compounds contained more, and half less, of the respective element. The large molecular diversity and the molecular mass range of DOM in hailstones were comparable to DOM in aquatic systems [26] as well as to water-soluble compounds in aerosols [27]. Forty-four percent of all identified compounds contained one or two nitrogen atoms. All nitrogen was associated to phenolic and unsaturated compounds, whereas peptides and proteins were not present in detectable concentrations. While the molecular diversity of nitrogen-containing compounds was high (1242 compounds contained nitrogen), their abundance in terms of relative concentration was low. Thirteen percent of all compounds contained one sulfur atom. Most sulfur containing compounds were sulfonic acids, some of which are common synthetic products.

The molecular composition of higher molecular mass range DOM is indicative of its history. A few compounds contained less than 10 carbon atoms (Figure 2) and were potentially volatile, but most compounds were too large to be volatile and must have reached the atmosphere as particles. As aromatic compounds in dissolved organic matter are very susceptible to photochemical

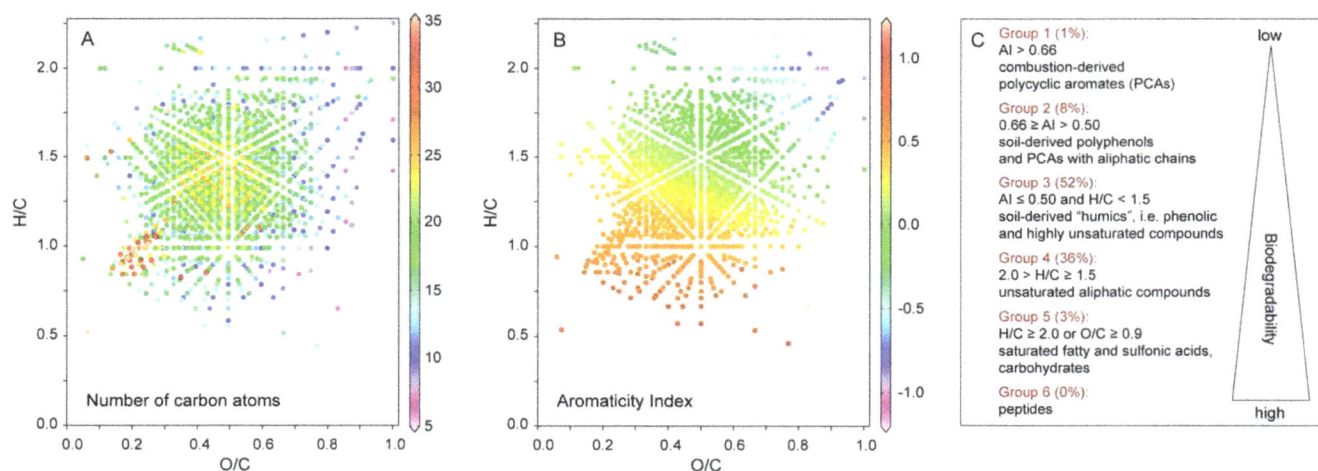

Figure 2. The molecular composition of dissolved organic matter in hail. The element ratio H/C is plotted as a function of O/C for each detected molecular formula detected by ultrahigh-resolution mass spectrometry (FT-ICR-MS) in at least one of the three hailstones. Each dot in these plots represents the molecular formula of an intact molecule. Panel A: The number of carbons in each molecular formula is displayed as a color code in the third dimension. Most compounds are large (C>10) and polar (O/C>0) and are likely not volatile. Panel B: The aromaticity index (AI-mod, [29]) of each molecular formula is displayed as a color code in the third dimension. An aromaticity index 0.5<AI is unambiguous evidence for aromatic compounds and an aromaticity index 0.66<AI is unambiguous evidence for condensed aromatics. Panel C: Compound groups were assigned to molecular formulae based on their aromaticity and element ratios [29], [39]. The biodegradability roughly increases from group 1 to group 6, e.g., polycyclic aromates are among the most stable compounds in the environment, whereas most peptides are quickly decomposed in the environment. Peptides (group 6) have the same characteristics as group 4, but contain nitrogen.

decay [28], the dominance of this compound class in the hailstones is indicative of a fast transfer from soil to atmosphere and hail on the time scale of less than one day.

The aromaticity of each molecular formula was assessed with help of the Aromaticity Index (AI [29]. An aromaticity index 0.66<AI is an unambiguous criterion for polycyclic aromatic structures that are produced during combustion, but not by organisms. An aromaticity index 0.5<AI is an unambiguous criterion for the presence of aromatic compounds that are abundant in vascular plant debris (e.g. lignin). Thus compounds having an aromaticity index 0.5<AI<0.66 are plant derived material that has undergone microbial degradation in soils. By using AI and element ratios of molecular formulae (H/C and O/C), all detected molecules were grouped according to their molecular structure. The majority of the compounds (60%) were highly unsaturated or phenolic organic acids, typical for soil-derived organic matter in rivers (groups 2 and 3 in Figure 2) (e.g. [30], [31]). Less than 3% of the identified compounds were plant waxes, fatty acids or carbohydrates (group 5 in Figure 2), and less than 1% of the compounds were unambiguously combustion-derived (group 1 in Figure 2). As the majority of compounds were soil-derived (groups 2, 3 and 4 in Figure 2) and leaf surface compounds were present to a lesser degree (part of group 5 in Figure 2), the most likely scenario that explains the molecular composition of DOM in the hailstones is dissolution and desorption of organic matter from soil particles that were mobilized from a local source. A fraction of these particles most probably carried bacterial cells, which often are found aerosolized attached to particles [2]. Peptides were not present in detectable concentrations, thus the only highly biodegradable group of compounds were plant waxes, fatty acids and carbohydrates (group 5 in Figure 2), which represented less than 3% of the identified compounds. Due to the short lifetime of the storm cloud, from which the hailstones were obtained [6], only group 5 compounds are likely to be metabolized by bacteria during their residence time in the atmosphere. As 80% of the cloud water

remains airborne after the hail event, the hailstones do not represent the eventual fate of all cloud water, containing organic matter and bacteria. When the cloud dissipates and the droplets evaporate, the aerosols remain airborne and can serve as condensation nuclei for future cloud events. Thus, the detected trace amounts of easily degradable organic compounds, may serve as bacterial substrate even after the cloud has dissipated.

The analysis of 12 individual hailstones by flow cytometry revealed total bacterial numbers ranging from 778 to 21 321 cells per ml (Me = 1973, Q1–Q3 = 1485–2960, Figure 3). The bacterial densities in the storm cloud are in the lower range of previously reported cell numbers in cloud water, which ranged between 1500 [13] and 430 000 [11] bacteria per ml. Based on the average bacterial density in hailstones and an assumed initial cloud droplet diameter of 10 μm [32], we can calculate that on average only 1 out of 10^6 storm cloud droplets carried a bacterial cell. Thus, cloud droplets are sparsely populated environments, where competition for nutrients and space between bacterial cells is likely insignificant. In addition, we can conclude that cloud water is a nutrient-rich microbial environment, in which significant increase in cell numbers would be possible even if only 3% of the high molecular mass DOM is readily biodegradable.

The median cultivability of bacteria was 0.8% (Q1–Q3 = 0.2%–1.5%), with high variability characteristic for individual hailstones. The reported range of cultivable bacteria in the atmosphere is between 0.01% and 75% [2]. However, cloudborne bacterial communities have previously been found to be characterized by a lower cultivability of between <1% and 2% [10], [33]. Up to 10.5% of all storm cloud bacteria were cultivable on nutrient agar plates (Figure 3), a property that is consistent with an opportunistic ecologic strategy. Lower cultivability of cells from some hailstones was probably a result of stress factors that the cells were subjected to during hailstone formation. E.g. these cells could have been subjected to several cycles of freeze-thawing during hailstone formation, which may cause cultivable bacteria to die or develop into a viable but non-cultivable state [34]. The fact than an

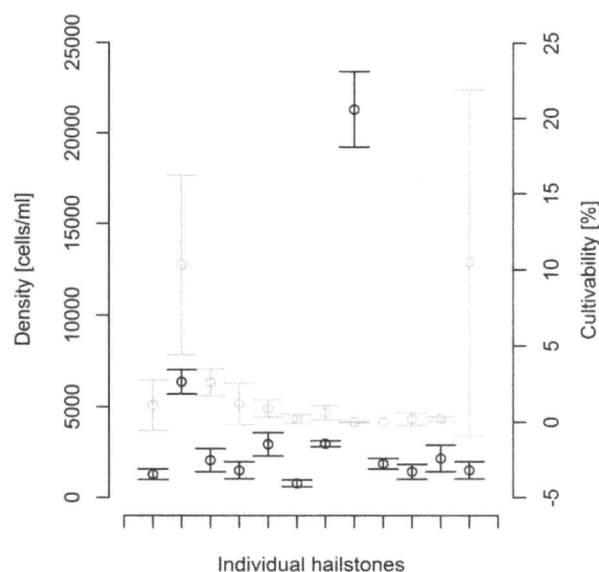

Figure 3. Bacterial density and proportion of cultivable cells. The mean density of bacterial cells as determined with flow cytometry in individual hailstones is presented (gray lines). The proportion of cultivable cells is shown for the same hailstones (dark lines). Error bars denote the standard deviation.

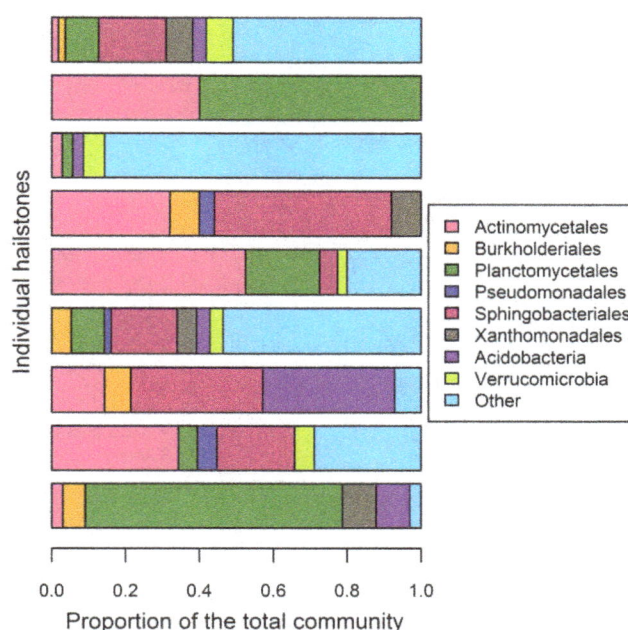

Figure 4. Total community composition in the storm cloud. Proportion of characteristic orders and phyla in 9 out of 12 hailstones, from which clone libraries were made. Characteristic orders and phyla are defined as the ones detected in ≥3 hailstones.

unusually high cultivability has recently been described for epiphytic bacterial community [35], might point to an epiphytic origin of a part of storm cloud community [36], although high cultivability is not a strong proof by itself.

The total bacterial community was highly variable in individual hailstones (Figure 4), which is likely a consequence of storm clouds being a highly dynamic and temporary environment. However, there were taxa that were found in ≥3 hailstones and may represent typical cloud inhabitants. Bacterial orders and phyla, of which representatives were found in at least three individual hailstones, are termed characteristic and are presented in Figure 4. Sequences from *Actinobacteria* (23% of all sequences) were found in all investigated hailstones and representatives from *Plantomycetes* (11%), the *Bacteroidetes* (14%) as well as the γ-*Proteobacteria* (12%) were found in ≥5 hailstones. Also, sequences belonging to *Acidobacteria* (3%), *Verrucomicrobia* (3%), α- (5%) and β-*Proteobacteria* (8%) were present in ≥3 individual hailstones. Terrestrial habitats, and plants in particular, have been identified as major sources of bacteria in the atmosphere [2]. When investigating the spatial variability of airborne bacterial communities, Bowers et al [3] found that the community composition depended on the type of their source environment. Considering the high concentrations of *Actinobacteria* and *Bacteroides* and the low concentrations of *Rhizobiales* (Figure 4), the storm cloud community of our study resembles most closely the airborne communities found by Bowers et al [3] at suburban locations. This agreement fits well with the location of the storm cloud formation, which was over a city. The indicative bacterial taxa of our study pointed to a predominant terrestrial source of the atmospheric bacterial community [3]. The low abundance of *Acidobacteria* and *Rhizobiales* (Figure 4) indicated that soil bacteria were not dominant in the community, whereas the high relative abundance of γ-*Proteobacteria* and *Sphingobacteriales* (Figure 4) suggested that the microbiota in cloud droplets is to a large extent influenced by bacteria of epiphytic origin. This fits well with the high fraction of cultivable cells found for the storm

cloud community and is consistent with the results obtained by others (e.g. [3]).

Common cultivable genera, found in individual hailstones, are presented in Figure 5. Bacterial genera of the cultivable community, which were isolated from ≥3 individual hailstones, were considered characteristic for the cloud. In contrast to the community represented by the clone library, the cultivable bacterial community had a higher proportion (43.5%) of characteristic genera (Figure 5). Some of the characteristic genera (*Bacillus*, *Paenibacillus*, *Bradyrhizobium*) that were represented by isolates are consistent with soil origin of the cultivable community, but there was a remarkable (22%) contribution of typical plant-surface bacteria belonging to the genus *Methylobacterium*. They are adapted to a number of stress factors common for plant surfaces and the atmosphere [6], and therefore predestinated to remain active in the airborne state. We found, for example, that about 90% of the *Methylobacterium* isolates produced reddish, most likely carotenoid-type pigments, which can protect the cells against UV-induced cell damage [37]. In addition to being adapted to atmospheric stress, several *Methylobacterium* isolates have a wide substrate range, which is consistent with an opportunistic ecological strategy [6] and would predispose these cells to growth in the atmosphere. On the contrary, members of typical soil inhabiting genera, *Bacillus* and *Paenibacillus*, most likely get airborne as endospores, which hinders their growth in the atmosphere.

Despite the fact that the total and cultivable bacterial community composition as well as high cultivability all indicated the dominance of plant-associated bacterial groups, the molecular characteristics of DOM pointed to a soil origin of most aerosol particles in the cloud droplets. In fact, very few molecules suggested direct plant-surface origin, as the plant-derived compounds showed the chemical signature of decomposition in soil prior to aerosolization. A likely explanation for the discrepancy between chemical and microbial data is that bacteria originating

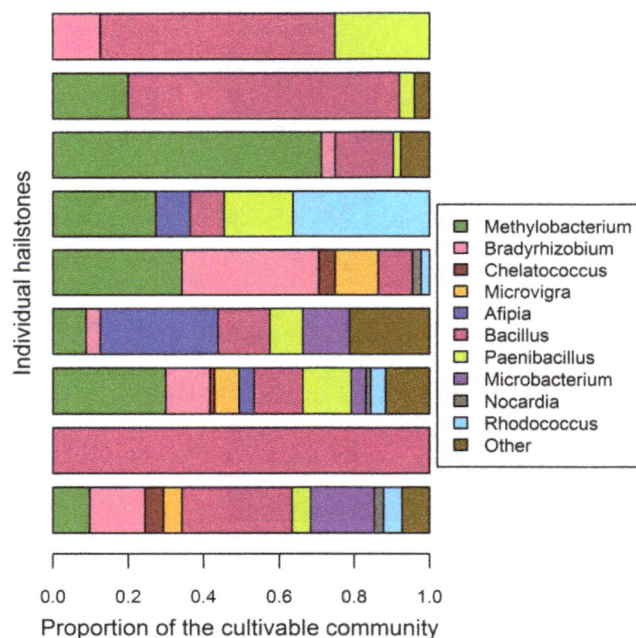

Figure 5. Cultivable genera in the storm cloud. Proportion of characteristic cultivable genera in 9 out of 12 hailstones, which contained cultivable bacteria. Characteristic genera are defined as the ones isolated from ≥3 hailstones.

from plant surfaces are better adapted to survival and growth in the atmosphere, whereas the stressors encountered in the atmosphere such as desiccation and UV radiation act as strong selective barriers against the soil inhabiting bacteria. Consequently, epiphytic bacteria may get enriched in the atmosphere, which could not only affect the chemical composition of the atmosphere,

but also impact precipitation patterns more strongly than previously thought, as most INA bacteria stem from plant surfaces [38].

Conclusions

The unique data sets that we obtained by analyzing individual hailstones provide us with unprecedented insight into the microbial and chemical inventory of storm clouds. They allow us to conclude that while the majority of aerosols were mainly soil-derived, the storm cloud contained microbial communities with a strong plant-surface signature, which links the troposphere to the phyllosphere. Many plant-associated bacteria are efficient in utilizing variable substrates on short timescales as well as in coping with atmospheric stress. Growth of these bacteria can be supported by the trace amounts of carbohydrates, lipids and some nitrogen-containing compounds that we detected among the high molecular mass DOM. The accumulating evidence strongly points to a selection process of bacterial cells in the course of cloud formation, which likely impacts the long-distance transport and the global distribution of bacteria. Our study on hailstones indicates that storm clouds are among the most extreme habitats on Earth, where microbial life can exist.

Acknowledgments

We thank Marijan Govedič for sample collection and Tina Thane, Kathrin Klaproth and Matthias Friebe for excellent technical assistance. We appreciate the helpful advice of Mark A. Lever and Kasper U. Kjeldsen regarding the molecular work with low density bacterial environments.

Author Contributions

Conceived and designed the experiments: TŠT KF BMH UGK. Performed the experiments: TŠT TD RT. Analyzed the data: TŠT TD RT NWN. Contributed reagents/materials/analysis tools: TŠT TD RT KF BMH UGK. Wrote the paper: TŠT KF UGK TD.

References

1. Delort AM, Vaïtilingom M, Amato P, Sancelme M, Parazols M, et al. (2010) A short overview of the microbial population in clouds: potential roles in atmospheric chemistry and nucleation processes. Atmos Res 98(2–4): 249–260.
2. Burrows SM, Elbert W, Lawrence MG, Pöschl U (2009) Bacteria in the global atmosphere – Part 1: Review and synthesis of literature data for different ecosystems. Atmos Chem Phys Discuss 9: 10777–10827.
3. Bowers RM, McLetchie S, Knight R, Fierer N (2011) Spatial variability in airborne bacterial communities across land-use types and their relationship to the bacterial communities of potential source environments. ISME J 5:601–612.
4. Bowers RM, Lauber CL, Wiedinmyer C, Hamady M, Hallar AG, et al. (2009) Characterization of Airborne Microbial Communities at a High-Elevation Site and Their Potential To Act as Atmospheric Ice Nuclei. Appl Environ Microbiol 75(15): 5121–5130.
5. Kourtev PS, Hill KA, Shepson PB, Konopka A (2011) Atmospheric cloud water contains a diverse bacterial community. Atmos Environ 45: 5399–5405.
6. Temkiv TŠ, Finster K, Hansen BM, Nielsen NW, Karlson UG (2012) The microbial diversity of a storm cloud as assessed by hailstones. FEMS Microbiol Ecol 81(3): 684–695. DOI: 10.1111/j.1574-6941.2012.01402.x
7. Sun J, Ariya P (2006) Atmospheric organic and bio-aerosols as cloud condensation nuclei (CCN): A review. Atmos Environ 40: 795–820.
8. Möhler O, DeMott PJ, Vali G, Levin Z (2007) Microbiology and atmospheric processes: the role of biological particles in cloud physics. Biogeosciences 4: 1059–1071.
9. Govindarajan AG, Lindow SE (1988) Size of bacterial ice-nucleation sites measured in situ by radiation inactivation analysis. Proc Nat Acad Sci USA 85(5): 1334–1338.
10. Bauer H, Kasper-Giebl A, Löflund M, Giebl H, Hitzenberger R, et al. (2002) The contribution of bacteria and fungal spores to the organic carbon content of cloud water, precipitation and aerosols. Atmos Res 64: 109–119.
11. Hill KA, Shepson PB, Galbavy ES, Anastasio C, Kourtev PS, et al. (2007) Processing of atmospheric nitrogen by clouds above a forest environment. J Geophys Res 112: 1–16.

12. Herlihy LJ, Galloway JN, Mills AL (1987) Bacterial utilization of formic and acetic acid in rainwater. Atmos Environ 21(11): 2397–2402.
13. Sattler B, Puxbaum H, Psenner R (2001) Bacterial growth in supercooled cloud droplets. Geophys Res Lett 28(2): 239–242.
14. Amato P, Demeer F, Melaouhi A, Fontanella S, Martin-Biesse AS, et al. (2007) A fate for organic acids, formaldehyde and methanol in cloud water: their biotransformation by micro-organisms. Atmos Chem Phys 7(15): 4159–4169.
15. Vaïtilingom M, Amato P, Sancelme M, Laj P, Leriche M, et al. (2010) Contribution of Microbial Activity to Carbon Chemistry in Clouds. Appl Environ Microbiol 76(1): 23–29.
16. Reasoner DJ, Geldreich EE (1985) A new medium for the enumeration and subculture of bacteria from potable water. Appl Environ Microbiol 49: 1–7.
17. Stubbins A, Dittmar T (2012) Low volume quantification of dissolved organic carbon and dissolved nitrogen. Limnol Oceanogr. Methods 10, 347–352.
18. Dittmar T, Koch BP, Hertkon N, Kattner G (2008) A simple and efficient method for the solid-phase extraction of dissolved organic matter (SPE-DOM) from seawater. Limnol Oceanogr-Meth 6: 230–235.
19. Marie D, Brussaard CPD, Thyrhaug R, Bratbak G, Vaulot D (1999) Enumeration of marine viruses in culture and natural samples by flow cytometry. App Environ Microbiol 65(1): 45–52.
20. CD-HIT Suite: Biological Sequence Clustering and Comparison website. Available: http://weizhong-lab.ucsd.edu/cdhit_suite/cgi-bin/index.cgi. Accessed 2012 July 10.
21. Wang Q, Garrity GM, Tiedje JM, Cole JR (2007) Naive Bayesian classifier for rapid assignment of rRNA sequences into the new bacterial taxonomy. Appl Environ Microbiol 73(16): 5261–5267.
22. Marinoni A, Laj P, Sellegeri K, Mailhot G (2004) Cloud chemistry at the Puy de Dôme: variability and relationships with environmental factors. Atmos Chem Phys 4: 715–728.
23. Willey JD, Kieber RJ, Eyman MS, Avery GB Jr (2000) Rainwater dissolved organic carbon: Concentrations and global flux. Global biochem cy 14(1): 139–148.

24. Cornell SE, Jickells TD, Cape JN, Rowland AP, Duce RA (2003) Organic nitrogen deposition on land and costal environments: a review of methods and data. Atmos Environ 37: 2173–2191.

25. Maita Y, Yanada M (1990) Vertical distribution of total dissolved nitrogen and dissolved organic nitrogen in seawater. Geochem J 24: 245–254.

26. Dittmar T, Paeng J (2009) A heat-induced molecular signature in marine dissolved organic matter. Nature Geosci 2: 175–179.

27. Wozniak AS, Bauer JE, Sleighter RL, Dickhut RM, Hatcher PG (2008) Technical Note: Molecular characterization of aerosol-derived water soluble organic carbon using ultrahigh resolution electrospray ionization Fourier transform ion cyclotron resonance mass spectrometry. Atmos Chem Phys 8: 5099–5111.

28. Stubbins A, Spencer RGM, Chen H, Hatcher PG, Mopper K, et al.(2010) Illuminated darkness: molecular signatures of Congo River dissolved organic matter and its photochemical alteration as revealed by ultrahigh precision mass spectrometry. Limnol Oceanogr 55: 1467–1477.

29. Koch BP, Dittmar T (2006) From mass to structure: An aromaticity index for high-resolution mass data of natural organic matter. Rapid Commun Mass Spectrom 20: 926–932.

30. Stenson AC, Marshall AG, Cooper WT (2003) Exact Masses and Chemical Formulas of Individual Suwannee River Fulvic Acids from Ultrahigh Resolution Electrospray Ionization Fourier Transform Ion Cyclotron Resonance Mass Spectra. Anal Chem 75(6): 1275–1284.

31. Tremblay LB, Dittmar T, Marshall AG, Cooper WJ, Cooper WT (2007) Molecular characterization of dissolved organic matter in a north Brazilian mangrove porewater and mangrove-fringed estuary by ultrahigh resolution Fourier transform-ion cyclotron resonance mass spectrometry and excitation/emission spectroscopy. Mar Chem 105: 15–29.

32. Ahrens CD (2009) Meteorology Today. An Introduction to Weather, Climate, and the Environment. Ninth Edition. Belmont: Brooks/Cole. 549 p.

33. Amato P, Menager M, Sancelme M, Laj P, Mailhot G, et al. (2005) Microbial population in cloud water at the Puy de Dome: Implications for the chemistry of clouds. Atmos Environ 39(22): 4143–4153.

34. Oliver JD (2005) The Viable but Nonculturable State in Bacteria. J Microbiol 43: 93–100.

35. Niwa R, Yoshida S, Furuya N, Tsuchiya K, Tsushima S (2011) Method for simple and rapid enumeration of total epiphytic bacteria in the washing solution of rice plants. Can J Microbiol 57: 62–67.

36. Garland JL, Cook KL, Adams JL, Kerkhof L (2001) Culturability as an Indicator of Succession in Microbial communities. Microbiol Ecol 42(2): 150–158.

37. Jacob JL, Carroll TL, Sundin GW (2004) The Role of Pigmentation, Ultraviolet Radiation Tolerance, and Leaf Colonization Strategies in the Epiphytic Survival of Phyllosphere Bacteria. Microbial Ecol 49: 104–113.

38. Morris CE, Sands DC, Vinatzer BA, Glaux C, Guilbaud C, et al. (2008) The life history of the plant pathogen *Pseudomonas syringae* is linked to the water cycle. ISME J: 1–14.

39. Perdue EM (1984) Analytical constraints on the structural features of humic substances. Geochim Cosmochim Ac 48: 1435–1442.

17

Do Clouds Save the Great Barrier Reef? Satellite Imagery Elucidates the Cloud-SST Relationship at the Local Scale

Susannah M. Leahy[1,2]*, **Michael J. Kingsford**[1,2], **Craig R. Steinberg**[3]

1 School of Marine and Tropical Biology, James Cook University, Townsville, Queensland, Australia, **2** Australian Research Council Centre of Excellence for Coral Reef Studies, James Cook University, Townsville, Queensland, Australia, **3** Australian Institute of Marine Science, Townsville, Queensland, Australia

Abstract

Evidence of global climate change and rising sea surface temperatures (SSTs) is now well documented in the scientific literature. With corals already living close to their thermal maxima, increases in SSTs are of great concern for the survival of coral reefs. Cloud feedback processes may have the potential to constrain SSTs, serving to enforce an "ocean thermostat" and promoting the survival of coral reefs. In this study, it was hypothesized that cloud cover can affect summer SSTs in the tropics. Detailed direct and lagged relationships between cloud cover and SST across the central Great Barrier Reef (GBR) shelf were investigated using data from satellite imagery and *in situ* temperature and light loggers during two relatively hot summers (2005 and 2006) and two relatively cool summers (2007 and 2008). Across all study summers and shelf positions, SSTs exhibited distinct drops during periods of high cloud cover, and conversely, SST increases during periods of low cloud cover, with a three-day temporal lag between a change in cloud cover and a subsequent change in SST. Cloud cover alone was responsible for up to 32.1% of the variation in SSTs three days later. The relationship was strongest in both El Niño (2005) and La Niña (2008) study summers and at the inner-shelf position in those summers. SST effects on subsequent cloud cover were weaker and more variable among study summers, with rising SSTs explaining up to 21.6% of the increase in cloud cover three days later. This work quantifies the often observed cloud cooling effect on coral reefs. It highlights the importance of incorporating local-scale processes into bleaching forecasting models, and encourages the use of remote sensing imagery to value-add to coral bleaching field studies and to more accurately predict risks to coral reefs.

Editor: Juan A. Añel, University of Oxford, United Kingdom

Funding: Work supported by:-Australian Research Council Centre of Excellence for Coral Reef Studies, funding to MJK-Q-IMOS Satellite Remote Sensing funding via CRS' relationship with the Australian Government's National Collaborative Research Infrastructure Strategy, the Super Science Initiative, the Australian Institute of Marine Science, and the Queensland State Government The funders had no role in study design, data collection and analysis, decision to publish, or preparation of the manuscript.

Competing Interests: The authors have declared that no competing interests exist.

* E-mail: Susannah.Leahy@my.jcu.edu.au

Introduction

The reality of climate change is now well established [1,2], and there is strong consensus that anthropogenic changes in carbon dioxide, methane, and nitrous oxide are contributing to global warming [2]. In the context of coral reefs, global warming has been implicated in the consistent rise in sea surface temperatures (SSTs) over the past 45 years [3,4], and consequently to thermal bleaching events [5]. The extent and intensity of the 1998 bleaching event in the Indo-Pacific in particular has drawn attention to the future of coral reefs on a warming planet [5–8].

Extensive analyses of bleaching patterns [5,8,9], together with comprehensive laboratory experimentation [10–12], have dramatically improved our understanding of coral thermal tolerances and our ability to predict future bleaching occurrences [13,14]. They have also led to the development of a simple and straightforward paradigm regarding critical temperature thresholds: multi-day exposure to SSTs 1–2°C above the long-term local average will cause mass coral bleaching [15–17].

While the general mechanisms of anthropogenic global warming are straightforward, feedback loops, particularly involving water vapour, can confound predictions of patterns of warming at the regional and local scales [9,18]. Atmospheric warming is caused by incident shortwave solar radiation and trapping of the longwave radiation re-emitted by planetary surfaces and "greenhouse gases" [19]. A warm atmosphere promotes evaporation, increasing atmospheric water vapour content while cooling SSTs via the transfer of sensible and latent heat from the ocean's surface into the atmosphere. At this point, a positive feedback mechanism may be generated, with the "greenhouse" properties of water vapour trapping more heat energy, serving to raise air temperatures and perpetuating a cycle of evaporation and rising air temperatures [20,21]. Alternatively, a negative feedback mechanism may occur in which increased atmospheric water vapour condenses into clouds; these reflect incident solar radiation, preventing further surface warming [19,22,23]. This mechanism has been implicated in distinct events in which coral bleaching thresholds were not attained due to local cloud cover [24–30], and has been suggested as a key mechanism constraining tropical SSTs in a warming world [22,23]. There is merit to both hypotheses, in that observed cloud build-up and temperature responses depend on a number of physical parameters including the strength of convective activity [20,21] and cloud parameters such as cloud height, altitude, spatial distribution, optical depth, liquid water content, and particle size and state [31–33]. These factors determine to what extent local clouds will reflect or transmit incident shortwave radiation, and reflect or transmit outgoing longwave radiation, with measurable consequences for local air and sea temperatures [21,22,32,34,35].

The complexity of the water vapour feedback mechanisms makes cloud processes one of the major confounding factors in climate models, with a significant proportion of variation between models directly attributable to differences in parametrisation of cloud phenomena [32,36]. The quality of climate models is further constrained by the scale at which relevant processes are forced, with many models produced on a global, or at best, an ocean basin scale ([13], but see [14,23]). The output of these models can be downscaled for relevance to ecological management, such as coral reef areas, but the downscaling process is known to reduce the certainty associated with climate predictions, and inaccuracies can be high [37]. Capturing the full range of physical processes involved in climatology, while at the same time producing realistic predictions for local management authorities, is a major challenge in climate modelling; the result is often a disconnect between the predicted regional conditions and the observed local conditions (e.g. [9,14,18]).

It is therefore of key importance to collect empirical evidence of atmospheric feedback processes at local (10s of kilometres) to regional (100s to 1,000s of kilometres) scales, and to quantify their effects on incident solar radiation, and subsequently on SST, a direct causative agent of coral bleaching. This study aimed to find empirical evidence of local atmospheric processes, in particular cloud cover, affecting SST on the central Great Barrier Reef (GBR), and to quantify its effect during the vulnerable summer months, when high temperatures and generally low convective activity increase the probability of mass thermal bleaching events. Records of SST, incident solar radiation, and cloud cover during the multiyear period from 2004 to 2008 were retrieved from a combination of in situ loggers and satellite imagery and were successfully used to identify relationships between the variables, including responses that were temporally lagged, at different positions across the GBR shelf.

Materials and Methods

Study Period

The Austral summer on the GBR extends from October to March, with incident solar radiation peaking around December and SSTs peaking between December and February [38]. The atmospheric circulation of the Australian summer monsoon increases cloud cover and brings in weaker and moister surface winds during this time [38], such that cloud cover and rainfall are highest in February [39]. Thermal bleaching risk is highest during this period, with particularly dramatic mass bleaching events recorded in the Austral summers of 1998 [5] and 2002 [9].

Study Area

It was hypothesized that cloud cooling effects would vary with distance from shore, potentially due to orographic effects [40], as well as a coastal-to-ocean gradient in both bathymetry and exposure, in which features an inner shelf open coastal lagoon, a mid shelf complex reef matrix, and an outer shelf exposed to the Coral Sea [41,42]. Environmental data was therefore collected from Australian Institute of Marine Science (AIMS) island and buoy weather stations deployed at sites at inner (Orpheus Island, 18°36′46.08″S, 146°28′59.16″E; Cleveland Bay, 19°8′27.6″S, 146°53′23.4″E; and Middle Reef, 19°11′40.2″S, 146°48′36.72″E), mid (Davies Reef, 18°49′53.82″S, 147°38′4.2″E; John Brewer Reef, 18°37′15.24″S, 147°3′13.68″E; and Kelso Reef, 18°26′42.84″S 146°59′32.06″E), and outer shelf

(Dip Reef, 18°24′5.33″S, 147°27′3.67″E; Chicken Reef, 18°39′17.57″S, 147°43′15.49″E; and Myrmidon Reef, 18°16′27.29″S, 147°22′54.25″E) positions (Fig. 1). These distance strata were also of biological interest, as substantial variation in marine assemblages are found cross-shelf (e.g. soft corals [43], sponges [44], hard corals [45], herbivorous fishes [46]).

"Regions of interest" that equated to major cross-shelf positions on the central GBR were produced in ESRI ArcGIS 10.0, using shapefiles of GBR features and management regions defined and provided by the Great Barrier Reef Marine Park Authority (GBRMPA) as UTM Zone 55 projections, GDA94. Management regions were modified using expert opinion to produce inner (6,734 km²), mid (10,417 km²), and outer shelf (5,471 km²) study regions, extending along the coastline from Hinchinbrook Island (18°21′0.00″S, 146°17′49.20″E) south to Cape Bowling Green (19°24′46.80″S, 147°28′8.40″E) and covering the full cross-shelf area (Fig. 1).

Collection and Processing of SST Data

Multi-year SST data was available from multiple near-surface temperature loggers in each shelf position (inner: Orpheus Island, Cleveland Bay, Middle Reef; mid: Davies Reef, John Brewer Reef; outer: Dip Reef, Chicken Reef). As SST trends were consistent within each shelf position, further analyses were conducted using data from loggers with the longest uninterrupted reports between 1 October 2004 and 31 March 2008 at each shelf position (inner: Orpheus Island, mid: Davies Reef, outer: Dip Reef). Multi-year comparisons were produced from expert quality-controlled data from loggers at similar and biologically relevant depths (3–6 m). Thermal readings were taken every 30 minutes. Datasets were trimmed to include only peak heating hours (1100–1600 h, comparable to [40]), and were averaged by day, producing a "mean daily daytime SST." Within-day variation was low, and was generally attributable to tidal fluxes. Preliminary analyses indicated that two summers in the study period were particularly warm (2005 and 2006), while two were relatively cooler (2007 and 2008). These are distinguished graphically for all relevant analyses.

The full SST datasets were detrended using a 21 day moving average in order to remove the seasonal component of variation in the data. Further analyses were conducted on the residuals or "white noise" in the dataset as per Chatfield [47].

Collection and Processing of Cloud Cover Data

Pre-processed remotely-sensed cloud cover information was used to address the question of cloud cooling effects on the central GBR. Cloud imagery was collected using the Moderate Resolution Imaging Spectroradiometer (MODIS) mounted on board the Terra and Aqua satellites. Images were pre-processed to Level 2 Cloud Product state using MOD06 cloud retrieval algorithms as described in King et al. [48]. Images were downloaded from NASA's LAADS server (ladsweb.nascom.nasa.gov, collection 5.1) at a resolution of 5×5 km and projected as a Universal Transverse Mercator (UTM) Zone 55, WGS84.

For each image, the cloud fraction layer was extracted from the original Level 2 Cloud Product file and reclassified from a 0–255 RGB range to a binomial "yes/no" cloud present in each 5×5 km pixel. The reclassified layer was then trimmed to the regions of interest defined above, and the number of yes-cloud and no-cloud pixels in each study region was recorded and converted to a percent cloud cover. Information from satellite imagery that provided only partial coverage of the study region was discarded if it contained <50% coverage of each shelf position, i.e. <130.5, 208.5, or 110 informative pixels for the inner, mid, or outer shelf study areas, respectively. Only daytime imagery was used for

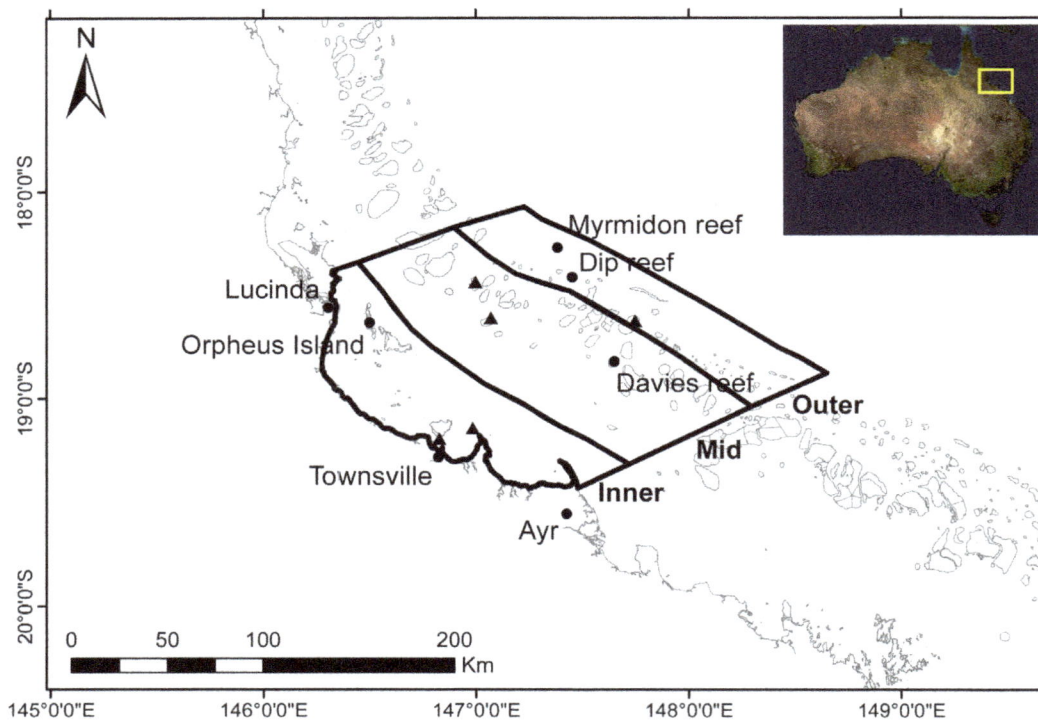

Figure 1. Study area. Location and extent of cross-shelf study regions and AIMS monitoring loggers used in this study. Named reefs indicate loggers used in this study. Black-fill triangles indicate loggers used to validate the generality of SST trends for each study region. Inset: approximate location of the study area; composite satellite image of Australia courtesy of NASA (2002) MODIS technology.

analyses presented here, i.e. one Terra pass at ~1200–1300 h and one Aqua pass at ~1500–1600 h.

Collection and Processing of Radiation Data

Records of incident Photosynthetically Active Radiation (PAR) were collected to serve as proxies for the amount of light reaching the ocean's surface. Measurements of PAR indicate incident visible light in the 400–700 nm range, which is only a fraction of total incident solar radiation ("insolation," ~100–14,000 nm) and is monitored for its relevance to primary productivity [49]. The ratio of PAR to insolation is approximately constant across a range of cloud conditions [31,50] and so is considered a reliable proxy for insolation.

Measurements of PAR, integrated over 10 minutes, were recorded in $\mu mol \cdot s^{-1} \cdot m^{-2}$ every 30 minutes between 1 October 2004 and 1 March 2008 from AIMS monitoring stations at all three shelf positions (inner: Orpheus Island; mid: Davies Reef; outer: Myrmidon Reef). PAR datasets were trimmed to include only peak heating hours (1100–1600 h) and were then averaged by day, producing a "mean daily daytime PAR."

Data Analysis

Daytime daily average SST, cloud cover, and PAR during the peak bleaching period (January to March of each study year) were analysed using a 3-way fixed factor ANOVA to test for the effects of study summer (2005 to 2008), month (January to March), shelf position (inner, mid, or outer shelf), and their interactions. The daily average value of each variable at each shelf position was treated as an independent replicate. The cloud cover dataset was arcsine transformed ($asin(\sqrt{value})*180/\pi$) to meet the assumptions of ANOVA [51]. The SST and PAR datasets violated the

assumption of normality despite transformations; therefore a more conservative critical p value ($p<0.01$) was applied to these data [52].

To test the hypothesis that cloud cover intercepts incident solar radiation, the smoothed time series (five-day moving averages) of PAR was regressed against smoothed arcsine transformed cloud cover for each study summer in each study region.

Overlayed smoothed time series (five-day moving average) of SST and cloud cover were used for qualitative and quantitative assessments of the relationship between the two variables and identification of potential lags between cloud phenomena (increase or decrease in cloud cover) and SST responses. The pattern and strength of the relationship between arcsine transformed daily cloud cover and detrended daily SST (i.e. SST residuals) was then quantified for ±10 time lags using a cross-correlation (SYSTAT 12.02.00). The lags producing the strongest relationship between the two variables were incorporated into a linear regression for each shelf position in each study summer. When detrended SST was lagged *after* the cloud cover dataset, it was considered to be the dependent variable and was regressed against transformed cloud cover. When detrended SST was lagged *before* the cloud cover dataset, cloud cover was considered the dependent variable and was regressed against detrended SST. Directionality of all regressed relationships is indicated in-text and in relevant tables.

Results

Spatial and Temporal Variation in SST

There was great temporal variation in SST at the inner and mid shelf positions both within and among years, serving to identify two years with generally warmer summers (2005 and 2006) and two years with generally cooler summers (2007 and 2008, Fig. 2A,

B). Summer SSTs also varied significantly by shelf position. Variation by shelf position by summer and month resulted in a significant three-way interaction (Table 1). The pattern of SST buildup, peak, and decline between January and March of each summer demonstrated a clear seasonal trend, with SST maxima generally attained in February of each study summer (Fig. 2A, B). In contrast, in the outer shelf region, there was one cool (2007) and three warm summers (2005, 2006, and 2008), and there was a less distinct temperature peak in February (Fig. 2). Summer 2007 was consistently the coolest across all shelf positions.

Spatial and Temporal Variation in Cloud Cover

Cloud cover did not vary significantly by shelf position, but did vary significantly by summer and by month, and by summer*-month interaction (Fig. 3, Table 1). Cloud cover was consistently high in January, February, and early March of 2007 and 2008, and declined in mid-March of both 2007 and 2008 (Fig. 3A–C). The trend was different in 2005 and 2006, with high cloud observed across all shelf positions in January and early March, but not in February (Fig. 3A–C). The nature of cloud cover also differed between study summers, with typically patchier cloud cover noted in 2005 and 2006, and consistently higher and more extensive cloud cover observed in 2007 and 2008 (Fig. 4).

Spatial and Temporal Variation in Incident Radiation

PAR varied significantly among shelf positions, study summers, and months, resulting in a significant three-way interaction (Table 1). However, clear patterns were found among summers at all shelf positions, with the greatest differences among summers occurring in late January and early to mid February (Fig. 5). In the inner shelf region, mean daytime PAR was high across all four study summers, with the exception of a large (500–1,000 $\mu mol \cdot s^{-1} \cdot m^{-2}$) drop in late January 2007 and 2008, which persisted throughout the month of February (Fig. 5A). The same relative difference in PAR between warm (2005 and 2006) and cool (2007 and 2008) years was observed in the mid shelf region, but details of the pattern differed. In the mid shelf region, consistently low summer PAR values gave way to a distinct rise (of approximately 1,100 $\mu mol \cdot s^{-1} \cdot m^{-2}$) in early February 2005 and late February 2006, which persisted for approximately two weeks (Fig. 5B). A similar, albeit weaker, pattern occurred in the outer shelf region, where PAR values rose approximately 700 $\mu mol \cdot s^{-1} \cdot m^{-2}$ higher in late February 2005 and 2006, as compared to the same period in 2007 and 2008 (Fig. 5C).

Incident solar radiation in the PAR range was generally a strong proxy for cloud cover, although the strength of the relationship varied in both time and space (Table 2). The relationship between cloud cover and PAR was consistently strong and negative across all shelf positions in almost all study summers; cloud cover accounted for up to 55.76% of the variation in PAR (summer

Table 1. ANOVAs that tested sea surface temperature (SST), photosynthetically active radiation (PAR), and arcsine transformed cloud cover by shelf position, summer, month, and their interactive effects.

	Factor	d.f.	MS	F	p
SST	Shelf position	2	26.194	181.17	**<0.001**
	Summer	3	25.323	175.146	**<0.001**
	Month	2	55.389	383.097	**<0.001**
	Shelf position * Summer	6	10.959	75.798	**<0.001**
	Shelf position * month	4	1.949	13.482	**<0.001**
	Summer * month	6	20.06	138.744	**<0.001**
	Shelf position * summer * month	12	0.817	5.652	**<0.001**
	Error	1,047	0.145		
Cloud cover	Shelf position	2	1,553.80	2.395	0.092
	Summer	3	15,817.78	24.382	**<0.001**
	Month	2	4,551.47	7.016	**0.001**
	Shelf position * Summer	6	49.439	0.076	0.998
	Shelf position * month	4	109.143	0.168	0.955
	Summer * month	6	10,106.87	15.579	**<0.001**
	Shelf position * summer * month	12	123.409	0.19	0.999
	Error	1,047	648.748		
PAR	Shelf position	2	35,126,017.63	167.626	**<0.001**
	Summer	3	23,773,571.26	113.451	**<0.001**
	Month	2	4,387,579.55	20.938	**<0.001**
	Shelf position * Summer	6	575,370.67	2.746	0.012
	Shelf position * month	4	3,479,080.00	16.603	**<0.001**
	Summer * month	6	8,987,451.70	42.889	**<0.001**
	Shelf position * summer * month	12	879,638.44	4.198	**<0.001**
	Error	1,047	209,549.94		

SST and PAR violated the assumption of normality and are therefore interpreted with a more conservative $p<0.01$.

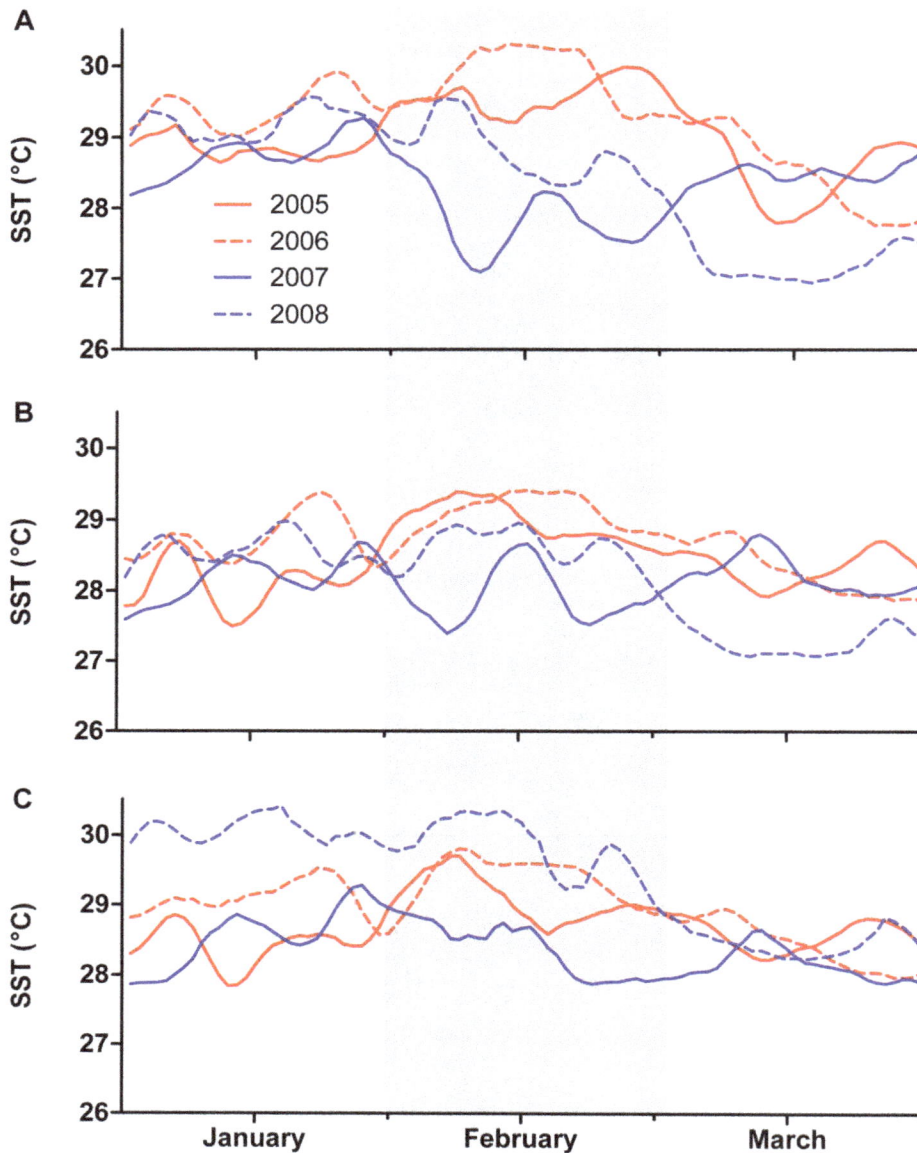

Figure 2. SST time series. Summer daytime SSTs in the (A) inner, (B) mid, and (C) outer shelf study regions. SSTs are smoothed using five-day moving averages. Red lines denote summers with warmer thermal profiles (2005 and 2006); blue lines denote summers with cooler thermal profiles (2007 and 2008). The period of greatest difference in SST, cloud cover, and PAR between the warmer and cooler summers (discussed in-text) is highlighted in grey.

2006, inner shelf, Table 2). Among summers, the fraction of PAR explained by cloud cover was generally highest in the inner shelf region, was lowest mid shelf, and was intermediate in the outer shelf. There is no clear pattern between summers (Table 2).

Relationship between Cloud Cover and SST

A direct comparison of cloud cover and SST time series indicated a lagged relationship between the two variables, with distinct cloud events (increases or decreases) followed by inverse SST changes (decrease or increase) several days later (Fig. 6). For example, two relatively short "pulses" of cloud in early January 2006 were associated with subsequent declines in SST of approximately 0.5°C, while oscillating low and medium cloud cover in early February was associated with high SST in mid to late February (Fig. 6). Formal analysis of the lagged relationship using a cross-correlation indicated two major peaks in the correlation between cloud cover (arcsine transformed) and SST (detrended; residuals only, Fig. 7). The first peak indicated a strong negative correlation between cloud cover and SST residuals three days later (lag: −3). The second peak indicated a strong positive correlation between cloud cover and SST residuals three days prior (lag: +3), which can instead be expressed as a positive correlation between SST residuals and cloud cover three days later (Fig. 7).

Regressing cloud cover (arcsine transformed) against SST residuals lagged three days later produced significant relationships, with cloud cover explaining between 2.1 and 32.1% of the variation in SST residuals three days later (Table 3). This was true over most of the study region in almost all study summers (Table 3).

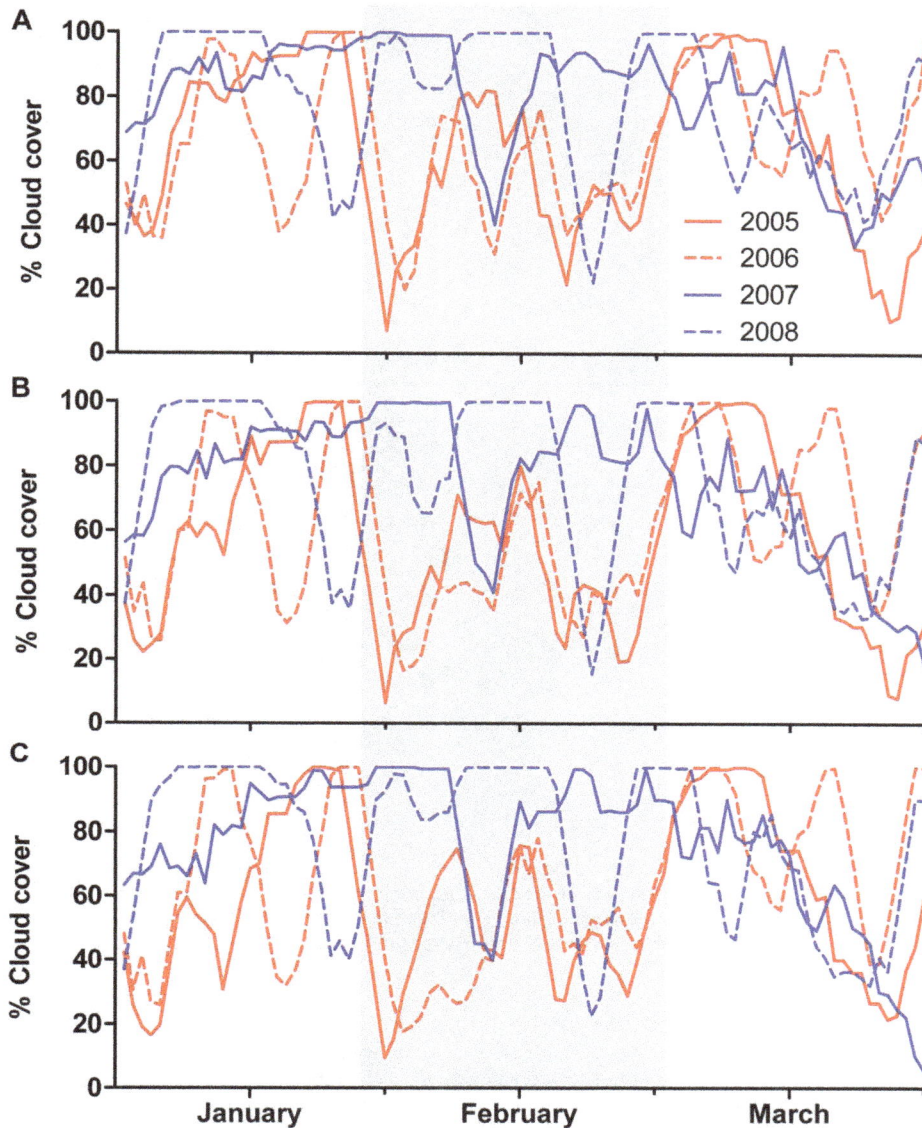

Figure 3. Cloud time series. Summer daytime cloud cover in the (A) inner, (B) mid, and (C) outer shelf study regions. Cloud cover is smoothed using five-day moving averages. Red lines denote summers with warmer thermal profiles (2005 and 2006); blue lines denote summers with cooler thermal profiles (2007 and 2008). The period of greatest difference in SST, cloud cover, and PAR between the warmer and cooler summers (discussed in-text) is highlighted in grey. The spike in cloud cover in late March 2006 is due to the path of Tropical Cyclone Larry.

The relationship was significant in 2005 at all shelf positions, explaining between 15.4 and 32.1% of the variation in lagged SST. Cloud cover significantly affected SST residuals across all shelf positions in 2008, but to a lesser extent than in 2005, explaining only 6.6 to 13.1% of the variation (Table 3). The relationship between cloud cover and lagged SST residuals was fairly weak in both 2006 and 2007. The regression was significant in the mid shelf region in all study summers, explaining between 5.5 and 17.7% of the variation. The strongest regression occurred in the inner shelf region in 2005 (32.1%, table 3).

Regressions of SST residuals against arcsine transformed cloud cover lagged three days later (i.e. SST affecting cloud cover) identified a much weaker overall relationship, with fewer significant relationships within each study summer, and a totally non-significant relationship in 2005 (Table 4). The relationship was strongest in 2008, with SST residuals explaining between 9.1

and 21.6% of the variation in cloud cover three days later. The overall significance of the relationship in 2006 and 2007 was driven by individual shelf positions in each study summer (9.9% in the mid shelf in 2006, 12.6% in the inner shelf in 2007, Table 4). The regressions were most frequently significant in the inner shelf region, and appear to show a cross-shelf gradient in 2008, with SST residuals explaining the highest variation in lagged cloud cover in the outer shelf region (21.6%).

Discussion

Cloud Cooling Effects on Lagged SST

The combination of *in situ* SST data and remotely-sensed cloud imagery used in this study indicated a significant, albeit variable, relationship between cloud cover and lagged SST, with cloud cover explaining up to 32.1% of the variation in SST three days

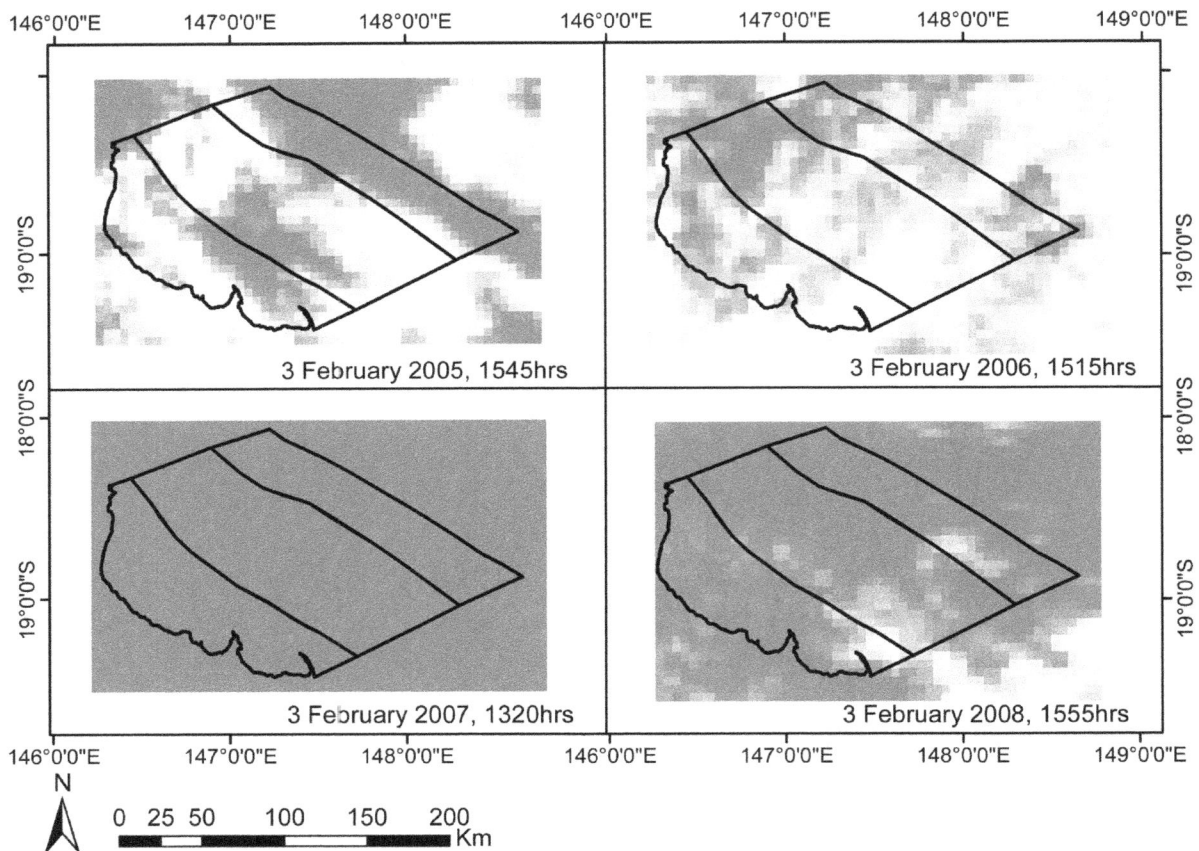

Figure 4. Sample MODIS imagery. Images were taken on the same day in each of the four study summers. Images have been reclassified to Y/N cloud present (grey: yes cloud; white: no cloud). Black polygons indicate study regions (inner, mid, and outer shelf). Cloud cover illustrated in each image is approximately representative of mean and median cloud cover for February of each study summer.

later (Table 3). Our data suggest that the SST response to cloud cover is primarily a result of cloud interception of solar radiation (Table 2). This work provides the first empirical evidence of a phenomenon that has until now been reported only anecdotally in the literature on coral bleaching [9,24–30,53]. While the capacity for specific cloud types to limit or reduce SSTs is well known, particularly in the physical modelling literature [2,21,22,54,55], to our knowledge, this study provides the first quantitative assessment of cloud cooling effects in coral reef systems.

While this study focused on the effects of cloud cover alone, other cloud parameters (e.g. cloud height, altitude, optical depth, liquid water content, and particle size and state [31,32]) and other environmental variables (e.g. evaporation from wind and lagoonal effects such as tidal mixing, wind mixing, and local currents [56] [41]) are also known to affect SSTs at spatial scales of kilometres to hundreds of kilometres. These factors can alter patterns in the interception of solar radiation, the mixing of water masses, or air-sea heat exchange processes. Furthermore, some unexplained variation in our study may be a result of the insensitivity of the regressions to short "pulse" cloud events that may have a cumulative effect on SST (e.g. Fig. 6). The addition of these factors to our analysis could explain a greater proportion of the variation in SST and lagged SST.

However, given the number of variables involved in thermal forcing on coral reefs [41,42], the extent to which cloud cover alone explained variation in lagged SST is surprisingly large. Furthermore, the peak in the strength of the relationship with a

three day lag was not particularly distinct, suggesting a great deal of inertia in the system, with cloud build-up inducing a slow, gradual, and persistent SST response. Indeed, the cross-correlation indicated a weaker, but still significant relationship between cloud cover and subsequent SST responses with lags anywhere between 0 and 8 days, which indicated both a rapid SST response to changes in cloud cover (0 day lag), and a persistent effect (up to 8 day lag, Fig. 7).

The greatest variation in the strength of the cloud-SST relationship was among study summers. However, the strength of the regression was not associated with summers with similar thermal profiles (i.e. "warmer" 2005 and 2006 versus "cooler" 2007 and 2008), as was originally expected. Instead, cloud cooling effects were found to be consistently strongest in 2005 and 2008, which exhibited strongly differing thermal profiles (warmer and cooler, respectively). The El Niño Southern Oscillation (ENSO) strongly influences regional atmospheric circulation during the austral summer [57] and is frequently associated with sustained SST anomalies on the GBR [39,58]. The ENSO phenomenon may influence the variation in the strength of the cloud cover-SST relationship identified in this study, as the strongest cloud cooling effects were noted in strongly ENSO summers: 2005 experienced a strong El Niño summer, and 2008 a strong La Niña summer. The other study summers (2006 and 2007) experienced neutral Southern Oscillation Indices [59]. Our results indicated that the extent to which cloud cover was responsible for cooling SST was strongest in a hot, dry El Niño summer (2005), but was also

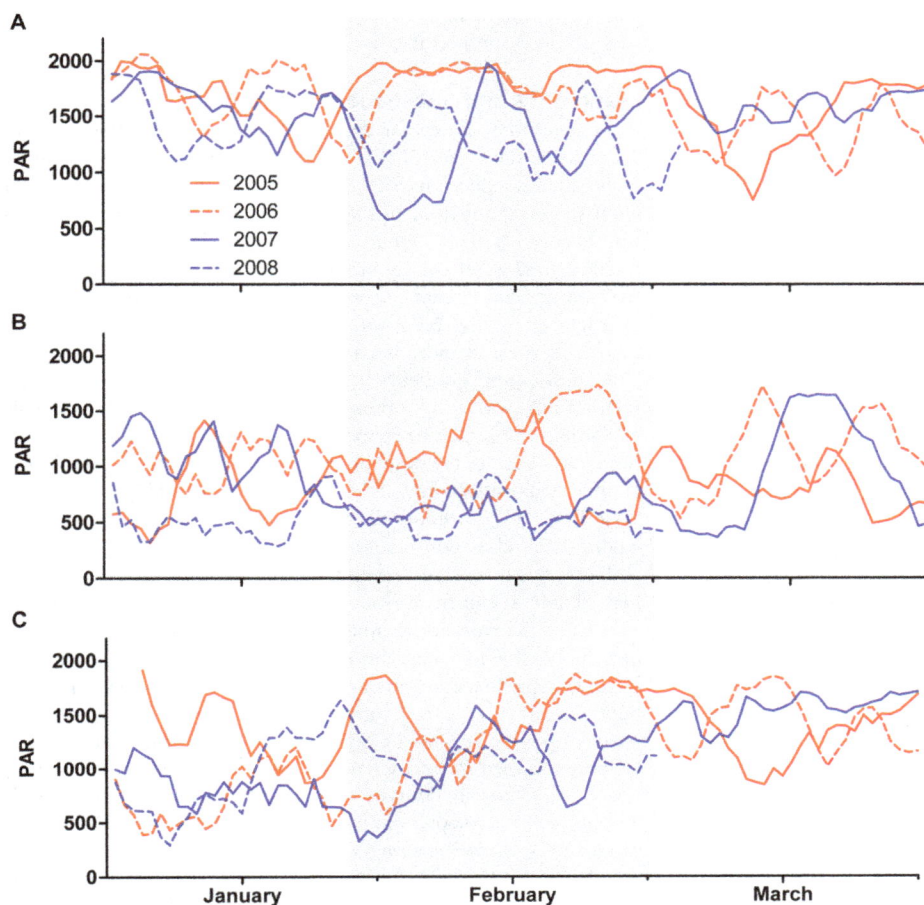

Figure 5. PAR time series. Summer daytime PAR (Photosynthetically Active Radiation, in $\mu mol \cdot s^{-1} \cdot m^{-2}$) in the (A) inner, (B) mid, and (C) outer shelf study regions. PAR is smoothed using five-day moving averages. Red lines denote summers with warmer thermal profiles (2005 and 2006); blue lines denote summers with cooler thermal profiles (2007 and 2008). The period of greatest difference in SST, cloud cover, and PAR between the warmer and cooler summers (discussed in-text) is highlighted in grey.

important during an overcast, cool La Niña summer (2008, Table 3). The importance of the cloud cooling effect in the El Niño summer may be due to the reduced activity of other cooling mechanisms such as surface winds [39,57] and higher initial SSTs (both from local heating and from advection of warm water via the South Equatorial Current [42]). The strong cloud cooling effect observed in the La Niña summer may in part be due to the extent

Table 2. Regressions between 5-day moving averages of PAR and arcsine transformed cloud cover across all shelf positions in all study summers.

Shelf position		Summer 2005	Summer 2006	Summer 2007	Summer 2008
Inner	R^2	0.471	0.558	0.343	0.465
	p	**<0.001**	**<0.001**	**<0.001**	**<0.001**
	N	90	90	90	63
Mid	R^2	0.026	0.125	0.088	0.085
	p	n.s.	**<0.001**	**0.005**	**0.023**
	N	90	90	90	61
Outer	R^2	0.378	0.007	0.468	0.231
	p	**<0.001**	n.s.	**<0.001**	**<0.001**
	N	87	90	90	61

All significant relationships were negative.

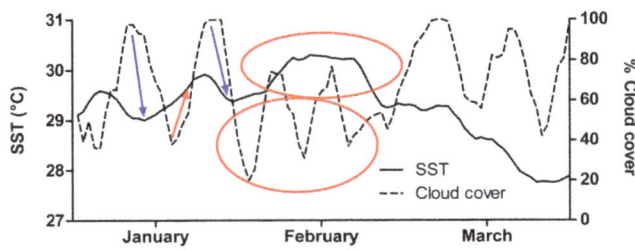

Figure 6. Sample cloud cover and SST overlay. Visualization of lags between cloud cover phenomena and SST responses; summer 2006, inner shelf position. Blue arrows indicate suggested causative lagged cloud cooling events; red arrows and circles indicate suggested causative lagged warming events.

of cloud cover in that study summer (Fig. 3 and 4), as well as the effect of cloud cover covariates such as wind speed [60].

The spatial scale at which the study was carried out did not indicate consistent cross-shelf patterns; however, cloud cover explained the greatest proportion of SST variation in the inner shelf in both 2005 and 2008, which were the strongly ENSO summers in which the overall cloud cooling effect was greatest. This may be a result of local orographic effects [40], in which clouds "pile up" against the Great Dividing Range, a coastal mountain range in this region. This explanation is supported by the strength of the observed relationship between cloud cover and PAR, which was generally highest in the inner shelf region, indicating the presence of optically thicker clouds in this area. Reduced transmittance of solar radiation in the PAR range is indicative of low-altitude clouds [50], which have high albedo, and therefore strong cooling effects, due to characteristics such as small-radius, high particle density, liquid-phase water droplets [33]. Greater radiation reflection or interception by clouds in the inner shelf region may be particularly beneficial to coral reefs in this area, where bleaching patterns are best described by radiation patterns rather than by SSTs [61].

Cloud cooling effects were weaker in the mid shelf region than in the inner shelf region, but were significant across all four study summers (Table 3), suggesting that cloud cover represents a natural means of regulating SST in this region, where water

column mixing is limited by the reduced tidal currents characteristic of the dense reef matrix [41].

Rising SSTs and Subsequent Cloud Buildup

Changes in SST explained up to 21.6% of the variation in cloud cover three days later (Table 4). Taking into account subsequent cloud cooling effects on SST, this is indicative of a negative cloud feedback mechanism in this system. The strong positive relationship between SST and cloud cover was only consistently observed in 2008, the La Niña study summer. This may be a consequence of the lower barometric pressure and greater wind strength characteristic of La Niña summers in this part of the world [39] increasing local evaporative activity.

A cross-shelf gradient in the strength of the SST-cloud relationship was only observed in 2008, with SST residuals explaining the greatest proportion of the variation in cloud cover three days later in the outer shelf region. This may be a result of a greater evaporation rate offshore, where warm, highly saline water "puddles" [41], forming clouds which are then pushed inshore, where their subsequent cooling effect is strongest.

The increase in cloud cover following the rises in SST reported here (Table 4) can be attributed to increased evaporative activity, as is likely the case in the inner shelf region [41]. However, other potential influences may be operating in this system, including biological feedback mechanisms. For example, observed cloud build-up may be a response to local increases in the aerosol particles serving as Cloud Condensation Nuclei. The waters of the GBR are known to be a significant source of these aerosols [62], which are primarily composed of dimethylsulfide (DMS) [63,64]. The DMS precursor, dimethylsulphoniopropinate (DMSP), is produced by marine phytoplankton [33] and coral-symbiont dinoflagellates [65–67]. Atmospheric DMS production by corals increases with increased light intensity and SST, but declines if corals are exposed to prolonged thermal and light stress [68]. On a global scale, DMS is involved in a negative feedback loop with incident solar radiation and cloud cover [69]. Part of the SST-cloud relationship observed here may therefore be a result of DMS production by corals on the central GBR, particularly in the coral-rich mid shelf region [68,70]. This coral-produced atmospheric DMS may increase local densities of cloud condensation nuclei

Table 3. Transformed cloud cover regressed against SST residuals three days later, across all shelf positions and study summers.

Shelf position		Summer 2005	Summer 2006	Summer 2007	Summer 2008
Inner	R^2	0.321	0.020	0.012	0.131
	p	**<0.001**	n.s.	n.s.	**<0.001**
	N	87	87	87	88
Mid	R^2	0.177	0.081	0.055	0.066
	p	**<0.001**	**0.008**	**0.030**	**0.015**
	N	87	87	87	88
Outer	R^2	0.154	0.103	0.006	0.100
	p	**<0.001**	**0.002**	n.s.	**0.003**
	N	87	86	87	88
Overall	R^2	0.208	0.061	0.021	0.096
	p	**<0.001**	**<0.001**	**0.020**	**<0.001**
	N	261	260	261	264

All significant relationships were negative, i.e. increased cloud cover was associated with subsequent decreases in SST.

Table 4. SST residuals regressed against transformed cloud cover three days later, across all shelf positions and study summers.

Shelf position		Summer 2005	Summer 2006	Summer 2007	Summer 2008
Inner	R^2	0.001	0.044	0.126	0.091
	p	n.s.	**0.051**	**<0.001**	**0.004**
	N	87	87	87	88
Mid	R^2	0.003	0.099	0.037	0.174
	p	n.s.	**0.003**	0.073	**<0.001**
	N	87	87	87	88
Outer	R^2	0.016	0.040	0.018	0.216
	p	n.s.	0.075	n.s.	**<0.001**
	N	87	86	87	88
Overall	R^2	0.004	0.058	0.049	0.152
	p	n.s.	**<0.001**	**<0.001**	**<0.001**
	N	261	260	261	264

All significant relationships were positive, i.e. rising SSTs were associated with subsequent increases in cloud cover.

enough to promote the formation of high albedo clouds (composed of high-density small-radius liquid-phase droplets [33]).

Implications

We have demonstrated that cloud cover is an important determinant of SST, which has direct implications for thermal coral bleaching. From a computational or modelling perspective, our results highlight the importance of incorporating local sources of variation in order to better align model predictions with the reality of coral bleaching patterns on a local scale (10s to 100s of kilometres). The current push towards incorporating local physical processes and increasing the spatial resolution of models is producing dramatic improvements in model predictive abilities [14,71]. We recommend the inclusion of quantitative cloud parameters (e.g. cloud cover, but also optical depth) in order to maximize model relevance for reef managers and coral reef biologists.

From a biological perspective, our work underscores the importance of quantifying local environmental conditions within the context of basin-scale physical processes (e.g. ENSO) when assessing imminent, current, or historical bleaching records. Thermal bleaching reports frequently note the occurrence of "doldrum" periods prior to and during a mass bleaching event [5,26,30]. Despite the fact that the direct (decreased UV stress [15,34]) and indirect (decreased SST [this paper, 9,24]) benefits of cloud cover are well-recognized, the information available from satellite imagery remains a largely underutilised resource. The use of remote sensing technologies such as satellite imagery can be used for both atmospheric forecasting and hindcasting, substantiating observations made in the field, value-adding to bleaching reports, and improving our predictive abilities and a posteriori understanding of bleaching events.

Acknowledgments

We acknowledge the use of data products from the Australian Institute of Marine Science, of GIS layers from the Great Barrier Reef Marine Park Authority, and of imagery from the Level 1 and Atmosphere Archive and Distribution System operated by NASA. We thank Wayne Mallett (JCU) and Scott F. Heron (NOAA) for assistance with image processing techniques, and Graham B. Jones (SCU), Billy D. Causey (NOAA) and two anonymous reviewers for suggestions for the improvement of the manuscript.

Author Contributions

Conceived and designed the experiments: SML MJK CRS. Performed the experiments: SML. Analyzed the data: SML. Contributed reagents/materials/analysis tools: MJK CRS. Wrote the paper: SML MJK CRS.

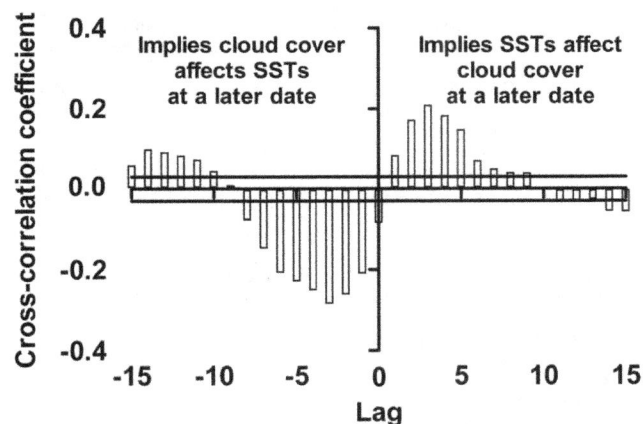

Figure 7. Cross-correlation output. Cross-correlation of cloud cover against SSTs, across all study regions and summers. Note the negative peak in the relationship at −3 days of lag, and the positive peak in the relationship at +3 days lag.

References

1. Brierley AS, Kingsford MJ (2009) Impacts of Climate Change on Marine Organisms and Ecosystems. Curr Biol 19: R602–R614.
2. Solomon S, Qin D, Manning M, Chen Z, Marquis M, et al (2007) Climate Change 2007: The Physical Science Basis. Cambridge, UK and NY, USA: Cambridge University Press.
3. Veron JEN, Hoegh-Guldberg O, Lenton TM, Lough JM, Obura DO, et al. (2009) The coral reef crisis: The critical importance of <350 ppm CO2. Mar Pollut Bull 58: 1428–1436.
4. Barnett TP, Pierce DW, AchutaRao KM, Gleckler PJ, Santer BD, et al. (2005) Penetration of Human-Induced Warming into the World's Oceans. Science 309: 284–287.
5. Hoegh-Guldberg O (1999) Climate change, coral bleaching and the future of the world's coral reefs. Mar Freshwater Res 50: 839–866.

6. Goreau T, McClanahan T, Hayes R, Strong A (2000) Conservation of Coral Reefs after the 1998 Global Bleaching Event. Conserv Biol 14: 5–15.

7. Wilkinson CR (1999) Global and local threats to coral reef functioning and existence: review and predictions. Mar Freshwater Res 50: 867–878.

8. Hoegh-Guldberg O, Mumby PJ, Hooten AJ, Steneck RS, Greenfield P, et al. (2007) Coral Reefs Under Rapid Climate Change and Ocean Acidification. Science 318: 1737–1742.

9. Berkelmans R, De'ath G, Kininmonth S, Skirving W (2004) A comparison of the 1998 and 2002 coral bleaching events on the Great Barrier Reef: spatial correlation, patterns, and predictions. Coral Reefs 23: 74–83.

10. Berkelmans R, Willis BL (1999) Seasonal and local spatial patterns in the upper thermal limits of corals on the inshore Central Great Barrier Reef. Coral Reefs 18: 219–228.

11. Ulstrup KE, Berkelmans R, Ralph PJ, van Oppen MJH (2006) Variation in bleaching sensitivity of two coral species across a latitudinal gradient on the Great Barrier Reef: the role of zooxanthellae. Mar Ecol Prog Ser 314: 135–148.

12. Middlebrook R, Hoegh-Guldberg O, Leggat W (2008) The effect of thermal history on the susceptibility of reef-building corals to thermal stress. J Exp Biol 211: 1050–1056.

13. Liu G, Strong AE, Skirving W, Arzayus LF (2006) Overview of NOAA Coral Reef Watch Program's near-real time global satellite coral bleaching monitoring activities. Proceedings of the 10th International Coral Reef Symposium. Okinawa, Japan. 1783–1793.

14. Maynard JA, Turner PJ, Anthony KRN, Baird AH, Berkelmans R, et al. (2008) ReefTemp: An interactive monitoring system for coral bleaching using high-resolution SST and improved stress predictors. Geophys Res Lett 35.

15. Jokiel PL (2004) Temperature Stress and Coral Bleaching. In: Rosenberg E, Loya Y, editors. Coral Health and Disease. 401–425.

16. Fitt WK, Brown BE, Warner ME, Dunne RP (2001) Coral bleaching: interpretation of thermal tolerance limits and thermal thresholds in tropical corals. Coral Reefs 20: 51–65.

17. Coles SL, Brown BE (2003) Coral bleaching - capacity for acclimatization and adaptation. Adv Mar Biol. 46: 183–223.

18. McClanahan T, Ateweberhan M, Ruiz Sebastián C, Graham N, Wilson S, et al. (2007) Predictability of coral bleaching from synoptic satellite and in situ temperature observations. Coral Reefs 26: 695–701.

19. Newell RE (1979) Climate and the Ocean: Measurements of changes in sea-surface temperature should permit us to forecast certain climatic changes several months ahead. Am Sci 67: 405–416.

20. Tompkins AM (2001) On the Relationship between Tropical Convection and Sea Surface Temperature. J Clim 14: 633–637.

21. Williams IN, Pierrehumbert RT, Huber M (2009) Global warming, convective threshold and false thermostats. Geophys Res Lett 36.

22. Ramanathan V, Collins W (1991) Thermodynamic regulation of ocean warming by cirrus clouds deduced from observations of the 1987 El Niño. Nature 351: 27–32.

23. Kleypas JA, Danabasoglu G, Lough JM (2008) Potential role of the ocean thermostat in determining regional differences in coral reef bleaching events. Geophys Res Lett 35.

24. Mumby PJ, Chisholm JRM, Edwards AJ, Andrefouet S, Jaubert J (2001) Cloudy weather may have saved Society Island reef corals during the 1998 ENSO event. Mar Ecol Prog Ser 222: 209–216.

25. Turner J, Hardman E, Klaus R, Fagoonee I, Daby D, et al. (2000) The reefs of Mauritius. Stockholm, Sweden: Coral Reef Degradation in the Indian Ocean Status Report 2000.

26. Glynn PW (1993) Coral reef bleaching: ecological perspectives. Coral Reefs 12: 1–17.

27. Jokiel PL, Brown EK (2004) Global warming, regional trends and inshore environmental conditions influence coral bleaching in Hawaii. Global Change Biol 10: 1627–1641.

28. Celliers L, Schleyer MH (2002) Coral bleaching on high-latitude marginal reefs at Sodwana Bay, South Africa. Mar Pollut Bull 44: 1380–1387.

29. Obura DO (2005) Resilience and climate change: lessons from coral reefs and bleaching in the Western Indian Ocean. Estuar Coast Shelf Sci 63: 353–372.

30. Causey BD (2001) Lessons learned from the intensification of coral bleaching from 1980–2000 in the Florida Keys, USA. In: Salm RV, Coles SL, editors. Proceedings of the Workshop on Mitigating Coral Bleaching Impact through MPA Design. Honolulu, Hawaii. 60–66.

31. Frouin R, Pinker RT (1995) Estimating Photosynthetically Active Radiation (PAR) at the earth's surface from satellite observations. Remote Sens Environ 51: 98–107.

32. Stephens GL (2005) Cloud feedbacks in the climate system: a critical review. J Clim 18: 237–273.

33. Charlson RJ, Lovelock JE, Andreae MO, Warren SG (1987) Oceanic phytoplankton, atmospheric sulphur, cloud albedo and climate. Nature 326: 655–661.

34. Barton TJ, Paltridge GW (1979) The Australian Climatology of Biologically Effective Ultraviolet Radiation. Aust J Dermatol 20: 68–74.

35. Takahashi Y, Okazaki Y, Sato M, Miyahara H, Sakanoi K, et al. (2010) 27-day variation in cloud amount in the Western Pacific warm pool region and relationship to the solar cycle. Atmos Chem Phys 10: 1577–1584.

36. Webb M, Senior C, Sexton D, Ingram W, Williams K, et al. (2006) On the contribution of local feedback mechanisms to the range of climate sensitivity in two GCM ensembles. Clim Dyn 27: 17–38.

37. Jones RN (2000) Managing Uncertainty in Climate Change Projections – Issues for Impact Assessment. Clim Change 45: 403–419.

38. Lough JM (2007) Chapter 2: Climate and Climate Change on the Great Barrier Reef. In: Johnson JE, Marshall PA, editors. Climate Change and the Great Barrier Reef: a Vulnerability Assessment. Australia: Great Barrier Reef Marine Park Authority and Australian Greenhouse Office. 15–50.

39. Lough JM (1994) Climate variation and El Niño-Southern Oscillation events on the Great Barrier Reef: 1958 to 1987. Coral Reefs 13: 181–185.

40. Barry RG, Chorley RJ (2003) Atmosphere, weather, and climate. London and New York: Routledge. 460 p.

41. Wolanski E (1994) Physical oceanographic processes of the Great Barrier Reef. Boca Raton, Florida, USA: CRC Press, Inc. 194 p.

42. Steinberg C (2007) Chapter 3 Impacts of climate change on the physical oceanography of the Great Barrier Reef. In: Johnson JE, Marshall PA, editors. Climate Change and the Great Barrier Reef: a Vulnerability Assessment. Australia: Great Barrier Reef Marine Park Authority and Australian Greenhouse Office. 51–74.

43. Dinesen ZD (1983) Patterns in the distribution of soft corals across the central Great Barrier Reef. Coral Reefs 1: 229–236.

44. Wilkinson CR, Cheshire AC (1989) Patterns in the distribution of sponge populations across the central Great Barrier Reef. Coral Reefs 8: 127–134.

45. Done TJ (1982) Patterns in the distribution of coral communities across the central Great Barrier Reef. Coral Reefs 1: 95–107.

46. Russ G (1984) Distribution and abundance of herbivorous grazing fishes in the Central Great Barrier Reef. I. Levels of variability across the entire continental shelf. Mar Ecol Prog Ser 20: 23–34.

47. Chatfield C (1980) The Analysis of Time Series: An Introduction. London and New York: Chapman and Hall. 268 p.

48. King MD, Tsay S-C, Platnick SE, Wang M, Liou K-N (1997) Cloud Retrieval Algorithms for MODIS: Optical Thickness, Effective Particle Radius, and Thermodynamic Phase. MODIS Algorithm Theoretical Basis. Document No. ATBD-MOD-05, MOD06 - Cloud product.

49. Jokiel PL, Lesser MP, Ondrusek ME (1997) UV-Absorbing Compounds in the Coral Pocillopora damicornis: Interactive Effects of UV Radiation, Photosynthetically Active Radiation, and Water Flow. Limnol Oceanogr 42: 1468–1473.

50. Grant RH, Heisler GM, Gao W (1996) Photosynthetically-active radiation: sky radiance distributions under clear and overcast conditions. Agric For Meteorol 82: 267–292.

51. Quinn G, Keough M (2002) Experimental Design and Data Analysis for Biologists. New York: Cambridge University Press. 537 p.

52. Underwood AJ (1997) Environmental decision-making and the precautionary principle: what does this principle mean in environmental sampling practice? Landsc Urb Plann 37: 137–146.

53. Wilkinson C, Olof L, Cesar H, Hodgson G, Jason R, et al. (1999) Ecological and Socioeconomic Impacts of 1998 Coral Mortality in the Indian Ocean: An ENSO Impact and a Warning of Future Change? Ambio 28: 188–196.

54. Lau K-M, Sui C-H (1997) Mechanisms of Short-Term Sea Surface Temperature Regulation: Observations during TOGA COARE. J Clim 10: 465–472.

55. Bony S, Lau KM, Sud YC (1997) Sea Surface Temperature and Large-Scale Circulation Influences on Tropical Greenhouse Effect and Cloud Radiative Forcing. J Clim 10: 2055–2077.

56. Skirving W, Guinotte J (2001) Chapter 18: The Sea Surface Temperature Story on the Great Barrier Reef during the Coral Bleaching Event of 1998. In: Wolanski E, editor. Oceanographic Processes of Coral Reefs. Boca Raton, Florida, USA: CRC Press LLC. 301–314.

57. Redondo-Rodriguez A, Weeks SJ, Berkelmans R, Hoegh-Guldberg O, Lough JM (2012) Climate variability of the Great Barrier Reef in relation to the tropical Pacific and El Niño-Southern Oscillation. Marine and Freshwater Research 63: 34–47.

58. Lough JM (1999) Sea Surface Temperature on the Great Barrier Reef: a Contribution to the Study of Coral Bleaching. Australia: Great Barrier Reef Marine Park Authority.

59. Oliver JK, Berkelmans R, Eakin CM (2009) Chapter 3: Coral Bleaching in Space and Time. In: van Oppen MJH, Lough JM, editors. Coral Bleaching: Patterns, Processes, Causes and Consequences. Berlin, Heidelberg: Springer-Verlag. 21–39.

60. Mann KH, Lazier JRN (2006) Dynamics of Marine Ecosystems. Carlton, Victoria, Australia: Blackwell Publishing. 394 p.

61. Masiri I, Nunez M, Weller E (2008) A 10-year climatology of solar radiation for the Great Barrier Reef: implications for recent mass coral bleaching events. Int J Remote Sens 29: 4443–4462.

62. Modini RL, Ristovski ZD, Johnson GR, He C, Surawski N, et al. (2009) New particle formation and growth at a remote, sub-tropical coastal location. Atmos Chem Phys 9: 7607–7621.

63. Andreae MO, Crutzen PJ (1997) Atmospheric aerosols: Biogeochemical sources and role in atmospheric chemistry. Science 276: 1052–1058.

64. Vallina SM, Simó R (2007) Re-visiting the CLAW hypothesis. Environ Chem 4: 384–387.

65. Swan HB, Jones GB, Deschaseaux E (2012) Dimethylsulfide, Climate and Coral Reef Ecosystems. In: Proceedings of the 12th International Coral Reef Symposium. Cairns, Australia.

66. Jones GB, Curran MAJ, Broadbent AD (1994) Dimethylsulphide in the South Pacific. In: Bellwood O, Choat H, Saxena N, editors. Recent Advances in

Marine Science and Technology '94: Pacon International and James Cook University of North Queensland. 183–190.

67. Broadbent AD, Jones GB, Jones RJ (2002) DMSP in corals and benthic algae from the Great Barrier Reef. Est Coast Shelf Sci 55: 547–555.

68. Fischer E, Jones G (2012) Atmospheric dimethysulphide production from corals in the Great Barrier Reef and links to solar radiation, climate and coral bleaching. Biogeochemistry 110: 31–46.

69. Vallina SM, Simó R (2007) Strong relationship between DMS and the solar radiation dose over the global surface ocean. Science 315: 506–508.

70. Deschaseaux E, Jones G, Miljevic B, Ristovski Z, Swan H, et al. (2012) Can corals form aerosol particles through volatile sulphur compound emissions? In: Proceedings of the 12th International Coral Reef Symposium. Cairns, Australia.

71. Eakin CM, Liu G, Rauenzahn JL, Burgess T, Christensen TR, et al. (2012) NOAA Coral Reef Watch's Decision Support System: A Global view to help managers protect coral reefs in a changing climate. In: 2012 Ocean Sciences Meeting. Salt Lake City, Utah, USA.

18

Patterns and Variability of Projected Bioclimatic Habitat for *Pinus albicaulis* in the Greater Yellowstone Area

Tony Chang*, Andrew J. Hansen, Nathan Piekielek

Department of Ecology, Montana State University, Bozeman, Montana, United States of America

Abstract

Projected climate change at a regional level is expected to shift vegetation habitat distributions over the next century. For the sub-alpine species whitebark pine (*Pinus albicaulis*), warming temperatures may indirectly result in loss of suitable bioclimatic habitat, reducing its distribution within its historic range. This research focuses on understanding the patterns of spatiotemporal variability for future projected *P.albicaulis* suitable habitat in the Greater Yellowstone Area (GYA) through a bioclimatic envelope approach. Since intermodel variability from General Circulation Models (GCMs) lead to differing predictions regarding the magnitude and direction of modeled suitable habitat area, nine bias-corrected statistically down-scaled GCMs were utilized to understand the uncertainty associated with modeled projections. *P.albicaulis* was modeled using a Random Forests algorithm for the 1980–2010 climate period and showed strong presence/absence separations by summer maximum temperatures and springtime snowpack. Patterns of projected habitat change by the end of the century suggested a constant decrease in suitable climate area from the 2010 baseline for both Representative Concentration Pathways (RCPs) 8.5 and 4.5 climate forcing scenarios. Percent suitable climate area estimates ranged from 2–29% and 0.04–10% by 2099 for RCP 8.5 and 4.5 respectively. Habitat projections between GCMs displayed a decrease of variability over the 2010–2099 time period related to consistent warming above the 1910–2010 temperature normal after 2070 for all GCMs. A decreasing pattern of projected *P.albicaulis* suitable habitat area change was consistent across GCMs, despite strong differences in magnitude. Future ecological research in species distribution modeling should consider a full suite of GCM projections in the analysis to reduce extreme range contractions/expansions predictions. The results suggest that restoration strageties such as planting of seedlings and controlling competing vegetation may be necessary to maintain *P.albicaulis* in the GYA under the more extreme future climate scenarios.

Editor: Ben Bond-Lamberty, DOE Pacific Northwest National Laboratory, United States of America

Funding: This work was supported by the National Aeronautics and Space Administration Applied Sciences Program (Grant 10-BIOCLIM10-0034); Funder URL: http://www.nasa.gov (AJH TC). It also received support from the National Science Foundation Experimental Program to Stimulate Competitive Research (EPSCoR) Track-I EPS-1101342 (INSTEP 3); Funder URL: http://www.nsf.gov/div/index.jsp?div=EPSC (NBP TC); and the North Central Climate Science Center (G13AC00392-G-8829-1); Funder URL: http://www.doi.gov/csc/northcentral/index.cfm (AJH NBP). The funders had no role in study design, data collection and analysis, decision to publish, or preparation of the manuscript.

Competing Interests: The authors have declared that no competing interests exist.

* Email: tony.chang@msu.montana.edu

Introduction

Over the next century, it is expected that most of North America will experience climate changes related to increased concentrations of anthropogenic greenhouse gas emissions and natural variability [1]. At regional scales these changes are highly variable and can result in areas of increased mesic, xeric, or even hydric habitat conditions relative to present day. These shifting climates in turn also transform the suitable habitat for individual species that may result in changes in species composition and dominant vegetation types.

Whitebark pine (*Pinus albicaulis*) is a native conifer of the Western U.S. that is considered a keystone species in the sub-alpine environment. It provides a food source for animals such as the grizzly bear (*Ursus arctos*), red squirrel (*Tamiasciurus hudsonicus*), and Clark's nutcracker (*Nucifraga columbiana*) [2]. It also serves the ecosystem functions of stabilizing soil, moderating snow melt and runoff, and facilitating establishment for other species [2,3]. Whitebark pine has experienced a notable decline in

the past two decades within the U.S. Northern Rockies due to high rates of infestation from the mountain pine beetle (*Dendroctonus ponderosae*) and infections from white pine blister rust (*Cronartium ribicola*), resulting in an 80% mortality rate within the adult population [4–7]. Given the potential loss of important ecosystem functions that whitebark pine contribute to the landscape under this mortality event, there is an emphasis to understand the climate characteristics of its habitat to identify the restoration strategies and locations that may aid the persistence of the species under future climates.

One method of understanding species response to climate change is through bioclimate niche modeling, which has become a common practice for assessing potential vegetation shifts under new environmental conditions [8–13]. Ecological niche theory proposes there exists some range of bioclimatic conditions within which a species can persist [14]. In bioclimatic niche modeling, the realized niche is modeled by empirical relationships between the presence or absence of a species and the associated abiotic, and

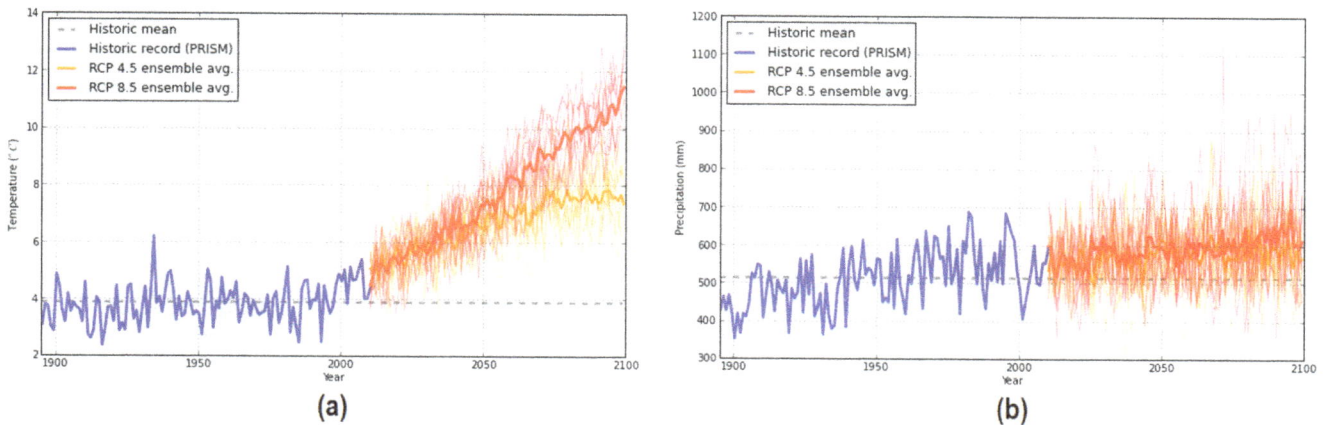

Figure 1. Historic and projected climates variables for the GYA from 1895–2099 under RCP 4.5 and 8.5 scenarios. Light shaded orange and red lines represent individual GCMs for RCP 4.5 and 8.5 respectively. Bold lines represent GCM ensemble average. (a) Mean annual temperature (b) Mean annual precipitation.

sometimes biotic, variables that describe the niche space. Bioclimatic models assume that species are in equilibrium with their environment and that the current abiotic relationships reflect a species environmental preferences which may be retained into the future [15,16]. At macro scales, bioclimatic approaches have demonstrated success at predicting current distributions of species [17,18]. Most bioclimatic models do not explicitly consider the many additional ecological factors that ultimately influence a

species distribution such as dispersal, disturbance, or biotic interaction. Thus the approach does not predict where a species will actually occur in the future, but rather it predicts locations where climatic conditions will be suitable for the species.

Bioclimatic niche methodology has demonstrated utility in modeling historic ranges of species for conservation and management applications. By modeling the present day suitable habitat and then projecting those habitats into the future, bioclimatic

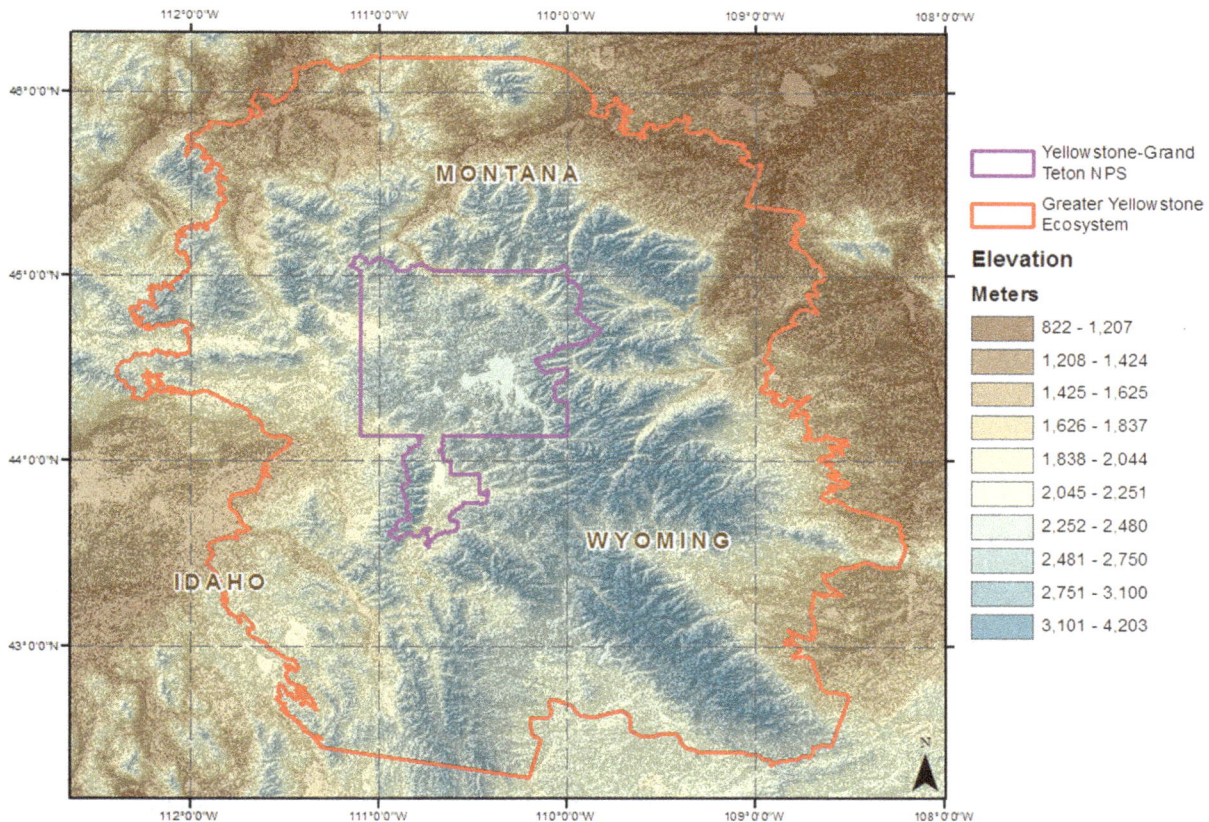

Figure 2. The Greater Yellowstone Area, representing an area of 150,700 km² with an elevational gradient from 522–4,206 m.

Table 1. General Circulation Models for analysis.

Name	Institute	Country
CESM1-CAM5	National Center for Atmospheric Research	US
CCSM4	National Center for Atmospheric Research	US
CESM1-BGC	National Center for Atmospheric Research	US
CNRM-CM5	Centre National de Recherche Meteorologiques	FR
HadGEM2-AO	Met Office Hadley Centre Climate Programme	UK
HadGEM2-ES	Met Office Hadley Centre Climate Programme	UK
HadGEM2-CC	Met Office Hadley Centre Climate Programme	UK
CMCC-CM	Centro Euro-Mediterraneo per Cambiamenti Climatici	ITA
CanESM2	Canadian Centre for Climate Modelling and Analysis	CAN

Selection of AR5 GCMs that represent historic climate in the U.S. Pacific Northwest region for future bioclimate habitat modeling.

niche models can serve as the first step filter for conservation action plans, such as mapping suitable species reintroduction sites or habitat reserve selection [19–21]. For *P.albicaulis*, McLane and Aitken [22] utilized bioclimate niche models to successfully implement experimental assisted migration on persisting climate habitat in British Columbia. Additionally, models of *Pinus flexis*, a closely related species of five needle pine, have been used to evaluate management options in Rocky Mountain National Park [23]. Given these examples, an effort to model and projected suitable climate habitat for *P.albicaulis* within a regional domain can provide valuable insight to land resource managers.

In this study, we present a bioclimatic habitat model for *P.albicaulis* within the Greater Yellowstone Area (GYA). Although *P. albicaulis* has a range-wide distribution that is split into two broad sections, one along Western North America: the British Columbia Coast Range, the Cascade Range, and the Sierra Nevada; and the other section in the Intermountain West that covers the Rocky Mountains from Wyoming to Alberta [2,24]; the GYA was selected as the primary geographic modeling domain for three reasons: 1) evidence that the *P. albicaulis* sub-population in the GYA is genetically distinct from other regional populations with different climate tolerances [25]; 2) the high regional investment in *P. albicaulis* conservation in the area [6]; 3) the high density of climate stations within the region. Climate within the GYA is highly heterogenous due to complex topography, and sharp elevational gradients. Current knowledge of the region expects climate to shift towards increased mean annual temperatures and earlier spring snowmelt [26,27]. This shift is expected to have an impact on the total suitable habitat area for *P. albicaulis*. Modeling at a regional scale can provide a finer resolution spatially explicit description of the bioclimatic envelope of *P. albicaulis* in the GYA.

Here we also present an opportunity to investigate the effect of future climate variability on projected species distributions. In 2013, the World Climate Research Programme Coupled Model released the new generation General Circulation Model (GCM) projections through the Coupled Model Intercomparison Project Phase 5 (CMIP5) [28]. These new GCM projections also include four possible climate futures are modeled with each GCM under the Representative Concentration Pathways (RCP) of greenhouse gas/aerosol. These RCP scenarios designate four different levels of radiative forcing (2.6, 4.5, 6.0 and 8.5 W/m^2) that may occur by the year 2099 [29]. In practice, research of future species suitable climate generally use a small suite of GCM/RCP combinations to project future climate [8,11,30]. However, internal variability in

these GCMs that arise from modeled coupled interactions among the atmosphere, oceans, land, and cryosphere can result in atmospheric circulation fluctuations that are characteristic of a stochastic process [31]. Such intrinsic atmospheric circulation variations from model structure induce regional changes in air temperature and precipitation on the multi-decadal time scale [31]. For the GYA specifically, this GCM variability has be observed with mean annual temperatures projected to increase by $2-9°C$ and mean annual precipitation to change by -50 to $+225$ mm (Fig. 1). This suggests that magnitude and direction of projected species distributions at a regional scale can vary depending on the GCM selected and the modeled species response to more xeric or mesic future climate conditions [32].

To summarize, this study presents a bioclimate niche model for *P. albicaulis* based on historic climate observations and field sampling of *P. albicaulis* presence and absence. Using this modeled bioclimate envelope, projections of future total climate suitable habitat area under nine GCMs and two RCP scenarios will be measured. Since different GCMs may project a diverging spectrum of climates, it is expected that measures of total suitable habitat will reduce with varying degrees of area loss. It is also expect that number and size of continous patches of *P. albicaulis* habitat will reduce due to the limited available number of subalpine areas distributed within the landscape. This research provides an analysis of the variability of biotic response under a large suite of GCMs to provide managers/researchers with a measure of the uncertainty associated with future species distribution models. Furthermore, this analysis explicitly describes the spatial patterns of bioclimatic niches for *P. albicaulis* to gain a better understanding of topographic characteristics, such as elevation, on suitable habitat. Changes in these spatial patterns are examined through quantifying landscape patch dynamic that may result from GCM projections to understand the species trends for persistence on the landscape.

Methods

Study area

The GYA, which includes Yellowstone National Park, Grand Teton National Park, and a number of state and federally managed forests, is a mid- to high-latitude region in the Northern Rocky Mountains of western North America. Conifers are dominant in the range, with forest types composed of *Pinus contorta*, *Abies lasiocarpa*, *Pseudotsuga menziesii*, *Pinus albicaulis*, *Juniperus scopulorum*, *Pinus flexis* and *Picea engelmannii*, although the deciduous hardwood *Populus tremuloides*, is also wide spread. Plateaus and lowlands are dominated by species of *Artemisia tridentata* and open grasslands of mixed composition. The GYA study area encompasses 150,700 km^2 with an elevational gradient from 522–4,206 m that represents 14 surrounding mountain ranges (Fig. 2).

Data

Biological data. Field observations of *P. albicaulis* adult presences and absences were compiled from three data sources. First, 2,545 observations from the Forest Inventory and Analysis (FIA) program were assembled. FIA plots are located on a regular

gridded sampling design with one plot at approximately every 2,500 forested hectares, with swapped and fuzzed exact plot locations within 1.6 km to protect privacy [33]. Gibson et al. [34] found that model accuracy to not be dramatically affected by data fuzzing, but to provide the most spatial accuracy, this study culled FIA field points where measured elevation were > 300 m different from a 30 m USGS DEM [35]. To capitalize on additional field observations of *P. albicaulis* within the study area, and because false absences are one of the most problematic data issues in constructing bioclimatic niche models [36]; supplementary points were drawn from the Whitebark/Limber Pine Information System (WLIS) [37], and long-term monitoring plots established by the

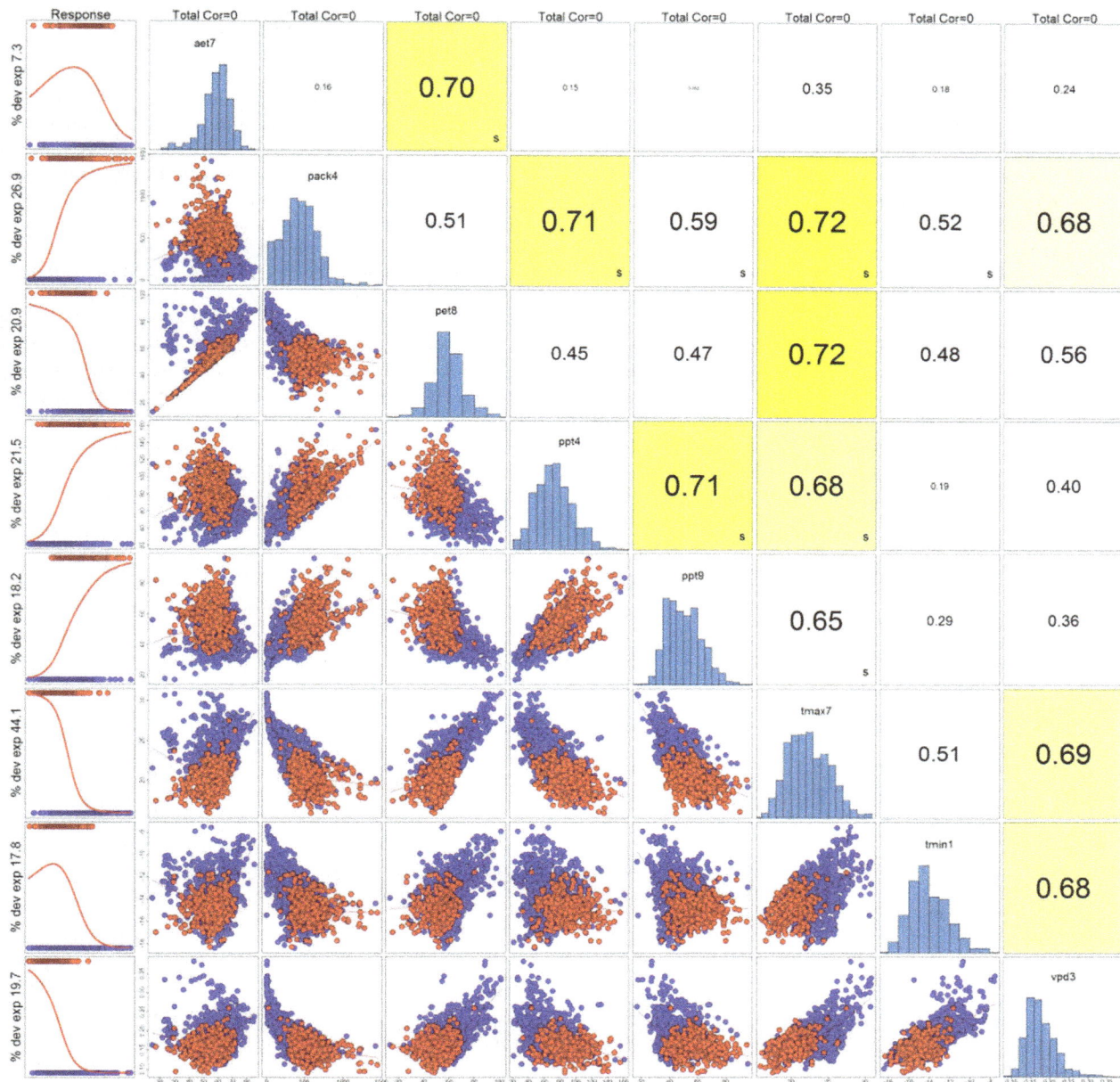

Figure 3. Selected predictor variables based on Principal Component Analysis and a maximum correlation filter of $\leq$ 0.75. Scatter plots represent one-to-one covariate plots where red points represent *P. albicaulis* presence, and blue points represent absence from field data. Far-left columns display logistic-regression of covariates from Generalized Additive Modeling using the Software for Assisted Habitat Modeling (SAHM [59]).

Table 2. Bioclimatic predictor variable list.

Code	Predictor Variable
tmin1	Minimum Temperature January
vpd3	Vapor Pressure Deficit March
ppt4	Precipitation April
pack4	Snow Water Equivalent April
tmax7	Maximum Temperature July
aet7	Actual Evapotranspiration July
pet8	Potential Evapotranspiration August
ppt9	Precipitation September

Final predictor variable set for Random Forest modeling. All variables were calculated as a 30-year climate mean from 1950–1980.

National Park Service Greater Yellowstone Inventory and Monitoring Network (GYRN) [38]. The presences in these two additional datasets were collocated within predictor pixels of FIA absence to correct for false absences. In doing so, only one *P. albicaulis* presence or absence record was associated per predictor pixel, thereby avoiding issues associated with sampling bias that are common when building bioclimate niche models with data from targeted surveys [39]. This compilation of data represents an effort for "completeness" as described by Kadmon et al. [40] and Franklin [36], to capture all climate conditions where a species does exist. New data sources added 119 *P. albicaulis* presences that would have been missed by using FIA data alone, for a total of 938 presences and 1,633 absences.

"Adult" class *P. albicaulis* were selected for modeling based on a recorded diameter at breast height (DBH) >20 cm. *P. albicaulis* within the Central Montana are reported to reach 100 years of age at approximately 8–12 m in height with DBHs between 15–20 cm

[41]. Given previous silvicultural studies, it was assumed that 20 cm DBH *P. albicaulis* represent adult class individuals for the GYA, with potential to reproduce [24]. Furthermore, this study focused on adult size class due to difficulties distinguishing younger age class *P. albicilus* from *P. flexis*.

Historic climate data. Climate inputs for modeling were acquired from the 30-arc-second (~800 m) monthly Parameter-elevation Regressions on Independent Slopes Model (PRISM), a derived product that interpolates local station measurements across a continuous grid [42]. PRISM data includes monthly average minimum temperatures (T_{min}), maximum temperature (T_{max}), mean temperature (T_{mean}), and mean precipitation (Ppt). All monthly data were averaged for the temporal extent of 1950–1980 for bioclimatic niche model fitting. The 1950–1980 temporal extent was selected for modeling since: 1) a sufficient density of weather stations were operating by 1950 to provide a reasonable network; 2) evidence of anthropogenic warming that begins in the

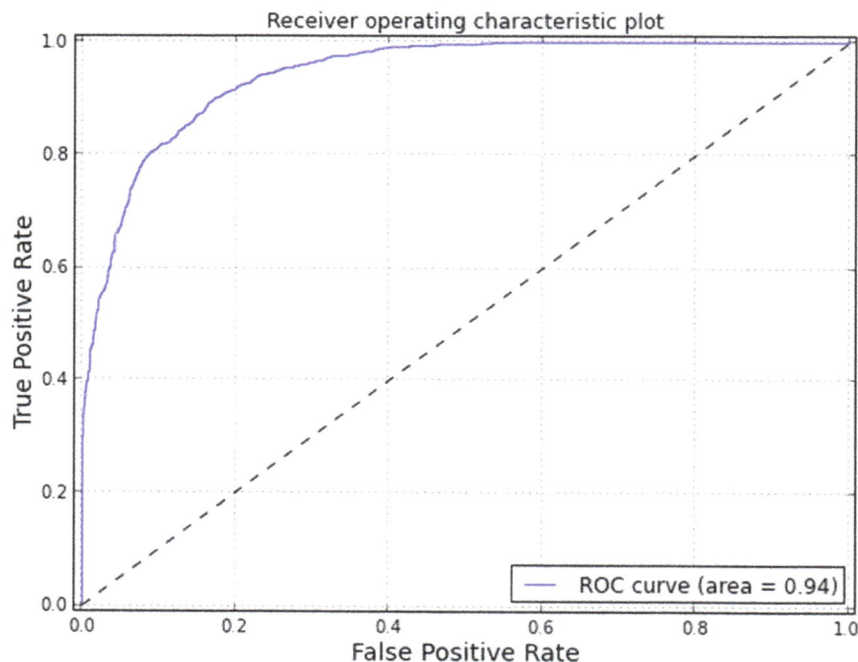

Figure 4. Area under curve for the receiver operating characteristic plot suggests adequate performance from the Random Forest modeling.

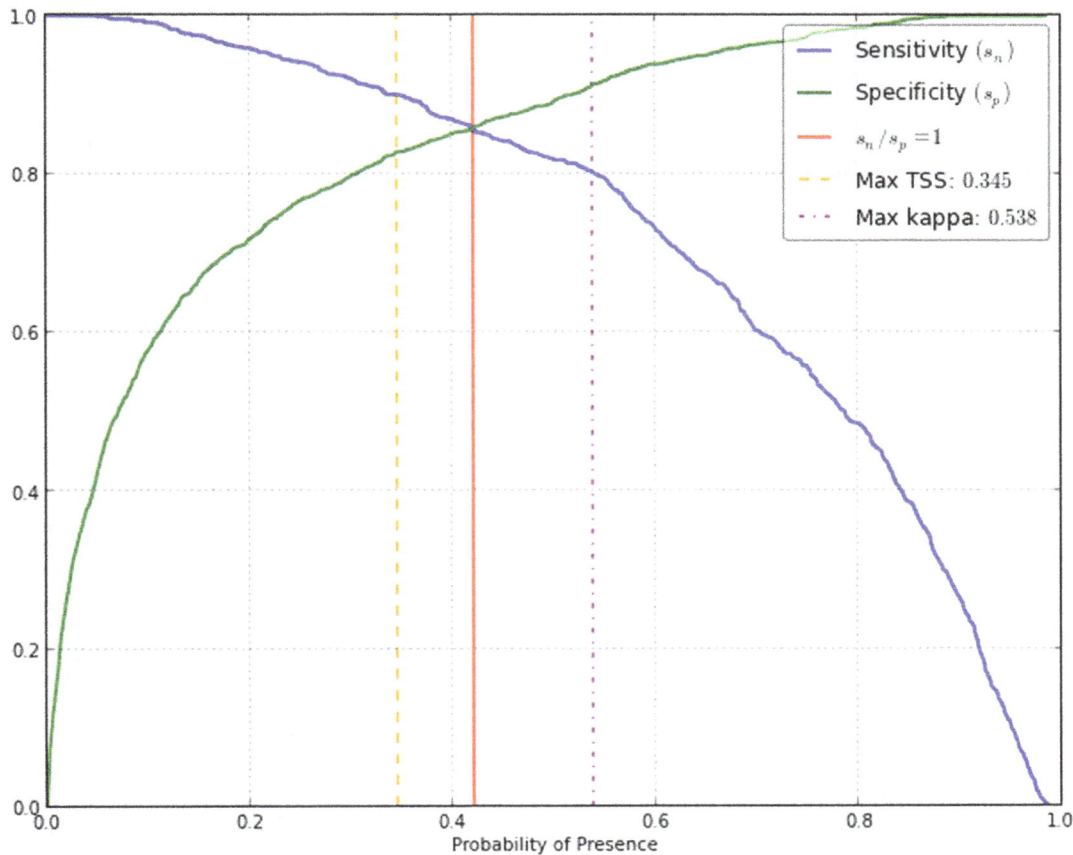

Figure 5. Threshold for probability of presence of 0.421 determined at the intersection of true positive rate (TPR) and true negative rate (TNR). Equivalent TPR and TNR, displayed a compromise between the maximum true skill statistic (TSS : 0.345) and maximum Kappa statistic (κ : 0.538).

late 1980s; 3) trees old enough to bear seeds today likely established under a similar climates to the 1950–1980 period.

Water balance. A Thornthwaite-based dynamic water balance model was used to estimate a number of variables that include actual evapotranspiration (AET) and potential evapotranspiration (PET) [43–45]. The model required only monthly mean temperatures, dew point temperatures, and precipitation (see Text S1). Water was stored as soil moisture or in surface snowpack, with the excess taking the form of evaporated vapor or loss through seepage/runoff. In addition to the climatic variables, latitude and physical characteristics of the soil were required to define water holding capacity. Soil attributes assigned by the Soil Survey Geographic (STATSGO) datasets were allocated from the Natural Resource Conservation Service at a 30-arc-second

resolution to determine soil water holding capacity and estimates for soil depth [46]. All water balance variables, which include PET, AET, soil moisture, vapor pressure deficit (vpd), and snow water equivalent (pack), were averaged by month over 1950–1980 to match with historic climate data for bioclimate model fitting.

GCM data. The general circulation model (GCM) experiments conducted under CMIP5 for the Intergovernmental Panel on Climate Change Fifth Assessment Report provided future projected climate data sets for assessing the effects of global climate change. Using a Bias-Correction Spatial Disaggregation (BCSD) approach, an archive of statistically down-scaled CMIP5 climate projections for the conterminous United States at 30-arc-second spatial resolution was assembled by the NASA Center for Climate Simulation NEX-DCP30 [47]. For this analysis, a subset of the

Table 3. Confusion matrix from out-of-bag analysis.

		Validation data set	
		Presence	**Absence**
Model	Presence	763 (81.9%)	169 (13.1%)
	Absence	176 (10.9%)	1437 (89.1%)

Random Forest tree estimators displays higher OOB specificity than sensitivity. Area Under Curve (AUC) value of 0.94 suggests model has high predictive capacity for projecting future suitable bioclimate habitat.

Feature importance plot

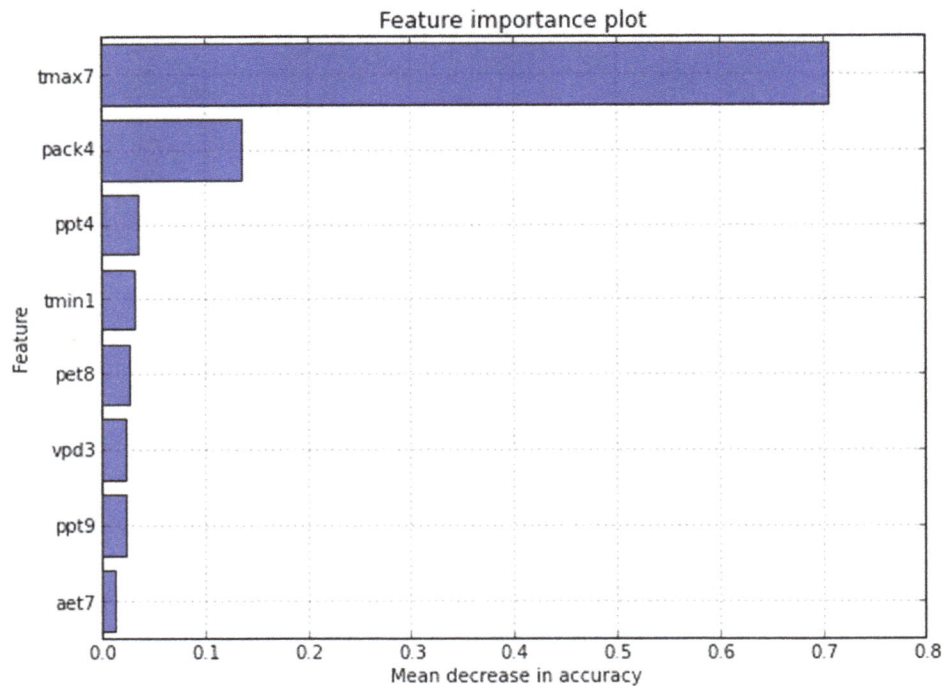

Figure 6. Random Forest out-of-bag variable importance plots find removal of maximum temperatures for July and April snow water equivalent to create the greatest reducing in model accuracy.

total GCM models available from NASA were selected that best represent the Northwestern US. Rupp et al. [48] recently presented an analysis of GCM performance versus the observed historic climate in the U.S. Pacific Northwest under 18 specified climate metrics. In their analysis, Rupp et al. ranked GCMs for accuracy using an empirical orthogonal function (EOF) analysis of the total normalized error compared to reference data. This analysis selected models with a normalized error score <0.5 as a threshold to cull the full suite of GCMs to the top nine models. These GCMs were used to project modeled *P. albicaulis*

distributions into the future (Table 1). Two RCP scenarios were selected to understand effects of differing carbon futures under climate change from 2010 to 2099. RCP 4.5 was the first, representing increased radiative forcing until stabilization of greenhouse emissions between 2040–2050 and total radiative forcing of 4.5 W/m^2 by 2099. RCP 8.5 was the second, representing the ''business as usual'' scenario, with uncontrolled radiative forcing increasing with stabilization of 8.5 W/m^2 by 2099 [49,50].

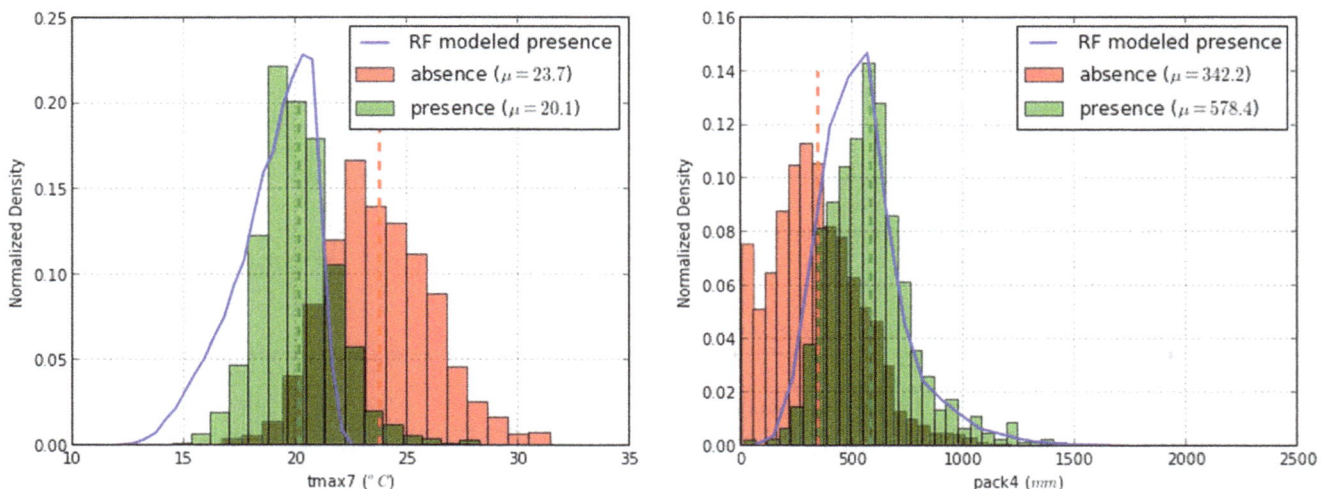

Figure 7. Modeled binary presence for *P.albicaulis* under 1980–2010 mean July maximum temperatures and mean April snow water equivalent bioclimate variables shows agreement with field presence data. Dotted lines designate climate means for corresponding *P. albicaulis* field points. Blue lines represent the distribution of Random Forest modeled presence within the GYA.

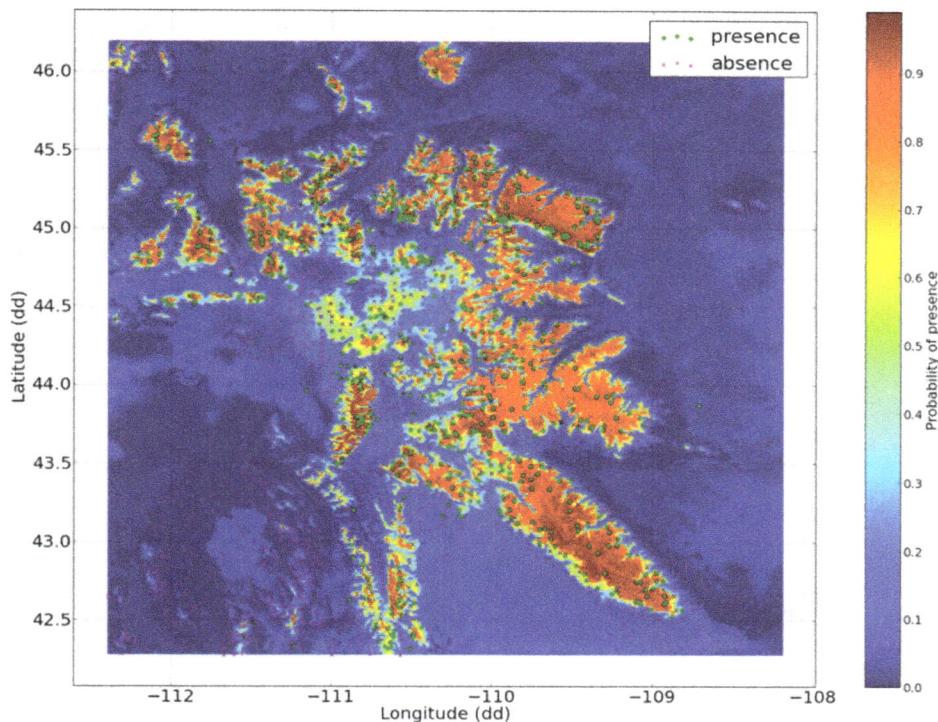

Figure 8. Probabiliy presence for *P.albicaulis* ≥ **20 cm DBH within the GYA for the 2010 climate period.**

Modeling methods

A random forest (RF) [51] algorithm was used to create a bioclimate niche model of *P. albicaulis* in the GYA. Random forest is an ensemble learning technique that generates independent random classification trees using a subset of the total predictor variables and classifies a bootstrap random subsample of the data. These trees are aggregated and a majority vote over all trees in the random forest defines the resulting response class. This method of random trees with subsampling ensures a robust ensemble classification reducing overfitting and collinearity issues, especially with a large number of trees [9,51–53]. The python programming language (Python 3.3) and the Scikit-Learn library was used to fit the random forest model and predict current habitat niche, with parameters for number of trees ($n_{estimators} = 1000$), number of variables ($max_{features} = 4$), and node size ($min_{samplesleaf} = 20$) [54].

First pass filtering of environmental covariates was performed using Principal Component Analysis (PCA) to generate proxy sets [55–57]. After initial list was constructed, an additional filter was imposed on the variables with a 0.75 maximum correlation threshold to avoid collinearity issues (Fig. 3) [55]. Physiologically relevant variables to *P. albicaulis* presence were given precidence in final culling in cases of correlation above the specified maximum threshold. The final variable list selected were tmin1, vpd3, ppt4, pack4, tmax7, aet7, pet8, ppt9 (Table 2). The Software for Assisted Habitat Modeling (SAHM) was used to visualize correlations with the pairs function embeded in the VisTrails scientific workflow management system [58,59].

Model evaluation was performed under a variety of methods. An out-of-bag (OOB) error estimate was calculated by comparing the modeled probability of presence using approximately two-thirds of the field data, while withholding a subset of the remainder. Accuracy was evaluated by calculating: 1) the sensitivity, representing the true positive rate (TPR), 2) the specificity, representing the true negative rate (TNR), 3) the receiver operator characteristic curve (AUC). Importance of a specific predictor variable was calculated by examination of the increase in prediction error within the OOB sample when the predictor variable was permuted while others were held constant [54,60]. The rate of prediction error with permutation of a specified variable can be interpreted as the level of dependence of presence or absence response to that variable [61].

Projections for *P. albicaulis* were computed using 30 year moving climate averages for the period from 2010–2099 for both RCP 4.5 and 8.5 climate scenarios. Changes of suitable habitat area were determined using a binary classification of expected presence and absence. Binary class assignment was made under a probability of presence threshold where the ratio of sensitivity and specificity equalled 1. This method ensured an equal ability of the model to detect presence and absence. The Kappa and True Skill Statistic (TSS) were also calculated to observe how sensitivity and specificity responded under differing probability thresholds [62]. Survey plots predicted as suitable under climatic conditions in 2010 served as a reference for projections. The presence classifications were evaluated as the amount of suitable habitat changed over time, confined within specified elevational limits. To account for the need for a minimum patch size, total number of patches and median sizes using the an eight-neighbor rule (see Text S1) for patch identification were tracked over time [63].

Results

Model evaluation

The random forest model displayed an out-of-bag (OOB) error rate of 16.1% with greater errors of commission (13.1%) than omission (10.9%) (Table 3). The AUC was 0.94, displaying high

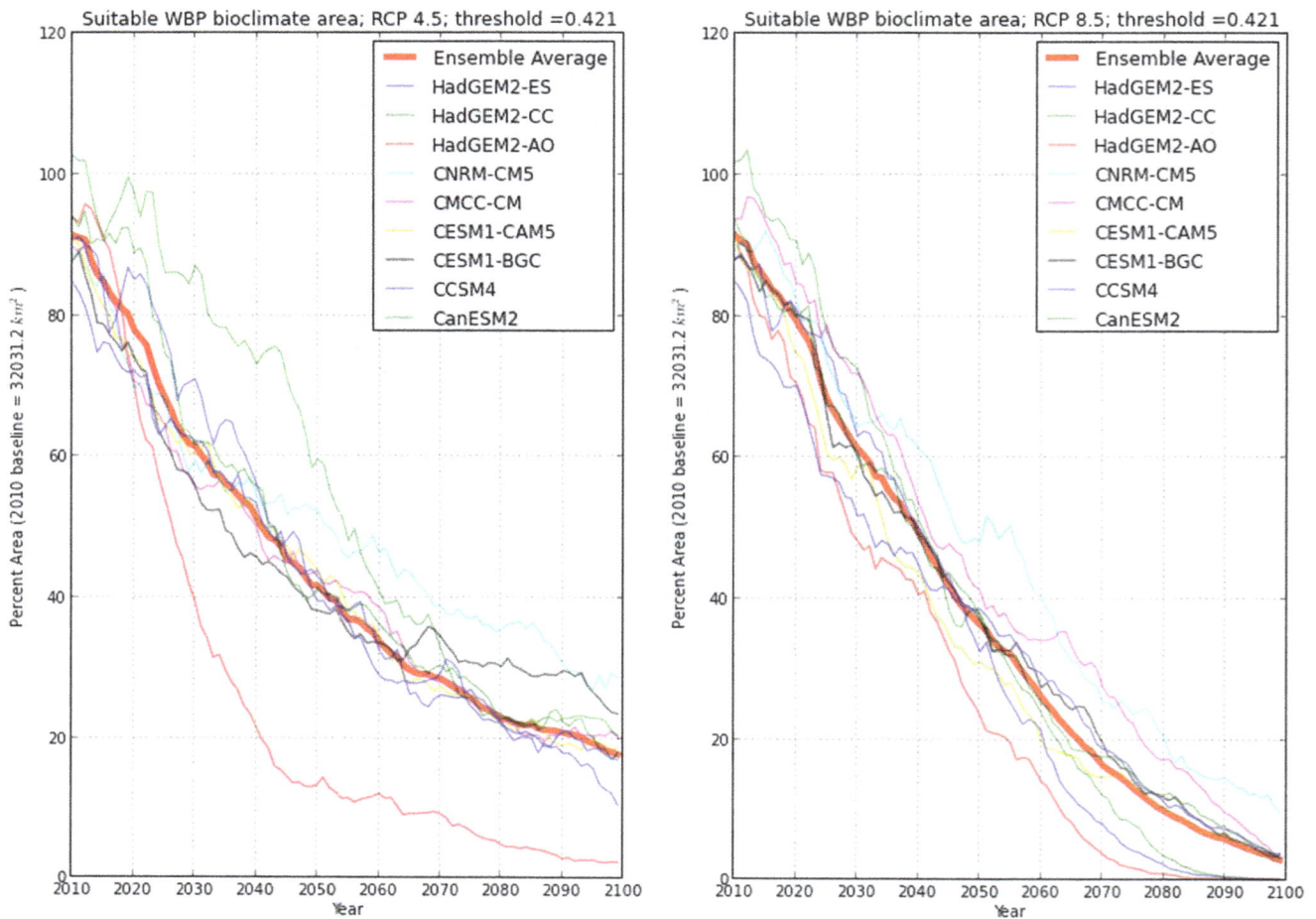

Figure 9. Bioclimate projections for *P.albicaulis* for 2010 to 2099 under 30-year moving averaged climates.

specificity and sensitivity (Fig. 4). Threshold probability of presence for a binary classification was selected at 0.421 (i.e where sensitivity = specificity). A probability threshold where TPR and TNR were equal was compared to the maximum Kappa statistic (0.538) and the maximum True Skill Statistic (TSS) (0.345) and found to be a compromise between the diagnostics (Fig. 5).

Estimates of variable importance plots revealed that permutation of maximum temperatures of summer months from all random trees resulted in a large drop in mean accuracy for distinguishing presence and absence of *P. albicaulis* (0.706 decrease in mean accuracy). This was followed by spring time snowpack (0.137 decrease in mean accuracy) (Fig. 6). Histogram plots of July maximum temperatures and April snowpack provided evidence of discrimination for presence and absence that are consistent with the modeled probability of presence for the year of calibration (Fig. 7).

Spatially explicit probability plots for the 2010 climate displayed highest probability of presence values within the ≥ 2500 m mountain ranges of the GYA in agreement with studies employing aerial imagery and remote sensing [4,5] (Fig. 8). Assuming that the modeled suitable bioclimate for *P. albicaulis* remains similar in the next century, the model demonstrated capacity to predict probable future *P. albicaulis* suitable habitat under projected climate conditions.

Model projections

Under both RCP 4.5 and 8.5, there was a predicted steady reduction of suitable bioclimate habitat for *P. albicaulis* over the course of this century, with RCP 8.5 displaying steeper declines than RCP 4.5 (Fig. 9). Under the RCP 4.5 and 8.5 scenarios, suitable habitat shifts from 100–85% to 2–29% by 2099, and 100–85% to 0.04–10% by 2099 respectively (Table 4).

CNRM-CM5, CMCC-CM, and CESM1-BGC projections showed the highest probabilities for suitable habitat area at the end of the century, while HadGEM2-AO, HadGEM2-ES, and HadGEM2-CC indicated the lowest probabilities. The standard deviations per year for both RCPs progressively decreased over time (Fig. 10). Among climate scenarios, standard deviations for both RCPs display low variability for the first five projection years and a rapid increase of variability peaking at 2043. For RCP 4.5, high variability existed primarily due to differing climate projections by models HadGEM2-AO and HadGEM2-CC, resulting in uncertainties in probabilities of presence fluctuating between 8 and 15% until 2068, after which variability was between 6–8%. Under RCP 8.5, standard deviations between GCMs were consistently lower than RCP 4.5. Regardless of the GCM, by 2079 the areas of suitable habitat converged to similar values.

Spatially explicit mapping of probability surfaces presented similar contractions of *P. albicaulis* habitat suitability toward the

Table 4. Projected binary *P. albicaulis* presence area within GYA to 2099.

Ensemble Average RCP 4.5	2010	2040	2070	2099
Area (km²)	29250.9	16381.2	9151.1	5685.9
	(27134–32858)	(6918–23359)	(2962–12477)	(763–9194)
% Total Threshold Area*	91.3	51.1	28.6	17.8
	(85–103)	(22–73)	(9–39)	(2–29)
Mean Elevation (m)	2875.7	3020.2	3128.0	3217.9
	(2842–2895)	(2938–3182)	(3055–3297)	(3114–3471)
2.5 Percentile Elevation (m)	2356.3	2494.2	2595.0	2691.5
	(2320–2376)	(2433–2656)	(2506–2758)	(2571–3041)
97.5 Percentile Elevation (m)	3521.9	3603.5	3677.8	3734.6
	(3507–3530)	(3551–3701)	(3636–3783)	(3673–3905)
Ensemble Average RCP 8.5	**2010**	**2040**	**2070**	**2099**
Area (km²)	29259.3	15746.0	5271.5	960.0
	(27188–32604)	(12985–19581)	(1247–8850)	(13–3105)
% Total Threshold Area*	91.3	49.2	16.5	3.0
	(85–102)	(40–61)	(4–28)	(0–10)
Mean Elevation (m)	2874.7	3022.5	3225.5	3470.5
	(2845–2893)	(2974–3061)	(3116–3412)	(3255–3749)
2.5 Percentile Elevation (m)	2353.1	2492.2	2691.3	3001.5
	(2322–2369)	(2436–2547)	(2553–2934)	(2622–3401)
97.5 Percentile Elevation (m)	3522.1	3605.7	3739.5	3908.7
	(3508–3530)	(3576–3631)	(3677–3866)	(3775–4063)

Summary of projection outputs under RCP 4.5 and 8.5 climate scenarios displays loss of bioclimate habitat from 2010 to 2099 (low and high probability of presence GCM summaries displayed in parentheses). Projections into 2099 under all 9 GCMs suggest rapid loss of suitable bioclimate habitat to below 70% of the current modeled distribution and shifts towards the limited high elevation zones (>3000 m). *(Percent threshold areas calculated from the 2010 PRISM reference probabilities of presence.)*

upper elevation zones of the GYA that included the Beartooth Plateau and Wind River Ranges (Fig. 11). This implied that rapid warming may lead to conditions outside of the *P. albicaulis* niche in lower elevation areas, and limiting the species to the alpine zones. Elevational analysis of cells within threshold presence probabilities over time observed mean elevations of suitable bioclimates shifting from 2,875 to 3,218 m and 2,875 to 3,470 m for RCP 4.5 and 8.5 respectively. By 2099, ensemble averaged GCM projections displayed over 70% loss of habitat under both scenarios.

P. albicaulis patches from the 2010 baseline observed 202 patches with median patch size of ~180 km². Projected patch dynamics analysis denoted a quadratic relationship of patch size over time. Patch dynamics displayed a slow increase in number of *P. albicaulis* patches to a maximum at 2074 and 2057 for RCP 4.5 and 8.5 respectively, followed by a decreasing trend. RCP 4.5 patch numbers were more sporadic, displaying fluctuations across the time period compared to RCP 8.5 associated with the greater interannual climate variability amongst GCM models. Median patch size saw a steady decrease from 72–65 km² to 21–8 km² for RCP 4.5 and 8.5 respectively, for the projection period, suggesting habitat loss through fragmentation (Fig. 12).

Discussion

In this analysis, the spatiotemporal patterns for *P. albicaulis* distributions were assessed under nine climate models and two emissions scenarios. Bioclimate modeling of *P. albicaulis* illustrat-

ed that presence and absence were strongly separated by summer temperatures and spring snowpack. This was in agreement with empirical findings of *P. albicaulis* presence in cool summertime environments where July temperatures range between 4–18°C [64]. Concordantly, these cool summer regions were synonymous with late snow melt, supporting snowpack as an important feature in distinguishing presence and absence.

Future projections by all nine GCMs suggested a contraction in suitable *P. albicaulis* climate area by the end of the century to <30% of current conditions. This was consistent with the results from various other research using either niche models or hybrid process models, predicting similar amounts of *P. albicaulis* contraction [8,9,65]. Variability among projected suitable habitat areas under differing GCMs decreased as all projected maximum temperatures increased above 1°C from the 100 year historic mean. This pattern of warming convergence occurred earlier for the GCMs under the RCP 8.5 scenario than those under RCP 4.5, resulting in the observed low variability of *P. albicaulis* suitable habitat area under RCP 8.5. Despite temperature variability remaining relatively constant amongst GCMs within a RCP, once mean annual temperatures increased beyond *sim* 1°C from the historic average, all bioclimatic habitat models exhibited a pattern of contracting total area and variability. These results lead to the conclusion that explicit selection of a GCM to model under may not necessarily matter for *P. albicaulis* bioclimatic niche modeling studies, especially if the direction of change is solely of concern. However, if investigation of the magnitude of change is relevant,

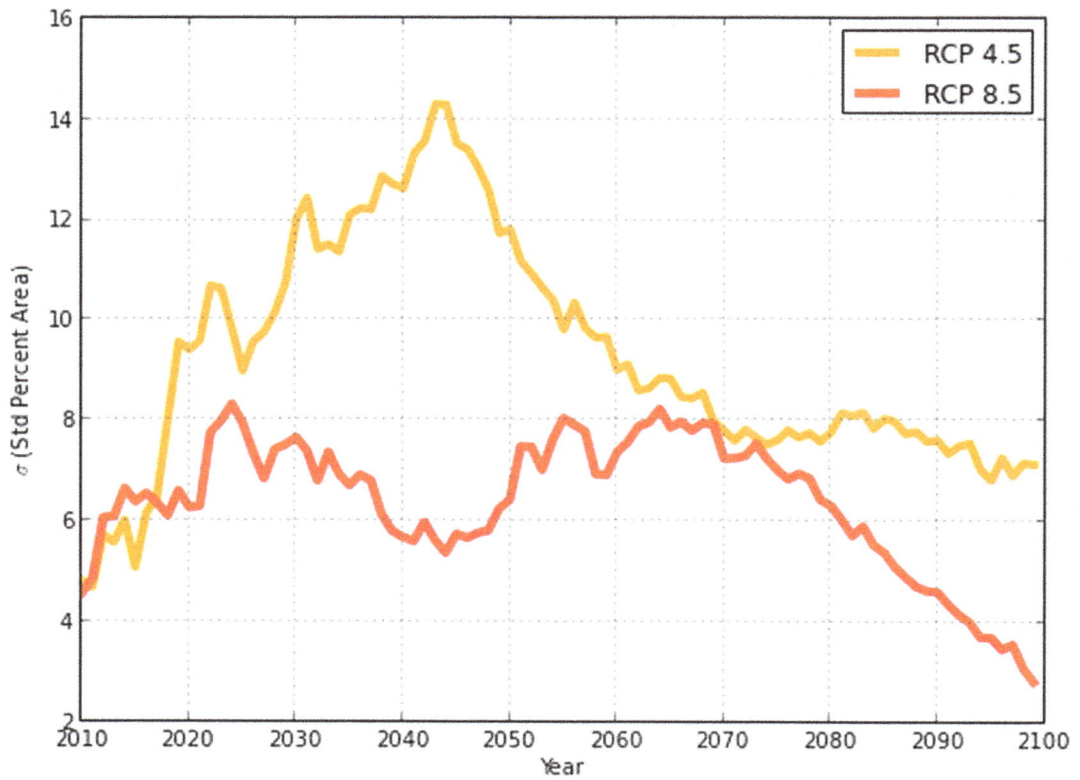

Figure 10. Evaluation of the standard deviation σ **for percent suitable habitat area by RCP scenario.**

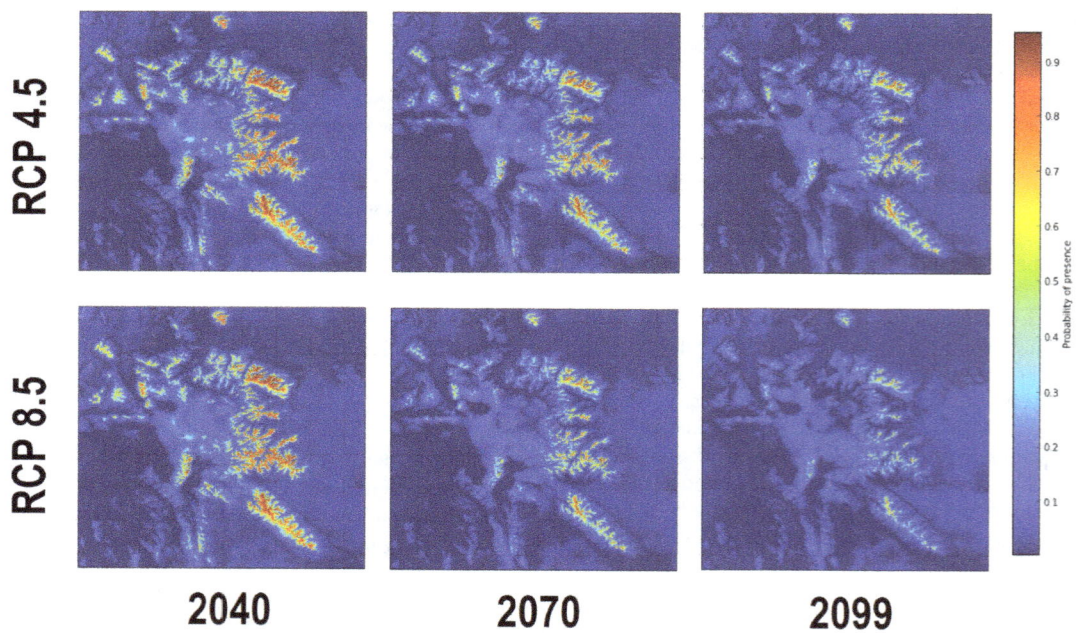

Figure 11. Spatially explicit probabilty surfaces for 2040 to 2099 suggest contraction of suitable bioclimatic habiatat for *P.albicaulis* **into the** $\geq$ **2500 m elevation zones.**

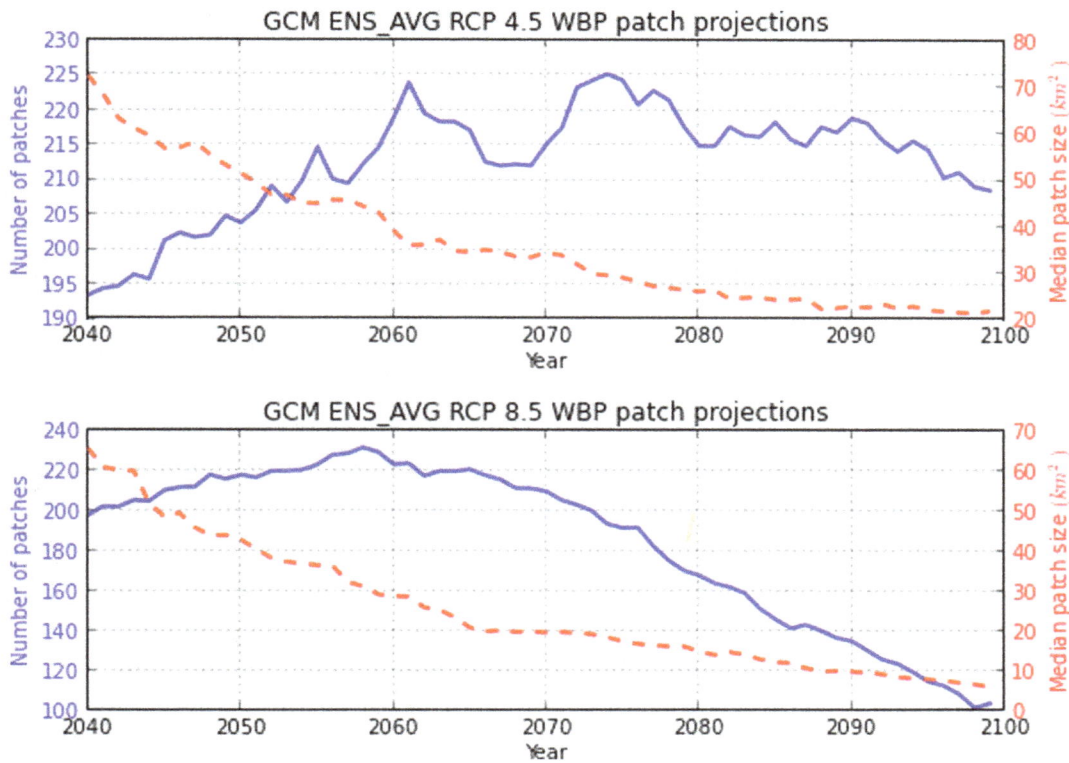

Figure 12. Patch dynamics of modeled *P.albicaulis*. Time series of *P.albicaulis* patch projections for number of patches and median patch size to 2099.

then GCM selection may directly influence the projected total suitable habitat area. This can be observed with RCP 4.5 habitat projection models differing by as much as 27% total suitable habitat area by the year 2099. Therefore arbitrary selection of a GCM for future projection modeling is likely inappropriate since it could lead to overly optimistic/pessimistic results for the species of concern.

Temporal patch dynamic analysis present an increase in fragmentation of the larger *P. albicaulis* suitable habitats over the next five decades, suggested through an increase in the total number of continuous patches but decreases in median size. This was followed by a contraction of small patches until they were almost absent from the system. Remaining habitat patches were smaller and less prevalent on the landscape by the end of the century. Reduced habitat patch size and density may reduce the likelihood for *N. columbiana* to disperse successful germinating seed caches, due to the limited size and area of suitable patch space. If changing climate habitats result in mortality within adult patches, genetic diversity may be lost resulting in a population bottleneck, thus reducing the robustness of the species to adapt to future disturbances. Experimental trials of P. albicaulis survival and fecundity under warmer and drier conditions outside the currently known range would provide greater confidence of the species ability to persist under future change. Limited analysis on seedling environmental conditions would also elucidate spatially explicit dispersal ranges and greater understanding of probable ranges for future establishment and survivorship.

Projected distributions of persistent *P. albicaulis* patches displayed a strong trend towards contraction into high elevation zones. Physiologically, there does not appear to be any upper elevation limit for *P. albicaulis* in the GYA. *P. albicaulis* in the

region has been reported to survive in absolute temperatures as low as $-36°C$ [64]. Lab experiments performed on *Pinus cembra*, a related five-needle pine residing in similar climates, were able to endure cold temperature extremes as low at $-70°C$ without cellular tissue damage [66]. Considering the current absolute minimum temperatures the species resides in and cold tolerance of its relatives suggests that *P. albicaulis* treeline in the GYA are not limited by lower temperatures. Controlled laboratory experimentation on *P. albicaulis* tolerances to temperatures would greatly improve this physiological understanding of cold tolerance.

Elevational habitat constriction do not imply that *P. albicaulis* will be completely gone from the region, but merely the loss of suitable climate habitat. Currently pre-established adult age class individuals will likely persist, since projected conditions of increased temperatures and CO_2 concentrations physiologically indicate increased growth rates of *P. albicaulis* [67]. Furthermore, micro-refugia sites may exist in the GYA that support *P. albicaulis* survival into the future, but were failed to have been modeled due to the coarseness of 30-arc-second climate data resolution. Since this bioclimatic envelope modeling approach was parameterized by the realized niche from in-situ data, it was difficult to determine if lower elevation limits are driven by warmer climate conditions or competition for light, water, or nutrients [15,17]. For example, lower treeline limits for *P. albicaulis* maybe driven primarily by competitive exclusion from late seral species *A.lasiocarpa*, *P.contorta*, and *P.engelmannii*. This follows from paleoecological pollen records of competitor migration during the Early Holocene (9000–5000 yr B.P), when climate conditions were warmer and drier. Longer growing seasons allowing competitors to invade likely drove *P. albicaulis* communities +500 m in elevation [68–70]. If future climate conditions become analogous to this Early

Holocene period, invasion of competitor species will likely contract *P. albicaulis* habitat to the limited high elevation zones of the GYA, specifically the Beartooth Plateau region and Wind River range [71].

Conclusion

This analysis examined the future of *P. albicaulis* suitable climate in the GYA and explicitly addressed the question of distribution variability under 9 representative GCMs and 2 emission scenarios. Increases in temperature within the GYA will likely result in a high level of contraction of suitable climate habitat for *P. albicaulis* over the next century. This contraction was consistent for all GCM projections, with approximately 20% uncertainty in total probable area. This analysis recommends that care be taken for species distribution modeling in future studies during the selection of GCMs due to their relevance for magnitudes of change. GCM ensemble averaging may be a solution to this issue, however it should be noted that averaging should take place after an individual GCM is projected in order to maintain interannual variability.

Although other studies have examined *P. albicaulis* species distribution models [8,65,72], this study is a step forward through its focus on relevant regional scale design, expansive local datasets, inclusion of high resolution climate and dynamic water balance variables, and selective projection under the latest AR5 GCMs. It is reiterated that the bioclimate niche model approach has high utility for understanding habitat conditions through correlative relationships with environmental variables, however, it may fail to explicitly model competitive exclusion, disturbance, phenotypic plasticity, and other complex interactions that are vital in determining a species' actual presence as it experiences changes in climate [15,17,73,74]. These unmodeled factors create uncertainties suggesting that this modeling effort does not identify the full potential climatic range of *P. albicaulis* in the future.

Uncertainties also exist regarding new suitable climates that may occur outside the current species range. Despite most rangewide studies confirming our results of total suitable habitat area reduction, there is potential for previously unsuitable habitat to become available under future climate change in the Northern regions [8,22,65]. Caution is therefore advised to individuals interpreting these findings. Changing climate will inevitably result in impacts on biomes and community structures. As such, mitigation and adaptation for potential futures are vital to conservation of climate sensitive species [75]. Future research that combines bioclimatic niche modeling with a mechanistic based disturbance, dispersal, and competition model will likely provide greater insight to the potential range of *P. albicaulis* in a climate changing world [76,77]. It would furthermore provide insight towards informing management options for restoration that may include controlled fire, selected thinning of competitor species, or assisted migration.

Acknowledgments

We are indebted for the insightful reviews from Richard Waring, William Monahan, and Tom Olliff. Many thanks to Marian and Colin Talbert, and Mark Greenwood for the software and statistical consultation.

Author Contributions

Conceived and designed the experiments: TC AJH NBP. Performed the experiments: TC NBP. Analyzed the data: TC. Contributed reagents/ materials/analysis tools: TC AJH NBP. Wrote the paper: TC AJH NBP.

References

1. Intergovernmental Panel on Climate Change (2007) Fourth Assessment Report: Climate Change 2007: The AR4 Synthesis Report. Geneva: IPCC.

2. Tomback DF, Arno SF, Keane RE (2001) Whitebark pine communities: ecology and restoration. Island Press.

3. Callaway RM (1998) Competition and facilitation on elevation gradients in subalpine forests of the Northern Rocky Mountains, USA. Oikos 82: pp. 561–573.

4. Macfarlane WW, Logan JA, Kern W (2012) An innovative aerial assessment of greater yellowstone ecosystem mountain pine beetle-caused whitebark pine mortality. Ecological Applications.

5. Jewett JT, Lawrence RL, Marshall LA, Gessler PE, Powell SL, et al. (2011) Spatiotemporal relationships between climate and whitebark pine mortality in the greater yellowstone ecosystem. Forest Science 57: 320–335.

6. Logan JA, Macfarlane WW, Willcox L (2010) Whitebark pine vulnerability to climate-driven mountain pine beetle disturbance in the greater yellowstone ecosystem. Ecological Applications 20: 895–902.

7. Logan JA, Bentz BJ (1999) Model analysis of mountain pine beetle (coleoptera: Scolytidae) seasonality. Environmental Entomology 28: 924–934.

8. Rehfeldt GE, Crookston NL, Sáenz-Romero C, Campbell EM (2012) North American vegetation model for land-use planning in a changing climate: a solution to large classification problems. Ecological Applications 22: 119–141.

9. Rehfeldt GE, Crookston NL, Warwell MV, Evans JS (2006) Empirical analyses of plant-climate relationships for the western United States. International Journal of Plant Sciences 167: 1123–1150.

10. Thuiller W (2004) Patterns and uncertainties of species' range shifts under climate change. Global Change Biology 10: 2020–2027.

11. Iverson LR, Prasad AM, Matthews SN, Peters M (2008) Estimating potential habitat for 134 eastern US tree species under six climate scenarios. Forest Ecology and Management 254: 390–406.

12. Guisan A, Theurillat JP, Kienast F (1998) Predicting the potential distribution of plant species in an alpine environment. Journal of Vegetation Science 9: 65–74.

13. Busby J (1988) Potential impacts of climate change on Australias flora and fauna. Commonwealth Scientific and Industrial Research Organisation, Melbourne, FL, USA.

14. Hutchinson GE (1957) Concluding remarks. Cold Spring Harbor Symposia on Quantitative Biology 22: 415–427.

15. Austin M (2007) Species distribution models and ecological theory: a critical assessment and some possible new approaches. Ecological Modelling 200: 1–19.

16. Austin M (2002) Spatial prediction of species distribution: an interface between ecological theory and statistical modelling. Ecological Modelling 157: 101–118.

17. Pearson RG, Dawson TP (2003) Predicting the impacts of climate change on the distribution of species: are bioclimate envelope models useful? Global Ecology and Biogeography 12.

18. Willis KJ, Whittaker RJ (2002) Species diversity–scale matters. Science 295: 1245–1248.

19. Araújo MB, Cabeza M, Thuiller W, Hannah L, Williams PH (2004) Would climate change drive species out of reserves? an assessment of existing reserve-selection methods. Global Change Biology 10: 1618–1626.

20. Ferrier S (2002) Mapping spatial pattern in biodiversity for regional conservation planning: where to from here? Systematic Biology 51: 331–363.

21. Pearce J, Lindenmayer D (1998) Bioclimatic analysis to enhance reintroduction biology of the endangered helmeted honeyeater (lichenostomus melanops cassidix) in Southeastern Australia. Restoration Ecology 6: 238–243.

22. McLane SC, Aitken SN (2012) Whitebark pine (pinus albicaulis) assisted migration potential: testing establishment north of the species range. Ecological Applications 22: 142–153.

23. Monahan WB, Cook T, Melton F, Connor J, Bobowski B (2013) Forecasting distributional responses of limber pine to climate change at management-relevant scales in Rocky Mountain National Park. PloS ONE 8: e83163.

24. Arno SF, Hoff RJ (1989) Silvics of whitebark pine (pinus albicaulis). Intermountain Research Station GTR-INT-253.

25. Mahalovich MF, Hipkins VD (2011) Molecular genetic variation in whitebark pine (pinus albicaulis engelm.) in the inland west. In: Keane RE, Tomback DF, Murray MP, Smith CM, editors, The future of high-elevation, five-needle white pines in Western North America: Proceedings of the High Five Symposium. 28–30 June 2010; Missoula, MT. Proceedings RMRS.

26. Pederson GT, Gray ST, Ault T, Marsh W, Fagre DB, et al. (2011) Climatic controls on the snowmelt hydrology of the Northern Rocky Mountains. Journal of Climate 24: 1666–1687.

27. Westerling AL, Hidalgo HG, Cayan DR, Swetnam TW (2006) Warming and earlier spring increase western US forest wildfire activity. Science 313: 940–943.
28. Taylor KE, Stouffer RJ, Meehl GA (2012) An overview of CMIP5 and the experiment design. Bulletin of the American Meteorological Society 93.
29. Hibbard KA, van Vuuren DP, Edmonds J (2011) A primer on representative concentration pathways (RCPs) and the coordination between the climate and integrated assessment modeling communities. CLIVAR Exchanges 16: 12–13.
30. Lutz JA, van Wagtendonk JW, Franklin JF (2010) Climatic water deficit, tree species ranges, and climate change in Yosemite National Park. Journal of Biogeography 37: 936–950.
31. Deser C, Phillips AS, Alexander MA, Smoliak BV (2014) Projecting North American climate over the next 50 years: Uncertainty due to internal variability. Journal of Climate 27: 2271–2296.
32. Beaumont LJ, Hughes L, Pitman A (2008) Why is the choice of future climate scenarios for species distribution modelling important? Ecology Letters 11: 1135–1146.
33. Smith WB (2002) Forest inventory and analysis: a national inventory and monitoring program. Environmental Pollution 116: S233–S242.
34. Gibson J, Moisen G, Frescino T, Edwards Jr TC (2014) Using publicly available forest inventory data in climate-based models of tree species distribution: Examining effects of true versus altered location coordinates. Ecosystems 17: 43–53.
35. Gesch D, Oimoen M, Greenlee S, Nelson C, Steuck M, et al. (2002) The national elevation dataset. Photogrammetric engineering and remote sensing 68: 5–32.
36. Franklin J (2009) Mapping species distributions: spatial inference and prediction. Cambridge University Press.
37. Lockman IB, DeNitto GA, Courter A, Koski R (2007) WLIS: The whitebark-limber pine information system and what it can do for you. In: Proceedings of the conference whitebark pine: a Pacific Coast perspective. US Department of Agriculture, Forest Service, Pacific Northwest Region, Ashland, OR. Citeseer, pp. 146–147.
38. Jean C, Shanahan E, Daley R, DeNitto G, Reinhart D, et al. (2010) Monitoring white pine blister rust infection and mortality in whitebark pine in the Greater Yellowstone Ecosystem. Proceedings of the future of high-elevation five-needle white pines in Western North America: 28–30.
39. Edwards Jr TC, Cutler DR, Zimmermann NE, Geiser L, Moisen GG (2006) Effects of sample survey design on the accuracy of classification tree models in species distribution models. Ecological Modelling 199: 132–141.
40. Kadmon R, Farber O, Danin A (2003) A systematic analysis of factors affecting the performance of climatic envelope models. Ecological Applications 13: 853–867.
41. Weaver T, Dale D (1974) Pinus albicaulis in central Montana: environment, vegetation and production. American Midland Naturalist: 222–230.
42. Daly C, Gibson WP, Taylor GH, Johnson GL, Pasteris P (2002) A knowledge-based approach to the statistical mapping of climate. Climate Research 22: 99–113.
43. Thornthwaite C (1948) An approach toward a rational classification of climate. Geographical Review 38: 55–94.
44. Thornthwaite C, Mather J (1955) The water balance. Publication of Climatology 8.
45. Dingman S (2002) Physical hydrology. Prentice Hall.
46. National Resources Conservation Service (2014) Available: http://soildatamart.nrcs.usda.gov. Accessed 2013 Apr 3.
47. Thrasher B, Xiong J, Wang W, Melton F, Michaelis A, et al. (2013) Downscaled climate projections suitable for resource management. Eos, Transactions American Geophysical Union 94: 321–323.
48. Rupp DE, Abatzoglou JT, Hegewisch KC, Mote PW (2013) Evaluation of CMIP5 20th century climate simulations for the Pacific Northwest USA. Journal of Geophysical Research: Atmospheres 118: 10–884.
49. Gent PR, Danabasoglu G, Donner LJ, Holland MM, Hunke EC, et al. (2011) The community climate system model version 4. Journal of Climate 24: 4973–4991.
50. Moss RH, Babiker M, Brinkman S, Calvo E, Carter T, et al. (2008) Towards new scenarios for analysis of emissions, climate change, impacts, and response strategies.
51. Breiman L (2001) Random forests. Machine Learning 45: 5–32.
52. Roberts DR, Hamann A (2012) Method selection for species distribution modelling: are temporally or spatially independent evaluations necessary? Ecography 35: 792–802.
53. Lawrence RL, Wood SD, Sheley RL (2006) Mapping invasive plants using hyperspectral imagery and breiman cutler classifications (randomforest). Remote Sensing of Environment 100: 356–362.
54. Pedregosa F, Varoquaux G, Gramfort A, Michel V, Thirion B, et al. (2011) Scikit-learn: Machine learning in Python. Journal of Machine Learning Research 12: 2825–2830.
55. Dormann CF, Elith J, Bacher S, Buchmann C, Carl G, et al. (2013) Collinearity: a review of methods to deal with it and a simulation study evaluating their performance. Ecography 36: 027–046.
56. Booth GD, Niccolucci MJ, Schuster EG (1994) Identifying proxy sets in multiple linear regression: an aid to better coefficient interpretation. Research paper INT.
57. Tabachnick B, Fidell LS (1989) Using multivariate statistics, 1989. Harper Collins Tuan, PD A comment from the viewpoint of time series analysis Journal of Psychophysiology 3: 46–48.
58. Freire J (2012) Making computations and publications reproducible with vistrails. Computing in Science & Engineering 14: 18–25.
59. Morisette JT, Jarnevich CS, Holcombe TR, Talbert CB, Ignizio D, et al. (2013) Vistrails SAHM: visualization and workflow management for species habitat modeling. Ecography 36: 129–135.
60. Liaw A, Wiener M (2002) Classification and regression by randomforest. R news 2: 18–22.
61. Cutler DR, Edwards Jr TC, Beard KH, Cutler A, Hess KT, et al. (2007) Random forests for classification in ecology. Ecology 88: 2783–2792.
62. Allouche O, Tsoar A, Kadmon R (2006) Assessing the accuracy of species distribution models: prevalence, kappa and the true skill statistic (tss). Journal of Applied Ecology 43: 1223–1232.
63. Turner MG, Gardner RH, O'Neill RV (2001) Landscape ecology in theory and practice: pattern and process. Springer.
64. Weaver T (2001) Whitebark pine and its environment. In: Tomback DF, Arno SF, Keane RE, editors, Whitebark pine communities: ecology and restoration, Washington D.C, USA: Island Press.
65. Waring RH, Coops NC, Running SW (2011) Predicting satellite-derived patterns of large-scale disturbances in forests of the pacific northwest region in response to recent climatic variation. Remote Sensing of Environment 115: 3554–3566.
66. Sakai A, Larcher W (1987) Frost survival of plants. Responses and adaptation to freezing stress. Springer-Verlag.
67. Chapin III FS, Chapin MC, Matson PA, Vitousek P (2011) Principles of terrestrial ecosystem ecology. Springer.
68. Whitlock C, Shafer SL, Marlon J (2003) The role of climate and vegetation change in shaping past and future fire regimes in the Northwestern US and the implications for ecosystem management. Forest Ecology and Management 178: 5–21.
69. Whitlock C (1993) Postglacial vegetation and climate of Grand Teton and southern Yellowstone national parks. Ecological Monographs: 173–198.
70. Bartlein PJ, Whitlock C, Shafer SL (1997) Future climate in the Yellowstone national park region and its potential impact on vegetation. Conservation Biology 11: 782–792.
71. Tausch RJ, Wigand PE, Burkhardt JW (1993) Viewpoint: plant community thresholds, multiple steady states, and multiple successional pathways: legacy of the quaternary? Journal of Range Management: 439–447.
72. Bell DM, Bradford JB, Lauenroth WK (2014) Early indicators of change: divergent climate envelopes between tree life stages imply range shifts in the western united states. Global Ecology and Biogeography 23: 168–180.
73. Keane B, Tomback D, Davy L, Jenkins M, Applegate V (2013) Climate change and whitebark pine: Compelling reasons for restoration. Whitebark Pine Ecosystem Foundation Whitepaper.
74. Guisan A, Thuiller W (2005) Predicting species distribution: offering more than simple habitat models. Ecology Letters 8: 993–1009.
75. Keane RE, Tomback DF, Aubry CA, Bower EM, Campbell CL, et al. (2012) A range-wide restoration strategy for whitebark pine (pinus albicaulis): General technical report. USDA FS, Rocky Mountain Research Station RMRS-GTR-279: 108.
76. Mathys A, Coops NC, Waring RH (2014) Soil water availability effects on the distribution of 20 tree species in Western North America. Forest Ecology and Management 313: 144–152.
77. Morin X, Thuiller W (2009) Comparing niche-and process-based models to reduce prediction uncertainty in species range shifts under climate change. Ecology 90: 1301–1313.

19

Evaluating Potential Spectral Impacts of Various Artificial Lights on Melatonin Suppression, Photosynthesis, and Star Visibility

Martin Aubé[1]*, Johanne Roby[2], Miroslav Kocifaj[3]

1 Département de physique, Cégep de Sherbrooke, Sherbrooke, Québec, Canada, **2** Département de chimie, Cégep de Sherbrooke, Sherbrooke, Québec, Canada, **3** Astronomical Institute, Slovak Academy of Sciences, Dúbravská 9, Bratislava, Slovak Republic

Abstract

Artificial light at night can be harmful to the environment, and interferes with fauna and flora, star visibility, and human health. To estimate the relative impact of a lighting device, its radiant power, angular photometry and detailed spectral power distribution have to be considered. In this paper we focus on the spectral power distribution. While specific spectral characteristics can be considered harmful during the night, they can be considered advantageous during the day. As an example, while blue-rich Metal Halide lamps can be problematic for human health, star visibility and vegetation photosynthesis during the night, they can be highly appropriate during the day for plant growth and light therapy. In this paper we propose three new indices to characterize lamp spectra. These indices have been designed to allow a quick estimation of the potential impact of a lamp spectrum on melatonin suppression, photosynthesis, and star visibility. We used these new indices to compare various lighting technologies objectively. We also considered the transformation of such indices according to the propagation of light into the atmosphere as a function of distance to the observer. Among other results, we found that low pressure sodium, phosphor-converted amber light emitting diodes (LED) and LED 2700 K lamps filtered with the new Ledtech's *Equilib* filter showed a lower or equivalent potential impact on melatonin suppression and star visibility in comparison to high pressure sodium lamps. Low pressure sodium, LED 5000 K-filtered and LED 2700 K-filtered lamps had a lower impact on photosynthesis than did high pressure sodium lamps. Finally, we propose these indices as new standards for the lighting industry to be used in characterizing their lighting technologies. We hope that their use will favor the design of new environmentally and health-friendly lighting technologies.

Editor: Shin Yamazaki, University of Texas Southwestern Medical Center, United States of America

Funding: This work was supported by Fond de recherche du Quebec - Nature et Technologie (FRQNT) Programme pour le dégagement d'enseignement des chercheurs de collège Regroupements stratégiques (http://www.fqrnt.gouv.qc.ca/) and Cegep de Sherbrooke - CERTEE (http://cegepsherbrooke.qc.ca). This work was partially supported by the Slovak National Grant Agency VEGA (grant No. 2/0002/12). This work was also partially supported by the Programme de soutien aux chercheurs et aux chercheuses du collégial from ministère de l'Enseignement supérieur, de la Recherche, de la Science et de la Technologie, Québec. The funders had no role in study design, data collection and analysis, decision to publish, or preparation of the manuscript.

Competing Interests: The authors have declared that no competing interests exist.

* E-mail: martin.aube@cegepsherbrooke.qc.ca

Introduction

2The development of artificial lighting technologies over the centuries has transformed human civilization and shaped the way we live. The world has become awash with artificial lighting both during the day and at night, indoors and outdoors, from office buildings to streetlights. Before Edison's invention of the light bulb (1879), people spent most of their time outdoors, receiving adequate daily doses of natural, full-spectrum sunlight during the day while spending their evenings and nights in relative darkness. With the growing availability of artificial lighting, people are spending an increasing amount of time inside under artificial lighting and consequently reducing the amount of time they are exposed to natural full-spectrum light during the day and darkness during the night. Around 99% of the population of the United States and Europe, and 62% of the world's remaining population, are exposed to artificial light at night (ALAN), the amount of which is increasing rapidly each year [1]. Not only are humans but also fauna and flora are exposed to ALAN, with ensuing environmental consequences. ALAN is one of the fastest growing and most common kinds of environmental pollution. The effects of ALAN on fauna have been well defined and documented, and almost only negative effects have been reported. ALAN affects behavior, foraging, reproduction, communication, breeding cycles and the habitat of many nocturnal species [2], [3], [4], including invertebrates [5] amphibians [6], birds [7], bats [8], turtles [9], [10], [11], fish [12] and reptiles [13]. On the other hand, the impact of ALAN on flora is less documented; a review on the topic is reported by Briggs [14]. Exposure to artificial light prevents many trees from adjusting to seasonal variations. The presence of ALAN stimulates photosynthesis at a time when photosynthesis does not normally occur. Similar to humans and animals, plants require a specific cycle of light/darkness in order to grow healthily. Light affects several plant processes, such as seed germination, stem elongation, leaf expansion, conversion from a vegetative to a flowering state, flower development, fruit development, cessation of leaf production (bud dormancy) and leaf senescence and abscission; for all these processes, the duration, wavelength and intensity of the light are crucial [14]. Some of this knowledge is commonly used by the greenhouse industry to promote

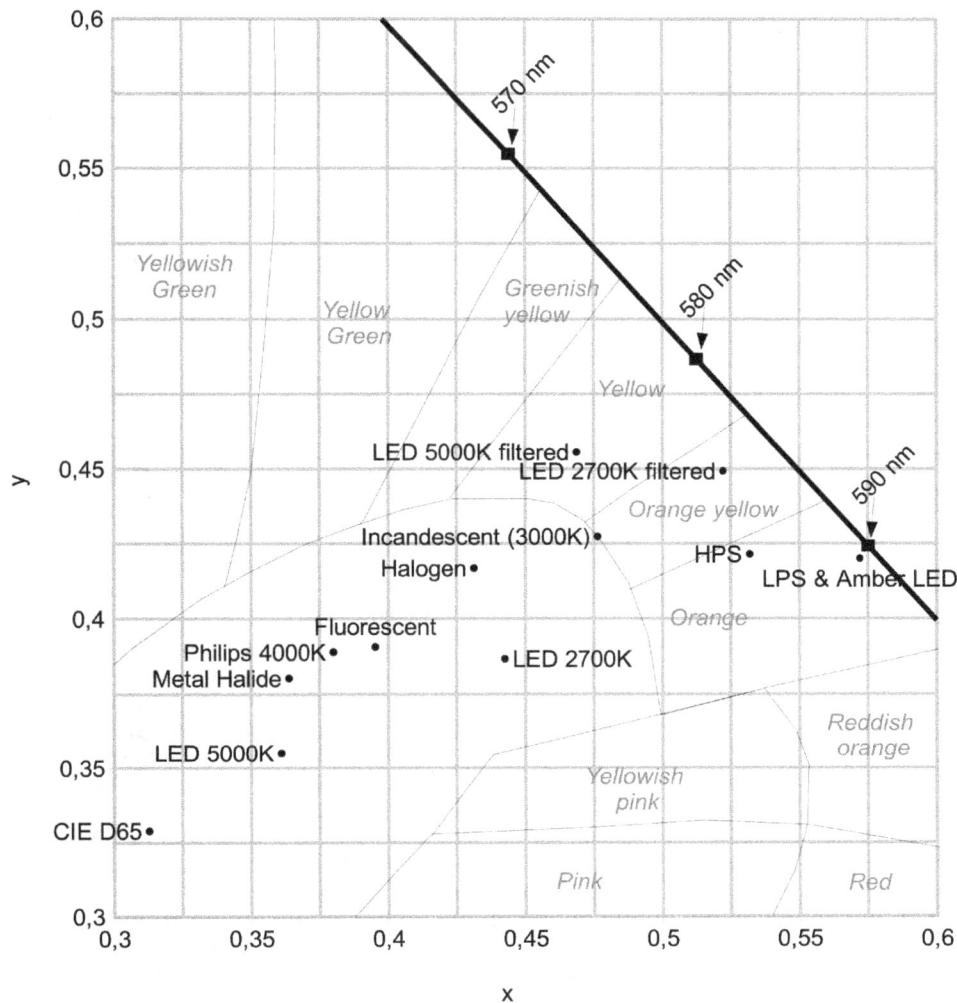

Figure 1. x and y chromaticity coordinates [52] for each lamp listed in Table 2. On that figure, the spectral locus, which is the line for monochromatic light, is shown by the thick black line. Thin black lines indicate color zones. Black squares show monochromatic values, while small black circles are lamps.

flowering and growth, and to stimulate fruit, vegetable and plant production. High intensity discharge (HID) lamps are popular for a large area of lighting applications in horticulture.

Artificial light is sometimes beneficial and sometimes detrimental to human health. Light has a profound impact on circadian systems and physiological functions. Because of this major impact and the growth in ALAN, there is evidence for a strong link between exposure to ALAN and disease. ALAN may be associated with an increased risk of breast cancer [15], [16], [17], [18], [19], [20], prostate and colorectal cancer [21], [22], and may also cause obesity [23], [24], diabetes [25] and depression [26]. Kloog et al. [20] report that women have 30–50% higher risk of breast cancer in the countries with the highest exposure to ALAN compared to those with the lowest exposure. One efficient way to monitor the effect of light on circadian systems is to evaluate melatonin suppression in biofluids. Melatonin, also called the sleeping hormone, is produced by the pineal gland and is released mainly during the night. Light induces a decrease in pineal melatonin hormone production and secretion; and it may also induce a phase shift in daily rhythms [27]. The discovery of a novel non-visual photoreceptor, with the photopigment melanopsin acting on our

circadian function [28], [29], has changed the understanding of that mechanism. Melanopsin responds to light by decreasing pineal melatonin hormone production, with a maximum spectral sensitivity at blue wavelengths. Two light variables, intensity and wavelength [30], are responsible for the suppression of melatonin, and an illuminance of only 1.5 lux may disrupt circadian rhythms [31]. According to our measurements, taken in Sherbrooke (Canada), an illuminance of ~2 lux is frequently encountered in urban bedrooms. Moreover, the human circadian system responds to millisecond flashes of light, delaying pineal melatonin production [32].

Indoor artificial lighting can be beneficial to humans. For example, light therapy is commonly prescribed by doctors against seasonal affective disorder [33]. It has also been shown that blue-enriched light during the day increases performance, vigilance and sleeping patterns [34]. Exposure to compact fluorescent light (CFL) at a correlated color temperature (CCT) of 6500 K (blue-enriched light) induced greater melatonin suppression, together with enhanced subjective alertness, well-being and visual comfort [35]. These results suggest that the selection of CFL with different CCT has a significant impact on circadian physiology and cognitive performance at home and at work. Finally, the availability of electronic devices with backlit screens, which are often used at night, is rapidly increasing throughout the world. In comparison with backlit liquid crystal display (LCD), evening exposure to a light emitting diode (LED)-backlit computer screen (blue-enriched light) resulted in attenuated salivary melatonin and sleepiness levels, with a concomitant increase in cognitive performance associated with sustained attention and with working and declarative memory [36]. With the progress of LED technologies, it will be important to build electronic device screens in accordance with the circadian cycle [36].

Since the 1960s, outdoor artificial lighting has progressively changed from incandescent-bulbs (orange-yellow color, see Fig. 1) to a high pressure sodium form (HPS, orange) and more recently to LED (blue-enriched white light). The indoor artificial lighting that is most used is cool-white fluorescent lighting (FL) for public areas and incandescent, halogen and CFL bulbs for private areas with a large span of CCT. In lighting engineering, lower CCT (CCT<5000 K) is often called warm white light while high CCT (CCT>5000 K) is called cool white light. The use of artificially generated full spectrum daylight for human activities is not common, but they are used in the field of light therapy, greenhouse lighting and for pet shops. This kind of light is reputed to mimic natural sunlight, but it is not exact in this, as will be discussed later.

Reduction in star visibility, one of the best known impacts of outdoor ALAN, has been identified by astronomers. A first abatement for the protection of night sky quality over professional astronomical observatories was adopted in 1958 in the vicinity of Flagstaff, AZ, USA. Astronomers have always preferred the use of low pressure sodium (LPS) lamps. This technology shows quasi-monochromatic spectral power distribution (SPD) in the orange part of the visible spectrum. This kind of SPD is easy to filter out using optical filters and its color is not very efficient in terms of atmospheric scattering. In fact, when light travels into the atmosphere, it is partly scattered by molecules and aerosols and can be redirected toward an observer looking at the stars. This astronomical light pollution is then competing with the faint light coming from the universe. According to the Mie and Rayleigh scattering theories, blue light is scattered more efficiently than

other colors (e.g. blue scattering is about one order of magnitude more efficient than red scattering).

The first main difference between daytime and nighttime natural light is the intensity level, since sunlight, starlight and moonlight are not so different in terms of their relative SPD. Sunlight is around five to nine orders of magnitude brighter than typical ancestral nighttime illumination (natural or human-made). In modern times, nighttime artificial illumination in lit areas is typically four orders of magnitude higher than illumination from a natural starry sky without moonlight, and around one to two orders of magnitude higher in comparison to full moonlight illumination. Light from wood/oil burning, which was the most intense source of human-made lighting for centuries, contains a very low blue light contribution in comparison to sunlight. Nowadays, human-made light shows important differences in comparison to wood/oil burning and the Sun's SPD. The most significant nighttime natural lights, such as those from the stars, moon and wood/oil burning, can be described as a quasi black body spectrum showing a predominant continuum SPD, while many modern artificial lights include the addition of discrete spectral lines with a very low continuum contribution. Natural light contains all wavelengths of the visible spectrum while some artificial lights contain only a subset or are dominated by a few spectral lines. Artificially generated full spectrum daylight lamps, which were designed to approximate to sunlight, contain essentially all the wavelengths of the visible spectrum but with relatively important discrete spectral lines superimposed to a continuum. Tungsten incandescent technology SPDs (halogen tungsten incandescent and standard tungsten incandescent) are similar to natural light in terms of the relative importance of the continuum, but with a lower CCT compared to the sun, moon and the brightest stars. In other words, tungsten-based SPD shows a higher relative red contribution compared to the sun. Even if, with passing time, humans are increasing their light spill into the environment, not enough consideration has been given to the development of lighting devices that have an SPD comparable to natural light (either daytime or nighttime).

In the field of lighting engineering, the parameters used to describe light spectra are very crude and do not characterize SPD in detail. As an example, CCT and the color rendering index

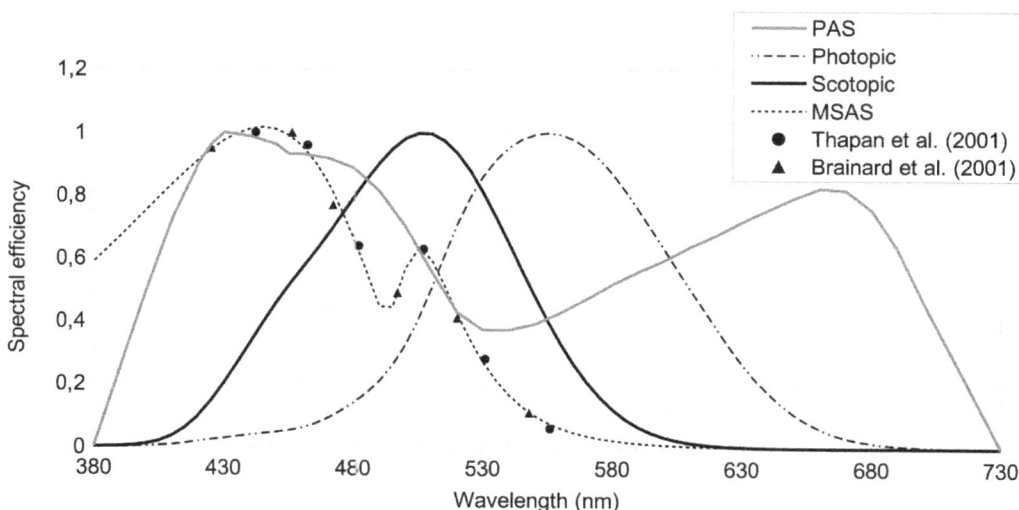

Figure 2. Various spectral sensitivities: human eye photopic [44] and scotopic [41] spectral sensitivity; MSAS: human melatonin suppression action spectrum; PAS: photosynthesis action spectrum.

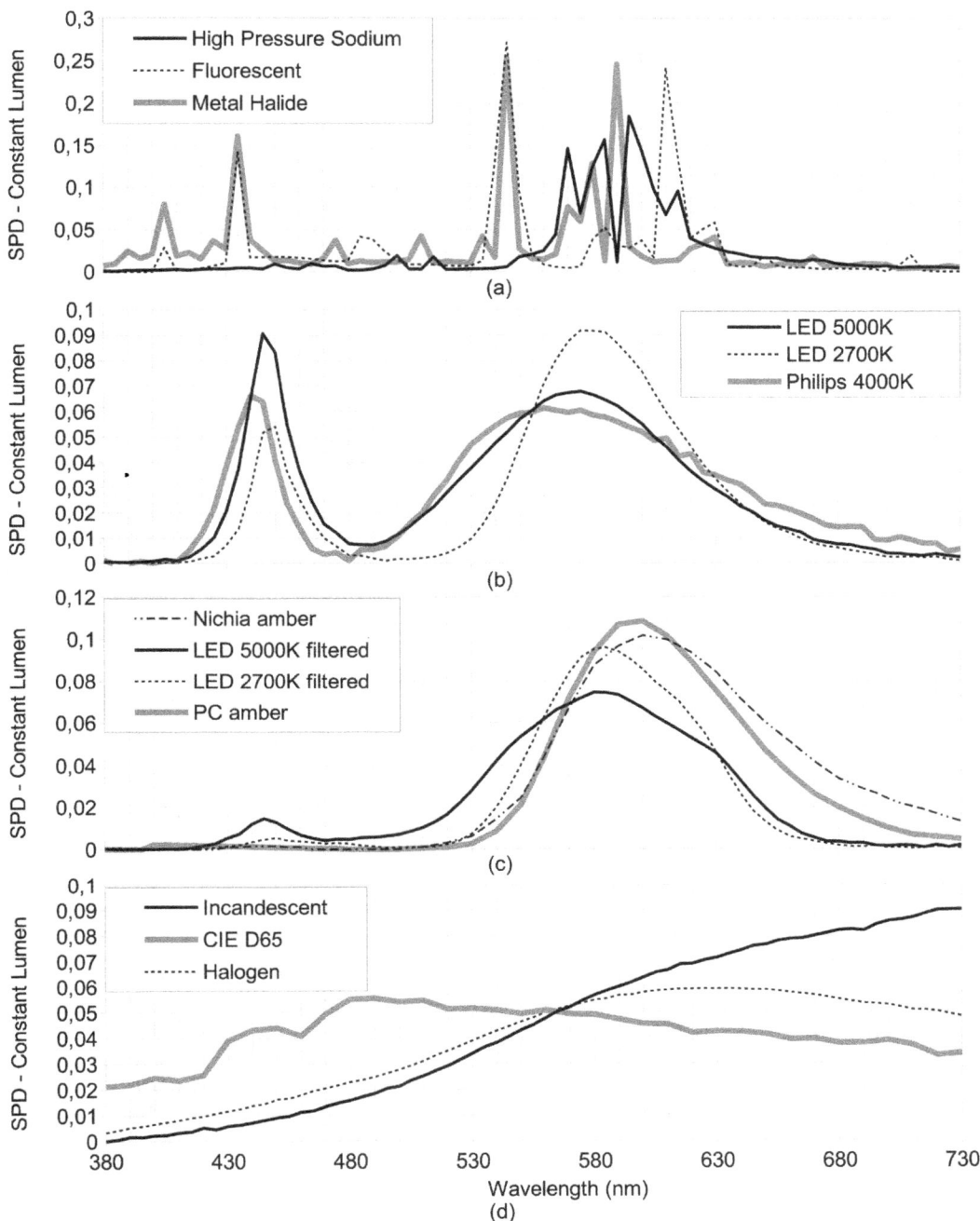

Figure 3. Constant lumen spectral power distributions. A subset of the spectrum used in this paper is shown here. Pane (a) shows HID lamps, pane (b) illustrates white LEDs, (c) shows low blue content broadband LEDs and (d) shows thermalized spectra such as for halogen and incandescent lamps and our reference D65 illuminant from the Commission Internationale de l'Éclairage (CIE). D65 illuminant corresponds to a midday Sun in Western/Northern Europe with CCT~6500 K. LED filtering was performed using the *Equilib* optical filter commercialized by Ledtech International.

(CRI) are often used, but both parameters refer to a black body style of spectrum. CCT is the black body temperature that gives the same color sensation to the average human eye. CRI gives information concerning to what extent a lamp spectrum can be compared to a black body SPD in terms of its color rendering. A CRI value of 100 means that the lamp SPD is a perfect black body spectrum. CRI can be evaluated by comparing the color rendering of the source with a black body of the same CCT. This is why tungsten halogen shows CRI~100, which is the optimal value of CRI. As stated above, most modern artificial light SPDs are very far from black body values. To obtain a good approximation of human eye color perception under sunlight, we need a CRI of 100 and a CCT of around 5800 K. One technology that is not far from this ideal target is the ceramic metal halide lamp (CCT~5400 and CRI~96), if we ignore spectral lines superimposed on the continuum spectrum.

Table 1. Parameters of the lognormal components of the MSAS fit.

Log normal #1		Log normal #2	
h_1	1.017	h_2	0.5239
λ_1	444.7	λ_2	509.3
w_1	111.9	w_2	30.18
b_1	−0.5785	b_2	0.6666

Recently, LED technology from the field of solid state physics has been introduced to the lighting industry. LED emits a quasi-monochromatic SPD with a typical full width at half maximum (FWHM) of the order of 30 nm and a nominal wavelength depending on the material used to make the diode junction. Nowadays, the most efficient LEDs are the blue ones with a nominal wavelength ranging from 440 nm to 480 nm. Such light have of course a CCT and CRI that are very far from natural solar radiation. To overcome this drawback, a phosphorous material is placed between the blue LED and the observer. The role of that phosphorous material is to expand the narrow SPD of the blue LED into a broad band SPD. The resultant light is almost white but, when observing it with a spectrometer, one can clearly see that the white LED SPD can be described as including the addition of a broadband yellowish SPD with a significant remaining narrow band blue SPD.

The impact of artificial light on photosynthesis, on star visibility and on melatonin suppression is closely related to the concordance of the given spectral sensitivity of the phenomena being considered with the spectrum of the light. As an example, the photosynthesis action spectrum (PAS), or $P(\lambda)$, which represents the efficiency of each wavelength in inducing photosynthesis for averaged vegetable species, shows two peaks: one in the blue region at around 450 nm and the other in the red part of the spectrum at around 660 nm (see Fig. 2). Basically this means that an artificial light having a significant emission around these wavelengths is more likely to interfere with photosynthesis, especially during the night when there is no solar light. White LEDs are somewhat problematic for nighttime photosynthesis because their blue peak fits almost perfectly with the blue sensitivity peak of PAS.

The same kind of analysis can be made to estimate: 1- the impact of ALAN on star visibility by considering the low illumination eye spectral sensitivity (scotopic response), and 2- the potential impact of artificial light on circadian cycle disruption using the melatonin suppression action spectrum (MSAS).

In this paper, we will introduce three new parameters or indices to characterize a light spectrum in terms of its potential impact on respective biological processes: 1- melatonin suppression, 2- photosynthesis, and 3- scotopic vision. Our indices are intended to separate SPD from other factors acting on the given biological process. As an example, a minimum illumination is required to induce circadian cycle disruption, but our new index ignores this minimum illumination level. By using such an index, we will therefore have to assume that all other variables known to have an impact on the given biological process are favorable. In this way, the indices only deal with the potential impact of SPD shape. After defining the indices, we apply them to a variety of existing lighting technologies. We finally calculate the impact of atmospheric light scattering on indices values as a function of the distance between the light source and the observer, with and without cloud cover.

All comparisons are made considering a constant lumen output for each lamp.

Materials and Methods

Lamp Spectral Power Distribution Data

This experiment was conducted on the basis of a Lamp SPD Database (LSPDD) available online [37] and maintained by our research group. This dataset aims to provide independent information about the spectral characteristics of commercial lamp products. Among other information, we distribute SPDs in ASCII text format, allowing any other researchers to use this data for their own research. LSPDD is released under Creative Commons BY-NC-ND license. Some examples of SPDs from this database are shown in Figure 3.

Spectral Sensitivities

Various spectral sensitivities have been used for this study. The first, photopic spectral sensitivity $V(\lambda)$, was used in order to normalize the spectrum (see section on constant lumen normalization). The three other curves, which are 1- the melatonin suppression action spectrum (MSAS), 2- the photosynthesis action spectrum (PAS) and 3- the scotopic spectral sensitivity $V'(\lambda)$, were used to calculate the spectral impact on related biological processes. In the following sections, we will detail the choice of spectral sensitivity curves with respect to each process.

Melatonin suppression action spectrum. In 2001, data concerning the melatonin suppression action spectrum in the spectral range from 425 nm to 560 nm were published [28], [29] (triangles and circles on Fig. 2). MSAS represents to what extent each wavelength is efficient in suppressing melatonin production. Unfortunately, the dataset provided is quite small (only 12 data points) and contains no information about MSAS values in the very deep blue and UV-A regions (no data below 425 nm). In order to generate MSAS all over our spectral range, from 380 nm to 730 nm, we tried to fit the data using a combination of two log normal curves (see Eq. 3). We chose a 2 lognormal curves to capture a breakdown in the slope from the point at 507 nm, assuming this data to be reliable. Previous authors (e.g. [29]) used a vitamin A1 retinaldehyde photo pigment template to fit the data. They obtained a correlation of $R^2 = 0.91$ with $\lambda_{max} = 464$nm. Our 2 lognormal fit is shown by the dotted curve on Figure 2. The correlation of that fit was very good with $R^2 = 0.997$ and we obtained $\lambda_{max} = 445$nm. The fitted function and parameters are given in Table 1 and in Equation 1 (wavelength λ is in nm):

$$MSAS(\lambda) = h_1 \exp\left(-\ln(2)\left(\frac{\ln\left(1+2b_1\frac{\lambda-\lambda_{max1}}{w_1}\right)}{b_1}\right)^2\right)$$
$$+ h_2 \exp\left(-\ln(2)\left(\frac{\ln\left(1+2b_2\frac{\lambda-\lambda_{max2}}{w_2}\right)}{b_2}\right)^2\right) \tag{1}$$

where b represents asymmetry parameters, w the profile widths, λ_{max} the maximum wavelengths and h the function heights.

Since no constraint was exerted on the fit for wavelengths lower than 425 nm, we were unable to determine if the fitted MSAS remained good in this part of the spectrum, but we assumed that to be the case.

Photosynthesis action spectrum. Photosynthesis is a process in which plants, algae and some bacteria transform solar light into organic compounds. To infer the possible impact of artificial

Table 2. Photometric characteristics of lamps under direct lighting.

	x	y	CCT	CRI	MSI	IPI	SLI	Typ. Lumen/W
LPS	0.58	0.42	1720	−47	0.017	0.380	0.088	100–200
HPS	0.53	0.42	2010	19	0.118	0.509	0.231	85
Metal Halide	0.36	0.38	4500	48	0.624	0.640	0.577	120
Halogen	0.43	0.42	3200	92	0.377	0.829	0.597	24
Incandescent	0.48	0.43	2600	93	0.255	0.923	0.490	15
Fluorescent T8 cool-white	0.40	0.39	3730	82	0.435	0.606	0.608	90
LED 5000 K	0.36	0.36	4440	61	0.542	0.636	0.617	100
Philips LED 4000 K	0.38	0.39	4100	63	0.452	0.623	0.563	75
LED 2700 K	0.44	0.39	2760	37	0.285	0.541	0.359	69
Nichia Amber	0.57	0.42	1720	47	0.043	0.682	0.170	53
Lumiled PC Amber	0.57	0.42	1720	36	0.046	0.610	0.154	59
LED 5000 K Filtered	0.47	0.46	2910	55	0.172	0.470	0.380	82
LED 2700 K Filtered	0.52	0.45	2260	34	0.077	0.443	0.230	61
CIE D65	0.31	0.33	6504	100	1.000	1.000	1.000	–

x and y are the chromaticity coordinates [52], CCT is the Correlated Color Temperature, CRI is the Color Rendering Index, MSI is the Melatonin Suppression Index, IPI is the Induced Photosynthesis Index and SLI is the Star Light Index.

lighting on photosynthesis, we needed to define the action spectrum to be used. Many authors have shown that action spectra change from one species to another (e.g. [38]). Chlorophyll a and chlorophyll b are the two most abundant pigments in plants but we know of four other structures of chlorophyll molecules that are adapted to various environmental characteristics. As an

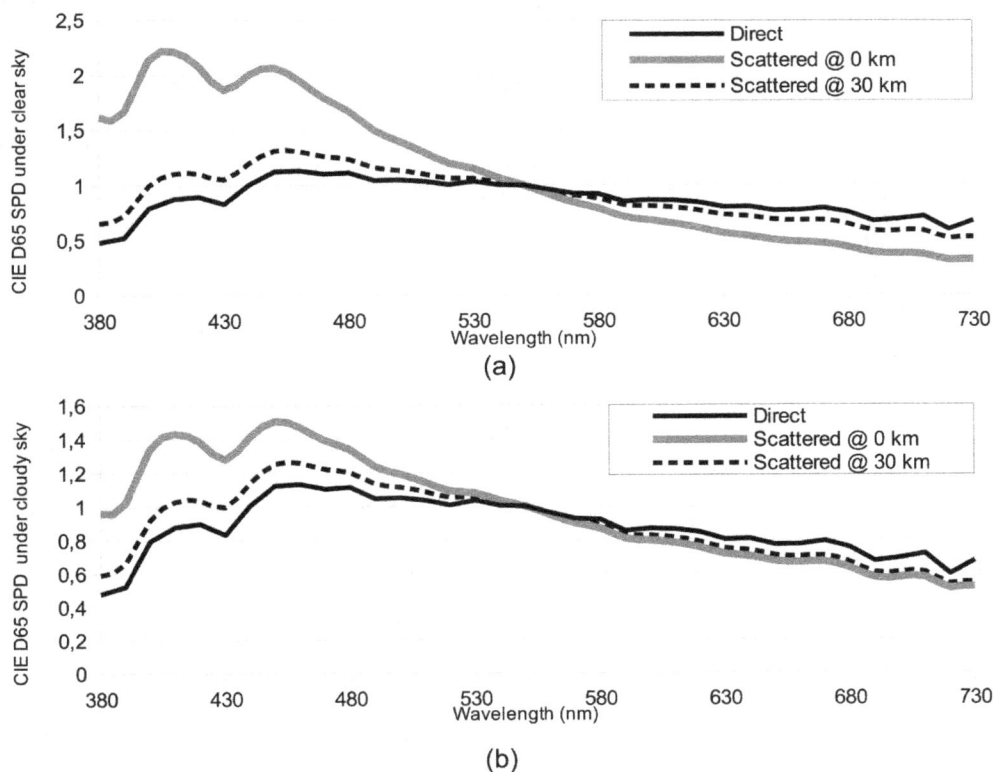

Figure 4. Spectral impact of the atmospheric transfer function as a function of the distance for the CIE D65 illuminant without cloud cover in (a) and with cloud cover in (b). Direct SPD relates to the original SPD before any atmospheric transformation. Compared to direct SPD, scattered SPD is bluer at short distances and redder at long distances.

Table 3. Photometric characteristics of lamps after scattering into the atmosphere.

Distance (km)	Clear sky						Cloudy sky					
	MSI		IPI		SLI		MSI		IPI		SLI	
	0	30	0	30	0	30	0	30	0	30	0	30
LPS	0.008	0.014	0.241	0.349	0.053	0.078	0.012	0.015	0.317	0.359	0.072	0.081
HPS	0.099	0.113	0.342	0.465	0.188	0.218	0.107	0.114	0.430	0.477	0.210	0.221
Metal Halide	0.663	0.634	0.678	0.651	0.547	0.567	0.647	0.631	0.658	0.648	0.563	0.570
Halogen	0.329	0.364	0.592	0.761	0.544	0.583	0.350	0.369	0.718	0.779	0.571	0.587
Incandescent	0.203	0.241	0.572	0.820	0.431	0.474	0.225	0.245	0.759	0.846	0.461	0.478
Fluorescent T8 cool-white	0.403	0.427	0.549	0.594	0.561	0.595	0.417	0.430	0.579	0.598	0.585	0.598
LED 5000 K	0.500	0.533	0.601	0.631	0.598	0.610	0.517	0.536	0.618	0.633	0.608	0.611
Philips LED 4000 K	0.422	0.445	0.555	0.607	0.521	0.550	0.435	0.448	0.591	0.611	0.543	0.553
LED 2700 K	0.261	0.280	0.439	0.516	0.330	0.349	0.271	0.282	0.492	0.524	0.346	0.352
Nichia Amber	0.026	0.038	0.384	0.598	0.12	0.156	0.034	0.040	0.543	0.620	0.146	0.159
Lumiled PC Amber	0.034	0.042	0.357	0.540	0.107	0.140	0.039	0.043	0.491	0.558	0.131	0.144
LED 5000 K Filtered	0.136	0.162	0.336	0.437	0.315	0.362	0.151	0.165	0.406	0.446	0.349	0.367
LED 2700 K Filtered	0.056	0.071	0.288	0.403	0.173	0.214	0.065	0.073	0.370	0.415	0.203	0.218
CIE D65	1.000	1.000	1.000	1.000	1.000	1.000	1.000	1.000	1.000	1.000	1.000	1.000

example, chlorophyll d is found in cyanobacteria and is more sensitive to far-red light ($\sim$710 nm). This particularity is useful, especially in a scattering medium, such as moderately deep water, where shorter wavelengths are removed by scattering along the light path. Chlorophyll a is common to all vegetable species, while chlorophyll b is mostly found in plants. These two pigments absorb more efficiently the blue and red part of the spectrum. This is in turn the reason why plants generally look green. However, along with these two important pigments, there exist some other accessory pigments, such carotenes, that can be involved to a lower extent in the process of photosynthesis. The action spectrum of carotenes shows a blue absorption peak many times larger than the red peak, leading to an orange tint. Depending on the plant species, the relative importance of each pigment differs and the final action spectrum can therefore be seen as a combination of each individual action spectrum of the various pigments. We decided to use a global action spectrum, which will fit an averaged vegetation type. Such an action spectrum was adopted by the German Institute for Standardization and numbered DIN 5031-10 [39]. This action spectrum is shown by the gray line in Figure 2.

Scotopic spectral sensitivity. Scotopic vision occurs when luminance is below 10^{-3} cd/m^2 [40]. This is clearly the case for stargazing conditions, and we decided to use scotopic spectral sensitivity as a basis for estimating star visibility. We retained the scotopic vision curve adopted by the commission international de l'éclairage (CIE) in 1951 [41]. This curve is shown in Figure 2: one important feature to notice on this curve is that the sensitivity peak is displaced towards the blue compared to the photopic curve. Basically, this means that for the same lumen, a lamp with a greater contribution in the blue part of the spectrum should generate more disturbance to stargazing and nocturnal vision of humans in general.

Constant Lumen Normalization

In this work, one key element of spectral data processing was the normalization of all SPDs to a constant lumen output. We chose this normalization because most lamps are intended to illuminate the environment in order to favor human activity. By normalizing the entire lamp SPD with the photopic spectral sensitivity, all the resultant SPDs produce the same stimuli to the human eye (ignoring color variations from one lamp to another). We assumed that lighting is designed in such a way that the desired luminance (L_v) reaching the eye is higher or equal to the minimum luminance required for the photopic vision regime of the human eye. The photopic vision luminance threshold can be set to 0.6 cd.m^{-2} [42]. In fact, when using a unified system of photometry [43] based on a combination of photopic and scotopic luminances, we obtained pure photopic luminance when $L_v = 0.6$ cd.m^{-2} or higher. Assuming a ground lambertian surface with a constant reflectance ρ, we were able to calculate the equivalent illuminance in lux (E_v) according to Equation 2:

$$E_v = \frac{L_v \pi}{\rho} \qquad (2)$$

Assuming that luminance is higher or equal to 0.6 cd.m^{-2} and that for a typical summer city ground reflectance was 0.08 (based on asphalt reflectance from the NASA's ASTER spectral library), at least $E_v \approx 23$ lux is required to be within the photopic vision regime. For a snow-covered surface this minimal illuminance is $E_v \approx 2$ lux (snow reflectance of 0.98). This range of illuminance is representative of what is found on typical roadways.

The normalization method given by Equation 3 below defines the constant lumen SPD ($\Phi_n(\lambda)$)

$$\Phi_n(\lambda) = \frac{\Phi(\lambda)}{\int\limits_{380nm}^{730nm} \Phi(\lambda) V(\lambda) d\lambda} \qquad (3)$$

where $\Phi(\lambda)$ is the unnormalized SPD of the lamp and $V(\lambda)$ is the photopic spectral sensitivity function [44]. This function was derived from corrections to a previous revision of CIE 1931 -

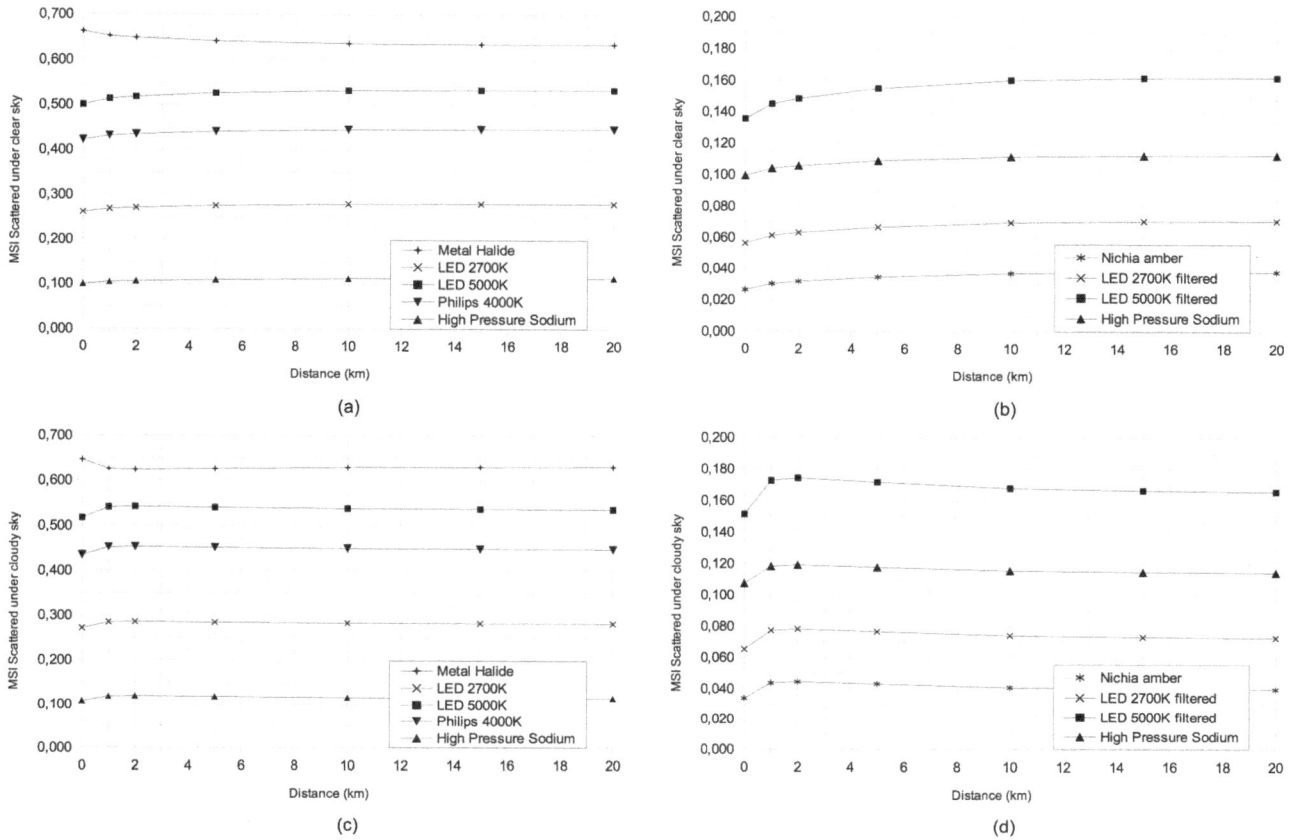

Figure 5. Atmospheric scattered Melatonin Suppression Index dependency, with distance between lamp and observer, (a) and (c) for usual street lamps and (b) and (d) for new technologies that reduce light pollution. The two upper plots (a) and (b) are for clear sky conditions, while (c) and (d) are for cloudy sky conditions.

2 deg color matching functions [45]. $V(\lambda)$ is shown on Figure 2. To calculate the constant lumen normalization, the integration spectral range (380–730 nm) was chosen to include significant spectral sensitivities of the biological processes covered by the present work. More specifically, we used the limits given by the PAS.

Impact of Atmospheric Optics

Light sources usually have complex spectra and emit radiation in all directions. A proportion of the electromagnetic energy that is propagated upwards undergoes scattering and absorption processes before it is detected at the ground as diffuse radiation. It should be noted that the optical behavior of the diffuse radiation depends strongly on both the original spectra of the light source and its radiant intensity distribution as a function of the zenith angle z' and azimuth angle A'. If $I_0(z', A', \lambda)$ is the spectral radiance of an elementary surface, than the amount of electromagnetic energy delivered to and removed by an elementary atmospheric volume dv is

$$d^2 E(h, z', A', \lambda) = I_0(z', A', \lambda) \, k_{ext}(h, \lambda)$$

$$\exp\left\{ -\int_0^h \frac{k_{ext}(h', \lambda)}{\cos z'(h')} dh' \right\} dv \, dt \, d\Omega' \quad (4)$$

where k_{ext} is the extinction coefficient of an atmospheric environment, dt is a time interval, λ is the wavelength, and $d\Omega'$ denotes an elementary solid angle in which the radiance of the ground-based light source is received at altitude h. In Equation (4), we assumed that z' depends on the altitude due to vertical stratification of the air refractive index. For zenith angles smaller than 80 degrees and/or for short optical paths such a dependency can be neglected, and thus the factor $1/\cos z'$ can be placed before the integral (in curly brackets). In all other cases a corresponding optical air mass should be used instead of $1/\cos z'$. The exponential function in Eq. (4) is also called the transmission function:

$$T(h1, h2, z') = \exp\left\{ -\int_{h1}^{h2} \frac{k_{ext}(h', \lambda)}{\cos z'(h')} dh' \right\} \quad (5)$$

and typically characterizes the attenuation of electromagnetic radiation along the beam path. It is evident that any increased atmospheric turbidity implies a more rapid intensity decay in the atmospheric environment. However, elevated turbidity conditions are closely related to the increased number of scattering domains, which also make the scattering processes more efficient. Since the elementary volume dv can collect light from all directions, the amount of energy scattered within the elementary solid angle

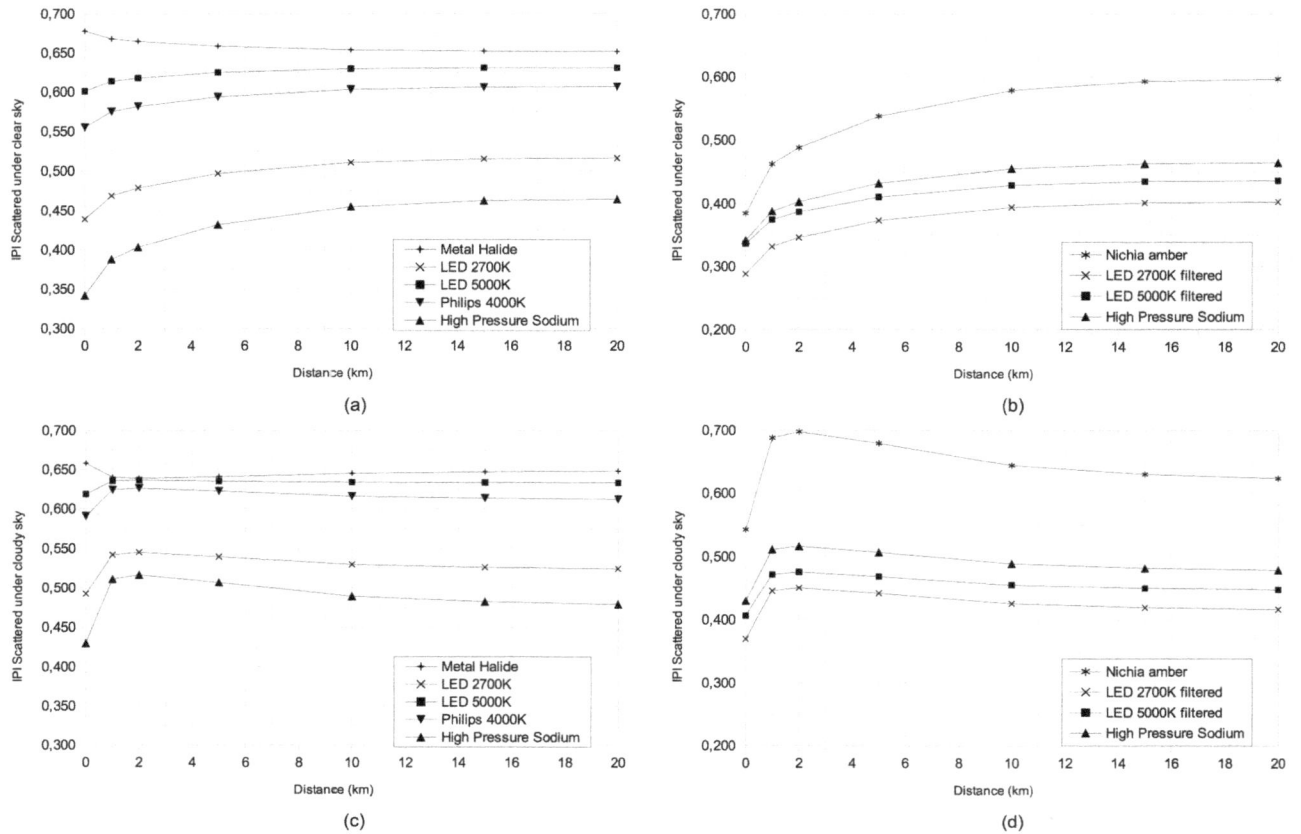

Figure 6. Same as Figure 5 but for Induced Photosynthesis Index.

$d\Omega = \sin z\, dz\, dA$ will be:

$$d^2 E^S(h,z,A,\lambda)$$

$$= \varpi k_{ext}(h,\lambda)\, dv\, dt\, d\Omega \int\limits_{4\pi} I_0(z',A',\lambda)$$

$$(6)$$

$$\exp\left\{ -\int\limits_0^h \frac{k_{ext}(h',\lambda)}{\cos z'(h')}dh' \right\} \frac{p(\theta,\lambda)}{4\pi}d\Omega'$$

where ϖ is the so-called atmospheric single scattering albedo [46], $p(\theta,\lambda)$ is the spectral scattering phase function, and z,A are the observational zenith and azimuth angles, respectively. The radiative energy scattered toward a hypothetical observer will be then given as:

$$dI^S(z,A,\lambda) = \frac{d^2 E^S(h,z,A,\lambda)}{dt\, d\Omega}\frac{dh}{dv}\frac{1}{\cos z} \qquad (7)$$

since the photometry law dictates that the elementary amount of electromagnetic energy crossing an elementary surface $d\sigma$ is by definition $d^2 E^S = dI^S\, d\sigma\, d\lambda\, d\Omega\, dt$. The scattered beam we considered in Equation (6) crosses the elementary volume $dv = d\sigma\, dh$ under the inclination angle z, so the projection area is $d\sigma \cos z$ instead of $d\sigma$ (consult Eq. 7). The total amount of scattered radiation received at a measuring point is obtained as an integral

product of the spectral transmission function and elementary radiance, so:

$$I^S(z,A,\lambda) = \int\limits_0^H T(0,h,z)\, dI^S(z,A,\lambda) \qquad (8)$$

where H approaches infinity in a cloudless atmosphere. Under overcast conditions, H appears to be the altitude of the cloud base. The contribution of clouds to the spectral radiance is computed in the form of an additional term [47] (not shown here). One of most important parameters affecting the spectral radiance under overcast conditions is the spectral reflectance of the cloud. Green et al. [48] have shown that the average reflectance of a cloud is about 0.46, and we have used the same value in our computations.

In principle, k_{ext}, introduced in Equation (4), is a sum over all the extinction coefficients of the atmospheric constituents. However, we considered only aerosol and molecular extinction coefficients since these are dominant in a cloudless or undercloud atmosphere. Water vapor absorption was not considered because it becomes important only for wavelengths larger that 755 nm; in this work, we do not use wavelengths larger than 730 nm.

The optical thickness τ is used in the radiative transfer computations, rather than the extinction coefficient k_{ext}. The relationship between τ and k_{ext} can be written as follows:

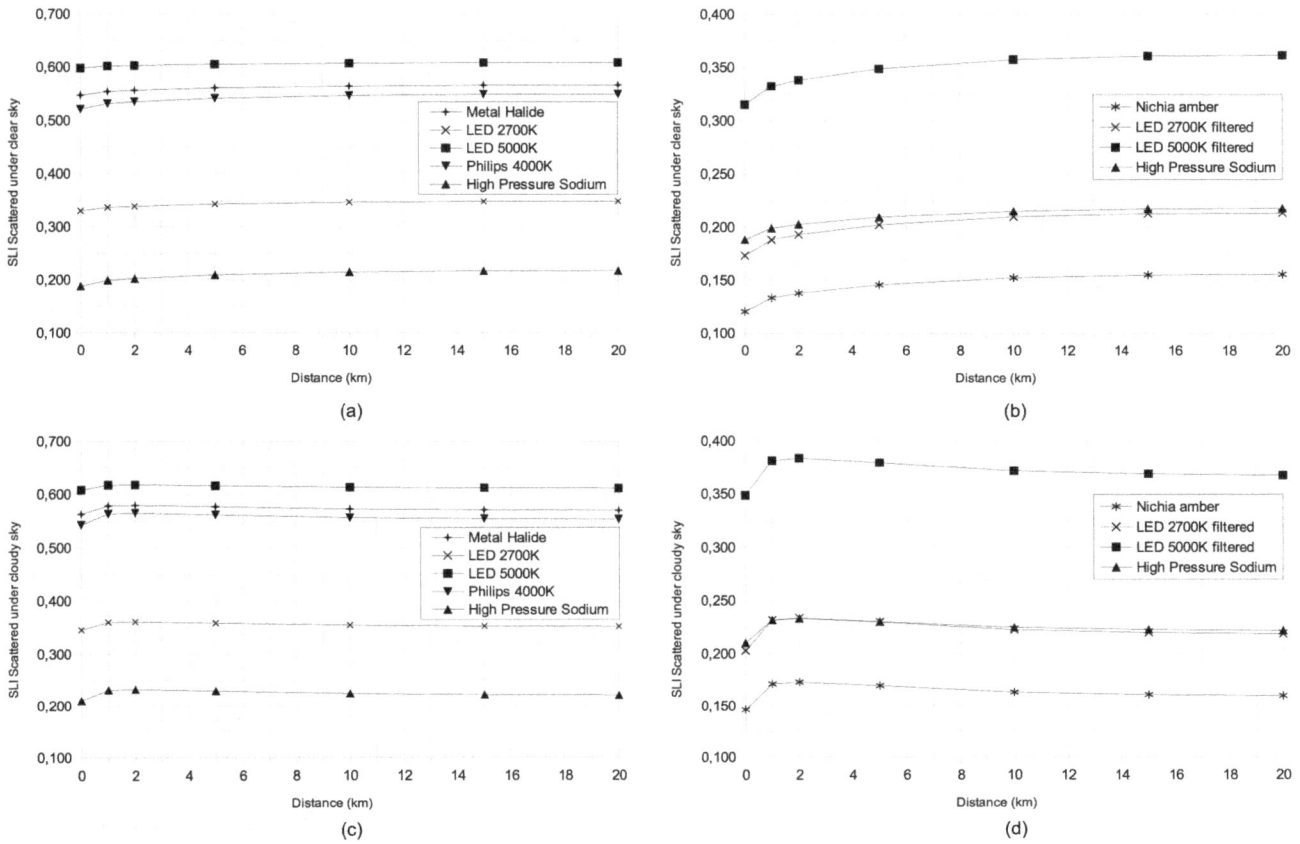

Figure 7. Same as Figure 5 but for Star Light Index.

$$\tau(h,\lambda) = \int_{h}^{\infty} k_{ext}(\xi,\lambda)\,d\xi \qquad (9)$$

Here, ξ is an integration variable that characterizes the altitude about ground level. It is commonly accepted that Rayleigh's optical thickness for the molecular atmosphere decreases exponentially with increasing altitude, h, i.e.

$$\tau_R(h,\lambda) = \tau_R(0,\lambda)\exp\left\{-\frac{h}{H_R}\right\} \qquad (10)$$

where $H_R \cong 8\ km$ is the so-called scale height of the molecular atmosphere: it is the altitude up to which a homogeneous molecular atmosphere would extend. For the molecular optical thickness at ground level we used the following formula [49]:

$$\tau_R(\lambda) = 0.008735\left(\frac{\lambda}{1000}\right)^{-4.08} \qquad (11)$$

with the wavelength measured in nm. The concept of scale height is also frequently applied for aerosols. However, the aerosol scale height H_A is much lower than that for the molecular atmosphere. For aerosols we can write:

$$\tau_A(h,\lambda) = \tau_A(0,\lambda)\exp\left\{-\frac{h}{H_A}\right\} \qquad (12)$$

where we employed $H_R \cong 1.5$km. This value correlates well with the results obtained e.g. by Horvath et al. [50]. For a typical clear atmosphere, we assumed values of $\tau_A(\lambda) = 0.2$ at $\lambda = 500$ nm with an Ångström coefficient of 1.0, because a set of experiments conducted in urban and suburban regions have shown that the Ångström exponent is close to unity (e.g. [51], [52]). The total optical thickness is computed as $\tau(h,\lambda) = \tau_R(h,\lambda) + \tau_A(h,\lambda)$.

The scattering angle θ used in Equation (6) is determined from spherical geometry in the form:

$$\cos\theta = \cos z\cos z' + \sin z\sin z'\cos(A - A') \qquad (13)$$

Once the scattering angle has been determined in this way, the scattering pattern for a turbid molecular-aerosol atmosphere can be expressed as follows:

$$\varpi k_{ext}(h,\lambda)\frac{p(\theta,\lambda)}{4\pi}$$

$$= \frac{1}{4\pi}[k_{ext,R}(h,\lambda)p_R(\theta) + \varpi_A k_{ext,A}(h,\lambda)p_A(\theta,\lambda)] \tag{14}$$

The expression given at the right hand side of Equation (14) has to be used in Equation (6). The molecular scattering phase function introduced with Equation (14) reads:

$$p_R(\theta) = \frac{3}{4}(1+\cos^2\theta) \tag{15}$$

while the aerosol scattering phase function is determined based on Mie theory:

$$p_A(\theta,\lambda) = \frac{\lambda^2}{\pi}\frac{1}{k_{sca,A}(\lambda)}$$
$$\int_{r=0}^{\infty}\left\{\frac{|S_1(\theta,\lambda,r)|^2 + |S_2(\theta,\lambda,r)|^2}{2}\right\}f(r)dr \tag{16}$$

In this equation, $k_{sca,A} = \varpi_A k_{ext,A}$, with ϖ_A being the aerosol single scattering albedo. The dimensionless Mie functions $S_1(\theta,\lambda,r)$ and $S_2(\theta,\lambda,r)$ are weighted by a particle size distribution, $f(r)$, where r is the particle radius. It is expected that both $k_{sca,A}$ and f are altitude-dependent. The computations based on Equation (16) could be CPU-intensive if used in vast modeling. Therefore, some convenient approximations, such as the Henyey–Greenstein function [53],

$$p_A(\theta,\lambda) = \frac{1-g^2(\lambda)}{[1+g^2(\lambda)-2g(\lambda)\cos\theta]^{3/2}} \tag{17}$$

are routinely applied in the use of radiative transfer tools. Although this function has no theoretical foundation, it can mimic the experimental scattering patterns quite well. The most significant advantage of this function is its analytical formulation. The asymmetry parameter $g(\lambda)$ used in Equation (17) is the cosine-weighted integral of the scattering phase function. In conjunction with the single-scattering albedo $\varpi_A(\lambda)$, the asymmetry parameter $g(\lambda)$ is the most important variable that is considered in any aerosol-related radiative problem. In principle, $\varpi_A(\lambda)$ can vary from 0 to 1, while the acceptable theoretical values of $g(\lambda)$ range from -1 to $+1$. Nevertheless, it has been well documented experimentally that typical values of $\varpi_A(\lambda)$ and $g(\lambda)$ range from 0.8 to 0.9 in urban and suburban regions [50], [54]. We have used $g = 0.8$ and $\varpi_A = 0.9$ for the visible spectrum.

We considered that photon removal from the light fixture to the observer can be calculated by assuming that direct light does not reach the observer because of obstacles such as trees, buildings or topography, and that the main source of illumination at remote sites is derived from the light dome. This assumption can of course be weak for an astronomical site located on top of a mountain from where a light fixture can be seen in a direct line of sight.

Reference Spectrum

In order to exclude any other parameter but wavelength, we decided to perform all calculations in reference to a standard SPD ($\Phi_{n(D65)}$). We chose the CIE D65 standard illuminant for this

purpose. This spectrum, or daylight illuminant, corresponds roughly to a midday sun in Western/Northern Europe. We also applied the constant lumen normalization to that spectrum, as explained by Equation 3. We considered that biological evolution occurred under sunlight, and assumed that the CIE D65 illuminant is closely related to the three biological processes studied here.

Spectral Indices Definition and Interpretation

Some attempt has been made in the past to define spectral indices related to photosynthesis and circadian action in the field of computer image rendering [55] and light and health research [56]. A melatonin suppression effect relative to HPS, and a ratio of the radiant power in a scotopic protected interval (440 nm - 540 nm) to photopic luminous flux, were also suggested in 2011 [30]. Starting from these anterior works, we set out to improve and standardize this concept, and to include the estimation of star visibility (scotopic vision). We also set out to explore the effect of the scattering of light through the atmosphere. In general, scattering processes may cause different effects in humans and other photosensitive species due to their different visual perceptions. Some species are even sensitive to the UV spectrum, as has been suggested in a recent study [57]. In this paper, the authors make a set of computations in order to estimate the real perception of three species by considering the spectral radiance produced by five different lamps. The findings are related specifically to the skyglow as perceived by humans (under scotopic vision) and two species of insects (one diurnal and one nocturnal). We suggested here three spectral indices ranging from 0 to $\sim$1 to characterize independently the potential spectral impact of a lamp SPD on the three following biological processes: 1- melatonin suppression, 2- photosynthesis, and 3- scotopic vision. To isolate the spectral properties from other variables, we first calculated a weighted lamp constant lumen SPD, where the weight function is the spectral sensitivity of the biological process considered. We then performed the same calculation for a CIE D65 constant lumen SPD in place of the lamp SPD. Finally, we computed the ratio of the weighted constant lumen lamp SPD over the weighted constant lumen CIE D65 SPD. When atmospheric effects were considered, the atmospheric transfer function was applied to the SPD of both constant lumens so that we could always compare the lamp with the CIE D65 illuminant under the same conditions. The integration limits were set to the spectral extent of the photosynthesis action spectrum (380 nm to 730 nm).

According to this, we can write the Melatonin Suppression Index (MSI), the Induced Photosynthesis Index (IPI) and the Star Light Index (SLI) in Equations 18, 19 and 20, respectively.

$$MSI = \frac{\int_{380nm}^{730nm}\Phi_{n_{(lamp)}}(r,\lambda)M(\lambda)d\lambda}{\int_{380nm}^{730nm}\Phi_{n_{(D65)}}(r,\lambda)M(\lambda)d\lambda} \tag{18}$$

$$IPI = \frac{\int_{380nm}^{730nm}\Phi_{n_{(lamp)}}(r,\lambda)P(\lambda)d\lambda}{\int_{380nm}^{730nm}\Phi_{n_{(D65)}}(r,\lambda)P(\lambda)d\lambda} \tag{19}$$

$$SLI = \frac{\int\limits_{380nm}^{730nm} \Phi_{n_{(lamp)}}(r,\lambda)V'(\lambda)d\lambda}{\int\limits_{380nm}^{730nm} \Phi_{n_{(D65)}}(r,\lambda)V'(\lambda)d\lambda} \qquad (20)$$

These indices can be easily calculated with the help of an OpenOffice spreadsheet file available at http://galileo.graphycs. cegepsherbrooke.qc.ca/lpds/index.php?n = Site.Products, if you have a SPD file to hand.

The three indices defined above enable the impact of the shape of the SPD to be evaluated. If we want to evaluate the global impact of a lamp we must also consider its illuminance. Since our indices use SPD normalized to a constant lumen (Eq. 3), we can thus obtain the global impact of a lamp by multiplying the indices by the illuminance of that lamp.

Results and Discussion

Spectral Indices for Direct Lighting

We first want to apply the above indices definitions to the special case of a direct sight towards a lighting device. By direct, we mean that we do not calculate any atmospheric effects. That special case does not only occur for direct sight, but can also be invoked when we observe a neutral reflecting surface lit by the lighting device. For example, when light trespass occurs, the light can enter a human eye after entering a bedroom by its window and then reflecting from a white wall. Table 2 shows the indices calculation results for a set of 13 lamps and for the CIE D65 illuminant. In our experiment, LED filtering was performed with the *Equilib* optical filter commercialized by Ledtech International. Together with the indices, we have provided in Table 2 the CCT approximation [58], CRI [59], x and y chromaticity coordinates [60], and typical spectral sensitivities (lumen/W). CCT, CRI, x and y were calculated for each lamp on the basis of their measured spectrum. The x and y chromaticity coordinates [60] for each lamp are given in Figure 1.

If we consider HPS as a reference device with MSI = 0.12, we can note that only LPS (0.02), Nichia amber (0.04), Lumiled Pc Amber (0.05) and LED 2700 K filtered (0.08) have a reduced potential impact on melatonin suppression (their MSI values are below the HPS value of 0.12, as shown in Table 2). So, these lamps should be favored for nighttime lighting in order to restrict the potential impact of ALAN on human health. It is important to note that, during the night, metal halides have an impact more than five times higher than HPS (0.64/0.12 = 5.33). This means that if the two lamps have the same lumen output, the metal halide will be 5.33x more efficient in suppressing melatonin, assuming that for both lamps the luminance is high enough to suppress melatonin. Similarly we can show that LED 5000 K, Philips LED 4000 K and fluorescent lamps are respectively 4.5x, 3.75x and 3.6x more efficient at suppressing melatonin than is HPS. If we need lighting during the daytime, LPS, phosphor-converted amber LEDs and LED 2700 K filtered lamps should not be used; during the daytime, we need to select the highest MSI values. In our lamp selection, the maximum MSI was obtained using the metal halide lamp (0.62), followed by the LED 5000 K (0.54), Philips LED 4000 K (0.45) and fluorescent (0.43) lamps. None of these are as efficient as the Sun itself, here represented by CIE D65 (1.0).

Making the same analysis for the impact on photosynthesis using IPI values, we determined that only LPS, LED 5000 K-filtered and LED 2700 K-filtered lamps had a lower impact than

HPS. The least beneficial lamp during the night is therefore clearly the incandescent lamp, with an index of 0.92, followed by the halogen lamp (0.83). However, during the daytime these lamps can be advantageously used for plant growth to compensate for shorter days and lower Sun illuminance in winter at high latitudes. This can also be helpful for greenhouse-based industries. Incandescent and halogen lamps showed a high IPI, but their lumen per watt ratio is low, so that even if they are efficient for photosynthesis per unit of lumen, they are not efficient for photosynthesis in terms of energy consumption. The IPI of phosphor-converted amber LEDs, metal halide and fluorescent lamps were found to be very similar (ranging from 0.61 to 0.68), even with very different SPDs (orange vs. white light). More specifically, phosphor-converted amber LEDs do not have any significant blue emission, in clear contrast to metal halide and fluorescent lamps (Fig. 3). However, if we look to the shape of PAS in Figure 2, we can see that both lamps have a similar IPI because phosphor-converted amber LEDs show a high emission in the region of the PAS red peak (~660 nm). For metal halide, the SPD shows an important line that coincides well with the blue PAS peak (~430 nm). Fluorescent light shows important lines in both the blue and red region.

The SLI is applicable to low illuminance conditions, which are basically satisfied during the nighttime. In this case, we used the scotopic spectral sensitivity to calculate the index so that it was not only applicable to the astronomical observation but more generally to low-light human vision. If again we take HPS as a reference (0.23), we can show that only LPS (0.09), phosphor-converted amber LEDs (0.17 and 0.15) and LED 2700 K filtered lamps (0.23) are better or equivalent to HPS for restricting the impact on observation of the night sky. LPS and phosphor-converted amber LEDs are better than HPS. The lamps with the most negative impact on night sky observation are, in decreasing impact order, LED 5000 K (0.62), fluorescent (0.61), halogen (0.6), metal halide (0.58) and Philips LED 4000 K (0.56). These lamps produce ~2.5x more astronomical light pollution than HPS for the same lumen output. It is of particular concern that white LEDs show such a high spectral impact on the dark sky. Groups and communities involved in dark sky protection initiatives should be aware of that, given the attractiveness of this technology to lighting professionals and local authorities. LED is a long lasting technology (50,000–100,000 hours), and is relatively efficient and easily controllable/dimmable. In order to constrain astronomical light pollution or to increase star visibility while changing HPS to cool white LED lighting, we recommend reducing newly installed LED lumens by a factor of 2.5x compared to the HPS lumen originally installed.

Spectral Indices for Indirect Lighting

We applied the atmospheric radiative transfer model described in the section 'Impact of Atmospheric Optics' to each lamp described in Table 2, and for a variety of distances from lamp to observer, ranging between 0 km and 30 km. This time, we did not observe the lamp directly but instead observed the light scattered by the surrounding atmosphere. It is important here to understand that the same model was applied to all SPDs, even the CIE D65. In other words, we compared the scattered light from a lamp with the reference CIE D65, both scattered to the same distance. In Table 3, we show the results for each index and for two distances (0 and 30 km). Atmospheric scattering increases the blue part of the SPD at 0 km. However, with increasing distance, the blue light is preferentially removed by extinction. This effect has a theoretical foundation and can be explained in terms of radiative transfer principles. If the distance from a hypothetical observer to the light source is small enough, the emitted light beams propagate through the atmosphere along short optical paths before they are backscattered toward the ground and detected by an optical device. Short optical paths imply a low attenuation due to extinction, and thus the scattering efficiency is by far the greatest modulator of downwelling radiation at surface level. Since an Ångström coefficient of 1.0 has been used in our numerical simulations, the aerosol optical thickness $\tau_A(\lambda)$ appears to be an inverse function of the wavelength. It is well-known that the Rayleigh optical thickness behaves like $\lambda^{-4.08}$ (Eq. 11), so molecular scattering definitely dominates in the blue part of the spectrum. In our model, the aerosol optical thickness is considered to be $\tau_A(500\ nm) = 0.2$ at the ground. Under these conditions, the Rayleigh optical thickness exceeds that for aerosols by 50% at 400 nm, but τ_R is almost three times smaller than τ_A at 650 nm. In addition, the aerosol backscatter is much weaker than Rayleigh theory dictates (compare e.g. Plate 10.7 b in [61] with Fig. 2.3 in [62]). Both factors, the low extinction at short optical paths and the fairly efficient molecular backscatter, result in a specific spectral behavior of diffuse radiation in which Rayleigh and aerosol optics appear to be important. This coincides very well with our numerical computations, which show that diffuse irradiance behaves like $\lambda^{-2.7}$ in the blue part of the visible spectrum. Thanks to a fairly steep decrease of $\tau_R(\lambda)$ with increasing wavelength, the spectral behavior of diffuse irradiance shows a functional dependency $\lambda^{-1.9}$ at the red edge of the visible spectrum. As the distance between an observer and light source increases, the situation changes significantly. The beams that form the light field near the measuring point are propagated at low elevation angles, thus traveling along long optical paths. In such a case, the aerosol optics become dominant due to the enhanced forward scatter and increased scattering efficiency. It should be noted that the ratio of forward to side scatter can exceed the value of 100 (or even more) for a polydispersed system of aerosol particles, while it is only 2:1 in the case of Rayleigh scattering. For these reasons, Rayleigh scatter is considered to have a marginal effect on sky glow if observations are made at larger distances from the light source. The spectral dependency of diffuse irradiance therefore mimics a function of λ^{-1} rather than λ^{-4}, because the large optical path means that the light beams interact with an elevated number of aerosol particles. On the one hand, the higher number of scattering domains translates to an increased scattering efficiency, while on the other hand it results in a more efficient intensity decay, due to the exponential form of the transmission function (Eq. 5). One could say these two effects act against each other. In a very simplified case, the energy of the scattered radiation or radiance (Eqs. 6 or 8) could behave like Xe^{-yX}, where X monotonically increases with growth in the optical path or number of scattering events (y is an arbitrary scaling parameter). This function peaks at $X_{peak} = y^{-1}$, so the extinction exceeds the scattering efficiency if $X > X_{peak}$, i.e. if the optical path is considerably large. The spectral dependency of diffuse irradiance is then further distorted, resulting in a functional behavior λ^{-a} with $a < 1$. The numerical experiments performed in this paper have shown that $a \cong 0.82$ for distances of about d = 15 km, $a \cong 0.77$ at d = 30 km, and $a \cong 0.75$ at d = 60 km. For very distant light sources, the coefficient a approaches its asymptotic behavior. The limit value of a depends on the mutual interaction of the aerosol phase function and the extinction and scattering efficiencies. Figure 4 illustrates the complexities of the atmospheric effects. This figure shows the SPD of illuminant CIE D65 as a function of distance for two different atmosphere ((a) clear sky and (b) cloudy

sky). All SPDs have been normalized at 550 nm to facilitate the comparison.

For most of the lamps shown in Table 3, MSI, IPI and SLI are lower in scattered light compared to direct lighting indices. The only exception is the metal halide lamp for which MSI and IPI are slightly higher after scattering. This is probably because of the presence of a significant deep blue line at around 405 nm. No other spectra in our selection show such an important emission at a very low wavelength where scattering is very efficient. The same phenomenon is not observed for SLI, because the scotopic spectral sensitivity curve is almost zero at 405 nm.

To obtain a better understanding of the distance dependency of each index, we produced plots showing the indices as a function of distance (Figs. 5, 6, 7) under clear and cloudy sky conditions. These figures show the complex behavior of indices with distance. We separated each figure into two cases, namely new technologies intended to reduce light pollution in the right-hand plots (b and d), and the more standard technologies that are shown in the left-hand plots (a and c). For all plots, we showed the HPS case as a common reference. The two upper plots (a and b) are for clear sky conditions, while the cloudy conditions are shown in the lower two plots (c and d). Only lamps used in street lighting are plotted. For all cases the scattered indices become closer to direct values with increasing distance. Under cloudy sky conditions, all indices show a peak located at a distance of approximately 2 km. This kind of curve inflexion is not observed under clear sky conditions.

Conclusions

In this paper, we introduced three new indices that can be used to characterize the SPD characteristics of any lighting device. These indices have been designed to allow a quick estimation of the potential impact of different lamp spectra on melatonin suppression, photosynthesis and star visibility. Indices have also been designed to separate the impact of the shape of the SPD from other factors, such as illuminance levels or the angular photometry of the lamp.

We used the indices to compare different lighting technologies in term of their spectral impacts. In particular we found that LPS, phosphor-converted amber LEDs and LED 2700 K-filtered lamps have a lower potential impact on melatonin suppression in comparison to HPS. LPS, LED 5000 K-filtered and LED 2700 K-filtered lamps show a lower impact on photosynthesis compared to HPS. Only LPS, phosphor-converted amber LEDs and LED 2700 K-filtered lamps are better than or equivalent to HPS in restricting the impact on star visibility. We also showed that atmospheric scattering under clear or cloudy skies generally reduces the value of the indices in comparison to direct light, except for metal halide lamps for which the effect is opposite.

During daytime, if we want to obtain the highest spectral impact, we should favor metal halide for increased melatonin suppression, while phosphor-converted amber LEDs and metal halide lighting is the best for stimulating photosynthesis.

It is important to realize that the spectral indices introduced here should not be considered alone in making a complete evaluation of the impact of a given lamp installation. As an example, a lamp with low MSI (low relative blue content) can impact significantly on melatonin suppression when the lamp illuminance is so high that the absolute blue flux becomes important. Basically, a global evaluation can be obtained by multiplying the index by the illuminance.

We hope that these indices will form some kind of standardization in characterization of lighting technologies, and that the lighting industry will provide such indices in the same way as they currently provide CCT or CRI, for example. Finally we consider that these indices will favor the design of new environmentally and health-friendly light devices, by providing a means of quickly evaluating the relative potential impact of a device at the design stage.

Acknowledgments

We want to thank Pierre Goulet, Mont-Mégantic National Park Director and responsible for the 1ˢᵗ International Dark Sky Reserve, for the fruitful conversations that have been helpful in rendering our work more attractive for potential end users. We also want to thank Ledtech International and Hydro Sherbrooke for providing some of the lamp fixtures used for this study

Author Contributions

Conceived and designed the experiments: MA JR. Performed the experiments: MA JR MK. Analyzed the data: MA JR MK. Contributed reagents/materials/analysis tools: MA JR MK. Wrote the paper: MA JR MK.

References

1. Cinzano P, Falchi PF, Elvidge CD (2001) The first world atlas of the artificial night sky brightness. Mon Not R Astron Soc 328: 689–707.
2. Longcore T, Rich C (2004) Ecological light pollution. Front Ecol Environ 2(4): 191–198.
3. Rich C, Longcore T (2006) Ecological consequences of artificial night lighting. Washington DC: Island Press. 478 p.
4. Navara KJ, Nelson RJ (2007) The dark side of light at night: physiological, epidemiological, and ecological consequences. J Pineal Res 43: 215–224.
5. Bruce-White C, Shardlow M (2011) A review of the impact of artificial light on invertebrates: Putting the backbone into invertebrate conservation. Peterborough, UK: Buglife – The Invertebrate Conservation Trust. 32 p.
6. Buchanan BW (2006) Observed and potential effects of artificial night lighting on Anuran amphibians. In: Richt C, Longcore T, editors. Ecological consequences of artificial night lighting. Washington DC: Island Press. 192–200.
7. Gauthreaux SA, Belser CG (2006) Effects of artificial night lighting on migrating birds. In: Rich C, Longcore T, editors. Ecological consequences of artificial night lighting. Washington DC: Island Press. 67–93.
8. Kuijper DPJ, Schut J, Van Dullemen D, Toorman H, Goossens N, et al. (2008) Experimental evidence of light disturbance along the commuting routes of pond bats (Myotis dasycneme). Lutra 51(1): 37–49.
9. Salmon M (2003) Artificial night lighting and sea turtles. Biologist 50: 163–168.
10. Calo F, Bardonnet C (2011) Les tortues marines et la pollutions lumineuses sur le territoire français. Rapport GTMF-SPN 2. Paris: MNHN-SPN. 40 p.
11. Fritsches KA (2012) Australian Loggerhead sea turtle hatchings do not avoid yellow. Mar Freshw Behav Physiol 45(2): 79–89.
12. Nightingale B, Longcore T, Simenstad A (2006) Artificial night lighting and fishes. In: Rich C, Longcore T, editors. Ecological consequences of artificial night lighting. Washington DC: Island Press. 257–276.
13. Perry G, Fischer NR (2006) Night lights and reptiles: Observed and potential effects. In: Rich C, Longcore T, editors. Ecological consequences of artificial night lighting. Washington DC: Island Press. 169–191.
14. Briggs WR (2006) Physiology of plant responses to artificial lighting. In: Rich C, Longcore T, editors. Ecological consequences of artificial night lighting. Washington DC: Island Press. 389–406.
15. Hansen J (2001) Light at night, shiftwork, and breast cancer risk. J Natl Cancer Inst 93(20): 1513–1515.
16. Blask DE, Brainard GC, Dauchy RT, Hanifin JP, Davidson LK, et al. (2005) Melatonin-depleted blood from premenopausal women exposed to light at night stimulates growth of human breast cancer xenografts in nude rats. Cancer Res 65(23): 11174–11184.
17. Stevens RG (2009) Light at night, circadian disruption and breast cancer: Assessment of existing evidence. Int J Epidemiol 38(4): 963–970.
18. Stevens RG (2011) Testing the light at night (LAN) theory for breast cancer causation. Chronobiol Int 28(8): 653–656.
19. Kloog I, Haim A, Stevens RG, Barchana M, Portnov BA (2008) Light at night co-distributes with incident breast but not lung cancer in the female population of Israel. Chronobiol Int 25(1): 65–81.
20. Kloog I, Stevens RG, Haim A, Portnoy BA (2010) Nighttime light level co-distributes with breast cancer incidence. Cancer Causes Control 21: 2059–2068.
21. Kloog I, Haim A, Stevens RG, Portnov BA (2009) Global co-distribution of light at night (LAN) and cancers of prostate, colon, and lung in men. Chronobiol Int 26(1): 108–125.
22. Haim A, Yukler A, Harel O, Schwimmer H, Fares F (2010) Effects of chronobiology on prostate cancer cells growth in vivo. Sleep Science 3(1): 32–35.
23. Spiegel K, Tasali E, Penev P, Van Cauter E (2004) Sleep curtailment in healthy young men is associated with decreased leptin levels, elevated ghrelin levels, and increased hunger and appetite. Ann Intern Med 141(11): 846–850.
24. Fonken LK, Workman JL, Walton JC, Weil ZM, Morris JS, et al. (2010) Light at night increases body mass by shifting the time of food intake. Proc Natl Acad Sci U S A 107(43): 18664–18669.
25. Spiegel K, Knutson K, Leproult R, Tasali E, Van Cauter E (2005) Sleep loss: A novel risk factor for insulin resistance and Type 2 diabetes. J Appl Physiol 99(5): 2008–2019.
26. Bedrosian TA, Fonken LK, Walton JC, Haim A, Nelson RJ (2011) Dim light at night provokes depression-like behaviors and reduces CA1 dendritic spine density in female hamsters. Psychoneuroendocrinology 36: 1062–1069.
27. Lockley SW, Brainard GC, Czeisler CA (2003) High sensitivity of the human circadian melatonin rhythm to resetting by short wavelength light. J Clin Endocrinol Metab 88(9): 4502–4505.
28. Thapan K, Arendt J, Skene DJ (2001) An action spectrum for melatonin suppression: Evidence for a novel non-rod, non-cone photoreceptor system in humans. J Physiol 535(1): 261–267.
29. Brainard GC, Hanifin JP, Greeson JM, Byrne B, Glickman G, et al. (2001) Action spectrum for melatonin regulation in humans: Evidence for a novel circadian photoreceptor. J Neurosci 21(16): 6405–6412.
30. Falchi F, Cinzano P, Elvidge CD, Keith DM, Haim A (2011) Limiting the impact of light pollution on human health, environment and stellar visibility. J Environ Manage 92: 2714–2722.
31. Wright Jr KP, Hughes RJ, Kronauer RE, Dijk DJ, Czeisler CA (2001) Intrinsic near-24-h pacemaker period determines limits of circadian entrainment to a weak synchronizer in humans. Proc Natl Acad Sci U S A 98(24): 14027–14032.
32. Zeitzer JM, Norman FR, Fisicaro RA, Heller HC (2011) Response of the human circadian system to millisecond flashes of light. PLoS ONE 6(7): e22078.
33. Rosenthal N, Blehar MC, editors (1989) Seasonal affective disorders and phototherapy. New York: Guilford Press. 350 p.
34. Viola AU, James LM, Schlangen LJM, Dijk DJ (2008) Blue-enriched white light in the workplace improves self-reported alertness, performance and sleep quality. Scand J Work Environ Health 34(4): 297–306.
35. Chellappa SL, Steiner R, Blattner P, Oelhafen P, Götz T, et al. (2011) Non-visual effects of light on melatonin, alertness and cognitive performance: Can blue-enriched light keep us alert? PLoS ONE 6(1): e16429.
36. Cajochen C, Frey S, Anders D, Späti J, Bues M, et al. (2011) Evening exposure to a light-emitting diodes (LED)-backlit computer screen affects circadian physiology and cognitive performance. J Appl Physiol 110: 1432–1438.
37. LS Data – Light Pollution Data Server. Available: http://galileo.graphycs.cegepsherbrooke.qc.ca/lpds/index.php?n = Site.LSData. Accessed 2013 May 30.
38. Clark JB, Lister GR (1975) Photosynthetic action spectra of trees - I. Comparative photosynthetic action spectra of one deciduous and four coniferous tree species as related to photorespiration and pigment complements. Plant Physiol 55: 401–406.
39. DIN 5031–10 (2000) Optical radiation physics and illuminating engineering - Part 10: Photobiologically effective radiation, quantities, symbols and action spectra. Germany: Deutsches Institut Fur Normung EV (German National Standard).
40. Bisketzis N, Polymeropoulos G, Topalis FV (2004) A mesopic vision approach for a better design of road lighting. WSEAS Transactions on Circuits and Systems 3(5): 1380–1385.
41. Wyszecki G, Stiles WS (1982) Color science: concepts and methods, quantitative data and formulae, second edition. New York: Wiley. 935 p.
42. Rea MS, Freyssinier JP (2009) Outdoor lighting: Visual efficacy (ASSIST recommends, Volume 6, Issue 2). New York: Lighting Research Center, Rensselaer Polytechnic Institute. 14 p.
43. Rea MS, Bullough JD, Freyssinier-Nova JP, Bierman A (2004) A proposed unified system of photometry. Light Res Technol 36(2): 85–111.
44. Vos JJ (1978) Colorimetric and photometric properties of a 2-deg fundamental observer. Color Res Appl 3: 125–128.
45. Judd DB (1951) Report of U.S. secretariat committee on colorimetry and artificial daylight. In: Proceedings of the twelfth session of the CIE, Stockholm. Paris: Bureau Central de la CIE. 1(7): 11 p.
46. Thomas GE, Stamnes K (2002) Radiative transfer in the atmosphere and ocean. Cambridge: Cambridge University Press. 548 p.
47. Leckner B (1978) The spectral distribution of solar radiation at the earth's surface – elements of a model. Sol Energy 20: 143–150.
48. Green RN, Wielicki BA, Coakley JA, Stowe LL, Hinton PO'R, et al. (1997) Clouds and the Earth's radiant energy system (CERES): Algorithm theoretical basis document. CERES inversion to instantaneous TOA fluxes, Release 2.2, June 2. Available: http://ceres.larc.nasa.gov/documents/ATBD/pdf/r2_2/ceres-atbd2.2-s4.5.pdf. Accessed : 2013 May 30.
49. Kocifaj M (2007) Light-pollution model for cloudy and cloudless night skies with ground-based light sources. Appl Optics 46: 3013–3022.
50. Horvath H, Arboledas LA, Olmo FJ, Jovanović O, Gangl M, et al. (2002) Optical characteristics of the aerosol in Spain and Austria and its effect on radiative forcing. J Geophys Res 107: 4386–4403.

51. Pesava P, Horvath H, Kasahara M (2001) A local optical closure experiment in Vienna. J Aerosol Sci 32: 1249–1267.

52. Gushchin GP (1988) The methods, instrumentation and results of atmospheric spectral measurements. Leningrad: Gidrometeoizdat. 200 p.

53. Braak JC, de Haan JF, Van der Mee CVM, Hovenier JW, Travis LD (2001) Parameterized scattering matrices for small particles in planetary atmospheres. J Quant Spectrosc Radiat Transf 69: 585–604.

54. Toublanc D (1996) Henyey-Greenstein and Mie phase functions in Monte Carlo radiative transfer computations. Appl Opt 35: 3270–3274.

55. Geisler-Moroder D, Dür A (2010) Estimating melatonin suppression and photosynthesis activity in real-world scenes from computer generated images. In: CGIV 2010/MCS'10 5th European Conference on Colour in Graphics, Imaging, and Vision and 12th International Symposium on Multispectral Colour Science, Joensuu, Finland. 346–352.

56. Gall D, Bieske K (2004) Definition and measurement of circadian radiometric quantities, light and health - non-visual effects. In: Proceedings of the CIE symposium '04 : 30 September–2 October 2004, Vienna, Austria. 129–132.

57. Lamphar HAS, Kocifaj M (2013) Light pollution in ultraviolet and visible spectrum: Effect on different visual perceptions. Plos One 8(2): e56563.

58. McCamy CS (1992) Correlated color temperature as an explicit function of chromaticity coordinates. Color Res Appl 17(2): 142–144.

59. CIE 13.3 (1995) Method of measuring and specifying colour rendering properties of light sources. Paris: Commission Internationale de l'Éclairage. 20 p.

60. CIE (1932) 1931 Commission Internationale de l'Éclairage Proceedings. Huitième session. Cambridge: Cambridge University Press, 19–29.

61. Mishchenko MI, Travis LD, Lacis AA (2002) Scattering, absorption, and emission of light by small particles. Cambridge: Cambridge University Press. 445 p.

62. Kokhanovsky AA (2004) Light scattering media optics: Problems and solutions. Berlin, Heidelberg, New York: Springer. 320 p.

20

A Unified Account of Perceptual Layering and Surface Appearance in Terms of Gamut Relativity

Tony Vladusich[1,2]*, **Mark D. McDonnell**[1]

1 Institute for Telecommunications Research, University of South Australia, Mawson Lakes, 5095, Australia, **2** Center for Computational Neuroscience and Neural Technology, Boston University, Boston, MA, United States of America

Abstract

When we look at the world—or a graphical depiction of the world—we perceive surface materials (e.g. a ceramic black and white checkerboard) independently of variations in illumination (e.g. shading or shadow) and atmospheric media (e.g. clouds or smoke). Such percepts are partly based on the way physical surfaces and media reflect and transmit light and partly on the way the human visual system processes the complex patterns of light reaching the eye. One way to understand how these percepts arise is to assume that the visual system parses patterns of light into layered perceptual representations of surfaces, illumination and atmospheric media, one seen through another. Despite a great deal of previous experimental and modelling work on layered representation, however, a unified computational model of key perceptual demonstrations is still lacking. Here we present the first general computational model of perceptual layering and surface appearance—based on a boarder theoretical framework called gamut relativity—that is consistent with these demonstrations. The model (a) qualitatively explains striking effects of perceptual transparency, figure-ground separation and lightness, (b) quantitatively accounts for the role of stimulus- and task-driven constraints on perceptual matching performance, and (c) unifies two prominent theoretical frameworks for understanding surface appearance. The model thereby provides novel insights into the remarkable capacity of the human visual system to represent and identify surface materials, illumination and atmospheric media, which can be exploited in computer graphics applications.

Editor: Hans A Kestler, University of Ulm, Germany

Funding: Mark D. McDonnell's contribution was supported by an Australian Research Fellowship from the Australian Research Council (http://www.arc.gov.au/), project number DP1093425. The funders had no role in study design, data collection and analysis, decision to publish, or preparation of the manuscript.

* Email: therealrealvlad@gmail.com

Introduction

The human visual system manifests the remarkable capacity to identify surface materials from the complex patterns of light reaching the eye [1,2]. This capacity is exploited in the computer graphics industry to create convincing renderings of surface materials based on physical models of 'light transport' [3–5]. The problem of understanding how the visual system represents surface materials (e.g. ceramic tiles or human skin), and related visual properties of illumination (e.g. shadows, shading and highlights) and atmospheric media (e.g. clouds, fog and smoke), is thus of immense practical importance in the field of computer graphics.

Models of physical light transport attempt to capture the immensely complicated ways in which physical surfaces and atmospheric media reflect, refract, scatter and transmit light [3–5]. The net result is that the light patterns reaching the eye from a rendered image consist of a mixture of physically modelled causes. Light 'reflected' from a rendered transparent surface using a standard α-blending model, for example, is combined with light 'transmitted' through the surface from the background [6]. Thus, even simple diffuse shading and/or blending models produce images that the human visual system parses into layered perceptual representations, one seen through another, as illustrated by the striking perceptual effects shown in Fig. 1. How the

human visual system parses such images into separate material, illumination and atmospheric layers remains a challenging problem in both human vision science and computer vision science.

In this article, we study the 'mid-level' computations that give rise to perceptual layering and related surface appearance properties, such as lightness and transparency, in images generated using simple diffuse shading and α-blending models [1,6–34]. Such mid-level computations evolved to process light associated with real physical sources, but in this article we will consider the more circumscribed issue of how the visual system represents surface materials, illumination and atmospheric media associated with graphically rendered physical sources. In this respect, the focus of this article will be the analysis of rendered images that elicit decomposition into surface and shadow/atmospheric layers (perceptual layering), rather than real physical scenes, which are known to sometimes elicit different perceptual interpretations when compared to rendered images [35–37]. We will also leave for future work the complex issue of how to model surface appearance in images that are difficult to interpret in terms of globally consistent perceptual layers, such as images containing certain types of gradients [37–39].

The perceptual effects shown in Fig. 1 are known as the Adelson checkerboard effect (Fig. 1A) [1] and the Anderson-

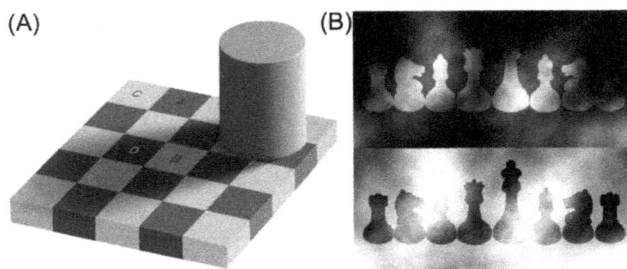

Figure 1. Two dramatic effects of perceptual layering and surface appearance. (A) Adelson checkerboard image [1] adapted from http://web.mit.edu/persci/people/adelson/checkershadow_illusion.html under the Creative Commons Attribution License: Checks labelled A and B (depicted as appearing in bright and dim illumination) have the same point-to-point luminance but check B appears light gray and check A dark gray. Checks B and D are seen through a 'transparent shadow layer', whereas checks A and C are seen in 'plain view' (without an accompanying transparent layer). Variations in illumination intensity level produce multiplicative changes in the luminance values depicted as being reflected from the checks in bright and dim illumination. (B) Anderson-Winawer effect reprinted from [12]: Chess pieces in the upper and lower rows have the same point-to-point luminance but appear white and black, respectively. The white pieces are seen through a blackish transparent 'atmosphere' whose transparency varies across space, while the black pieces are seen through a transparent whitish atmosphere. Variations in atmospheric transmittance levels produce additive changes in the luminance values depicted as being reflected from the black and white chess pieces. This article develops a model that aims to quantitatively predict surface lightness through transparent layers, irrespective of the physical source of the transparent layer.

Winawer effect (Fig. 1B) [12,13], respectively. In both effects, figure regions having the same point-to-point luminance distribution are perceived as having very different lightness due to variations in the surrounding 'ground' regions, which induce the impression of surfaces seen through different types of 'overlays'. In the Adelson checkerboard effect (Fig. 1A), grayish background checks are seen through a shadow cast over part of the display, whereas in the Anderson-Winawer effect (Fig. 1B), blackish or whitish chess pieces are seen through a cloud bank or wall of smoke that varies in its transparency at different points.

The demonstrations shown in Fig. 1 raise a number of important modelling challenges. First and foremost, a computational model is needed to explain how the human visual system represents different sources of physical variation—such as surfaces, illumination and atmospheric media—in terms of layered perceptual representations. Although much experimental and modelling work has been done on the topic of layered representations, and their relevance to surface material perception, a unified computational model of key perceptual layering effects is still lacking [1,6–18,26–34,39]. Second, the model must address the difficulty that variations in illumination intensity, such as shadows and shading, are associated with multiplicative changes in registered luminance, whereas variations in the transmittance of physical surfaces and atmospheric media are associated with additive changes in luminance [8,12,13,24,39]. Third, the model needs to incorporate an understanding of the manner in which the visual system represents the transparency of rendered physical surfaces and atmospheric media [6,9–13,15–18,29,32]. Fourth, the problem of separating an image region into perceptual layers is closely related to the problem of determining which surface regions appear in plain view and which appear through the transparent overlay, and thus requires an analysis in terms of figure-ground relationships [12,13,40].

Demonstrations of the sort illustrated in Fig. 1 also indirectly highlight the importance of considering stimulus- and task-driven constraints on surface appearance [37,39,41–50]. This is because stimulus- and task-driven constraints play a critical role in determining whether the visual system computes one or more perceptual layers [12,13]. In this article, we link stimulus- and task-driven constraints on the computation of perceptual layers to key perceptual matching data on the role of stimulus- and task-driven constraints on brightness (luminance) and lightness (reflectance) perception, respectively [25,30,41–51]. Of particular importance is the problem of teasing apart the complex relationship between the computational processes underlying different aspects of brightness and lightness perception. It is well known, for example, that human subjects adopt different strategies to perform matching tasks (e.g. brightness and lightness) under different stimulus conditions [41–45,48,49].

The following section of the article briefly reviews several key theoretical concepts underlying our model. The "Model" section then provides the detailed descriptions of empirical studies, mathematical equations, and computational specifications that are needed to explain perceptual data concerning the demonstrations shown in Fig. 1. The "Results" section provides conceptual analyses and computer simulations of the model under various stimulus- and task-driven constraints, demonstrating the model's capacity to quantitatively predict perceptual data. The "Discussion" section briefly explores some broader implications of the theoretical framework on which the current model is based.

A Brief Review of Gamut Relativity

The model we present is based on a recently introduced theoretical framework known as *gamut relativity* [52]. The interested reader can find detailed background information in several recent publications [52–57].

Blackness and whiteness are orthogonal dimensions

Our model explains how the visual system represents surfaces independently of variations in either illumination intensity (e.g. shadows; Fig. 1A) or atmospheric transmittance (e.g. clouds; Fig. 1B) in terms of computations performed in a *blackness-whiteness coordinate system* (Fig. 2). Roughly speaking, the whiteness coordinate value (ψ) increases with both increasing luminance and positive contrast magnitude, whereas the blackness coordinate value (ϕ) increases with decreasing luminance and increasing negative contrast magnitude. Blackness and whiteness are conceptualised as orthogonal dimensions of a two-dimensional (2-D) perceptual space [52–54] that can be 'sliced' in different ways, depending on stimulus conditions and task constraints.

Brightness and lightness are relative concepts

When illumination is perceived as uniform across a scene or object, luminance values corresponding to surfaces with different physical reflectance values are mapped to points falling on a single straight line ('slice') in blackness-whiteness space, termed the standard gamut line (Fig. 2A). We associate this mapping with the notion of 'brightness' perception. When illumination is perceived as non-uniform, by contrast, luminance values corresponding to different physical surfaces in bright illumination are mapped to points falling on the standard gamut line, whereas luminance values corresponding to different physical surfaces in dark illumination are mapped to points falling on one or more comparison gamut lines (Fig. 2B). The shifting of points from the standard to the comparison gamut line thus compensates for

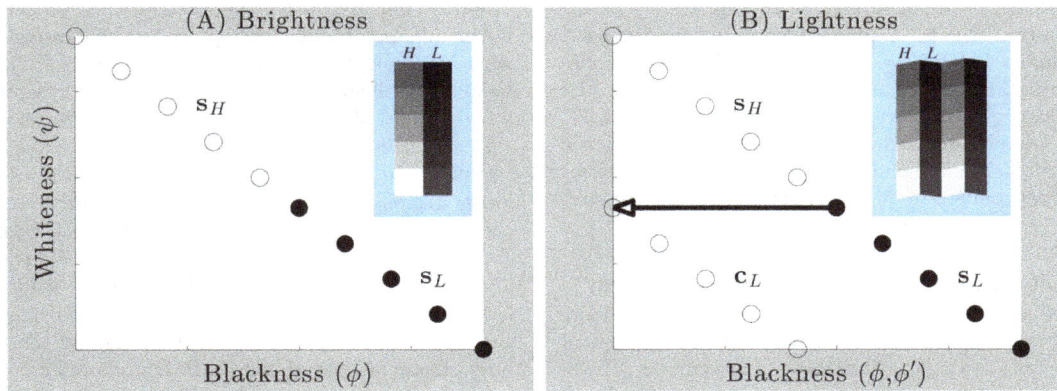

Figure 2. The representation of brightness and lightness in gamut relativity. (A) Surface regions represented under the assumption of a single illumination level and a planar arrangement of surfaces, such as co-ordinates $s_H = [\phi_H, \psi_H]$ and $s_L = [\phi_L, \psi_L]$, fall on a negatively sloped 'gamut' line in blackness-whiteness space, where $H \in \{1,2,3...,5\}$ and $L \in \{1,2,3...,5\}$ denote the columns of relatively higher and lower luminance squares depicted in the insets, respectively. (B) Surface regions represented under the assumption of two different illumination intensity levels and a corrugated arrangement of surfaces, such as co-ordinates $s_H = [\phi_H, \psi_H]$ and $c_L = [\phi'_L, \psi_L]$, fall on two different gamut lines (termed standard and comparison, respectively). The inset figures in (A) and (B) perceptually illustrate how identical sets of luminance values can be parsed according to the assumptions of uniform or variable illumination levels, respectively. In (A), pictorial image cues indicate that the bright and dark columns of squares (sets H and L) lie in the same depth plane, favoring the assumption of uniform illumination over all squares [25,89,90]. Horizontal pairs of squares are thus mapped to different blackness co-ordinates, $\phi_H \neq \phi_L$. As blackness co-ordinates constitute the computational correlate of diffuse reflectance in gamut relativity, squares in sets H and L appear to have different diffuse reflectance. In (B), the same sets of luminance values shown in the two columns in (A) are now pictorially depicted to lie in different depth planes (the repetition of rows here enhances this depiction), favoring the assumption of variable illumination [8,25,85,89,90]. Horizontal pairs of squares in this arrangement are mapped to the same blackness co-ordinates, $\phi_H = \phi'_L$, and thus appear to have the same diffuse reflectance. The horizontal vector depicts the shift of points from standard to comparison gamuts, which compensates for the presumptive illumination difference between sets H and L. Figure modified with permission from [56].

the difference in illumination levels between bright and dark. Vertically aligned points sharing the same blackness coordinates but falling on different gamut lines thus correspond to surfaces with the same physical reflectance [52]. We associate this mapping with the notion of 'lightness' perception.

In our model, then, it is the *relationships* between points lying on the standard gamut line—or between points lying on the standard and comparison gamut lines—that determine the properties characterising what we know as 'brightness' (Fig. 2A) and 'lightness' (Fig. 2B) perception, respectively. This emphasis on relationships between points lying on gamut lines is also the origin of the term 'gamut relativity'.

The reflectance-to-lightness mapping is relative

This distinction between our model and alternative models has a number of important correlates. Firstly, as blackness-whiteness space is two-dimensional, invariance along one dimension obviously does not imply invariance along the other dimension, meaning that surfaces sharing the same blackness coordinates needn't appear *identical*. Secondly, blackness coordinates vary from zero to some arbitrary upper bound, so do not themselves represent a range of 'lightness values' varying from black to white. Thirdly, different gamut lines represent different unique slices of blackness-whiteness space, with each line bookended by *different shades of black and white*. There thus exists no absolute mapping from reflectance to gray shades in gamut relativity—in the sense of an absolute scale of lightness values—and this proposal is consistent with a great deal of perceptual data that cannot be explained by classical approaches [52]. In short, our model underlies a more subtle relative account of the reflectance-to-lightness mapping than the classical absolute (scalar) reflectance-to-lightness mapping [25].

Luminance and contrast sum vectorially to facilitate figure-ground perception

The proposed illumination-shift process described above requires the visual system to compute the local luminance associated with each surface region [52] (and be capable of discriminating illumination edges from reflectance edges [22,25,46,58]). Another key idea in gamut relativity, then, is that luminance, in addition to contrast, plays a central role in determining surface appearance. This idea—as an anonymous reviewer of this article states—"flies against what we currently know about vision...current wisdom is that vision is not sensitive to luminance, only contrast." Our previous modelling successes—combined with the new analyses presented in this article—suggest that a modest revision to this conventional wisdom may be in order. In particular, we have previously shown how luminance and contrast can be represented as vectors that sum in blackness-whiteness space [54]; the proposed summation of luminance and contrast is consistent with recently reported cortical physiological data [59,60]. Here we show how this vector summation can facilitate perceptual layering and figure-ground perception by operating asymmetrically on figure and ground image regions (see Results).

Luminance is also important for ambient illumination perception

The sensitivity to luminance in our model also overcomes a key limitation of approaches based solely on contrast [58,61]; namely, how is it that we readily perceive variations in ambient (global) illumination? Psychophysical experiments showing that humans can distinguish light levels in *Ganzfeld* stimuli (i.e. containing no contrast) testify to the sensitivity of the visual system to global luminance [62,63]. Many classical and recent physiological studies [64–75] have, furthermore, revealed that both local and global luminance signals are present at early levels of both the cat and primate visual systems—although luminance signals are typically weaker than contrast signals, as documented in the classical early physiological studies of [76,77]—and recent studies have emphasised the functional importance of these signals in shaping the ON and OFF responses of visual cortical neurons [59,60,70,73,74,78–82]. Our model emphasises and interprets the available evidence concerning physiological luminance and contrast coding in terms of the relative contributions of these signals to surface appearance; see [52,54] for further discussion.

Gamut relativity is versatile and generalises effectively

A significant conceptual advantage of the gamut relativity framework is its ability to account for a wide range of perceptual phenomena in a parsimonious manner [6,8–13,22,23,25,29,32,83]. In addition to specifying brightness and lightness, for example, gamut relativity can also be used to specify the transparency level of a partially transmissive foreground surface or medium. The key idea is that the transparency level of the foreground layer is given by the distance between the standard and comparison gamut lines [55]. The equations of gamut relativity quantitatively explain some puzzling aspects of key demonstrations in classical studies of transparency perception [55], such as the observation that whitish transparent layers appear more opaque than blackish layers with the same physical transmittance [11,32]. This observation has proven difficult to explain in terms of classical transparency models [6]. Gamut relativity has also been extended to the domain of specularly reflecting surfaces to provide a unified account of layered perceptual representation in lightness and gloss perception [56].

Existing gamut relativity models need to be combined

The model presented in this article represents a unification of several previously published gamut relativity models that have dealt separately with aspects of brightness/lightness perception [52,54], lightness/transparency perception [55] and lightness/gloss perception [56], respectively. The latter two studies incorporated only luminance signals in the implemented models (e.g. the model depicted in Fig. 2). Here we show how these previous models can be combined—in a way that incorporates both luminance and contrast—in order to predict data on surface lightness perception through generically defined transparent overlays, whether they be associated with cast shadows, surface shading, atmospheric media or transmissive physical filters. The model goes beyond previous work by (a) qualitatively explaining some striking effects of perceptual transparency, figure-ground separation and lightness perception, (b) quantitatively accounting for the role of stimulus- and task-driven constraints on brightness/lightness matching performance, and (c) unifying two prominent theoretical frameworks for understanding surface appearance (see Discussion). The model thus provides the first quantitative account of perceptual data on the role of stimulus- and task-driven factors in brightness and lightness perception, in terms of a general theory of perceptual layering and surface appearance [25,30,41–52].

Materials and Methods

Perceptual data to be modelled

To motivate the computational modelling, consider the Adelson checkerboard effect (Fig. 1A), which is itself the product of two subtle image manipulations. Firstly, checks A and B—which have the same luminance but whose gray shades appear quite different—are seen against surrounding checks that themselves differ in luminance: check A is seen against checks of higher luminance (labeled check C), while check B is seen against checks

of lower luminance (labeled check D). This contextual difference induces the perceptual effect known as *simultaneous contrast* [25], whereby a target seen against a background of relatively higher luminance will appear relatively blacker than a target seen against a background of relatively lower luminance. Secondly, check A is seen in relatively bright illumination while check B is seen in relatively dim illumination, with an identifiable shadow separating image regions in relatively bright and dim illumination. This contextual difference induces the perceptual effect known as *discounting the illuminant*, whereby check B (and check D) in dim illumination is perceptually shifted in gray shade in order to compensate for the perceived illumination difference. This shift ensures that check B appears similar in gray shade to check C in bright illumination and that check D in dim illumination appears similar in gray shade to check A in bright illumination. This perceptual outcome is commonly termed *lightness constancy* [25]. The computational processes underlying simultaneous contrast and discounting the illuminant appear to combine to produce the dramatic perceptual difference that characterises Adelson's checkerboard display.

[58] sought to characterise the magnitude of perceptual shifts in variants of the Adelson checkerboard display [46] and a related display introduced by [22] among other displays. These authors had subjects adjust the luminance of a matching region, viewed against a black-and-white background, in order to make 'brightness' and 'lightness' matches to targets viewed within different versions of the checkerboard and simultaneous contrast displays. Two different stimulus conditions were examined. In the "Paint" conditions, all targets were viewed in the context of surfaces depicted as lying under uniform illumination (*without* a shadow overlay) but against surfaces appearing to have different reflectance ('paint jobs'). In the "Illumination" conditions, the targets were viewed under different depicted illumination levels (*with* a shadow overlay), seen against surfaces appearing to have the same or similar reflectance. Subjects adjusted the luminance of the matching region such that reference and matching regions either appeared to reflect the same "light intensity" (brightness match) or appeared "as if cut from the same paper" (lightness match). These task instructions had little or no influence in the "Paint" conditions, but had a dramatic influence in the "Illumination" conditions. The magnitude of the perceptual shift in the Adelson checkerboard display, for example, was much greater in the lightness matching task than in the brightness matching task. A key goal of the present study is to develop a model that quantitatively predicts how stimulus- and task-driven constraints control the computational processes that contribute to 'brightness' and 'lightness' matching behaviour [58].

The Anderson-Winawer effect (Fig. 1B)—in which physically identical textured surfaces are seen as either uniform black or white surfaces—depending on the surrounding context, has been theoretically analysed [12,13] as a perceptual decomposition, or 'scission' [6,9–11,32], into transparent foreground and opaque background layers. The computational process underlying this decomposition is sensitive to the spatial relationship between the target and background stimuli. Rotating the background textures by 90 degrees with respect to the target region, for example, eliminates the effect. According to [12,13], the visual system uses the fact that *figural contrast polarity* (black-to-white or white-to-black) is preserved around the entire perimeter of the target region to trigger the perceptual decomposition into surface layers. These authors proposed that, once decomposition is triggered, the visual system uses the surface region that appears in 'plain view'—that is, appearing without the intervening transparent medium—to compute the gray shade of the farther surface layer that is contained within the perimeter of the target region.

[12,13] provided lightness matching data to support this proposal and showed that the contribution of perceptual decomposition to the effect was far greater than the contribution attributable to simultaneous contrast. Another key goal of the present study is to demonstrate how the same model used to quantitatively predict the contributions of the computational processes underlying the Adelson checkerboard effect and brightness/lightness matching behaviour can also quantitatively predict the perceptual data on the decomposition and simultaneous contrast effects that contribute to the Anderson-Winawer effect.

Model overview

Two broad classes of computational processes work together to compute surface gray shades in the model: (A) vector summation of luminance and contrast, and (B) vector decompositions implementing the illuminant- and transmittance-shift processes to produce layered representations in different parts of the image.

General simulation methods

All software implementing the equations and algorithms defined below was written in MATLAB Version 8.0.0 (R2012b). Stimulus luminance values used in the computer simulations were taken from the published values given in [12,13] and [58].

Inputs to the model

In order to apply the model to arbitrary images, it would be necessary to solve the image segmentation problem, which generally involves parsing the retinal image into regions differing in either reflectance, illumination or transmittance [8–13,22,23,25,29,32,83]. A segmentation process is required in our model in order to (A) define an image region and its contrast with respect to immediately surrounding regions, and (B) divide the image into different regions upon which vector decomposition processes are differentially applied depending on stimulus- and task-driven constraints.

Fig. 3 illustrates how a standard segmentation algorithm from the computer vision literature [84] captures the intuition of a suitable segmentation to compute regional luminance and contrast in our analysis. The algorithm segments the Adelson checkerboard image and a simplified version of the Anderson-Winawer display into labelled regions in which mean pixel or luminance values are calculated. The segmented regions are thus characterised by differences in mean luminance, and each individual region is immediately surrounded by one or more regions containing a different mean luminance value.

In the present article, we adopt the following simplifying heuristic to extract predictions from the model. We assume that each check in the Adelson checkerboard image and each target region in the Anderson-Winawer display has been segmented into labelled regions whose mean luminance (more precisely, mean log luminance) we explicitly calculate based on stimulus specifications reported in relevant publications. This allows us to compute the luminance and contrast terms in the model equations, as described in detail below.

The segmentation algorithm can also sometimes produce region labels corresponding to different illumination and transmittance levels (e.g. the border between moons and surrounds in Fig. 3E)—particularly when the regional borders have high contrast—but such regional segmentations are often not computed (e.g. the shadow border in Fig. 3A). We thus explicitly set the values of the free parameters controlling the illuminant- and transmittance-shift

processes in a manner consistent with the stimulus-driven constraints (e.g. assuming the same or different illumination levels in different segmented regions), in addition to task-driven constraints (e.g. brightness or lightness matching tasks). In this way, we are able to extract predictions from the model without having to explicitly segment the image into regions differing in illumination or transmittance levels. We are currently developing a version of the model that will incorporate a sophisticated user-guided segmentation process to define regions differing in illumination and transmittance levels in a more general way.

In our analysis of the Adelson checkerboard (Fig. 4) and the related paint/transparency/shadow display of [58], we shall employ the following notation in order to define contrast in the equations below: A target check in relatively bright illumination will be labelled T for 'target' and surrounding checks of lower or higher luminance than the target will be labelled L for 'lower' or H for 'higher', respectively. The inputs to the model will then be luminance values labelled either ℓ_T, ℓ_L or ℓ_H. With reference to Fig. 1A, we explicitly define ℓ_A, ℓ_B, ℓ_C and ℓ_D as the luminance values of checks A, B, C and D, respectively. Thus, when $\ell_T = \ell_A$, then $\ell_H = \ell_C$ and $\ell_L = \ell_A$ (ensuring that the ratio ℓ_T/ℓ_L is unity and hence the log of this ratio is zero). Analogous specifications are applied to checks B, C and D in Fig. 1A. When the surround of target T has components that are both lower and higher in luminance than ℓ_T (e.g. a gray target seen against a black-and-white checkerboard, such as the test displays in [12,13] and [58]) then the ratios of ℓ_T/ℓ_L and ℓ_H/ℓ_T will both be positive.

In our analysis of the Anderson-Winawer display (Fig. 1B), individual pixels within regions T, L and H are indexed T, L and H, giving luminance values ℓ_T, ℓ_L and ℓ_H, respectively. We then define ℓ_T, ℓ_L or ℓ_H as the geometric mean luminance value of each region (e.g. $\ell_T = \Pi_{t=1}^{\mathcal{T}} \ell_t^{\frac{1}{\mathcal{T}}}$), where $\mathcal{T}$ denotes the number of pixels in region T. This choice is justified by the fact that these displays are characterised by luminance gradients, meaning that some method of averaging is required to compute contrast. Our choice of the geometric mean luminance is consistent with the logarithmic transformation applied in our model. In the case of the Adelson checkerboard (Fig. 1A), it is the case that $\ell_t = \ell_T$, $\ell_l = \ell_L$ and $\ell_h = \ell_H$. For greatest generality, we write the model equations in terms of these individually indexed luminance values. In general, therefore, we write the luminance of pixel t in region T as

$$\ell_t = \ell_T \forall t = 1, 2, \ldots, \mathcal{T}, \qquad (1)$$

where for reasons explained below, we label pixel indices in a sequential manner such that $\ell_1 \leq \ell_2 \leq \ldots \leq \ell_{\mathcal{T}}$.

Outputs of the model

We now describe the computational model itself, which specifies the algorithmic mapping of image luminance values specified at the pixel level into vector-valued surface representations characterised by 'blackness' (ϕ) and 'whiteness' (ψ) coordinates. In particular, the model maps scalar-valued image representations into vector-valued surface representations. A vector decomposition process produces surface representations that are used to predict human behavioural performance under various stimulus- and task-driven constraints. The output of the model is the vector-valued surface representation, given for each pixel t by the equation

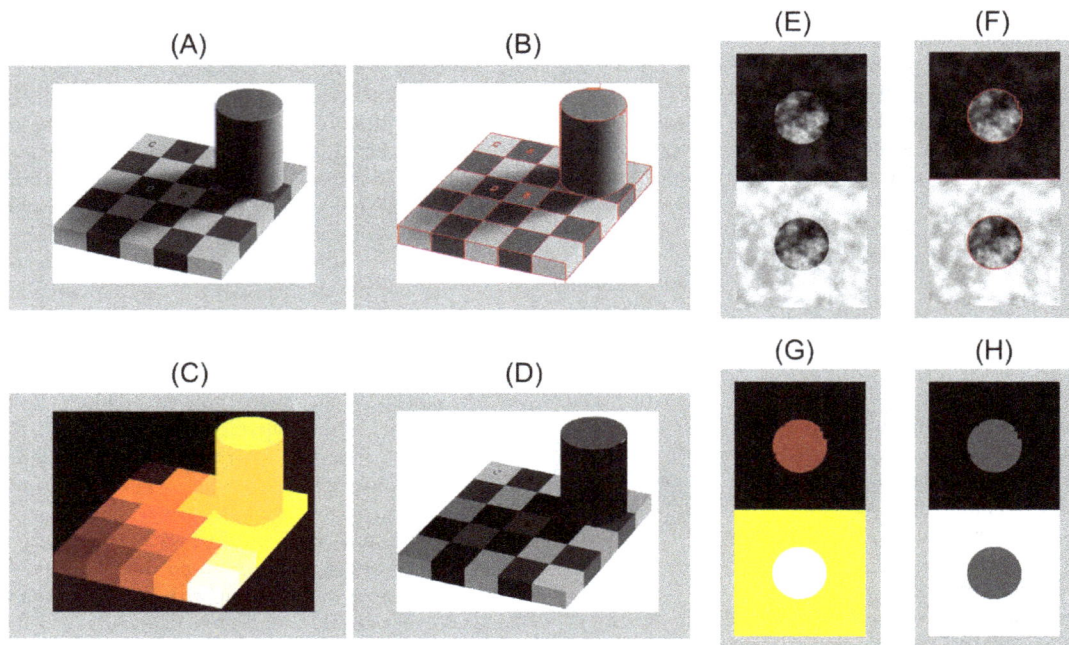

Figure 3. Two examples of image segmentations used to guide the computation of region luminance and contrast. (A) Adelson checkerboard image [1], modified with permission under the Creative Commons Attribution License. (B) Segmentation computed with a standard computer vision algorithm [84] (parameters: $\beta = 120$, stop $= 10^{-4}$). (C) The algorithm returns region labels for each image region. (D) Region labels enable the calculation of mean pixel or luminance values within each segmented region. (E-H) Same as above, except applied to a simple version of the Anderson-Winawer display (adapted from http://www.psy.ritsumei.ac.jp/~akitaoka/AIC2009.html with permission).

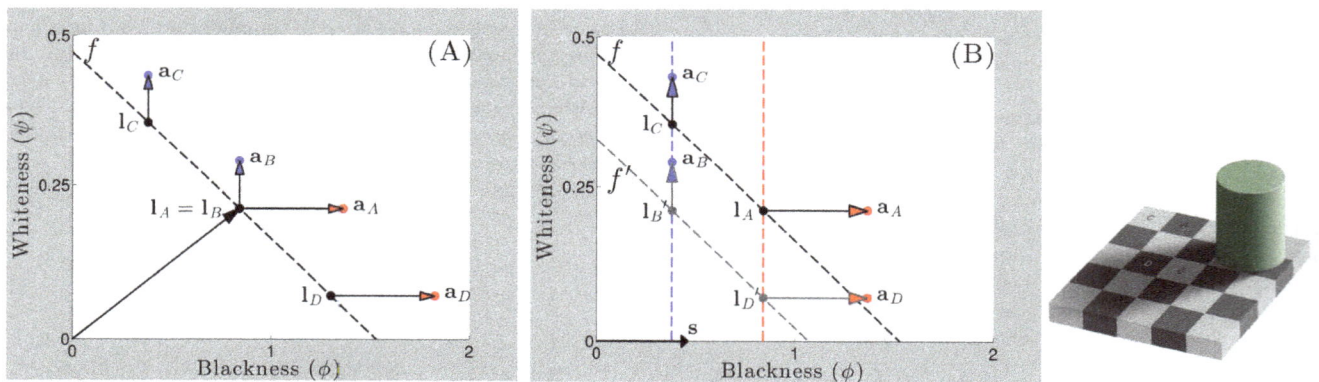

Figure 4. Adelson checkerboard display parsed in the brightness and lightness modes. The model explains the key perceptual properties implied by the Adelson checkerboard display shown in Fig. 1A. Surface gray shades are specified in a perceptual blackness-whiteness space given by the coordinates (ϕ, ψ). The free parameter $\zeta \in [0,1]$ controls the balance between so-called brightness ($\zeta = 0$) and lightness ($\zeta = 1$) 'modes' that represent the respective assumptions of spatially uniform or variable illumination. (A) **Brightness mode**: According to the model, the summation of luminance and contrast vectors ensures that check B in the Adelson checkerboard display has higher whiteness than check A ($\psi_B > \psi_A$ with respect to $\mathbf{a}_B$ and $\mathbf{a}_A$) and check A has higher blackness than check B ($\phi_A > \phi_B$ with respect to $\mathbf{a}_A$ and $\mathbf{a}_B$), consistent with various data on the simultaneous contrast effect [54]. (B) **Lightness mode**: According to the model, an illuminant-shift process combines with the vector summation underlying simultaneous contrast to produce the Adelson checkerboard effect, i.e. $\mathbf{a}_B = \mathbf{l}_B + \mathbf{c}_B - \mathbf{s} = \mathbf{l}'_B + \mathbf{c}_B$, where $\mathbf{s}$ is a 'shadow vector' with non-zero blackness and zero whiteness components that introduces the comparison luminance gamut, f'. The illuminant-shift process transforms the blackness coordinates of checks B and D in relatively dim illumination towards the blackness axis, e.g. ϕ_B is smaller in lightness mode than it is in the brightness mode example illustrated in subfigure (A). Checks with the same reflectance thus share the same blackness coordinates ($\phi_B = \phi_C$), and checks with different reflectance but the same luminance have very different blackness coordinates ($\phi_A \gg \phi_B$ with respect to $\mathbf{a}_A$ and $\mathbf{a}_B$). Due to the asymmetrical scaling of blackness coordinates relative to whiteness coordinates, blackness plays the dominant role in determining the surface gray shade [54]. The model thus explains both the independence of surface gray shades with respect to variable illumination intensity levels and the large magnitude of the Adelson checkerboard effect relative to simultaneous contrast alone. Adelson checkerboard image adapted from http://web.mit.edu/persci/people/adelson/checkershadow_illusion.html under the Creative Commons Attribution License.

$$\mathbf{a}_t = (\phi_t, \psi_t) = \mathbf{l}_t + \mathbf{c}_t - (\mathbf{s} + \mathbf{u}_t), \qquad (2)$$

where the vector components are defined below.

Note that, although model outputs can be displayed as image pairs (i.e. corresponding to ϕ and ψ coordinates), we find that displaying outputs in blackness-whiteness coordinate space (e.g. Fig. 4) at selected pixels t provides greater insight into the model computations. We therefore eschew the common practice of displaying model outputs as images, while still acknowledging that such representations can be useful in certain contexts.

Model equations

The various vectors comprising Eqn. (2) are defined as follows.

- A *luminance vector* is given by

$$\mathbf{l}_t = \left(\alpha\mu \, \log_{10} \frac{k_\phi}{\ell_t}, (1-\alpha)v \, \log_{10} \frac{\ell_t}{k_\psi} \right), \qquad (3)$$

where ℓ_t is defined above, $k_\phi = \max(\bar{\ell}, \bar{k}_\phi)$ and $k_\psi = \min(\bar{\ell}, \bar{k}_\psi)$ are 'anchoring' parameters, $\bar{\ell}$ is the highest luminance value in the entire display, $\bar{k}_\phi = 100$ and $\bar{k}_\psi = 1$ are constants, and $\alpha = 0.87$, $\mu = 0.88$ and $v = 1.81$ are estimated constants based on psychophysical data [54]. We term the blackness and whiteness components of the luminance vector *luminance blackness* and *luminance whiteness*, respectively.

The anchoring scheme defined above implies that scenes with luminance values below $\bar{k}_\phi$ will contain no white surfaces, but scenes with luminance values above this threshold will contain one white surface corresponding to the highest luminance value in the scene. We have found that this rule—coupled with our choice of value for $\bar{k}_\phi$ and $\bar{k}_\psi$—is suitable to model the perception of diffusely reflecting surfaces rendered on low dynamic range displays viewed under typical daylight adaptation conditions. See [56] and [52] for discussions of more complex anchoring rules in the context of brightness, lightness and gloss perception.

- A *contrast vector* is given by

$$\mathbf{c}_t = \left(\alpha\beta_T \, \log_{10} \frac{\ell_H}{\ell_T}, (1-\alpha)(1-\beta_T) \, \log_{10} \frac{\ell_T}{\ell_L} \right), \qquad (4)$$

where β_T represents the proportion of the surrounding region with luminance higher than the target, as in the equation $\beta_T = \mathcal{H}/(\mathcal{H}+\mathcal{L})$. In practice, we set β_T by hand in a manner consistent with this equation. We refer to the blackness and whiteness components of the vector specified in Eqn. (4) as *contrast blackness* and *contrast whiteness*, respectively. Note that the individual scalar components defining the contrast and luminance vectors above are summed to give the values of ϕ_t and ψ_t defined in Eqn. (2).

- An *illuminant-shift* vector ($\mathbf{s}$) specifies the magnitude of 'illuminant-discounting' in a manner that depends on the ratio of the highest-luminance regions designated as appearing in relatively bright illumination (labelled $\hat{\mathrm{H}}$) and dim illumination (labelled $\bar{\mathrm{H}}$), respectively. The illuminant-shift vector is expressed as

$$\mathbf{s} = \zeta \left(\alpha\mu \, \log_{10} \frac{\ell_{\hat{\mathrm{H}}}}{\ell_{\bar{\mathrm{H}}}}, 0 \right), \qquad (5)$$

where $\zeta \in [0,1]$ is a free parameter representing various stimulus- and task-driven constraints [52], as discussed below. In the perceptual demonstrations of surface and shadow perception analysed in this article (Fig. 1A, Fig. 5B), the illuminant-shift vector is applied asymmetrically; namely, only to those target regions in dim illumination (e.g. checks A, C in Fig. 1A), not bright illumination (e.g. checks B, D in Fig. 1A). The illuminant-shift process constitutes a mathematical decomposition of the vector $\mathbf{i}_t = \mathbf{l}_t + \mathbf{c}_t$ into surface $\mathbf{a}_t$ and shadow $\mathbf{s}$ component vectors, such that $\mathbf{i}_t = \mathbf{a}_t + \mathbf{s}$.

- A *transmittance-shift* vector ($\mathbf{u}_t$) specifies the magnitude of 'transmittance-discounting' relative to pixels designated as appearing in 'plain view'

$$\mathbf{u}_t = \eta \left(\alpha\mu(1-\beta_T) \, \log_{10} \frac{\ell_{t=T}}{\ell_t}, (1-\alpha)v\beta_T \, \log_{10} \frac{\ell_t}{\ell_{t=1}} \right), \qquad (6)$$

where $\ell_{t=1}$ and $\ell_{t=T}$ equal the lowest and highest luminance values within the target region, respectively, and $\eta \in [0,1]$ is a free parameter representing figural-continuity (i.e. spatial continuity of contours across figure and ground regions) and contrast-polarity (i.e. continuity of border polarity between figure and ground regions) constraints that are known to characterise scission into transparent layers [6,8–13,85]. Note that $\beta_T = 0$ and $\beta_T = 1$ when the surround has higher and lower geometric mean luminance than the target region in the Anderson-Winawer display (Fig. 1B), respectively. By Eqn. (6), then, whiteness coordinates are shifted when the target is a decrement and blackness coordinates are shifted when the target is an increment, which is what is required to discount the physical transmittance shift in a manner consistent with figural contrast polarity (see Results). The transmittance-shift process defined above is proposed to underlie the separation of figural regions into figure and ground layers in a manner consistent with the figural-continuity and contrast-polarity constraints reported in extant perceptual studies [12,13,40]. Indeed, the transmittance-shift process with $\eta = 1$ constitutes a mathematical decomposition of the vector $\mathbf{i}_t = \mathbf{l}_t + \mathbf{c}_t$ into figure $\mathbf{f}_t$ and ground $\mathbf{g}_t$ component vectors, such that $\mathbf{i}_t = \mathbf{f}_t + \mathbf{g}_t + \mathbf{i}_o$, where $\mathbf{i}_o$ is the origin of the vector decomposition.

Model parameters and properties

We now highlight some key conceptual properties of the model, some of which have previously been detailed in recent publications [52,54–57]:

- We assume in what follows that $\ell_t \in [k_\psi, k_\phi]$. This assumption implies that the blackness and whiteness components of the luminance vector in Eqn. (3) are always non-negative. Likewise, the blackness and whiteness components of the contrast vector in Eqn. (4) are by the definitions of ℓ_H and ℓ_L also always non-negative. These constraints thereby ensure that the blackness and whiteness components of the sum of

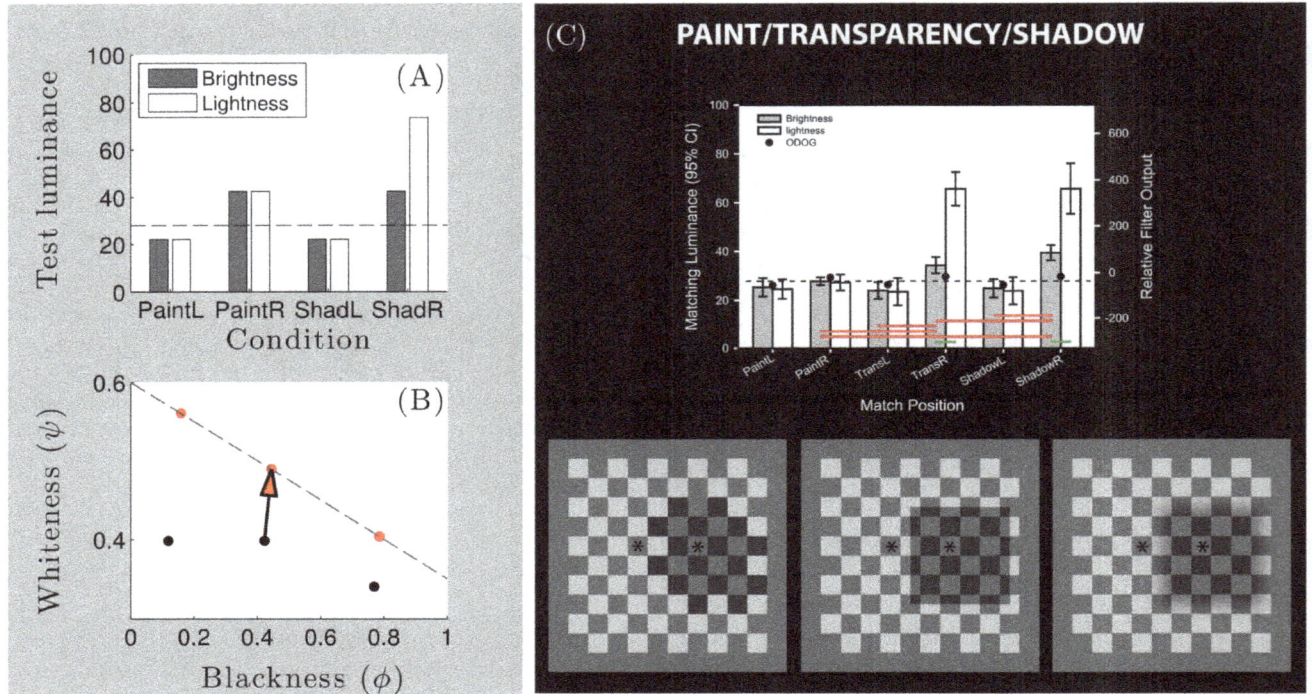

Figure 5. Model predictions of brightness and lightness matching data relating to the Blakeslee-McCourt paint/transparency/shadow display. (A) The model correctly predicts the influence of task instructions on perceptual matches made with surfaces seen under depicted uniform or variable illumination. The luminance of the target is shown by the dashed line, and predictions of luminance of the test target in each condition shown by the level of each bar. (B) Model luminance predictions shown in (A) were generated from minimal Euclidean distances between points representing the reference gray shades (black points, obtained from Eqn. (2) with $\zeta = 1$ and $\eta = 0$) and gamut lines representing the test display (red points on gray dotted line). The test display was assumed to have background luminance values equal to k_ϕ and k_ψ, and thus all grey shades in the test display fall on mixed gamut lines, since both blackness and whiteness coordinates have non-zero contrast components. (C) Data and depiction of stimuli reprinted from [58]. In total, there are 12 different test conditions: 6 of these are brightness tasks and 6 are lightness tasks. In (B), black dots indicate the blackness-whiteness coordinates, $\mathbf{a}_T$, for 8 of these 12 conditions. As the model predictions for the ShadowL and ShadowR conditions are equally applicable to the experimental TransL and TransR (transparency) conditions, we omit the 4 transparency conditions. There are only 3 unique coordinates, since the same blackness-whiteness coordinates at approximately ($\phi = 0.8$, $\psi = 0.3$) are obtained for all L conditions, and the same coordinates at approximately (0.4,0.4) are obtained for both PaintR conditions and ShadowR (labelled ShadR above) in the brightness task. The final black dot at approximately (0.1,0.4) occurs uniquely for ShadowR in the lightness matching task. The red arrow indicates the minimal perceptual match between reference and test coordinates for both PaintR conditions and ShadowR in the brightness matching task.

luminance and contrast vectors are always non-negative. This is why all points shown in Fig. 4A, for example, are restricted to the upper right quadrant of blackness-whiteness space.

- Assume $\zeta = 0$ in Eqn. (5) and $\eta = 0$ in Eqn. (6). Now consider an image with uniform luminance, $\ell_T = \ell_H = \ell_L$ (i.e. a Ganzfeld [62]). All pixel indices t in the image will have zero contrast values in both the blackness and whiteness coordinates. The blackness coordinate is then zero when $\ell_t = k_\phi$ and the whiteness coordinate is zero when $\ell_t = k_\psi$. We write the corresponding whiteness and blackness coordinates to these two cases as ψ_0 and ϕ_0. The *standard luminance gamut* is then defined as all points on a negatively sloped straight line in blackness-whiteness space defined between these two axis intercepts, $(0, \psi_0)$ and $(\phi_0, 0)$. This can be expressed as the equation

$$f := \psi = -\frac{\psi_0}{\phi_0} \phi + \psi_0. \qquad (7)$$

All luminance vectors, $\mathbf{l}_t$, are constrained to fall on the standard luminance gamut line, f; that is, letting $(\phi_t, \psi_t) = \mathbf{l}_t$

satisfies Eqn. (7). The black dotted lines in Fig. 4A,B, for example, represent the standard luminance gamut.

- In the case of a simple image with a single uniform target region on a *uniform* background region p, the blackness-whiteness coordinates corresponding to a pixel t within the target region will not fall on the standard luminance gamut, due to the contrast terms in the blackness-whiteness equations. As can be seen from Eqn. (2), the deviation from the standard luminance gamut is given by the contrast vector, $\mathbf{c}_t$. In the case of a contrast increment, the term $\mathbf{c}_t$ will have a non-zero whiteness co-ordinate and a zero blackness co-ordinate. This contrast whiteness component is added to the luminance vector to define a *standard increment gamut*. In the case of a contrast decrement, the term $\mathbf{c}_T$ will have a non-zero blackness co-ordinate and a zero whiteness co-ordinate. This contrast blackness component is added to the luminance vector to define a *standard decrement gamut*. For images containing both contrast increments and decrements (e.g. a checkerboard pattern), both contrast components will be non-negative. The contrast vector will then add both blackness and whiteness components to the luminance vector, defining a *standard mixed gamut*. In fact, in general it is possible to define families of gamut lines, both standard and comparison, each depending on stimulus- and task-related factors (see [52,55,56] for further details). One could, for example, draw separate comparison increment and decrement gamut lines through each individual blue and red point shown in Fig. 4B, but we shall omit these lines in order to maintain figural clarity.

- Now consider an image within which pixels indexed by t are identified as appearing in relatively dim illumination. In this case, $\zeta > 0$ and $\mathbf{s}$ becomes non-zero; a new gamut line representing surfaces appearing in the relatively dimmer illumination level is thus defined. We introduce $\mathbf{l}'_t = \mathbf{l}_t - \mathbf{s}$, such that the blackness-whiteness coordinate of pixel t is $\mathbf{a}_t = \mathbf{l}'_t + \mathbf{c}_t$. We then define a *comparison luminance gamut* as

$$f' := \psi = -\frac{\psi'_0}{\phi'_0}\phi + \psi'_0, \qquad (8)$$

which has a smaller whiteness intercept than the standard luminance gamut ($\psi'_0 < \psi_0$), indicating a relatively lower illumination level, but lies parallel to the standard luminance gamut, such that $\psi_0\phi_0^{-1} = \psi'_0\phi'_0^{-1}$. All luminance vectors, $\mathbf{l}'_t$, are constrained to fall on the comparison luminance gamut, f'; that is, letting $(\phi_t, \psi'_t) = \mathbf{l}'_t$ satisfies Eqn. (8). The light gray dotted line in Fig. 4B, for example, represents the comparison luminance gamut when $\zeta = 1$. According to the equation $\mathbf{a}_t = \mathbf{l}'_t + \mathbf{c}_t$, then, it is further possible to define increment, decrement and mixed *comparison* gamuts. For detailed discussion of the computational utility of the relationship between standard and comparison gamuts, see [52,55,56].

- Perceptual matches performed in psychophysical experiments generally correspond to *minimal perceptual mismatches* between points specified to lie along different gamut lines [52–54]. The minimum perceptual distance between a reference point lying on a gamut line specified for a reference display and the set of all points on another gamut line specified for the test (or matching) display determines the predicted luminance setting. It is calculated as the luminance value that minimises the Euclidean metric, $\sqrt{(\psi_r - \psi_t)^2 + (\phi_r - \phi_t)^2}$, where indices r and t denote reference and test targets, subject to the constraints imposed by the test gamut line. The blackness-whiteness plot shown in Fig. 5B provides an example of the manner in which the idea of minimal perceptual mismatches can help to account for perceptual data. Previous theoretical and experimental work also supports the idea that subjects' cannot generally make satisfactory brightness matches between targets viewed against backgrounds differing in luminance or perceived illumination level [52–54].

- The parameter ζ controls the balance between two perceptual 'modes' that each explain key properties of brightness and lightness perception, respectively (the parameter λ in [52] is equivalent to ζ here). Under the assumption that luminance variations between pixels are due entirely to reflectance variations, blackness coordinates are primarily correlated with local luminance ($\zeta = 0$; brightness mode). Under the assumption that luminance variations between pixels are due entirely to illumination variations, blackness coordinates are primarily correlated with diffuse surface reflectance ($\zeta = 1$; lightness mode). Intermediate values of ζ represent a 'balance of probability' [86,87] between these two extreme assumptions and thus represent linear combinations of presumptive illumination and reflectance variations. Here we generalise the distinction between brightness and lightness to describe the

surface perception under the assumption that $\zeta = 0$ and $\eta \in [0,1]$; that is, by generalising the definition to the case of surface perception through transmissive media (e.g. Fig. 6).

- The parameter ζ is itself a function of both the stimulus (ζ_s) and task (ζ_t), such that $\zeta = \zeta_s\zeta_t$, where $\zeta_s, \zeta_t \in [0,1]$. The assumption of uniform illumination corresponds to $\zeta_s = 0$ (e.g. the "Paint" condition of [58]). The assumption of variable illumination corresponds to $\zeta_s = 1$ (e.g. the "Illumination" condition of [58]). As ζ_t can only modify the value of ζ when $\zeta_s > 0$, this construction is consistent with psychophysical data reported in [58] showing that task-driven constraints on matching behaviour can only exert an influence when stimulus conditions support the perception of variable illumination. In the "Lightness" matching task of the "Illumination" conditions in [58], we assume that $\zeta = \zeta_s\zeta_t = 1$, whereas in the "Brightness" matching task we assume that $\zeta = \zeta_s\zeta_t = 0$. This construction reflects the fact that, under the assumption of uniform illumination across a scene, luminance and surface reflectance are correlated, whereas under the assumption of variable illumination, luminance and reflectance are uncorrelated. The capacity to flexibly switch between perceptual modes correlated with either luminance or reflectance thus underscores a key conceptual deviation of our model from the classical theory of surface perception as a problem of reflectance recovery.

- Blackness-whiteness space is asymmetrically scaled, meaning that a unit variation in physical luminance maps to a far greater variation in blackness coordinates than whiteness coordinates [52,54]. This proposal explains a wide range of otherwise puzzling data concerning asymmetries in the perception of contrast increments and decrements. The asymmetry can be appreciated, for example, by comparing the scales of the ϕ- and ψ-axes in Fig. 4. The precise ratio of blackness/whiteness variation depends on various factors, but has been estimated to be no less than approximately 3 [54]. Given the setting $\zeta = 1$ and $\eta = 0$, then under the assumption that surfaces seen under different illumination levels contain identically distributed sets of reflectance values, pairs of points associated with f' and f that have the same blackness coordinates (e.g. $\mathbf{a}_C$ and $\mathbf{a}_B$ in Fig. 4B) are perceptually more similar to one another than pairs of points associated with f' and f that have different blackness coordinates (e.g. $\mathbf{a}_C$ and $\mathbf{a}_A$ in Fig. 4B) [52].

Results

Surface perception under uniform and variable illumination

We now show how our model accounts for key properties of surface perception under uniform and variable illumination in the Adelson checkerboard effect. We claim that the effect actually consists of two distinct effects: simultaneous contrast and illuminant discounting. We first briefly recapitulate our previously published account of simultaneous contrast [54] in terms of the Adelson checkerboard display (Fig. 4A).

Our explanation of simultaneous contrast is most easily understood by assuming that the Adelson checkerboard display is parsed by the visual system such that only a single illumination level is perceived (i.e. by assuming that $\zeta = \eta = 0$). In other words, the 'shadow' region is actually perceived as having relatively lower reflectance than the 'brightly illuminated' region. As indicated above, the parameter setting of $\zeta = \eta = 0$ represents the brightness mode in gamut relativity. The luminance vector associated with

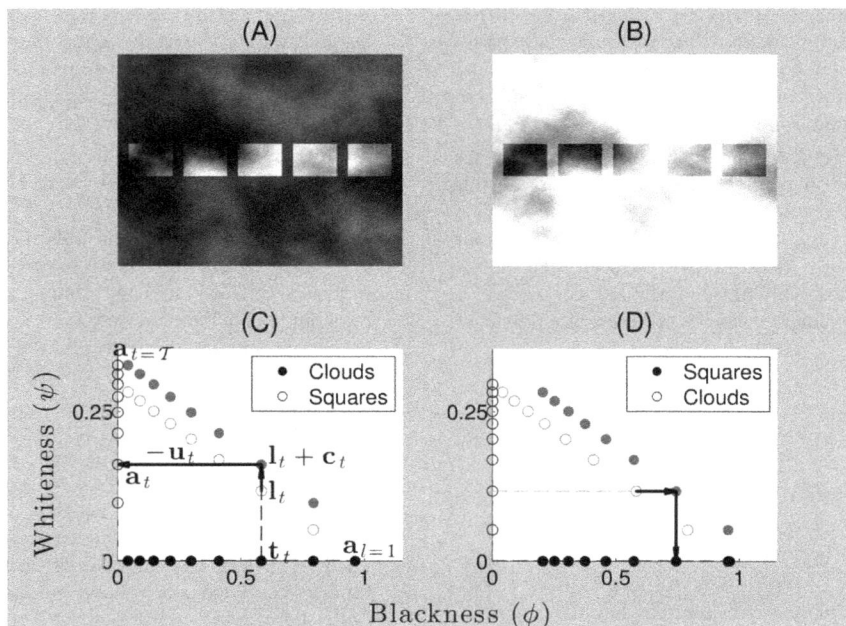

Figure 6. Anderson-Winawer display parsed in the brightness and lightness modes. (A,B) The Anderson-Winawer display with blackish and whitish backgrounds, respective y. (C,D) **Brightness mode**: The empty gray circles ($\mathbf{l}_t$ with $\eta = 0$) form the standard luminance gamut line for each pixel contained within each of the whitish or blackish squares shown in (A,B). The filled gray circles ($\mathbf{l}_t + \mathbf{c}_t$ with $\eta = 0$) form the standard increment and decrement gamut lines in (C) and (D), respectively, similar to Fig. 4A. These points, which are offset from the standard luminance gamut due to addition of the whiteness and blackness contrast vectors, would correspond to the perceived gray shades in (A,B) if the squares where rotated by $90°$ (rotation now shown here). **Lightness mode**: The model explains how the visual system computes separable whitish and blackish figural surface layers ($\mathbf{a}_t$) through blackish and whitish transparent 'ground' layers ($\mathbf{t}_t$) when $\eta = 1$. The transmittance-shift process subtracts the vector $\mathbf{u}_t$ from each filled gray circle to compute each $\mathbf{a}_t$ ($\mathbf{l}_t + \mathbf{c}_t - \mathbf{u}_t$ with $\eta = 1$). Surface layers are composed of the collection of every $\mathbf{a}_t$, represented here by the empty and filled black circles falling on the whiteness and blackness axes, respectively. The vertical and horizontal rows of empty and filled black circles thus correspond to the perceptually whitish and blackish layers evident in (A,B), respectively. The labelled vector corresponds to $t = 2$, $t = \mathcal{T}$ denotes the whitest pixel within the target region, and $l = 1$ denotes the blackest pixel in the surrounding region. Note that $\zeta = 0$.

each check (e.g. $\mathbf{l}_A$ and $\mathbf{l}_B$, where subscripts are used as labels rather than indices) are all constrained to fall on the standard luminance gamut, as defined by Eqn. (7), which is represented by the black dotted line in Fig. 4A. Checks with the same luminance (i.e. checks A and B) are thus mapped to identical points on the standard luminance gamut ($\mathbf{l}_A = \mathbf{l}_B$). The points $\mathbf{a}_A$, $\mathbf{a}_B$, $\mathbf{a}_C$ and $\mathbf{a}_D$ in Fig. 4A represent the blackness-whiteness coordinates of checks A, B, C and D following the addition of the contrast vector to the luminance vector (e.g. $\mathbf{l}_A + \mathbf{c}_A$). The coordinates $\mathbf{a}_A$ and $\mathbf{a}_C$ thus diverge, with a contrast blackness component added to $\mathbf{l}_A$, which is surrounded by brighter checks (check A) and a contrast whiteness component added to $\mathbf{l}_B$, which is surrounded by darker checks (check D). Checks A and B are thus mapped to blackness-whiteness coordinates that correspond to two different gray shades, $\mathbf{a}_B = \mathbf{l}_B + \mathbf{c}_B$ and $\mathbf{a}_A = \mathbf{l}_A + \mathbf{c}_A$.

Gamut relativity predicts that check B will be perceived as both 'blacker' and 'less white' than check A since the blackness coordinate of check A is larger than that of check B and the whiteness coordinate of check B is larger than that of check A. This prediction is generically consistent with the occurrence of the simultaneous contrast effect. As discussed in [54], moreover, this account of simultaneous contrast is quantitatively consistent with 'brightness matching' data and explains the inability of subject's to make satisfactory brightness matches when reference and test targets are viewed against backgrounds differing in luminance.

Our explanation of the large perceptual shift evident in the Adelson checkerboard display assumes that the display is parsed by the visual system into two different illumination levels (i.e. by

assuming that $\zeta = 1$). Fig. 4B illustrates the model account of the appearance of the Adelson checkerboard display when the illuminant-shift process is engaged. Given a parameter setting that represents the lightness mode in gamut relativity ($\zeta = 1$), the perceived difference in illumination level over the display is represented in the fact that $\mathbf{l}'_B$ and $\mathbf{l}_C$ now fall on separate luminance gamut lines, f' and f, respectively. Due to the process of discounting the illuminant, the blackness coordinates ($\phi_B = \phi_C$, $\phi_D = \phi_B$) of the vector pairs ($\mathbf{a}_B, \mathbf{a}_C$) and ($\mathbf{a}_D, \mathbf{a}_B$) remain invariant to differences in the depicted illumination intensity across the display. The perceptual shift between $\mathbf{l}'_B$ and $\mathbf{l}_B$ is equal in magnitude but opposite in sign to the physical shift in blackness induced by the illumination difference. The shift is given by the vector, $\mathbf{s}$, which specifies the magnitude of the discounting according to Eqn. (5), under the assumption that $\zeta = 1$. The shifted luminance vector coordinates are added to the contrast vectors to give $\mathbf{a}_B = \mathbf{l}_B + \mathbf{c}_B - \mathbf{s} = \mathbf{l}'_B + \mathbf{c}_B$. As the coordinates of check A in bright illumination remain unaffected by the discounting process, the magnitude of the difference between the vector $\mathbf{a}_B$ and $\mathbf{a}_A$ is much greater than the magnitude of the difference between the untransformed vectors in the brightness mode, given by $\mathbf{l}_B + \mathbf{c}_B$ and $\mathbf{l}_A + \mathbf{c}_A$ (Fig. 4A). The Adelson checkerboard display thus induces a far larger perceptual shift than would be expected on the basis of the processes underlying simultaneous contrast alone.

This perceptual shift can be understood as a manifestation of computational processes operating with the goal of parsing the retinal image into distinct surface and shadow layers. This goal can be clarified by first rewriting the equation $l'_B = l_B - s$ in the form $l_B = l'_B + s$. This equation says that the standard luminance vector associated with check B (l_B) is equal to the comparison luminance vector (l'_B) plus the shadow vector (s). In other words, the illuminant-shift process decomposes the standard luminance vector into surface and shadow vectors whose sum equals the original standard luminance vector. Due to this decomposition, s falls on the blackness axis and l'_B has been shifted towards the whiteness axis by the amount $|s|$. The decomposition thereby gives rise to the following property: The distance between the l_C and l'_B is less than than the distance between l_C and s; that is, the inequality $|l_C - l'_B| < |l_C - s|$ holds, given that $\zeta = 1$ and $\phi'_B > 0$. We claim that this inequality provides the basis for the capacity of the visual system to parse the Adelson checkerboard display into surface and shadow 'layers'. It ensures that points in backness-whiteness space representing physical surfaces in dim illumination can be unambiguously 'assigned' to corresponding points in bright illumination; that is, without interference from points representing shadows, which have been 'displaced' onto the blackness axis. These model properties thus explain the emergence of layered perceptual representations corresponding to surfaces and shadows.

To further emphasise the unique features of our model, we now analyse how the visual system might flexibly switch between brightness and lightness modes based on stimulus- and task-specific constraints. In this respect, we analyse data pertaining to the paint/transparency/shadow display used in [58]. In particular, we attempt to predict how stimulus- and task-driven constraints interact to determine brightness and lightness matches when the display appears either uniformly or variably illuminated (i.e. paint *versus* shadow, though the model predictions for the shadow condition apply equally well to the transparency condition). The model predictions are shown in Fig. 5A alongside the psycho-physical data in Fig. 5C, and agree reasonably well with the data. The model predicts the data well, with the main discrepancy being that the model predicts a slightly too strong simultaneous contrast effect with increments under uniform depicted illumination (c.f. condition PaintR) than is observed in the data. Of particular importance is to note that the model correctly predicts that lightness matching instructions have a disproportionately greater influence on contrast increments relative to decrements. This is because the model predicts that the increment region, which appears in dim illumination, undergoes the discounting, rather than the decrement region, which appears in bright illumination. Concordantly, the matching instructions have little influence in the latter case, but a large influence in the former case (i.e. condition ShadowR).

As discussed above, brightness and lightness matches are understood in the model as minimal perceptual mismatches between points lying on different gamut lines. The red test (or match) points lying on the dotted gray gamut line in Fig. 5B, for example, represent gray shades that minimal Euclidean distances with respect to the black points representing the gray shades of the target reference surfaces lying on different gamut lines (not shown). A key prediction of gamut relativity that sets it apart from alternative models [58] is thus that subjects cannot generally make satisfactory brightness or lightness matches [52–54]. Indeed, the model makes precise quantitative predictions that can be suitably compared against perceptual data obtained under conditions where subjects rate the perceptual similarity of their own matches [52–54,88]. The model is also consistent with perceptual data indicating that distinct computational processes subserve discrimination of targets against their local backgrounds ($\zeta = 0$) and lightness matching performance ($\zeta = 1$) [46].

Surface perception through atmospheric media

We now show how our model generalises to naturally account for properties of figure-ground separation and surface appearance through atmospheric media in terms of the Anderson-Winawer effect. We begin by illustrating the summation of luminance and contrast vectors in the brightness mode with $\eta = 0$ (Fig. 6C,D). In the absence of scission cues ($\eta = 0$), a single figural surface layer appears in plain view and is thus described as a surface brightness layer, according to the definition provided above. These vectors are given by the equation $l_t + c_t$. This latter situation occurs, for example, when the background regions of the Anderson-Winawer display are rotated by 90 degrees with respect to the target regions. The unfilled gray points shown in the blackness-whiteness plots of Fig. 6C,D correspond to a selection of pixels from within the square parts of the displays shown in Fig. 6A,B, and illustrate a mapping of physical luminance to standard luminance gamut. Note that these points are the same in Fig. 6C,D, since the physical luminance of all points in the squares in Fig. 6A,B are identical. The contrast vector c_t is associated with a pure whiteness 'boost' for figural contrast increments and a pure blackness 'boost' for figural contrast decrements. These contrast components are depicted as vertically and horizontally oriented whiteness and blackness vectors adding to the luminance vectors in Fig. 6C,D. For figural contrast increments (Fig. 6C), for instance, the boost shifts points on the standard luminance gamut upwards to form the standard increment gamut.

Given strong cues to the presence of transmissive media in an image, we assume that $\eta = 1$. The model equations then allow us to define $a_t = l_t + c_t - u_t$ to represent the underlying figural surface layer. The parameter β_T determines the orientation of the vector u_t; it is horizontal for target increments and vertical for target decrements. We may thus define a vector orthogonal to u_t and with different length using the definition $\beta_{\bar{T}} = 1 - \beta_T$; that is, we define the vector $t_t := l_t + c_t - u_{\bar{t}}$ to represent the transparent layer 'belonging' to the ground region surrounding the figure region. The transmittance-shift vectors u_t and $u_{\bar{t}}$ thus operate on each $l_t + c_t$ to compute each a_t and t_t. The shifts introduced by these vectors are equal in magnitude but opposite in sign to the physical shifts in blackness and whiteness induced by the transmittance difference between the ground medium and the underlying figural surface region seen in plain view (defined as $a_{t=T}$ in Fig. 6C). The application of these vectors implies that either blackness or whiteness coordinates always remains constant with respect to differences in the physical transmittance of the ground medium. This ensures, for example, that each a_T in Fig. 6C always lies closer to $a_{t=T}$ than does t_t; that is, the inequality $|a_t - a_{t=T}| < |t_t - a_{t=T}|$ holds for any $\eta > 0$.

This invariance is proposed to underlie the 'grouping' of vectors into perceptual layers characterising physically transmissive filters and media [55]. The net effect is to discount the transmittance of the ground medium in computing the underlying figural surface layer. The model thereby separates the figural image region into figure and ground layers, thereby accomplishing figure-ground separation. As indicated in the Model section, the transmittance-shift process with $\eta = 1$ is mathematically identical to a vector decomposition of the vector $i_t = l_t + c_t$ into figure $f_t = a_t$ and ground $g_t = t_t$ vectors corresponding perceptually to the figure and ground layers within the figural region.

In the Anderson-Winawer effect, this computational process generates the perceptual difference engendered by varying the mean luminance of the ground region outside the figure. In the case of the blackish ground region, vector decomposition operates to transform points on the standard increment gamut into a column of points lined up on the vertical constraint line provided by the whitest pixel of the underlying surface that appears in plain view. In the case of the whitish ground region, vector decomposition operates to transform points on the standard decrement gamut into a row of points lined up on the horizontal constraint line provided by the blackest pixel of the ground region that appears in plain view. The net effect is to produce separate sets of vectors corresponding to the ground and figural surface layers. As discussed above, each surface vector has the property that it lies closer to the surface vector appearing in plain view than does its 'partner' ground vector, allowing individual surface vectors to group together to form Gestalt-like representations of surface gray shades [25]. For figural contrast increments, for example, the vertical column of vectors lying on the whiteness axis form a whitish underlying figural surface layer by virtue of their relationship to the whitest figural pixel in plain view.

To quantitatively assess the predictions of the model with respect to perceptual data, we calculated predictions of 'lightness matches' ($\eta = 1$) for various Michelson contrast values of the target regions in the Anderson-Winawer display (Fig. 7A,C), as reported in [12,13]. Fig. 7B,D shows the model predictions alongside the perceptual data in Fig. 7E,F. The model correctly predicts that subjects' luminance settings always lie above the line indicating the luminance of the whitest or blackest pixels associated with figural contrast increments and decrements, respectively. This bias is a direct consequence of the asymmetric scaling of blackness-whiteness space, which forms a key computational feature of gamut relativity. In particular, the dominance of blackness with respect to whiteness ensures that the model weights the contrast blackness component more strongly than the contrast whiteness component. This causes a nominal gray shade seen against a black/white checkerboard, or black/white noise image, as used in the test displays reported in [12,13], to appear relatively whiter and less black than the reference region seen against more neutral backgrounds. The model thus compensates for this bias by selecting luminance values higher than those associated with pixels in plain view to produce the best 'lightness match'. Perceptual data on the Anderson-Winawer effect thus supports many of the key modelling postulates underlying gamut relativity. We leave to future work the goal of determining whether the model can accurately predict surface perception in the presence of simultaneous variations in both illumination and transmittance levels (i.e. with both $\zeta > 0$ and $\eta > 0$ [18]).

Discussion

We have presented a model that quantitatively accounts for perceptual data relating to some of the most striking and theoretically important effects of layered perceptual representation and surface appearance reported in the literature. In particular, the model reported in this paper documents four (4) key advances with respect to previously published work. The model: (1) provides the first unified analysis of how the visual system represents surfaces independently of shadows and atmospheric media, as exemplified in the Adelson checkerboard and Anderson-Winawer effects; (2) reconciles and unifies two prominent theories of surface lightness; (3) quantitatively predicts how stimulus- and task-driven factors combine to control brightness/lightness matching behaviours reported in published perceptual experiments; (4) unifies two previously published gamut relativity models, aimed at explaining properties of brightness/lightness perception [52,54], lightness/transparency perception [55] and lightness/gloss perception [56], respectively. The model thus provides the first unified account of the mid-level computations underlying layered perceptual repre-

sentation, which are believed to subserve the high-level computations involved in the identification of surface materials [1,2].

As indicated above, the model unifies two prominent theoretical approaches to surface lightness, known as the 'anchoring' and 'scission' theories [8–14,22–25,29,31–34], which have previously been applied separately to study the types of effects illustrated in Fig. 1. Lightness anchoring theory [25] posits that the visual system parses the scene into differentially illuminated regions, as in gamut relativity, before mapping relative reflectance values within each illumination level to absolute surface lightness values. This computation is captured in the current model in terms of the 'illuminant-shift' process applied to the blackness dimension. This process also generates a representation of the shadow layer. Scission theory [8–14,31–34] posits that the visual system parses the scene into layered representations, one seen through another, in order to disentangle the differential effects of surface reflectance and atmospheric media. This is accomplished by first estimating which surface regions appear in 'plain view' and which surface regions appear through atmospheric media of variable physical transmittance [12,13]. This computation is captured in the current model in terms of the 'transmittance-shift' process that is applied either to the blackness or whiteness dimensions, depending on figural contrast polarity. The computational outputs of the illuminant- and transmittance-shift processes are then combined in a single equation to compute layered representations. The current model thus mathematically unifies the central concepts in the lightness anchoring and scission theories.

The novel account of brightness and lightness perception embodied in gamut relativity may partially account for the wide range of behaviours observed when subjects perform perceptual matching tasks. At one extreme, task instructions to perform either brightness and lightness matches appear to have little or no influence on perceptual matches in the absence of a visible transparent layer. Such perceptual matches are associated with low intra- and inter-subject variability and tend to be subjectively relatively easy to make [58]. At the other extreme, lightness matches made under conditions where the task is largely underdetermined by stimulus-driven constraints—that is, in the absence of surface regions appearing in plain view—are associated with high intra- and inter-subject variability and tend to be subjectively relatively difficult to make. In such conditions, subjects may adopt a wide range of criteria to perform the matching task, such as attempting to 'infer' the surface appearance of the target under a certain illumination level [41–45,48,49]. In the middle ground, lightness matches made under conditions where stimulus-driven constraints are strongly present—that is, when surface regions appearing in plain view provide strong cue to the magnitude of the illuminant shift in shadow—are also associated with low intra- and inter-subject variability and tend to be subjectively relatively easy to make [58]. It is this class of lightness match that we have focused upon in this article. We expect to generalise our model to the more 'inferential' class of lightness match by demonstrating how subjects can 'infer' surface appearance under different gamut lines (i.e. by inferring the magnitude of the illumination or transmittance shift). The model thereby promises to provide a unified account of a wide range of matching strategies employed by human subjects in various experimental situations.

In providing a unified and general account of perceptual layering and surface appearance, our model provides crucial insights into the remarkable capacity of the human visual system to identify surface materials, illumination and atmospheric media. One potential application of this modelling framework involves the design of computer graphics software that allows a user to create

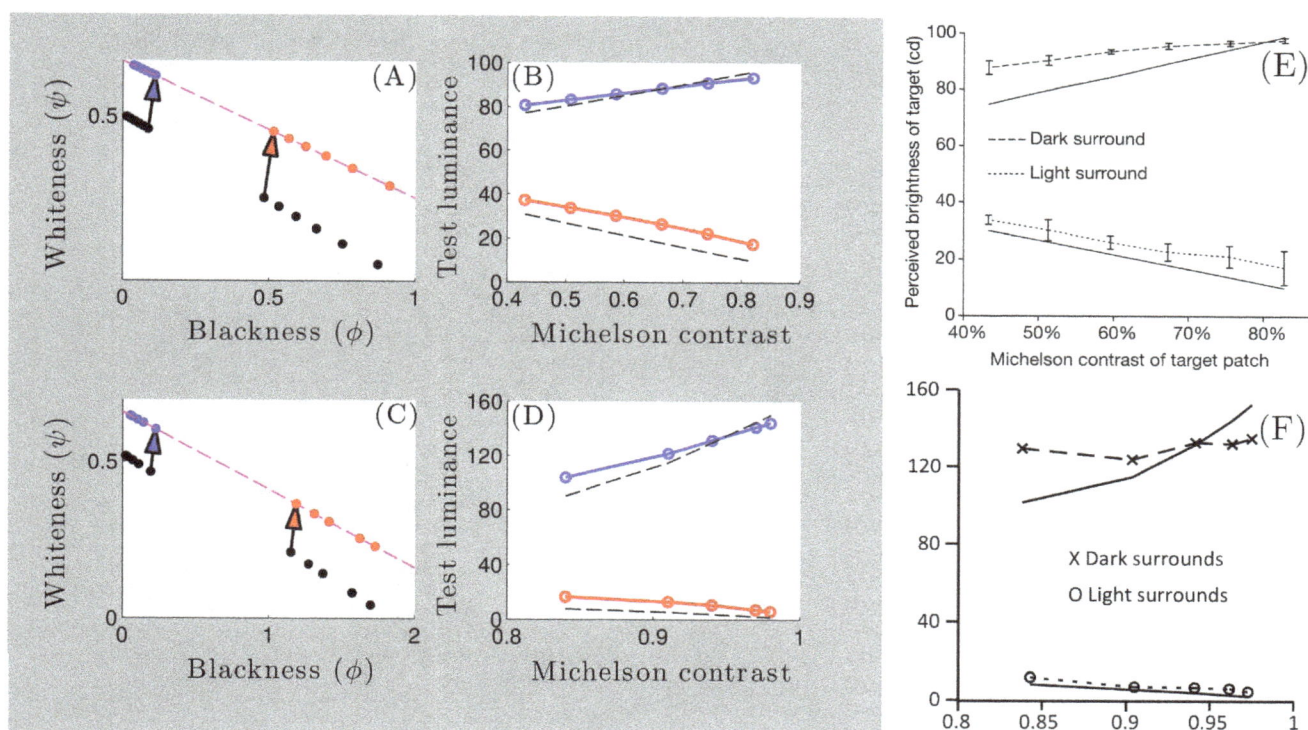

Figure 7. Model predictions of lightness matching data relating to the Anderson-Winawer effect. (A,C) Model luminance predictions were generated from minimal Euclidean distances between points representing the reference gray shades (black points) and gamut lines representing the test displays (red/blue points on purple dotted lines). Each black dot represents either the highest or lowest luminance value within the target reference region associated with each Michelson contrast level, depending on the figural contrast polarity of the reference region (i.e. highest for black dots matched to blue dots, and lowest for black dots matched to red dots) (B,D) The model correctly predicts that subjects set luminance values higher than the luminance values of the target reference region appearing in plain view (black dotted lines) for both figural contrast increments (blue points/lines) and decrements (red points/lines). Each empty blue and red dot corresponds to one of the filled blue or red dots in (A,C). Higher luminance values map to higher whiteness values and lower blackness values, respectively. (E) Data reprinted from [12]. (F) Data replotted from [13].

layered image representations by explicitly controlling perceptual variables (e.g. lightness and transparency), rather than indirectly specifying physical variables in models of light transport (e.g. reflectance and transmittance). We are also developing our modelling framework to leverage user-based image segmentation algorithms in a manner that will allow the user to predict brightness, lightness, transparency and gloss levels from arbitrary images.

Acknowledgments

We thank Fred Kingdom and an anonymous reviewer, whose comments led to significant improvements to the original manuscript.

Author Contributions

Conceived and designed the experiments: TV MDM. Performed the experiments: TV MDM. Analyzed the data: TV MDM. Contributed reagents/materials/analysis tools: TV MDM. Wrote the paper: TV MDM.

References

1. Adelson EH (2001) On seeing stuff: the perception of materials by humans and machines. In: Rogowitz BE, Pappas TN, editors, Proceedings of the SPIE, Human Vision and Electronic Imaging VI. pp. 1–12.
2. Fleming RW (2013) Visual perception of materials and their properties. Vision Research 94: 62–75.
3. Chandraker M, Bai J, Ng TT, Ramamoorthi R (2011) On the duality of forward and inverse light transport. IEEE Transactions on Pattern Analysis and Machine Intelligence 33: 2122–2128.
4. Donner C, Jensen HW (2005) Light diffusion in multi-layered translucent materials. ACM Transactions on Graphics (TOG) 24: 1032–1039.
5. Ghosh A, Hawkins T, Peers P, Frederiksen S, Debevec P, et al. (2008) Practical modeling and acquisition of layered facial reflectance. ACM Transactions on Graphics (TOG) 27: 1–10.
6. Metelli F (1974) The perception of transparency. Scientific American 230: 90–98.
7. Adelson EH (1995) Layered representations for vision and video. In: Proceedings of the IEEE Workshop on Representation of Visual Scenes. pp. 3–9.

8. Adelson EH (2000) Lightness perception and lightness illusions. In: Gazzaniga M, editor, The New Cognitive Neurosciences, Cambridge, MA: MIT Press. pp. 339–352.
9. Anderson BL (1997) A theory of illusory lightness and transparency in monocular and binocular images: the role of contour junctions. Perception 26: 419–453.
10. Anderson BL (1999) Stereoscopic surface perception. Neuron 24: 919–928.
11. Anderson BL (2003) The role of occlusion in the perception of depth, lightness, and opacity. Psychological Review 110: 785–801.
12. Anderson BL, Winawer J (2005) Image segmentation and lightness perception. Nature 434: 79–83.
13. Anderson BL, Winawer J (2008) Layered image representations and the computation of surface lightness. Journal of Vision 8: 1–22.
14. Barrow H, Tenenbaum J (1978) Recovering intrinsic scene characteristics from images. In: Hanson AR, Riseman EM, editors, Computer vision systems, Orlando, Fl: Academic Press. pp. 3–26.
15. Ekroll V, Faul F, Niederée R, Richter E (2002) The natural center of chromaticity space is not always achromatic: a new look at color induction.

Proceedings of the National Academy of Sciences of the United States of America 99: 13352–13356.

16. Faul F, Ekroll V (2002) Psychophysical model of chromatic perceptual transparency based on substractive color mixture. Journal of the Optical Society of America A 19: 1084–1095.

17. Faul F, Ekroll V (2011) On the filter approach to perceptual transparency. Journal of Vision 11: 1–33.

18. Faul F, Ekroll V (2012) Transparent layer constancy. Journal of Vision 12: 1–26.

19. Fleming RW, Dror RO, Adelson EH (2003) Real-world illumination and the perception of surface reflectance properties. Journal of Vision 3: 347–368.

20. Fleming RW, Torralba A, Adelson EH (2004) Specular reflections and the perception of shape. Journal of Vision 4: 798–820.

21. Fleming RW (2012) Human perception: Visual heuristics in the perception of glossiness. Current Biology 22: R865–6.

22. Gilchrist AL, Delman S, Jacobsen A (1983) The classification and integration of edges as critical to the perception of reflectance and illumination. Perception & Psychophysics 33: 425–436.

23. Gilchrist AL, Jacobsen A (1983) Lightness constancy through a veiling luminance. Journal of Experimental Psychology: Human Perception and Performance 9: 936–944.

24. Gilchrist AL (2005) Lightness perception: seeing one color through another. Current Biology 15: R330–2.

25. Gilchrist AL (2006) Seeing black and white. New York, NY: Oxford University Press.

26. Kim J, Anderson BL (2010) Image statistics and the perception of surface gloss and lightness. Journal of Vision 10: 1–17.

27. Kim J, Marlow P, Anderson BL (2011) The perception of gloss depends on highlight congruence with surface shading. Journal of Vision 11: 1–19.

28. Kim J, Marlow PJ, Anderson BL (2012) The dark side of gloss. Nature Neuroscience 15: 1590–1595.

29. Kingdom FAA (2008) Perceiving light versus material. Vision Research 48: 2090–2105.

30. Kingdom FAA (2011) Lightness, brightness and transparency: a quarter century of new ideas, captivating demonstrations and unrelenting controversy. Vision Research 51: 652–673.

31. Mausfeld R (2003) The dual coding of colour: 'surface colour' and 'illumination colour' as constituents of the representational format of perceptual primitives. In: Mausfeld R, Heyer D, editors, Colour perception: Mind and the physical world, New York: Oxford University Press. pp. 381–430.

32. Singh M, Anderson BL (2002) Toward a perceptual theory of transparency. Psychological Review 109: 492–519.

33. Sinha P, Adelson E (1993) Recovering reflectance and illumination in a world of painted polyhedra. In: Proceedings of the Fourth International Conference on Computer Vision. pp. 156–163.

34. Tappen MF, Freeman WT, Adelson EH (2005) Recovering intrinsic images from a single image. IEEE Transactions on Pattern Analysis and Machine Intelligence 27: 1459–1472.

35. Logvinenko AD, Kane J (2004) Hering's and Helmholtz's types of simultaneous lightness contrast. Journal of Vision 4: 1102–1110.

36. Logvinenko AD, Petrini K, Maloney LT (2008) A scaling analysis of the snake lightness illusion. Perception & Psychophysics 70: 828–840.

37. Todorović D (2006) Lightness, illumination, and gradients. Spatial Vision 19: 219–261.

38. Bressan P (2001) Explaining lightness illusions. Perception 30: 1031–1046.

39. Logvinenko AD (1999) Lightness induction revisited. Perception 28: 803–816.

40. Poirier FJAM (2009) The Anderson-Winawer illusion: it's not occlusion. Attention, Perception & Psychophysics 71: 1353–1359.

41. Arend LE, Goldstein R (1987) Simultaneous constancy, lightness, and brightness. Journal of the Optical Society of America A 4: 2281–2285.

42. Arend LE, Goldstein R (1990) Lightness and brightness over spatial illumination gradients. Journal of the Optical Society of America A 7: 1929–1936.

43. Arend LE, Spehar B (1993) Lightness, brightness, and brightness contrast: 1. Illuminance variation. Perception & Psychophysics 54: 446–456.

44. Arend LE, Spehar B (1993) Lightness, brightness, and brightness contrast: 2. Reflectance variation. Perception & Psychophysics 54: 457–468.

45. Blakeslee B, Reetz D, McCourt ME (2008) Coming to terms with lightness and brightness: effects of stimulus configuration and instructions on brightness and lightness judgments. Journal of Vision 8: 1–14.

46. Hillis JM, Brainard DH (2007) Distinct mechanisms mediate visual detection and identification. Current Biology 17: 1714–1719.

47. Logvinenko AD, Tokunaga R (2011) Lightness constancy and illumination discounting. Attention, Perception & Psychophysics 73: 1886–1902.

48. Schirillo JA (1999) Surround articulation. I. Brightness judgments. Journal of the Optical Society of America A 16: 793–803.

49. Schirillo JA (1999) Surround articulation. II. Lightness judgments. Journal of the Optical Society of America A 16: 804–811.

50. Schirillo J, Reeves A, Arend LE (1990) Perceived lightness, but not brightness, of achromatic surfaces depends on perceived depth information. Perception & Psychophysics 48: 82–90.

51. Gilchrist AL (2007) Lightness and brightness. Current Biology 17: R267–9.

52. Vladusich T (2013) Gamut relativity: A new computational approach to brightness and lightness perception. Journal of Vision 13: 1–21.

53. Vladusich T, Lucassen MP, Cornelissen FW (2007) Brightness and darkness as perceptual dimensions. PLoS Computational Biology 3: e179.

54. Vladusich T (2012) Simultaneous contrast and gamut relativity in achromatic color perception. Vision Research 69: 49–63.

55. Vladusich T (2013) A reinterpretation of transparency perception in terms of gamut relativity. Journal of the Optical Society of America A 30: 418–426.

56. Vladusich T (2013) A unified account of gloss and lightness perception in terms of gamut relativity. Journal of the Optical Society of America A 30: 1568–1579.

57. Vladusich T (2014) Brightness scaling according to gamut relativity. Color Research & Application 39: 463–465.

58. Blakeslee B, McCourt ME (2012) When is spatial filtering enough? Investigation of brightness and lightness perception in stimuli containing a visible illumination component. Vision Research 60: 40–50.

59. Geisler WS, Albrecht DG, Crane AM (2007) Responses of neurons in primary visual cortex to transient changes in local contrast and luminance. Journal of Neuroscience 27: 5063–5067.

60. Zurawel G, Ayzenshtat I, Zweig S, Shapley R, Slovin H (2014) A contrast and surface code explains complex responses to black and white stimuli in V1. Journal of Neuroscience 34: 14388–14402.

61. Rudd ME (2013) Edge integration in achromatic color perception and the lightness-darkness asymmetry. Journal of Vision 13: 1–29.

62. Barlow RB, Verrillo RT (1976) Brightness sensation in a ganzfeld. Vision Research 16: 1291–1297.

63. Knau HH (2000) Thresholds for detecting slowly changing Ganzfeld luminances. Journal of the Optical Society of America A 17: 1382–1387.

64. Barlow HB, Levick WR (1969) Changes in the maintained discharge with adaptation level in the cat retina. The Journal of Physiology 202: 699–718.

65. Barlow RB, Snodderly DM, Swadlow HA (1978) Intensity coding in primate visual system. Experimental Brain Research 31: 163–177.

66. Brown TM, Gias C, Hatori M, Keding SR, Semo M, et al. (2010) Melanopsin contributions to irradiance coding in the thalamo-cortical visual system. PLoS Biology 8: e1000558.

67. Dacey DM, Liao HW, Peterson BB, Robinson FR, Smith VC, et al. (2005) Melanopsin-expressing ganglion cells in primate retina signal colour and irradiance and project to the LGN. Nature 433: 749–754.

68. Ecker JL, Dumitrescu ON, Wong KY, Alam NM, Chen SK, et al. (2010) Melanopsin-expressing retinal ganglion-cell photoreceptors: cellular diversity and role in pattern vision. Neuron 67: 49–60.

69. Kayama Y, Riso RR, Bartlett JR, Doty RW (1979) Luxotonic responses of units in macaque striate cortex. Journal of Neurophysiology 42: 1495–1517.

70. Kinoshita M, Komatsu H (2001) Neural representation of the luminance and brightness of a uniform surface in the macaque primary visual cortex. Journal of Neurophysiology 86: 2559–2570.

71. Odermatt B, Nikolaev A, Lagnado L (2012) Encoding of luminance and contrast by linear and nonlinear synapses in the retina. Neuron 73: 758–773.

72. Papaioannou J, White A (1972) Maintained activity of lateral geniculate nucleus neurons as a function of background luminance. Experimental Neurology 34: 558–566.

73. Peng X, Van Essen DC (2005) Peaked encoding of relative luminance in macaque areas V1 and V2. Journal of Neurophysiology 93: 1620–1632.

74. Rossi AF, Paradiso MA (1999) Neural correlates of perceived brightness in the retina, lateral geniculate nucleus, and striate cortex. Journal of Neuroscience 19: 6145–6156.

75. Schmidt TM, Do MTH, Dacey D, Lucas R, Hattar S, et al. (2011) Melanopsin-positive intrinsically photosensitive retinal ganglion cells: from form to function. The Journal of Neuroscience 31: 16094–16101.

76. Hubel DH, Wiesel TN (1960) Receptive fields of optic nerve fibres in the spider monkey. The Journal of Physiology 154: 572–580.

77. Hubel DH, Wiesel TN (1961) Integrative action in the cat's lateral geniculate body. The Journal of Physiology 155: 385–398.

78. Mante V, Frazor RA, Bonin V, Geisler WS, Carandini M (2005) Independence of luminance and contrast in natural scenes and in the early visual system. Nature Neuroscience 8: 1690–1697.

79. Vladusich T, Lucassen MP, Cornelissen FW (2006) Do cortical neurons process luminance or contrast to encode surface properties? Journal of Neurophysiology 95: 2638–2649.

80. Xing D, Yeh CI, Shapley RM (2010) Generation of black-dominant responses in V1 cortex. Journal of Neuroscience 30: 13504–13512.

81. Xing D, Yeh CI, Gordon J, Shapley RM (2014) Cortical brightness adaptation when darkness and brightness produce different dynamical states in the visual cortex. Proceedings of the National Academy of Sciences of the United States of America 111: 1210–1215.

82. Yeh CI, Xing D, Shapley RM (2009) "Black" responses dominate macaque primary visual cortex V1. Journal of Neuroscience 29: 11753–11760.

83. Anderson BL, Kim J (2009) Image statistics do not explain the perception of gloss and lightness. Journal of Vision 9: 1–17.

84. Grady L, Schwartz EL (2006) Isoperimetric graph partitioning for image segmentation. IEEE Transactions on Pattern Analysis and Machine Intelligence 28: 469–475.

85. Adelson EH (1993) Perceptual organization and the judgment of brightness. Science (New York, NY) 262: 2042–2044.

86. Allred SR, Brainard DH (2013) A Bayesian model of lightness perception that incorporates spatial variation in the illumination. Journal of Vision 13: 1–18.

87. Corney D, Lotto RB (2007) What are lightness illusions and why do we see them? PLoS Computational Biology 3: 1790–1800.

88. Logvinenko AD, Maloney LT (2006) The proximity structure of achromatic surface colors and the impossibility of asymmetric lightness matching. Perception & Psychophysics 68: 76–83.

89. Gilchrist AL (1977) Perceived lightness depends on perceived spatial arrangement. Science (New York, NY) 195: 185–187.

90. Gilchrist A, Kossyfidis C, Bonato F, Agostini T, Cataliotti J, et al. (1999) An anchoring theory of lightness perception. Psychological Review 106: 795–834.

21

Revisiting the Two-Layer Hypothesis: Coexistence of Alternative Functional Rooting Strategies in Savannas

Ricardo M. Holdo*

Divison of Biological Sciences, University of Missouri, Columbia, Missouri, United States of America

Abstract

The two-layer hypothesis of tree-grass coexistence posits that trees and grasses differ in rooting depth, with grasses exploiting soil moisture in shallow layers while trees have exclusive access to deep water. The lack of clear differences in maximum rooting depth between these two functional groups, however, has caused this model to fall out of favor. The alternative model, the demographic bottleneck hypothesis, suggests that trees and grasses occupy overlapping rooting niches, and that stochastic events such as fires and droughts result in episodic tree mortality at various life stages, thus preventing trees from otherwise displacing grasses, at least in mesic savannas. Two potential problems with this view are: 1) we lack data on functional rooting profiles in trees and grasses, and these profiles are not necessarily reflected by differences in maximum or physical rooting depth, and 2) subtle, difficult-to-detect differences in rooting profiles between the two functional groups may be sufficient to result in coexistence in many situations. To tackle this question, I coupled a plant uptake model with a soil moisture dynamics model to explore the environmental conditions under which functional rooting profiles with equal rooting depth but different depth distributions (*i.e.*, shapes) can coexist when competing for water. I show that, as long as rainfall inputs are stochastic, coexistence based on rooting differences is viable under a wide range of conditions, even when these differences are subtle. The results also indicate that coexistence mechanisms based on rooting niche differentiation are more viable under some climatic and edaphic conditions than others. This suggests that the two-layer model is both viable and stochastic in nature, and that a full understanding of tree-grass coexistence and dynamics may require incorporating fine-scale rooting differences between these functional groups and realistic stochastic climate drivers into future models.

Editor: Robert Planque, Vrije Universiteit, The Netherlands

Funding: This work was funded by the University of Missouri. The funders had no role in study design, data collection and analysis, decision to publish, or preparation of the manuscript.

Competing Interests: The author has declared that no competing interests exist.

* E-mail: holdor@missouri.edu

Introduction

The distribution of most terrestrial biomes can be derived from climatic variables [1], but savannas challenge this model, often occurring under conditions that can theoretically support forests [2]. Why do trees fail to competitively exclude grasses in many ecosystems, and *vice versa*? Walter's two-layer model [3] proposes that differences in soil moisture use as a function of depth results in niche partitioning (and therefore coexistence) between trees and grasses [3,4,5,6]. The role of this mechanism as a general explanation for tree-grass coexistence has not been comprehensively tested. Despite this, it has gradually fallen out of favor in the savanna literature [7,8,9,10,11], giving way to demographic models, which assume that trees and grasses essentially compete for the same resources, but periodic disturbances prevent trees from completely excluding grasses [3,12].

As originally proposed by Walter, vertical resource partitioning interacted with other tree-grass trait differences (e.g., in root morphology and water use) to promote coexistence, but only in certain savanna types [6,13]. Subsequent work somewhat simplified the niche differentiation model by focusing on rooting separation alone, and over time this vertical resource partitioning model has become a general hypothesis for explaining tree-grass coexistence [4,6,7,14,15]. The evidence for or against such vertical partitioning in savannas has been mixed [16,17,18,19,20], and this may explain why demographic explanations have become more dominant over the past decade or so [8,12,21]. Fire and herbivory are well-known to be strong drivers of tree cover change in many savanna ecosystems [21,22,23,24], but it is still far from clear how important or general the vertical resource partitioning mechanism is for allowing coexistence and for determining tree-grass ratios in the absence of fire. In fact, global and continental-scale studies suggest that water availability is the ultimate factor determining the upper boundary of tree cover over a wide precipitation range [2,12,21], and that, below a certain threshold of mean annual precipitation, resource availability is the driving force behind tree-grass coexistence. Ultimately, it still has not been satisfactorily demonstrated that niche partitioning mechanisms are incapable of explaining the savanna state, even under quite mesic conditions. In other words, are aboveground drivers such as fire and herbivory necessary for tree-grass coexistence, or are they modifiers acting upon a system that is ultimately made possible by resources alone? Second, how pervasive is niche partitioning likely to be as a viable mechanism of coexistence across broad edaphic and climatic gradients?

The empirical case against the vertical resource partitioning model includes the observation that trees and grasses sometimes show substantial rooting overlap [20,25,26], the suggestion that water may rarely infiltrate to deeper soil layers during the growing season [27], and the fact that grasses have deep roots and therefore possibly the same access to deep water as trees [28]. Against this, it must be considered that rooting differences may be subtle but important [7,29], that the degree of deep infiltration during the growing season may vary systematically with climate and soils [30], and that deep grass roots may play little functional significance, except as a survival mechanism during drought [31]. Grass and tree roots have very distinct morphologies; for example, grass roots have no secondary growth or central taproot, but rather are fine and adventitious, with extremely high total root length and surface area. These morphological differences translate into functional differences in terms of the relative ability of trees and grasses to extract water as a function of depth, independently of maximum rooting depth [31].

From the theoretical perspective, few models have explicitly investigated the importance of vertical rooting separation for tree-grass coexistence. It is unclear, for example, how much rooting separation is necessary to allow coexistence, and under what conditions competition is likely to be most intense. Early theoretical models based on the two-layer model suggested that trees and grasses could coexist stably under certain conditions [4,14], but these models relied on extreme rooting separation between the two functional groups, assuming that grasses and trees have exclusive access to topsoil and subsoil moisture, respectively. This extreme assumption, unsupported by empirical evidence [28], may have contributed to the gradual loss of support for the resource partitioning hypothesis. Models that relaxed this extreme assumption failed to predict coexistence in the absence of disturbance [10,32]. Over the past decade or so, a number of savanna models have tended to focus on the role of disturbance (primarily fire) as a requirement for the stable maintenance of the savanna state [8,9]. Other theoretical studies that have focused primarily on hydrological mechanisms [27,33,34] have, on the other hand, shown convincingly that water availability alone can predict the savanna state, but these models, motivated perhaps by a lack of apparent empirical support for vertical niche differentiation [27], have side-stepped the issue of rooting separation between trees and grasses. A common feature of these models is the stochastic treatment of rainfall inputs [33,35]. This stochasticity, in combination with a spatially-explicit representation of local vegetation interactions and ecophysiological differences or tradeoffs between trees and grasses, allows the savanna state to persist.

In this paper I present a simple model of competition among alternative water-use strategies, defined by contrasting rooting profiles (i.e., patterns of root biomass allocation as a function of depth) to test the role of vertical resource partitioning as a coexistence mechanism in savannas. The model puts aside some of the mechanisms that have already been shown by theory to promote coexistence between trees and grasses, such as phenological differences, differences in maximum rooting depth [36], and lateral competition for resources [27]. I assume that variation in root shape along the vertical axis is the key trait differentiating trees from grasses (or other functional groups, or even species within a functional group), so that the model is spatially-implicit on the horizontal axis but spatially-explicit on the vertical one. I first investigate whether the model predicts coexistence of two or more rooting profiles for a single, well-studied site (Nylsvlei, South Africa). I then test the model across a broad rainfall gradient and for two very different types of soil texture, to ask the following questions: 1) can vertical resource partitioning mechanisms provide a sufficient explanation for coexistence across a wide range of conditions, and 2) are such mechanisms more likely to offer an explanation for coexistence under some climatic and edaphic conditions than others?

Materials and Methods

Soil moisture dynamics and plant transpiration model

I modeled the biomass and water uptake dynamics of four competing plant rooting profiles. I assumed that different rooting profiles correspond to distinct water-acquisition strategies defined by particular root shapes or depth distributions. I assume that functional root mass (the ability of roots to acquire water), which is the key variable here, can be represented by root biomass. It should be noted that physical rooting profiles (whether measured by root length, width, or mass) do not necessarily show a direct correspondence with functional profiles [37], but I make this simplifying assumption to keep the model tractable (i.e., to avoid having to allocate biomass increases to roots that absorb water and those that do not). I assume throughout the paper that the term "rooting profile" corresponds to functional root activity.

To model soil moisture dynamics, I used a modified version of Rodriguez-Iturbe et al.'s [38] soil moisture balance equation as the starting point for a vertically-resolved 'bucket-type' soil moisture model. I assumed a maximum rooting depth of 70 cm and divided the soil space into eight layers, as follows: 0–5 cm, 5–10 cm, and in 10-cm increments thereafter. The soil moisture dynamics in each layer i are determined by an ordinary differential equation:

$$\frac{dS_i}{dt} = \left[I_i(t) - K_{sat} \left[\frac{S_i}{S_f} \right]^{2\tau+2} - E_i - \sum_k T_{i,k} \right] / (nD_i) \quad (1)$$

where S_i is the relative moisture saturation of soil layer i, I_i is infiltration (in mm d^{-1}) from the layer above (in the case of layer 1, I is rainfall input), K_{sat} (in mm d^{-1}) and τ are texture-dependent parameters that determine the rate of water infiltration to deeper layers [39], E_i is evaporation, $T_{i,k}$ is transpiration (both in mm d^{-1}) of rooting profile k from soil layer i, n is soil porosity, D_i is the depth of soil layer i (in mm), and S_f is the field capacity of the soil. The second term on the r.h.s. of eq. 1 serves as an input for the layer below, and losses from the last layer (layer 8, 60–70 cm) result in deep drainage. I assumed that evaporation only occurs in the top 5 cm, i.e., layer 1 [40].

To model plant transpiration, I assumed a pipe model in which roots allocated to a particular soil layer retain vascular independence (Fig. 1a). This pipe model framework has previously been used to inform theoretical studies of plant ecophysiology and allometry [41,42], and is supported by empirical studies [43,44]. The amount of water that originates from a specific soil layer and that is then transpired depends on the moisture saturation of the soil layer and the root mass allocated to it. The total transpiration for any given profile is equal to the sum of the fractions contributed by each layer. The resulting net biomass gain resulting from carbon assimilation is then distributed across the profile so as to maintain the (fixed) root shape, i.e., transpiration results in biomass increases or declines but plant architecture is conserved. As a result, root biomass in a poorly-performing layer can still increase as a result of a subsidy from a layer with high uptake. I assumed that transpiration from each layer follows a Michaelis-Menton (MM) function:

$$T_{i,k}^{\max} = \frac{T_{atm} B_k \times RMR \times R_{k,i}}{B_{hsat} + \sum_k B_k \times RMR \times R_{k,i}} \quad (2)$$

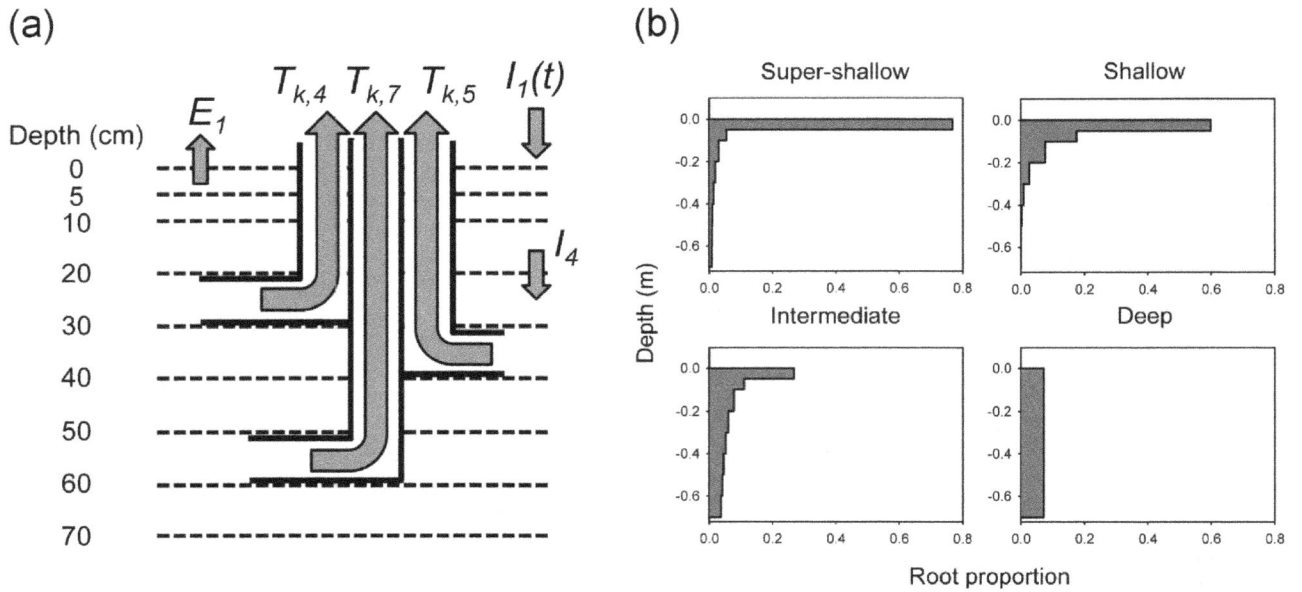

Figure 1. Water balance terms and rooting profiles used in the model. (a) The soil profile is partitioned into eight discrete layers for simulation of soil moisture dynamics. Transpiration is modeled according to a pipe model, with independent vessels linking each soil layer with the atmosphere. A generic functional group k with all root biomass allocated to soil layers 4 (20–30 cm), 5 (30–40 cm) and 7 (50–60 cm) is depicted. Water fluxes shown are rainfall inputs to layer 1 ($I_1(t)$), infiltration from layer 3 to layer 4 (I_4), evaporation from layer 1 (E_1), and transpiration from layers 4, 5 and 7. (b) Rooting profiles for four functional groups; the maximum rooting depth is 70 cm in all cases and cumulative relative root mass equals 1.

Here, $T_{i,k}^{\max}$ is the maximum transpiration of profile k from soil layer i, assuming that soil moisture is not limiting, B_k is the total biomass (in g m^{-2}) of profile k, RMR is the root mass ratio (the proportion of the total biomass that is comprised by roots, assumed equal across functional groups), $R_{k,i}$ is the proportional root mass allocation of profile k to layer i, and B_{hsat} (in g m^{-2}) is the half-saturation value of root biomass in the MM function. As root biomass increases, transpiration $T_i^{\max}$ saturates at T_{atm}, the maximum atmospheric evapotranspirational demand. This occurs as long as soil moisture S_i exceeds S^*, the relative soil moisture content below which stomatal conductance declines as a result of moisture stress [35,45]. The choice of the MM was somewhat arbitrary – any saturating function might prove acceptable. The rationale for a saturating function is that the ability to transport water is limited by atmospheric demand, regardless of root conducting capacity. The nonlinear function minimizes the need to truncate transpiration rates at T_{atm} (see below, and p. 71 in [28]). The MM function in particular is widely used in biology, both as a mechanistic and phenomenological representation of rate limitation, and has been used to model water uptake [24,46]. Below S^*, transpiration declines as follows:

$$T_{i,k} = \begin{cases} 0 & if \quad S_i \leq S_w \\ \frac{S_i - S_w}{S^* - S_w} T_{i,k}^{\max} & if \quad S_w < S_i < S^* \\ T_{i,k}^{\max} & if \quad S_i \geq S^* \end{cases} \quad (3)$$

This assumes that transpiration declines in a linear fashion as a function of S_i between S^* and S_w, the wilting point [38,45]. Below S^*, transpiration is no longer solely under the control of atmospheric demand. The total transpiration is summed across soil layers and profiles. Because the soil layers and their respective roots are treated independently, total uptake could theoretically exceed T_{atm} despite eq. 2. To prevent this from occurring, when

$\sum_i \sum_k T_{i,k} > T_{atm}$, $T_{i,k}$ values are rescaled proportionately so that the sum equals T_{atm} [28]. The rescaled values are used for the transpiration term in eq. 1.

The biomass dynamics for each functional group are then given by:

$$\frac{dB_k}{dt} = WUE \sum_i^8 T_{i,k} - RESP \times B_k \quad (4)$$

Where WUE (in g m^{-2} per kg of water transpired) and $RESP$ (in g g^{-1} m^2) are the whole-plant water use efficiency and mass-specific respiration costs, respectively, and the $T_{i,k}$ values are the rescaled values.

I obtained most model parameters from published estimates (Table 1). Most of the parameters apply to soil texture and vegetation data from the Nylsvley site, an extensively-studied tropical savanna ecosystem in South Africa [28] that has often been used as a model system for savanna ecohydrology [27,33,34,45]. I used published estimates of total transpiration and net primary productivity (NPP) for the site [28] to derive an aggregate value of WUE. I solved eq. 4 using a mean annual transpiration estimate and estimates of standing biomass obtained from [28] to obtain a value for $RESP$. I assumed that the maximum standing biomass estimates of B for Nylsvlei represent a steady state, and that mean respiration costs balance assimilation gains, which allowed me to set the l.h.s. in eq. 4 to 0 and solve for $RESP$. There was no parameter tuning or fitting.

Rainfall

To generate rainfall scenarios that are consistent with observed patterns across mean annual precipitation (MAP) gradients, I relied on North American daily climate data from Long Term Ecological Research (LTER) sites (given the lack of comparable

Table 1. Parameters for the soil moisture and biomass dynamics model.

Symbol	Soil texture		Units[†]	Description	Source(s)
	Coarse	Fine			
S_f	0.30	0.50		Field capacity	[45,61]
S_w	0.05	0.1		Wilting point	[45,61]
S^*	0.1	0.2		Soil moisture saturation leading to reduced transpiration	[45,61]
K_s	1100	300	mm d^{-1}	Saturated conductivity	[39,61]
τ	4.5	6.0		Exponent for conductivity function	[39,45]
n	0.42	0.45		Soil porosity	
E	1.5	1.5	mm d^{-1}	Maximum evaporative demand	[45]
T_{atm}	3.25	3.25	mm d^{-1}	Maximum atmospheric demand in mm d^{-1}	[45]
B_{hsat}	800	800	g m^{-2}	Half-saturation value of B	Author's estimate
RMR	0.5	0.5		Root Mass Ratio	[28]
WUE	4.0	4.0	g m^{-2} kg^{-1} H$_2$O	Water-use efficiency	Derived from data in [28]
$RESP$	0.001	0.001	g g^{-1} m^2	Respiratory cost	Derived from data in [28] and eq. 4

[†]Empty spaces signify dimensionless parameters.

data for African sites): Jornada (long-term MAP: 276 mm), Shortgrass (329 mm), Cedar Creek (778 mm) and Kellog (896 mm). I treated rainfall as a Poisson process, with exponentially distributed interarrival times and lognormally distributed depth. I was therefore able to summarize long-term daily rainfall across sites using three parameters: a rate parameter (λ) for the interarrival time and mean (μ) and standard deviation (σ) parameters for event depths. These parameters showed strong linear correlations with MAP (Fig. S1). This allowed me to generate time series of rainfall realizations for specific MAP values with a stochastic simulator. I assumed a mean annual rainfall of 650 mm for Nylsvlei. Although more realistic tools are available for modeling Southern African rainfall [47], the approach I used enabled me to derive rainfall sequences with reasonable event depths and interarrival times as a function of a single independent variable (MAP).

Rooting profiles

I used a Beta distribution, rescaled to vary between 0 and 70 cm (the maximum soil depth in the model) to generate the four rooting profiles. I generated distributions in R [48] by using the *pbeta* function to obtain discretized rooting fractions in each of the eight soil layers used by the model (Fig. 1b). The distributions ranged from a 'super-shallow' profile (with Beta distribution parameters shape 1 = 0.1 and shape 2 = 1) where >75% of root mass was concentrated in the top 5 cm of soil, to a 'deep' profile (shape 1 = 1; shape 2 = 1) with a uniform distribution along the depth axis (Fig. 1b). The 'super-shallow' and 'shallow' (shape 1 = 0.5; shape 2 = 5) profiles might be thought of as a grassy/herbaceous functional group, and the 'intermediate' (shape 1 = 0.5; shape 2 = 1) and 'deep' profiles as contrasting deeper-rooted forms typical of shallow- vs. deep-rooted tree species (Fig. 1b), such as *Terminalia sericea* and *Pterocarpus angolensis*, respectively [49]. None of these profiles were fit to real rooting data, but rather represent a range of possible shapes that are qualitatively consistent with profiles reported in the literature [28,50].

Model testing

I solved the model equations numerically in Microsoft Visual C++ using a daily time step (the full code is available in Source

Code S1). I ran the simulations to steady state in every case (100–1000 years). I initially ran simulations with a number of preliminary scenarios designed to test the robustness and soundness of the model. In these simulations, I assumed that parameters unrelated to root shape, *i.e.*, *RMR*, *WUE*, and *RESP* were identical across rooting profiles. I conducted the following simulations:

i. A single 1000-year run with each rooting profile as the sole strategy, to establish that each profile was viable in isolation.

ii. Simulations with all four profiles competing. For each run (N = 5, 500 years), the starting biomass of each profile was chosen from a uniform random distribution (range: 1000–2000 g m^{-2}).

iii. To test the robustness of the model to the particular uptake function chosen, I also conducted a pair of 1000-year runs, the first assuming the default MM uptake function and the second assuming a linear uptake function (i.e., with $B_{hsat} = 0$).

iv. To investigate the role of the stochastic nature of rainfall inputs, I conducted 100-year simulations with all functional groups present, but assumed a constant, deterministic rainfall input, equal to MAP divided by 365 to give an input of 1.78 mm d^{-1}. I set initial biomass values for all rooting profiles to 1000 g m^{-2}.

v. I conducted a global sensitivity analysis (GSA), consisting of 400 runs (of 500 years each), in which 13 parameters were assigned offset values (multiplied by their default values from Table 1) drawn from a uniform random distribution with range 0.8–1.2 (i.e., each parameter deviated by a maximum of ±20% of its default value). I included all 12 model parameters from Table 1 in the analysis, plus an additional parameter (default value = 0.5 g m^{-2} kg^{-1} H$_2$O) that determined the *WUE* advantage (in relation to the *WUE* of the other three profiles) of the super-shallow profile. I chose this profile for testing the effect of varying *WUE* because it was the most likely to go extinct during simulations, and I wanted to explore the ability of *WUE* to 'rescue' this strategy. The primary objective of the GSA was to explore the parameter space of coexistence. I calculated the proportion of runs for which two or more profiles coexisted after 500 years, and the median number of coexisting profiles.

In all of the above simulations, I assumed equal *WUE* (set at the default value, Table 1) for all rooting profiles, and assumed the environmental conditions of the Nylsvlei site (650 mm MAP and coarse-textured soils).

Single site coexistence. I next ran a series of simulations in which each rooting profile was assigned a slightly higher WUE than the other three (4.5 instead the of the default value of 4.0 g m^{-2} kg^{-1} H$_2$O). The competitive exclusion principle states that two species with identical traits cannot coexist stably [51]; over the long term, a slight advantage in one species (e.g., a marginally higher intrinsic population growth rate) will lead to competitive exclusion. In most of the simulations in this paper, therefore, I parted from the assumption that one of the four profiles under consideration has a slight advantage in *WUE* (*i.e.*, higher biomass growth per g of water transpired), and then tested the ability of the other profiles to persist and invade despite this *WUE* disadvantage. For each set of *WUE* conditions (out of four), I ran 500-year simulations (N = 5 runs) for the Nylsvlei environmental conditions and tested for coexistence at the end of each run. In these and other cases I treated a final biomass of <1 g m^{-2} as extinction. To test the robustness of the coexistence results, I repeated the simulations under the following conditions: the strategy with the *WUE* advantage was run alone to steady state (from an initial biomass of 4000 g m^{-2}) for 100 years, and the remaining strategies were then introduced to the system with small initial biomasses (10 g m^{-2}) so as to test their ability to invade the established profiles (*i.e.*, increase over 100 additional years. For completeness (and to further test the model), I also ran simulations in which the established profile tried to invade itself, but with a lower (default) value of *WUE*.

Coexistence across rainfall and edaphic gradients. To test the robustness of the coexistence results across a range of rainfall and soil conditions, I ran the model (500-year simulations, N = 5 runs, initial biomass drawn randomly from the interval 1000–2000 g m^{-2}) across a simulated rainfall gradient spanning 300–1500 mm y^{-1} (in intervals of 300 mm y^{-1}) under two soil texture scenarios: a coarse-textured soil substrate (based on Nylsvlei) consisting of sand/loamy sand soils, and a fine-textured substrate consisting of a higher clay fraction. These simulations build on the GSA, with the difference that this exercise varied external environmental conditions rather than intrinsic biological parameters. Variation in rainfull inputs was achieved purely by varying MAP in the rainfall generator; simulating contrasting soil substrates required adjusting some of the model parameters (Table 1).

Results

Model testing

All rooting profiles were viable when run in isolation under the conditions present at Nylsvlei (Fig. 2a), rapidly converging on a standing biomass of around 4500 g m^{-2} in all cases. This represents total (above plus belowground) biomass, and given the assumed *RMR* of 0.5 represents about 2250 g m^{-2} of aboveground biomass. This value is compatible with the estimate of tree standing biomass of 1627 g m^{-2} (which excludes grasses, so is an under-estimate of total aboveground biomass) reported by Scholes and Walker [28] for this site. When all profiles were included, the shallow and intermediate profiles achieved steady-state (nonzero) biomass values, with the super-shallow and the deep profiles going extinct rapidly and gradually, respectively (Fig. 2b). The runs with

linear uptake did not differ qualitatively from those with MM uptake, suggesting that the model is robust to the particular uptake function chosen (Fig. 2c). In the deterministic rainfall scenario, the super-shallow rooting profile excluded all others within 100 years (Fig. 3a). I repeated the simulation by excluding the winning strategy after each run. In every case, a single profile (always the shallowest one) excluded all others (Fig. 3b–d), suggesting that the

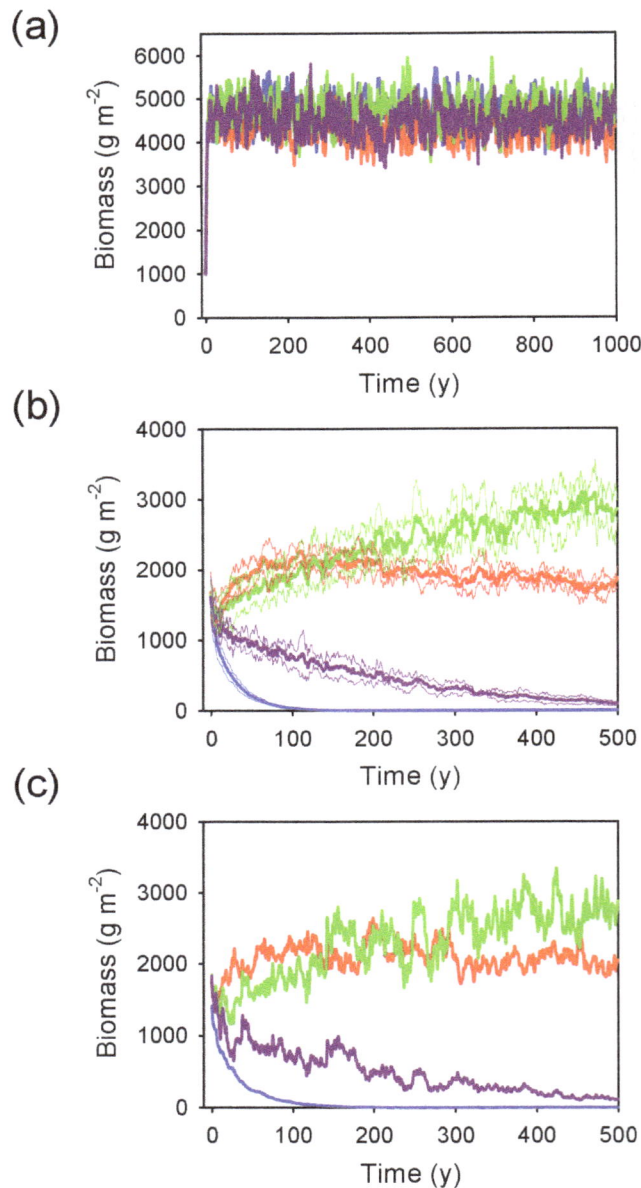

Figure 2. Time series of biomass dynamics for four rooting profiles assuming coarse-textured soils and MAP = 650 mm y^{-1}. Model outcomes with (a) each profile in isolation (note: profiles are plotted together for comparision); (b) all profiles initially present, with equal *WUE*; (c) the same scenario as b, but assuming a linear uptake function instead of the default Michaelis-Menton function. Thick lines show mean values and thin lines (where present) show the mean ± 1 SD across runs. Profile key: blue = super-shallow, red = shallow, green = intermediate, purple = deep.

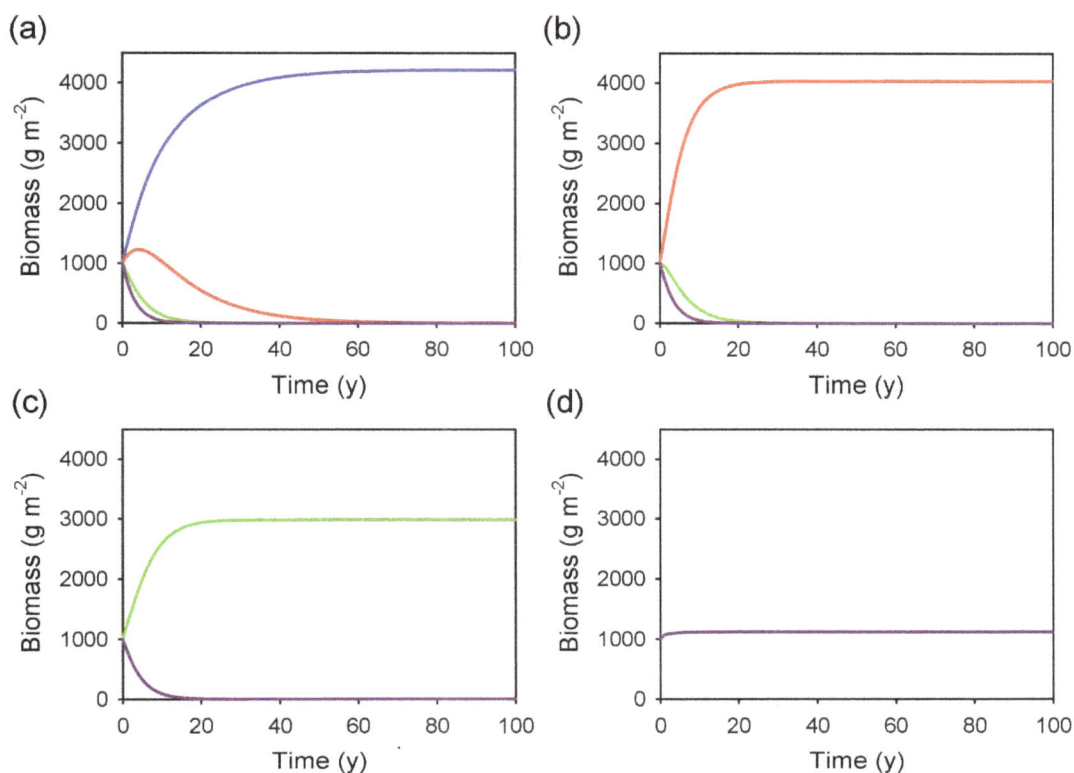

Figure 3. Time series of biomass dynamics for four rooting profiles on coarse-textured soil under a deterministic (constant) rainfall regime. Model outcomes (a) with all profiles included; (b) after removal of the winning profile (*i.e.*, super-shallow) from a; (c) after removal of the winning profile (*i.e.*, shallow) from b; (d) with the deep profile in isolation. MAP was fixed at $= 650$ mm y^{-1} and divided by 365 to generate a constant daily rainfall input. Profile key: blue = super-shallow, red = shallow, green = intermediate, purple = deep.

stochastic nature of the rainfall inputs is a key ingredient for the coexistence of alternative profiles.

In the sensitivity analysis, coexistence of two or more strategies (median = 2) persisted after 500 years in 56% of the 400 runs. The outcome of the GSA was strongly influenced by the value of a single parameter, however. Variation in S_f (field capacity) accounted for 70% of the variance in total biomass after 500 years. Given that field capacity is both a readily-measurable biophysical parameter and one that is closely correlated (*i.e.*, not independent of) with other ecohydrological parameters such as the wilting point (S_w) and saturated conductivity (K_s), it was clear that the GSA was too conservative (in other words, "generous" in its exploration of parameter space) in its assumption of independent parameter variation. To restrict this dominant effect of S_f, I examined a subset of results in which S_f was restricted to values that fell within 5% of its default value of 0.30 (104 of the 300 runs). For this subset of the GSA runs, coexistence of two or more strategies (median = 3) persisted in 96% of the runs, suggesting that the basic coexistence result is robust.

Single site coexistence

Two or more rooting profiles coexisted under each of the four *WUE* conditions (Fig. 4). When the super-shallow profile had a slightly higher *WUE* than the other profiles, all profiles were still present after 500 years (although the deep profile was still declining; Fig. 4a). When the shallow profile had higher *WUE* than the others, it coexisted with the deep profile (but not the other two; Fig. 4b). A higher *WUE* for the intermediate profile allowed coexistence with the shallow one (Fig. 4c), and a higher *WUE* for

the deep profile allowed it to coexist with the shallow profile, but not the other two (Fig. 4d). Two general results were that: i) the super-shallow profile was not viable unless it had a *WUE* advantage; and ii) profiles that were dissimilar (shallow and deep) were more likely to coexist than those that were more similar (shallow and intermediate or intermediate and deep), as predicted by the competitive exclusion principle. The coexistence results were robust, as suggested by the ability of alternative profiles to invade the dominant one when the latter was at steady state (Fig. 5).

Coexistence across rainfall and edaphic gradients. The coexistence results were also robust to variation in rainfall and soils (Figs. 6 and 7). Two or more rooting profiles (usually the shallow and deep ones) were able to coexist (depending on which of the four was assumed to have a *WUE* advantage) across a broad precipitation gradient on both coarse-textured (Fig. 6) and fine-textured (Fig. 7) soils. Overall, the combined biomass of all profiles (those that remained viable) increased as a function of MAP (Figs. 6 and 7), and increases in MAP led to sequential switches in dominance from shallower to deeper profiles. Therefore, as profile depth increases (shallow to intermediate to deep), the model predicted that the conditions that tend to favor it tend to shift to the moister end of the MAP gradient. The model results suggest that each profile has an optimum set of conditions that will allow it to dominate. In Fig. 6a, the shallow profile has a biomass peak at about 600 mm y^{-1}, whereas the intermediate profile peaks at about 1200 mm y^{-1}. The deep profile continues to increase in peak biomass beyond this range (Fig. 6). The location of the peaks appears to depend on the relative *WUE* values of the different

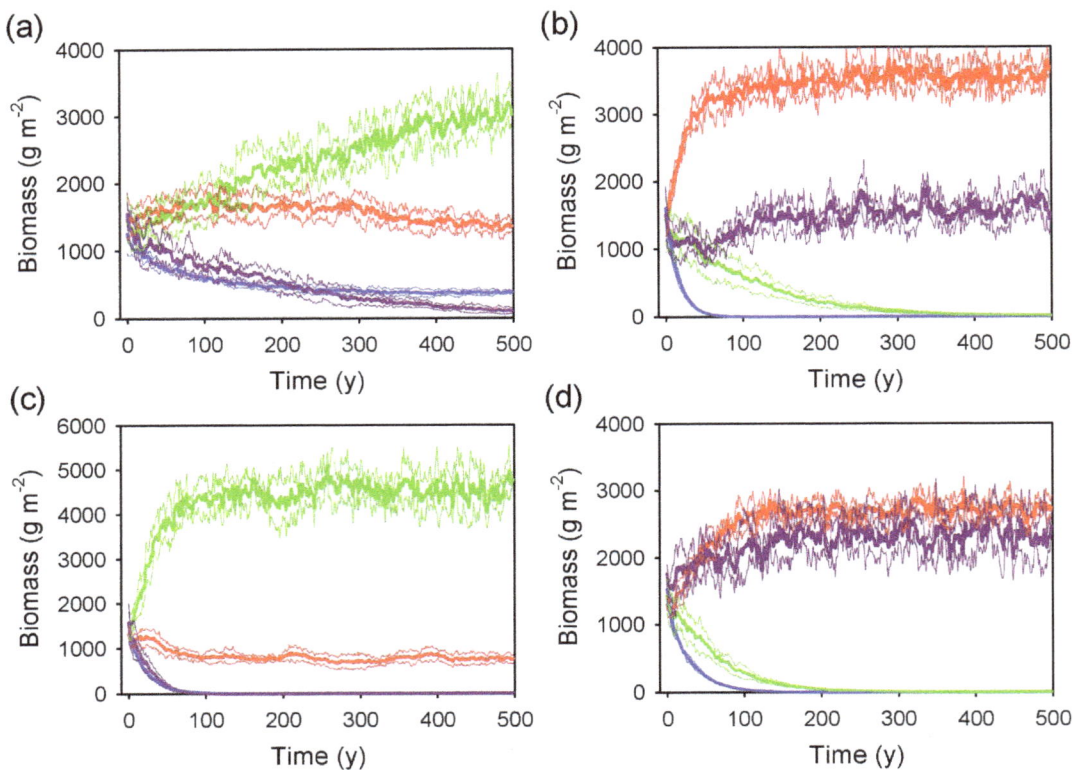

Figure 4. Time series of biomass dynamics for four rooting profiles assuming coarse-textured soils and MAP = 650 mm y^{-1}. Each panel assumes a water use efficiency advantage (WUE = 4.5 for the target profile vs. 4.0 g m^{-2} kg^{-1} for the other three) for a different target profile: (a) super-shallow, (b) shallow, (c) intermediate, and (d) deep. Thick and thin lines represent the mean ± 1 SD (N = 5 runs) of total (above and belowground) biomass. Profile key: blue = super-shallow, red = shallow, green = intermediate, purple = deep.

groups, and to a lesser extent, on the soil substrate (Figs. 6 and 7). The simulations also suggested that the fine-textured soil conditions are relatively more favorable to shallower rooting profiles than to deeper ones (Fig. 7), as expected by the slower rate of infiltration in fine-textured soils.

Discussion

The model results suggest that differences in plant rooting profiles are sufficient to explain coexistence when water is the limiting resource under realistic conditions, without the need to invoke the 'exclusive-use' premise [3,14]. It goes further, moreover, by suggesting that multiple rooting strategies may coexist when a single resource is limiting for extended periods of time, even when rooting profiles show quite subtle differences in allocation as a function of depth (compare the shallow and intermediate strategies in Fig. 1b). The two-layer model, as previously interpreted, largely relied on the notion of 'exclusive access' to subsoil water on the part of trees, and the superior performance of grasses in topsoil layers, at least in theoretical models [3,14]. Though these characteristics may well apply in many situations—in fact, we do know that grasses are able to outcompete trees in upper soil layers [31]—, their absence does not necessarily preclude coexistence. Beyond the question of tree-grass coexistence, the results have implications for interactions among tree species that coexist locally, given the diversity in root shape and allocation as a function of depth encountered in this growth form [15,49]. The results suggest that the original premise of the Walter two-layer model, i.e., that niche differentiation can

explain coexistence, is valid. Coexistence is possible even with substantial rooting overlap and quite strong competitive interactions, provided that the shapes of the functional rooting profiles differ sufficiently. The underlying mechanism proposed here is not new: it is well known that spatial and temporal heterogeneity in resource supply can promote species coexistence [52]. Plant species and functional groups exhibit clear differences in depth-specific rooting profiles [49,53], and substantial spatiotemporal heterogeneity in soil moisture availability occurs in real systems. The relative competitive advantage gained by allocating roots to alternative depths therefore shifts over time, resulting in a dynamic equilibrium for competing strategies.

Previous efforts to model tree-grass dynamics by accounting for rooting differences alone have often failed to predict coexistence [10,32]. A key requirement for coexistence is the occurrence of deep infiltration events. As demonstrated here, the stochastic nature of precipitation inputs (which lead to these events) is critical for coexistence, but a reasonably realistic representation of the temporal dynamics of the rainfall function may also be critical [33,38]. Jeltsch *et al.* [32] used both deterministic and stochastic rainfall functions in a tree-grass dynamics model, but found that stochasticity alone was insufficient to generate coexistence. A better characterization of interarrival times and storm depths allows for infrequent but intense rainfall events (which result in deep recharge, and may therefore be critical for the success of deep-rooted strategies), unlike stochastic functions that simply 'spread out' the supply of water over time. When large precipitation events occur on sandy soils, significant infiltration can occur, opening up deeper soil layers as viable sources of

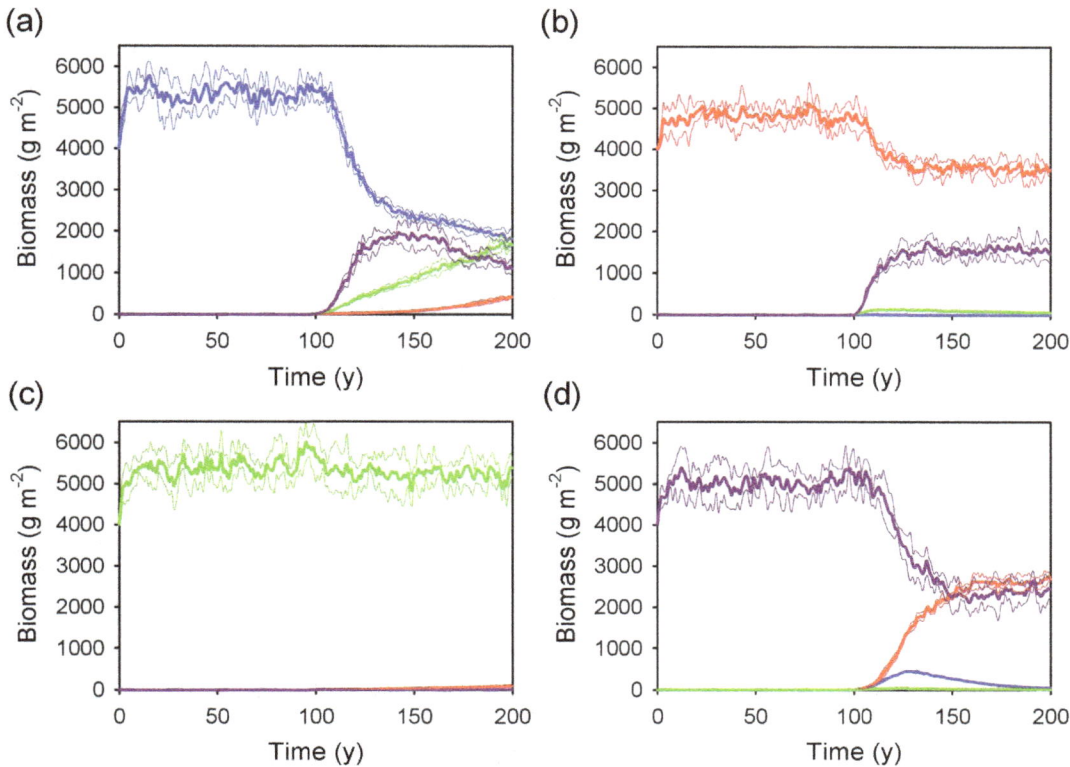

Figure 5. Invasibility of each rooting profile at steady state. Model outcomes are shown for (a) super-shallow, (b) shallow, (c) intermediate, and (d) deep profiles. Simulations are based on 100 years with no competition, followed by an introduction of 10 g m^{-2} of each of the other three profiles. Each profile being invaded has a *WUE* advantage over the others that matches what is shown in panels a–d of Fig. 4. Thick and thin lines represent the mean ± 1 SD (N = 5 runs) of total (above and belowground) biomass. Profile key: blue = super-shallow, red = shallow, green = intermediate, purple = deep.

moisture. Walter understood that, under certain conditions, the deeper roots of trees would allow them to capture this excess moisture, particularly under more mesic conditions, therefore allowing them to coexist with grasses despite the fact that grasses have higher transpiration rates than trees and are therefore the superior competitor of the two growth forms [5,6]. In addition to the timing and size of precipitation events, deep infiltration is a function of plant uptake, evaporative demand, and edaphic factors such as soil texture and pore size [30,38]. Rooting separation may therefore be the result of opportunities opening up for the exploitation of a new resource (*i.e.*, subsoil water) rather than a consequence of competition for a limiting resource.

Most authors will concede that niche-based explanations for the savanna state may well play an important role under certain conditions [7,11,27,28]. Many stable isotope-based studies have in fact shown clear differences in functional rooting profiles between trees and grasses and among functional groups, such as C$_3$ forbs and C$_4$ grasses [16,17,18,54,55], although it has been suggested that such partitioning tends to occur in systems where precipitation falls in the non-growing season, which is not the case in (for example) the extensive sub-Saharan African savannas [7]. The more limited amount of work conducted in African savannas does not appear to support the two-layer model [19,20,25], but upon closer scrutiny it appears that the approaches available may usually only be adequate for quantifying coarse differences in rooting profiles (a consequence of the difficulty of studying belowground patterns and processes). More painstaking studies have shown that trees and grasses may exhibit subtle rooting

differences as a function of depth [29,50], and that tree and grass profiles do differ, even when there is substantial overlap [19,28]. These small differences may be all that is required to allow long-term coexistence.

The fact that coexistence occurs across a wide environmental range does not necessarily mean that rooting separation is the primary factor responsible for the savanna state under all or even most conditions. The model results suggest that the viability of alternative rooting strategies (as measured by the functional diversity in Fig. 4c,d) varies as a function of rainfall and soil texture. The fact that the various rooting profiles exhibit biomass optima as a function of MAP suggests that any two strategies will tend to have equal biomass where their unimodal curves intersect. At this point, which will tend to occur in areas of intermediate MAP rather than at the extremes (Figs. 6 and 7), functional "diversity" is maximized. This suggests that niche partitioning generated by rooting differences may be a more plausible strategy under intermediate (*i.e.*, neither too dry nor wet) conditions. At the dry and wet ends of the spectrum, shallow-rooted and deep-rooted strategies are favored, respectively, as has been previously noted [36]. In arid and semi-arid savannas, this may contribute to explain the upper bound of tree cover that is associated with MAP [12,21]. Trees in semi-arid environments may have to 'choose' between two poor options: i) one in which they maintain a sub-optimal deep-rooted strategy (e.g., imposed by structural or architectural constraints), or ii) if rooting allocation is plastic, one in which they exploit shallow soil layers and compete directly with grasses. Either case is likely to lead to reduced tree biomass.

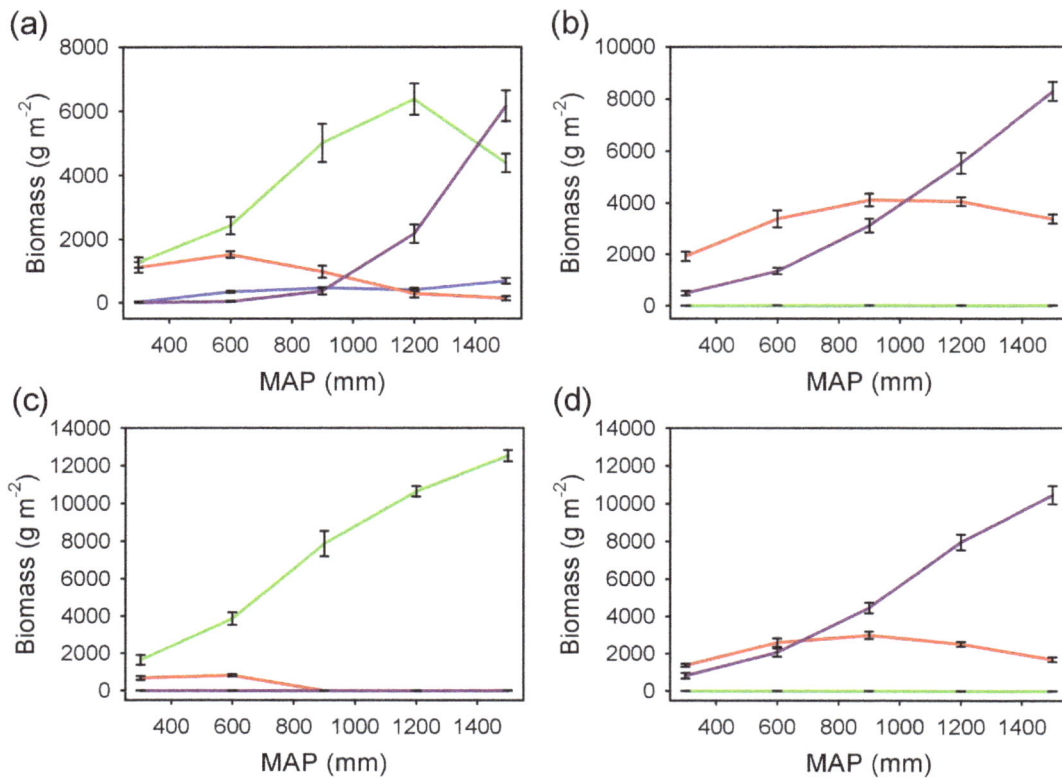

Figure 6. Steady-state biomass for the four rooting profiles as a function of MAP on coarse-textured soil. Each panel assumes a water use efficiency advantage ($WUE = 4.5$ for the target profile vs. 4.0 g m^{-2} kg^{-1} for the other three) for a different target profile: (a) super-shallow, (b) shallow, (c) intermediate, and (d) deep. The steady-state values (mean ± SD) are based on mean total (above and belowground) biomass across five runs. Profile key: blue = super-shallow, red = shallow, green = intermediate, purple = deep.

The reality probably lies somewhere between these two extremes: trees have been shown to be opportunistic and malleable in their ability to shift their functional rooting profile [50,55], but this ability has limits: the Serengeti plains, for example, support grasses but are too shallow to support trees [56]. In either case, trees and grasses are competing for a limiting resource: in the first case exploitative competition in topsoil leads to lower infiltration, and in the second case interference competition occurs within the soil layers where they overlap spatially. As MAP increases, spatial segregation of the resource increases, leading to greater opportunities for differentiation. At the upper end of the MAP boundary, high infiltration rates favor trees over grasses, and the savanna state is more likely to begin to give way to woodland or forest. At this point, other mechanisms are likely to begin to play a greater role, such as competition for light [9] and fire [12,21,24]. Over time, increasing tree and declining grass biomass in mesic sites may lead to fire suppression and species turnover from low-LAI (leaf area index) savanna species to high-LAI forest species, with light limitation eventually leading to a conversion to forest [23]. Although I do not include these mechanisms in this simple model, I propose that below-ground niche differentiation may continue to be an important component of savanna stability and species coexistence under quite mesic conditions.

Other considerations

As outlined above, the results of the model presented here (in combination with prior work [32]) suggest that the outcome of competitive interactions are influenced by the precise nature of the stochastic function describing precipitation events. Precipitation events can be described in terms of their timing and size [15], but also by the relationship between these two variables, given that times between storms may be correlated with the size of a preceding precipitation event. Alternative stochastic functions (including more realistic, site-specific rainfall generating functions, e.g., [47]) to the one used here and elsewhere need to be tested to explore the sensitivity of coexistence outcomes to variation in the timing and size of precipitation events. This may be particularly important at the dry end of the MAP spectrum, where plants can exhibit threshold responses [15] to precipitation pulses (and therefore water uptake becomes increasingly sensitive to the frequency distribution of event sizes).

For simplicity, I have omitted other key variables from this version of the model, most notably seasonality and plant phenology. Different function groups and species exhibit alternative responses to the onset of the dry season in savanna systems, trees often flushing before grasses and retaining their leaves later into the dry season, which allows them to opportunistically respond to late precipitation events [28,57]. Some savanna tree species are more drought-deciduous than others [28]. Niche partitioning along the temporal axis has been proposed as an alternative to the two-layer model [7,11]. I have ignored this seasonal axis here, but it should be incorporated into a more comprehensive model of tree-grass water use and partitioning, given that the vertical distribution of soil moisture (and therefore the relative advantage of shallow and deep-rooted species) varies seasonally. There are other elements I have left out in this model that have the potential to shift its quantitative conclusions. I ignore the role of stem flow for subsoil water recharge and changes in

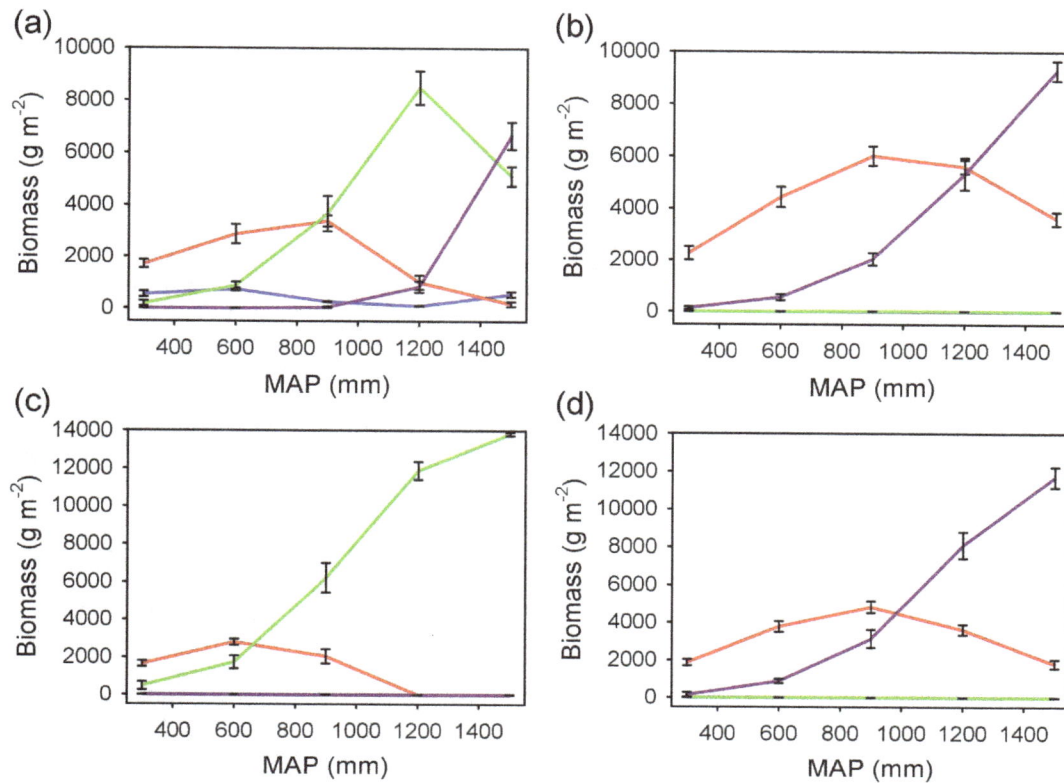

Figure 7. Steady-state biomass for four rooting profiles as a function of MAP on fine-textured soil. Each panel assumes a water use efficiency advantage ($WUE = 4.5$ for the target profile vs. 4.0 g m^{-2} kg^{-1} for the other three) for a different target profile: (a) super-shallow, (b) shallow, (c) intermediate, and (d) deep. The steady-state values (mean $\pm$ SD) are based on mean total (above and belowground) biomass across five runs. Profile key: blue = super-shallow, red = shallow, green = intermediate, purple = deep.

infiltration due to soil "capping" as the proportion of bare soil increases [4], the role of tree canopies on microclimate and evaporative demand in the grass layer [14], changes in root mass ratios along climate gradients [58], hydraulic redistribution by trees [59], changes in evaporative demand correlated with MAP, and plasticity in functional rooting profiles, particularly in trees [15,50]. I deliberately left out these factors to test the hypothesis that rooting separation can be a dominant explanation for the savanna state, although I note that the exploration of rooting niche space across environmental gradients requires further analysis.

Conclusions and broader implications

A simple vertically-explicit model suggests that the relative importance of tree-grass competition may vary systematically along environmental gradients, and this may be an important insight for developing a predictive understanding of tree-grass dynamics under a wide range of conditions. I suggest that explicitly incorporating a mechanistic representation of tree and grass vertical rooting profiles and modeling soil moisture dynamics as a function of depth may be important for improving predictive models of savanna dynamics across a wide range of edaphic and precipitation conditions, including novel climate regimes. Hydrologic templates are shifting rapidly as a result of climate change-induced shifts in the amount and distribution of rainfall inputs and evapotranspirational demand [60]. If there are indeed strong and consistent differences in rooting profiles among functional groups and species, who will be the winners and losers of such changes at

the community level? An important next step is to move beyond the debate about tree-grass coexistence and work towards the development of synthetic models that incorporate all the relevant factors necessary to move towards useful predictive model of tree-grass dynamics [7]. I propose that a systematic analysis of rooting profiles along key environmental gradients may be an important aspect of this process.

Supporting Information

Figure S1 Coefficients of stochastic rainfall generator as a function of mean annual precipitation (MAP) across four North American LTER sites. (a) rate λ of the exponential distribution describing interarrival tS1imes between precipitation events, (b) mean μ and (c) standard deviation σ of the lognormal distribution describing event size. Key to LTER sites: J = Jornada, S = Shortgrass, CC = Cedar Creek, and K = Kellog.

Acknowledgments

Data sets were provided by the Climate and Hydrology Database Projects, a partnership between the Long-Term Ecological Research program and the U.S. Forest Service Pacific Northwest Research Station, Corvallis, Oregon. I would like to acknowledge Jonathan Dushoff and Ben Bolker for assistance with modeling issues, and two anonymous reviewers for helpful comments on the manuscript.

Author Contributions

Conceived and designed the experiments: RMH. Performed the experiments: RMH. Analyzed the data: RMH. Contributed reagents/materials/analysis tools: RMH. Wrote the paper: RMH.

References

1. Whittaker RH (1975) Communities and Ecosystems. London: Collier Macmillan.
2. Bond WJ (2008) What limits trees in C4 grasslands and savannas? Annual Review of Ecology, Evolution and Systematics 39: 641–659.
3. Walker BH, Ludwig D, Holling CS, Peterman RM (1981) Stability of semi-arid savanna grazing systems. Journal of Ecology 69: 473–498.
4. Walker BH, Noy-Meir I (1982) Aspects of the stability and resilience of savanna ecosystems. In: Huntley BJ, Walker BH, editors. Ecology of Tropical Savannas. Berlin: Springer-Verlag. pp. 556–590.
5. Walter H (1971) Ecology of Tropical and Subtropical Vegetation. Endinburgh: Oliver and Boyd. 539 p.
6. Ward D, Wiegand K, Getzin S (2012) Walter's two-layer hypothesis revisited: back to the roots! Oecologia: 1–14.
7. Sankaran M, Ratnam J, Hanan NP (2004) Tree-grass coexistence in savannas revisited - insights from an examination of assumptions and mechanisms invoked in existing models. Ecology Letters 7: 480–490.
8. Higgins SI, Bond WJ, Trollope WSW (2000) Fire, resprouting and variability: a recipe for grass-tree coexistence in savanna. Journal of Ecology 88: 213–229.
9. Scheiter S, Higgins SI (2007) Partitioning of root and shoot competition and the stability of savannas. American Naturalist 170: 587–601.
10. Jeltsch F, Weber GE, Grimm V (2000) Ecological buffering mechanisms in savannas: A unifying theory of long-term tree-grass coexistence. Plant Ecology 150: 161–171.
11. House JI, Archer S, Breshears DD, Scholes RJ (2003) Conundrums in mixed woody-herbaceous plant systems. Journal of Biogeography 30: 1763–1777.
12. Sankaran M, Hanan NP, Scholes RJ, Ratnam J, Augustine DJ, et al. (2005) Determinants of woody cover in African savannas. Nature 438: 846–849.
13. Walter H (1973) Vegetation of the earth in relation to climate and the eco-physiological conditions. New York: Springer.
14. Eagleson PS, Segara RI (1985) Water-limited equilibrium of savanna vegetation systems. Water Resources Research 21: 1483–1493.
15. Ogle K, Reynolds J (2004) Plant responses to precipitation in desert ecosystems: integrating functional types, pulses, thresholds, and delays. Oecologia 141: 282–294.
16. Brown JR, Archer S (1990) Water relations of a perennial grass and seedling vs adult woody plants in a subtropical savanna, Texas. Oikos 57: 366–374.
17. Weltzin JF, McPherson GR (1997) Spatial and temporal soil moisture resource partitioning by trees and grasses in a temperate savanna, Arizona, USA. Oecologia 112: 156–164.
18. Dodd MB, Lauenroth WK, Welker JM (1998) Differential water resource use by herbaceous and woody plant life-forms in a shortgrass steppe community. Oecologia 117: 504–512.
19. Verweij RJT, Higgins SI, Bond WJ, February EC (2011) Water sourcing by trees in a mesic savanna: responses to severing deep and shallow roots. Environmental and Experimental Botany 74: 229–236.
20. Mordelet P, Menaut JC, Mariotti A (1997) Tree and grass rooting patterns in an African humid savanna. Journal of Vegetation Science 8: 65–70.
21. Staver AC, Archibald S, Levin SA (2011) The global extent and determinants of savanna and forest as alternative biome states. Science 334: 230–232.
22. Holdo RM, Sinclair ARE, Metzger KL, Bolker BM, Dobson AP, et al. (2009) A disease-mediated trophic cascade in the Serengeti and its implications for ecosystem C. PLOS Biology 7: e1000210.
23. Hoffmann WA, Geiger EL, Gotsch SG, Rossatto DR, Silva LCR, et al. (2012) Ecological thresholds at the savanna-forest boundary: how plant traits, resources and fire govern the distribution of tropical biomes. Ecology Letters 15: 759–768.
24. Holdo RM, Holt RD, Fryxell JM (2013) Herbivore–vegetation feedbacks can expand the range of savanna persistence: insights from a simple theoretical model. Oikos 122: 441–453.
25. Hipondoka MHT, Aranibar JN, Chirara C, Lihavha M, Macko SA (2003) Vertical distribution of grass and tree roots in arid ecosystems of Southern Africa: niche differentiation or competition? Journal of Arid Environments 54: 319–325.
26. February EC, Higgins SI (2010) The distribution of tree and grass roots in savannas in relation to soil nitrogen and water. South African Journal of Botany 76: 517–523.
27. Rodríguez-Iturbe I, D'Odorico P, Porporato A, Ridolfi L (1999) Tree-grass coexistence in savannas: The role of spatial dynamics and climate fluctuations. Geophysical Research Letters 26: 247–250.
28. Scholes RJ, Walker BH (1993) An African Savanna: Synthesis of the Nylsvlei Study. Cambridge: Cambridge University Press. 306 p.
29. Kulmatiski A, Beard KH, Verweij RJT, February EC (2010) A depth-controlled tracer technique measures vertical, horizontal and temporal patterns of water use by trees and grasses in a subtropical savanna. New Phytologist 188: 199–209.
30. Sperry JS, Hacke UG (2002) Desert shrub water relations with respect to soil characteristics and plant functional type. Functional Ecology 16: 367–378.
31. Nippert JB, Wieme RA, Ocheltree TW, Craine JM (2012) Root characteristics of C4 grasses limit reliance on deep soil water in tallgrass prairie. Plant and Soil 355: 385–394.
32. Jeltsch F, Milton SJ, Dean WRJ, VanRooyen N (1996) Tree spacing and coexistence in semiarid savannas. Journal of Ecology 84: 583–595.
33. Porporato A, D'Odorico P, Laio F, Ridolfi L, Rodriguez-Iturbe I (2002) Ecohydrology of water-controlled ecosystems. Advances in Water Resources 25: 1335–1348.
34. van Wijk MT, Rodriguez-Iturbe I (2002) Tree-grass competition in space and time: insights from a simple cellular automata model based on ecohydrological dynamics. Water Resources Research 38: 18-11–18-15.
35. Rodriguez-Iturbe I, Porporato A, Laio F, Ridolfi L (2001) Plants in water-controlled ecosystems: active role in hydrologic processes and response to water stress - I. Scope and general outline. Advances in Water Resources 24: 695–705.
36. Porporato A, Laio F, Ridolfi L, Caylor KK, Rodriguez-Iturbe I (2003) Soil moisture and plant stress dynamics along the Kalahari precipitation gradient. Journal of Geophysical Research 108: 4127.
37. Ogle K, Wolpert RL, Reynolds JF (2004) Reconstructing Plant Root Area and Water Uptake Profiles. Ecology 85: 1967–1978.
38. Rodriguez-Iturbe I, Porporato A, Ridolfi L, Isham V, Cox DR (1999) Probabilistic Modelling of Water Balance at a Point: The Role of Climate, Soil and Vegetation. Proceedings: Mathematical, Physical and Engineering Sciences 455: 3789–3805.
39. Moorcroft PR, Hurtt GC, Pacala SW (2001) A method for scaling vegetation dynamics: The ecosystem demography model (ED). Ecological Monographs 71: 557–585.
40. Xiao X, Horton R, Sauer TJ, Heitman JL, Ren T (2011) Cumulative soil water evaporation as a function of depth and time. Vadose Zone Journal 10: 1016–1022.
41. Valentine HT (1988) A carbon-balance model of stand growth: a derivation employing pipe-model theory and the self-thinning rule. Annals of Botany 62: 389–396.
42. West GB, Brown JH, Enquist BJ (1999) A general model for the structure and allometry of plant vascular systems. Nature 400: 664–667.
43. Nygren P, Rebottaro S, Chavarría R (1993) Application of the pipe model theory to non-destructive estimation of leaf biomass and leaf area of pruned agroforestry trees. Agroforestry Systems 23: 63–77.
44. Infante JM, Mauchamp A, Fernández-Alés R, Joffre R, Rambal S (2001) Within-tree variation in transpiration in isolated evergreen oak trees: evidence in support of the pipe model theory. Tree Physiology 21: 409–414.
45. Guswa AJ, Celia MA, Rodriguez-Iturbe I (2002) Models of soil moisture dynamics in ecohydrology: a comparative study. Water Resources Research 38: 1–15.
46. Pacala SW, Kinzig AP (2002) Introduction to theory and the common ecosystem model. In: Kinzig AP, Tilman D, Pacala SW, editors. Functional Consequences of Biodiversity: Empirical Progress and Theoretical Extensions. Princeton: Princeton University Press. pp. 169–174.
47. Zucchini W, Nenadić O (2006) A Web-based rainfall atlas for Southern Africa. Environmetrics 17: 269–283.
48. R Development Core Team (2011) R: A language and environment for statistical computing. Vienna, Austria: R Foundation for Statistical Computing.
49. Holdo RM, Timberlake J (2008) Rooting depth and above-ground community composition in Kalahari sand woodlands in western Zimbabwe. Journal of Tropical Ecology 24: 169–176.
50. Kulmatiski A, Beard K (2013) Root niche partitioning among grasses, saplings, and trees measured using a tracer technique. Oecologia 171: 25–37.
51. Hardin G (1960) The competitive exclusion principle. Science 131: 1292–1297.
52. Chesson PL (1985) Coexistence of competitors in spatially and temporally varying environments: A look at the combined effects of different sorts of variability. Theoretical Population Biology 28: 263–287.
53. Nippert J, Knapp AK (2007) Linking water uptake with rooting patterns in grassland species. Oecologia 153: 261–272.
54. Sala OE, Golluscio RA, Lauenroth WK, Soriano A (1989) Resource partitioning between shrubs and grasses in the Patagonian steppe. Oecologia 81: 501–505.
55. Nippert JB, Knapp AK (2007) Soil water partitioning contributes to species coexistence in tallgrass prairie. Oikos 116: 1017–1029.
56. Belsky AJ (1990) Tree-grass ratios in East African savannas - a comparison of existing models. Journal of Biogeography 17: 483–489.
57. Scholes RJ, Archer SR (1997) Tree-grass interactions in savannas. Annual Review of Ecology and Systematics 28: 517–544.

58. Tomlinson KW, Sterck FJ, Bongers F, da Silva DA, Barbosa ERM, et al. (2012) Biomass partitioning and root morphology of savanna trees across a water gradient. Journal of Ecology 100: 1113–1121.

59. Scholz FG, Bucci SJ, Goldstein G, Meinzer FC, Franco AC (2002) Hydraulic redistribution of soil water by neotropical savanna trees. Tree Physiology 22: 603–612.

60. Jung M, Reichstein M, Ciais P, Seneviratne SI, Sheffield J, et al. (2010) Recent decline in the global land evapotranspiration trend due to limited moisture supply. Nature 467: 951–954.

61. Laio F, Porporato A, Ridolfi L, Rodriguez-Iturbe I (2001) Plants in water-controlled ecosystems: active role in hydrologic processes and response to water stress - II. Probabilistic soil moisture dynamics. Advances in Water Resources 24: 707–723.

Permissions

List of Contributors

Yuying Wang, Chunsheng Hu, Hua Ming, Wenxu Dong, Yuming Zhang and Xiaoxin Li
Key Laboratory of Agricultural Water Resources, Center for Agricultural Resources Research, Institute of Genetics and Developmental Biology, Chinese Academy of Sciences, Shijiazhuang, Hebei, China

Oene Oenema
Department of Soil Quality, Wageningen University, Alterra, Wageningen, The Netherlands

Douglas A. Schaefer
Key Lab of Tropical Forest Ecology, Xishuangbanna Tropical Botanical Garden, Chinese Academy of Sciences, Menglun, Yunnan, China

William Carlisle Thacker
University of Miami and Atlantic Oceanographic and Meteorological Laboratory, Miami, Florida, United States of America

James R. Bell and Ka-Sing Lim
Department of Agro-Ecology, Rothamsted Research, Harpenden, Hertfordshire, United Kingdom

Prabhuraj Aralimarad
Department of Entomology, University for Agricultural Sciences, Raichur, Karnataka, India

Jason W. Chapman
Department of Agro-Ecology, Rothamsted Research, Harpenden, Hertfordshire, United Kingdom
Environment and Sustainability Institute, University of Exeter, Penryn, Cornwall, United Kingdom

Heikki Seppä
Key Laboratory of Humid Subtropical Eco-geographical Process (Ministry of Education), College of Geographical Sciences, Fujian Normal University, Fuzhou, Fujian province, China

Keyan Fang
Key Laboratory of Humid Subtropical Eco-geographical Process (Ministry of Education), College of Geographical Sciences, Fujian Normal University, Fuzhou, Fujian province, China
Department of Geosciences and Geography, University of Helsinki, Helsinki city, Helsinki, Finland

Fahu Chen and Wei Huang
Key Laboratory of Western Chinás Environmental Systems (MOE), Lanzhou University, Lanzhou, Gansu province, China

Asok K. Sen
Richard G. Lugar Center for Renewable Energy and Department of Mathematical Sciences, Indiana University, Indianapolis, Indiana, United States of America

Nicole Davi
Tree-Ring Lab, Lamont-Doherty Earth Observatory, Columbia University, Palisades, New York, United States of America

Jinbao Li
Department of Geography, University of Hong Kong, Hong Kong city, Hong Kong, China

Rachele Gallisai and Francesc Peters
Departament de Biologia Marina Oceanografia, Institut de Ciéncies del Mar, CSIC, Barcelona, Spain

Gianluca Volpe
Istituto di Scienze dell'Atmosfera e del Clima, Roma, Italy

Sara Basart
Istituto di Scienze dell'Atmosferae del Clima, Roma, Italy
Earth Sciences Department, Barcelona Supercomputing Center-Centro Nacional de Supercomputación, BSC-CNS, Barcelona, Spain

José Maria Baldasano
Earth Sciences Department, Barcelona Supercomputing Center-Centro Nacional de Supercomputación, BSC-CNS, Barcelona, Spain
Environmental Modelling Laboratory, Technical University of Catalonia, Barcelona, Spain

Bo-Hui Tang and Hua Wu
State Key Laboratory of Resources and Environment Information System, Institute of Geographic Sciences and Natural Resources Research, Chinese Academy of Sciences, Beijing, China

Si-Bo Duan
State Key Laboratory of Resources and Environment Information System, Institute of Geographic Sciences and Natural Resources Research, Chinese Academy of Sciences, Beijing, China
University of Chinese Academy of Sciences, Beijing, China
Laboratoire des sciences de l'ingenieur, de l'informatique et de l'imagerie, Université de Strasbourg, Centre National de la Recherche Scientifique, Illkirch, France

Zhao-Liang Li
Laboratoire des sciences de l'ingenieur, de l'informatique et de l'imagerie, Université de Strasbourg, Centre National de la Recherche Scientifique, Illkirch, France Key Laboratory of Agri-informatics,Ministry of Agriculture/Institute of Agricultural Resources and Regional Planning, Chinese Academy of Agricultural Sciences, Beijing, China

Lingling Ma and Chuanrong Li
Earth Observation Technology Application Department, Academy of Opto-Electronics, Chinese Academy of Sciences, Beijing, China

Enyu Zhao
University of Chinese Academy of Sciences, Beijing, China
Earth Observation Technology Application Department, Academy of Opto-Electronics, Chinese Academy of Sciences, Beijing, China

Tingting Cao and Jonathan E. Thompson
Department of Chemistry & Biochemistry, Texas Tech University, Lubbock, Texas, United States of America

Xurxo Costoya, Maite deCastro, Moncho Gómez-Gesteira and Fran Santos
EPHYSLAB, Environmental PHYsics LABoratory, Facultad de Ciencias, Universidad de Vigo, Ourense, Spain

Willow Hallgren, Udaya Bhaskar Gunturu and Adam Schlosser
The MIT Joint Program on the Science and Policy of Global Change, Massachusetts Institute of Technology, Cambridge, Massachusetts, United States of America

Toshihiko Ogura
Biomedical Research Institute, National Institute of Advanced Industrial Science and Technology (AIST), Umezono, Tsukuba, Ibaraki, Japan

Antonio Navarra and Giovanni Conti
Centro Euromediterraneo sui Cambiamenti Climatici, Bologna, Italy

Joe Tribbia
National Center for Atmospheric Research, Boulder, Colorado, United States of America

Jingbin Liu and Juha Hyyppä
Department of remote sensing and photogrammetry, Finnish Geodetic Institute, Masala, Finland

Ruizhi Chen
Conrad Blucher Institute for surveying & science, Texas A & M University Corpus Christi, Corpus Christi, United States of America

Zemin Wang and Jiachun An
Chinese Antarctic center of surveying and mapping, Wuhan University, Wuhan, China

Tianxing Wang, Jiancheng Shi, Yingying Jing, Tianjie Zhao, Dabin Ji and Chuan Xiong
State Key Laboratory of Remote Sensing Science, Institute of Remote Sensing and Digital Earth, Chinese Academy of Sciences. Beijing, China

Yaoming Ma
Key Laboratory of Tibetan Environment Changes and Land Surface Processes, Institute of Tibetan Plateau Research, Chinese Academy of Sciences, Beijing, China

Xuelong Chen
Key Laboratory of Tibetan Environment Changes and Land Surface Processes, Institute of Tibetan Plateau Research, Chinese Academy of Sciences, Beijing, China
Faculty of Geo-Information Science and Earth Observation, University of Twente, Enschede, The Netherlands

Zhongbo Su
Faculty of Geo-Information Science and Earth Observation, University of Twente, Enschede, The Netherlands

Juan A. Añel
Smith School of Enterprise and the Environment, University of Oxford, Oxford, United Kingdom
EPhysLab, Fac. of Sciences, Universidade de Vigo, Ourense, Spain

Laura de la Torre
EPhysLab, Fac. of Sciences, Universidade de Vigo, Ourense, Spain

Hennie Kelder
Eindhoven University, Eindhoven, The Netherlands

Jacob van Peet
Royal Netherlands Meteorological Institute (KNMI), Utricht, The Netherlands

Sergey Zimov and Nikita Zimov
Northeast Science Station, Pacific Institute for Geography, Russian Academy of Sciences, Cherskii, Russia

Bjarne Munk Hansen and Ulrich Gosewinkel Karlson
Department of Environmental Science, Aarhus University, Roskilde, Denmark

Tina Šantl-Temkiv
Department of Environmental Science, Aarhus University, Roskilde, Denmark

Microbiology Section, Department of Bioscience, Aarhus University, Aarhus, Denmark
Stellar Astrophysics Centre, Department of Physics and Astronomy, Aarhus University, Aarhus, Denmark

Kai Finster
Department of Environmental Science, Aarhus University, Roskilde, Denmark
Microbiology Section, Department of Bioscience, Aarhus University, Aarhus, Denmark
Stellar Astrophysics Centre, Department of Physics and Astronomy, Aarhus University, Aarhus, Denmark

Thorsten Dittmar
Max Planck Research Group for Marine Geochemistry, Institute for Chemistry and Biology of the Marine Environment, University of Oldenburg, Oldenburg, Germany

Runar Thyrhaug
Department of Biology, University of Bergen, Bergen, Norway

Niels Woetmann Nielsen
Danish Meteorological Institute, Copenhagen, Denmark

Susannah M. Leahy and Michael J. Kingsford
School of Marine and Tropical Biology, James Cook University, Townsville, Queensland, Australia
Australian Research Council Centre of Excellence for Coral Reef Studies, James Cook University, Townsville, Queensland, Australia

Craig R. Steinberg
Australian Institute of Marine Science, Townsville, Queensland, Australia

Tony Chang, Andrew J. Hansen and Nathan Piekielek
Department of Ecology, Montana State University, Bozeman, Montana, United States of America

Martin Aubé
Département de physique, Cégep de Sherbrooke, Sherbrooke, Québec, Canada

Johanne Roby
Département de chimie, Cégep de Sherbrooke, Sherbrooke, Québec, Canada

Miroslav Kocifaj
Astronomical Institute, Slovak Academy of Sciences, Dúbravská , Bratislava, Slovak Republic

Mark D. McDonnell
Institute for Telecommunications Research, University of South Australia, Mawson Lakes, 5095, Australia

Tony Vladusich
Institute for Telecommunications Research, University of South Australia, Mawson Lakes, 5095, Australia
Center for Computational Neuroscience and Neural Technology, Boston University, Boston, MA, United States of America

Ricardo M. Holdo
Divison of Biological Sciences, University of Missouri, Columbia, Missouri, United States of America

Index

www.ingramcontent.com/pod-product-compliance
Lightning Source LLC
Chambersburg PA
CBHW080533200326
41458CB00012B/4418